Students and naturalists are not only interested in which species live on the seashore but also about their biology. How does a particular species reproduce? What is its life-cycle? *A Student's Guide to the Seashore* is a unique, concise, illustrated guide to *both* the biology and identification of over 600 common and widespread shore animals and plants. In this new edition, for the first time, simple keys are included to allow accurate identification, and each species is beautifully illustrated by the authors' line drawings. Together with concise summaries of diagnostic features, and notes on biology, this is the first comprehensive guide to the seashore giving a fascinating insight into the diversity and complexity of life on the shore. An extensive glossary of scientific terms and a complete bibliography ensure that this book will be the premier biological text and identification guide for students at school and university, teachers and all those interested in the seashore for many years to come.

# A Student's Guide to the Seashore

SECOND EDITION

# A Student's Guide to the Seashore

SECOND EDITION

**J. D. Fish** *and* **S. Fish**
*Institute of Biological Sciences, The University of Wales, Aberystwyth*

CAMBRIDGE
UNIVERSITY PRESS

Published by the Press Syndicate of the University of Cambridge
The Pitt Building, Trumpington Street, Cambridge CB2 1RP
40 West 20th Street, New York, NY 10011-4211, USA
10 Stamford Road, Oakleigh, Melbourne 3166, Australia

First published by Unwin Hyman 1989
Second edition published by Cambridge University Press 1996

Printed in Great Britain at the University Press, Cambridge

*A catalogue record for this book is available from the British Library*

*Library of Congress cataloguing in publication data*

Fish, J. D.
    A students's guide to the seashore/J.D. Fish and S. Fish – 2nd
edn.
      p.   cm.
    Includes bibliographical references (p.    ) and index.
    ISBN 0 521 46279 7 (hc) – ISBN 0 521 46819 1 (pb)
    1. Seashore biology – Great Britain. 2. Seashore biology – Europe,
Northern.    I. Fish, S. (Susan)   II. Title.
    QH137.F52 1996
    574.94 – dc20   95–35162 CIP

ISBN 0 521 46279 7 hardback
ISBN 0 521 46819 1 paperback

'What branch of Natural History shall I begin to investigate, if it be but for a few weeks, this summer?'

To which I answer, Try 'the Wonders of the Shore.'

There are along every sea-beach more strange things to be seen, and those to be seen easily, than in any other field of observation which you will find in these islands. And on the shore only will you have the enjoyment of finding new species, of adding your mite to the treasures of science.

Charles Kingsley,
*Glaucus* or *The Wonders of the Shore*.
Macmillan and Co., 1890.
(By kind permission of Macmillan Publishers Ltd.)

# Contents

# Preface to the second edition

The welcome given to the first edition of *A Student's Guide to the Seashore* clearly indicated to us that the fusion of identification guide and biological text was much appreciated. This approach has been maintained and strengthened in the second edition. Simple dichotomous keys have been added, additional species have been included and the biological text has been updated and expanded.

We are most grateful to the many colleagues who have helped in one way or another during the production of this enlarged and revised edition. We have been sustained in this task by the support and encouragement received from readers of the first edition.

J. D. Fish
Susan Fish

*Aberystwyth, May 1995*

# Preface to the first edition

At one time or another, we have all been drawn by the fascination of the seashore. For the holiday maker, the relaxing day by the sea often turns out to be a most rewarding foray among rock pools and dense canopies of seaweed; for naturalists and students, the shore is one of the most challenging habitats. Whatever our interests and expertise, one of our first objectives when faced with the diversity of plant and animal life on the shore is to name the individual specimens and we quickly learn that this can be a difficult, though rewarding, occupation. Once an organism has been identified, a number of questions naturally follow. What is its life-cycle? How does it feed and reproduce? How long does it live? The answers to such questions give an insight into the lives of the plants and animals of the shore and are one of the first steps in an understanding of the complexity of the shore environment. However, the information required to answer such questions is not always easily accessible and even when it is known it is often scattered in various books and research journals, making it difficult and time consuming to find. Although a variety of identification keys and guides is available, some designed for the specialist, others for the amateur, such texts give little, if any, information on the biology of the organisms. This book has been designed as both a field guide and a biological text covering the common plants and animals of the shore. It is intended for undergraduates, sixth-form students and teachers but we hope that it will also appeal to the amateur naturalist and the occasional visitor to the shore.

The opening chapter deals with the shore environment. It is no more than an introduction to a complex subject. References to more advanced reading are given at the end of the section. Some basic tips as to methods of collecting intertidal organisms are given, but space does not permit a consideration of quantitative sampling methods. Reference is given to texts dealing with this subject. An outline of the general principles of scientific nomenclature and classification, and a glossary of scientific terms used in the text, have been included for the benefit of readers with little or no scientific training.

The bulk of the book is given over to a systematic coverage of the common plants and animals of the shore with emphasis on the biology of the species. Each group of organisms is prefaced by a statement of characteristics and an outline classification is included. For each species a statement of diagnostic features is given together with a line drawing (or photograph) and notes on the biology of the species. An elementary, illustrated guide to the identification of the different groups of plants and animals is included, but the book has not been designed as an identification key.

Drawing up a list of common species presented us with a number of problems. Clearly, there is no satisfactory definition of a 'common' species and one author's list will almost certainly differ from that of another. We have included species which, depending on locality, we think the student has a reasonably good chance of finding during the first few visits to the shore. A few species of restricted distribution but locally common or even abundant in some areas have been included as have plants and animals not normally found living between tidemarks but common on the strandline. The book covers the shores of north-west Europe (excluding the Baltic) but will be of particular use to those working in the British Isles.

The inspiration for writing this book came from undergraduate and postgraduate students at The University College of Wales, Aberystwyth, and members of extra-mural classes organized by the University. We are grateful to them all for their assistance. We are particularly indebted to the following for reading the first draft of selected chapters; Professor J. A. Allen, Dr J. Baxter, Professor A. D. Boney, Dr P. F. S. Cornelius, Dr J. H. Crothers, Dr P. R. Garwood, Dr Ruth Griffiths, Mr A. K. Jones, Dr A. M. Jones, Mr M. A. Kendall, Dr C. Mettam, Professor D. Nichols, Dr R. Seed, Dr T. E. Thompson, Dr. R. B. Williams and Dr R. J. Wootton. The text has been greatly improved by their constructive criticism. We are also indebted to friends and colleagues who provided specimens for drawing, in particular Mr Michael Bell, Mr John Dingley, Mr Gareth Owen, Mr Stephen Sankey and Mrs D. Slay of the Department of Zoology, Aberystwyth. The availability of this material has enabled us to make original drawings for the majority of specimens but in some cases drawings have been made from a number of existing sources which are acknowledged below.

In the preparation of the text we have of necessity drawn heavily on information given in a large number of books and, especially, research journals. Texts useful in identification are listed at the end of each section but in a book of this kind it is not possible to cite individually in the text all the many references used in writing the biological notes on the different species. A selection of such references is given in the general bibliography. We wish to acknowledge our indebtedness to the authors whose research has made this work possible.

Our thanks go to the publishers for their continued support and encouragement: Mr Miles Jackson guided us through the initial stages of preparation and Dr Clem Earl and Mr Andy Oppenheimer saw the project to fruition. Innumerable drafts of the manuscript were meticulously typed by Mrs Iris Thomas and Mrs Susan Davies. Their tolerance and understanding is greatly appreciated.

If this book in any way adds to the interest and pleasure of those visiting the shore, our objectives will have been met.

J. D. Fish

Susan Fish

*Aberystwyth, 1989*

# Acknowledgements for illustrations

We are grateful to the following authors and publishers for kind permission to use the figures listed:

Edward Arnold and J. R. Lewis for Figure 17 in *The ecology of rocky shores*. Hutchinson Educational Ltd and A. D. Boney for Figures 19, 32, 40(e) in *A biology of marine algae*. British Museum (Natural History) and N. Tebble for Figures 52A, B, C, 54A, B, 95C, 101B in *British bivalve seashells*; British Museum (Natural History) for Figures 70B, 239B in Newton, L., *A handbook of the British seaweeds*. Conseil International pour L'exploration de la Mer and W. Greve for Figure 5 in Zooplankton Sheet 146, *Fiches d'identification du zooplancton*. The Ray Society for Plate 2, Figures 1, 2; Plate 4, Figure 1; Plate 5, Figure 1; Plate 9, Figures 5, 6 in Allman, G. J., *A monograph of the gymnoblastic or tubullarian hydroids*; and Plate 4, Figure 2, Vol. 1; Plate 22, Figure 3, Vol. 2, in Stephenson, T. A., *The British sea anemones*; and Plate 1, Figures 1, 4; Plate 3, Figures 4, 6; Plate 5, Figure 1, Vol. 1, Part 1; Plate 25, Figure 1; Plate 26, Figure 1, Vol. 1, Part 2; Plate 114, Figure 2, Vol. 4, Part 1; Plate 116, Figures 2, 3, 5, Vol. 4, Part 2 in McIntosh, W. C., *A monograph of the British marine annelids*; and Fam. 1, Plate 5, Figure 1; Fam. 1, Plate 10, Figure 2; Fam. 1, Plate 11, Figures, 1, 8 in Alder, J. & Hancock, A., *A monograph of the British nudibranchiate Mollusca*; and Figures 97A, 98L, 102A, 103G, 107, 108I, 109A, 110E in Tattersall, W. M. & Tattersall, O. S., *The British Mysidacea*; and Plate 11, Figure 3, Vol. 1; Plate 51, Figure 1, Plate 52, Figure 3, Plate 54, Figure 6, Vol. 3 in Alder, J. & Hancock, A., *The British Tunicata*. Macmillan Publishers for Figure 42 in Bullough, W. S., *Practical invertebrate anatomy*. Cambridge University Press for Plates 3 and 4 in Russell, F. S., *The medusae of the British Isles*, Vol. 2; and for Plate 14, Figure 61; Plate 21, Figure 100a in Eales, N. B., *The littoral fauna of Great Britain*. Academic Press, The Estuarine and Brackish-water Sciences Association, The Linnean Society of London and R. L. Manuel for Figures 70, 78A, B in *British Anthozoa. Synopses of the British fauna*, No. 18. Academic Press, The Linnean Society of London and P. E. Gibbs for Figures 2A, 13A in *British sipunculans. Synopses of the British fauna*, No. 12. Academic Press, The Linnean Society of London and E. Naylor for Figure 10D, G, I in *British marine isopods. Synopses of the British fauna*, No. 3. Academic Press, The Linnean Society of London and R. H. Millar for Figures 10A, 55A in *British Ascidians. Tunicata: Ascidiacea. Synopses of the British fauna*, No. 1. E. J. Brill/ Dr W. Backhuys, The Estuarine and Brackish-water Sciences Association, The Linnean Society of London and P. J. Hayward for Figure 8A in *Ctenostome bryozoans. Synopses of the British fauna*, No. 33. S. A. Masson for Figure 669 in *Traité de Zoologie*, Vol. 5. The Linnean Society of London for Plate 44, Figures 1, 3 in Dyster, F. D., *Transactions of the Linnean Society of London*, Vol. 22. Penguin Books for Plate 5 in Stephenson, T. A., *Seashore life and pattern*. Editions Lechevalier for Figures 18i, 21b in Fauvel, P., *Faune de France 5. Polychètes errantes* and Figures 46b, 120a in Fauvel, P., *Faune de France 16. Polychètes sédentaires*. The Malacological Society of London for Figure 6B, D, G in Matthews, G., *Proceedings of the Malacological Society of London*, Vol. 29. The Freshwater Biological Association for Figure 15b, c in *A key to the British fresh-and brackish-water gastropods*, F.B.A. Scientific Publication No. 13. The Field Studies Council and J. H. and M. Crothers for Figure 9 in *Field Studies*, Vol. 5. The Zoological Society of London for Plate 1 in Bramble, F. W. R. & Cole, H. A., *Proceedings of the Zoological Society of London*, Vol.109. V. Driver (Du Heaume) for Figures on pp. 365, 424, 438 in Wheeler, A., *The fishes of the British Isles and north-west Europe*. Cammermeyers Forlag, Oslo, The Institute of Marine Research, Bergen, and The Zoologisk Museum, Bergen, for Plate 9 (part), Plate 10 (part), Plate 212 (part), Plate 225 (part), Plate 236 (part) in Sars, G. O., *An account of the Crustacea of Norway*. Vol. 1 Amphipoda; and Plate 31 (part), Plate 34 (part), Plate 43 (part) in Sars, G. O., *An account of the Crustacea of Norway*. Vol. 2 Isopoda. Collins for Figure 53 in Yonge, C. M., *The sea shore*. HMSO and The Natural History Museum, London for Figure 17E in Irvine, L. M. & Chamberlain, Y. M., *Seaweeds of the British Isles*. Vol. 1 *Rhodophyta* Part 2B. G. Rogerson for Figures 61A, C in Burrows, E. M., *Seaweeds of the British Isles*, Vol. 2. Cambridge University Press for Figure 2A, B in Cadman, P. S. & Nelson-Smith, A., *Journal of the Marine Biological Association of the United Kingdom*, Vol. 73 and Figure 50A in Gibson, R., *British nemerteans. Synopses of the British fauna*, No. 24. Academic Press for Figures 4E, 5B in *British marine isopods. Synopses of the British fauna*, No. 3. Academic Press and The Linnean Society of London for Figures 4C, 5C in Gibbs, P. E., *British sipunculans. Synopses of the British fauna*, No. 12 and Figures 4E, 5B in Naylor, E., *British marine isopods. Synopses of the British fauna*, No. 3. The Ray Society for Plate VI, Figure 3 in McIntosh, W. C., *A monograph of the British marine annelids*, Vol. 1, Part 1 and Figures 26A, I, 70G, 73F, 36A, 35 in Tattersall, W. M. & Tattersall, O. S., *The British Mysidacea*. Editions Lechevalier for Figures 30c, 41l, 150c in Fauvel, P., *Faune de France 5. Polychètes errantes* and Figures 72d, 73a in Fauvel P., *Faune de France 16. Polychètes sédentaires*. The Malacological Society of London for Figure 3D in Matthews, G., *Proceedings of the Malacological Society of London*, Vol. 29. Estuarine and Coastal Sciences Association and the Linnean Society of London for Figures 26B, 49 in Thompson, T. E., *Molluscs: benthic opisthobranchs. Synopses of the British fauna*, No. 8 (2nd edn).

# Introduction

## The seashore

One of the most striking features of the shore is the rich diversity of plant and animal life which is to be found there. A wide range of invertebrates, some highly mobile, others fixed or sedentary, and shore fishes, are a characteristic feature. Brightly coloured lichens often form distinct bands on the high shore; seaweeds may be present in abundance, and on mud flats flowering plants may dominate. Physical factors change rapidly and it is here that the student has the opportunity to observe and study some of the most fascinating adaptations shown by plants and animals.

The dominating force on the shore is the rise and fall of the tide. Tides result from the gravitational forces between the Moon and Sun, and the seas and oceans on the Earth's surface. The tides with which we are most familiar in north-west Europe are semi-diurnal: that is, we usually experience two high tides and two low tides each day. This can be appreciated if we picture the Earth revolving on its axis during the course of a day and passing through a water envelope which has been distorted by the gravitational forces of the Moon and Sun as in Fig. 1. In reality, however, the length of time between successive high tides is about 12 hours 25 minutes; subsequently the tides are approximately 50 minutes later each day, and there will not necessarily be two high tides and two low tides every day. Superimposed on the regular daily pattern of the tides are changes brought about by the relative positions of the Earth, Sun and Moon based on a

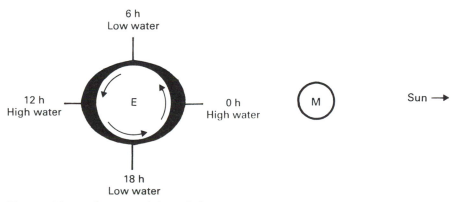

Figure 1   The Earth sectioned through the equator is rotating on its own axis every 24 hours. At 0 hours high water is experienced; 6 hours later low water; 12 hours later high water, and 18 hours later another low water. For detail see text.

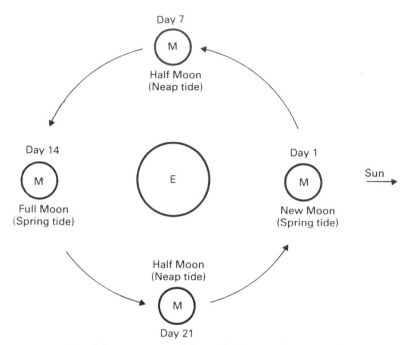

Figure 2  Orbit of the Moon around the Earth showing the times of spring and neap tides.

monthly and a yearly cycle. The Moon revolves around the Earth in approximately 28 days. At the times of full and new moon, the Earth, Moon and Sun are in line and the combined pull of the Moon and Sun on the seas produce tides with a large range. Such tides are referred to as spring tides (Fig. 2). At times of half moon when the Sun and Moon are at right angles to one another, the gravitational pull on the surface of the Earth is less, giving rise to tides with a smaller range. These tides are known as neap tides (Fig. 2).

The revolution of the Earth around the Sun in an elliptical orbit during the course of a year also affects the range of the tides. When the Sun is closest to the Earth its gravitational pull is greatest, and in March and September the combined pull of Sun and Moon results in very large spring tides, the spring and autumn equinoctial tides. Atmospheric pressure variation and wind speed can markedly alter the predicted height of the tide and the configuration of the coastline can also have a substantial effect. In the Bristol Channel, for example, where the tide is funnelled between narrowing headlands, a spring tide range of over 12 m is recorded.

The daily rise and fall of the tide results in different levels of the shore being covered (submersed) and uncovered (emersed) for varying periods of time. Specific tidal levels can be calculated: these prove to be useful reference points and five tidal levels are commonly referred to by shore ecologists. These are average levels and are: the mean high water level of spring tides (MHWS), mean low water level of spring tides (MLWS), mean high water level of neap tides (MHWN) and mean low water level of neap tides (MLWN) (Fig. 3). Mean tide level (MTL) is the average of these four tidal heights. In some texts,

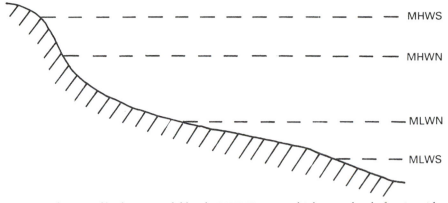

Figure 3   Shore profile showing tidal levels. MHWS – mean high water level of spring tides;
            MHWN – mean high water level of neap tides; MLWN – mean low water level of
            neap tides; MLWS – mean low water level of spring tides.

reference is made to 'extreme' levels (Fig. 4). Animals living at MHWS are covered for
only a short period of time at high water of spring tides and are not covered during neap
tides. Animals at MLWS are uncovered for only a brief period at low water of spring tides
and are permanently submersed at neap tides. These varying periods of submersion and
emersion lead to the development of a gradient of physical conditions such as temperature and desiccation, from high to low shore. The response of animals and plants to this
gradient, together with the effect of biological interactions between groups of organisms, such as competition for space and food, lead to zonation. Zonation can be seen on
rocky shores all around the world and is the occurrence of different species of plants and
animals at different levels on the shore.

    In his book on the ecology of rocky shores, Lewis (1964) divided the shore into three
major zones marked by the presence of conspicuous and widespread plants and animals.
The nomenclature proposed by him is shown in Fig. 4. It is important to note that the
zones are not defined by reference to the tidal levels described above. The highest zone
on the shore is the littoral fringe, the upper limit of which is marked by the upper limit
of the periwinkles and black lichens. The middle zone is the eulittoral zone, the upper
limit of which is marked by the upper limit of the acorn barnacles. The lowest zone is
the sublittoral zone which extends below low water but its upper limit, marked by the
presence of large laminarian seaweeds, can be explored at low water of spring tides. The
littoral fringe and the eulittoral zone together are known as the littoral zone.

    The upper limits of the zones on rocky shores are extended vertically on coasts which
are exposed to heavy wind and wave action (Fig. 4) as a result of spray being carried far
up the beach. In very exposed situations, the upper limit of the littoral fringe is raised by
many metres forming a broad zone extending well above the level of the highest tide. On
sheltered shores the littoral fringe is narrow. Exposure to wave action also affects the
range of plants and animals found on the shore, as some species are more tolerant of
exposure than others, and the density, diversity and form of the seaweeds on rocky
shores are useful indicators of exposure to wave action (see Ballantine, 1961). The upper

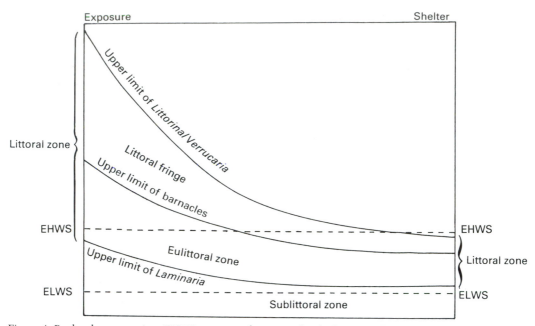

Figure 4  Rocky shore zonation. ELWS – extreme low water level of spring tides; EHWS – extreme high water level of spring tides (after Lewis, 1964).

reaches of the shore are harsh physical environments and it is not surprising to find that the number of different species of plants and animals is lower there compared with the middle and lower reaches where conditions are more favourable.

The distribution of littoral organisms rarely coincides with the boundaries of the defined zones explained above. As a result, the terms lower shore, middle shore and upper shore are often used by ecologists and have been used extensively in this text. The lower shore extends from the upper sublittoral to include the lower part of the eulittoral zone; the middle shore refers to the middle region of the eulittoral, and the upper shore to the upper eulittoral and littoral fringe.

Anyone walking or browsing on a rocky shore will immediately be aware of the plants and animals living there, but on sandy and muddy beaches the animals generally live buried in the sediment. The size of individual particles making up the sediment is of great importance as it determines the stability of the shore and the nature and abundance of the fauna. Coarse pebble and shingle deposits are unstable habitats supporting few species. When waves break on such beaches, the pebbles roll against one another and are likely to crush the fauna. When the tide retreats, the water drains quickly and little is retained in the spaces between individual particles. Very few animals are adapted to living in such harsh conditions, whereas sandy shores offer more stable habitats for burrowing animals and retain greater quantities of water during emersion. Generally, fine sands contain more species and individuals than coarse sediments. Sand is largely made up of quartz particles, the characteristic yellow-brown colour resulting from iron deposits on the grains. Some sands contain substantial quantities of calcareous material.

Sheltered sandy beaches support large numbers of animals but the number of different species is less than that found on rocky shores.

The finest particles, known as silt and clay, are deposited in sheltered areas; in such places mud-flats occur. Here, the increased stability of the sediment allows seaweeds and salt-tolerant flowering plants to become established, while their presence helps to stabilize the mud surface. Muddy shores are found most frequently in estuaries, where there is a heavy load of particulate matter in suspension, mainly brought down by rivers as runoff from the land. When fresh and salt water meet, the fine particles tend to flocculate and settle to form the mud banks so characteristic of estuaries. There is little circulation of water through mud deposits and this leads to stagnant and anaerobic conditions. Just under the surface of mud, oxygen may be completely lacking and in such situations anaerobic bacteria flourish, producing the characteristic smell of hydrogen sulphide so often detected when muddy deposits are disturbed. Anaerobic conditions are also found in sandy deposits, generally a few centimetres beneath the surface, and are marked by a conspicuous black layer in which the sediment is discoloured by sulphide deposits produced through the metabolism of anaerobic bacteria. The fauna of mud-flats survives in anaerobic sediments by a variety of morphological and physiological adaptations and can take advantage of the rich supplies of food available in the form of organic debris.

Organisms living in estuaries are further subjected to constantly changing salinities and because of the harshness of the physical conditions estuaries are frequently referred to as stress environments. The diversity of species is low compared with marine and freshwater habitats but abundance of individual species is high. Productivity of estuaries is high and the high densities of invertebrates makes intertidal mud-flats important feeding grounds for fishes such as flounders which migrate into the estuary on the tide, and for a wide variety of birds which feed as the tide recedes.

## Collection of specimens

One of the prime concerns of all who use the shore should be to cause as little disturbance as possible to the animals and plants living there. As far as possible specimens should be identified in the field, where necessary with the aid of a good-quality hand lens. Specimens should only be collected when it is essential to do so and even then as few as possible should be taken. It is as well to remember that many species will survive if returned to the beach from which they were collected after careful handling and observation in a laboratory. Before going to the shore it is essential to have information on any pecularities of the rate of ebb and flow of the tide and the times of high and low water for that area. Such data can be calculated for the British Isles and north-west Europe from Volume 1 of the *Admiralty Tide Tables* published by the Hydrographer of the Navy, but most coastal authorities now produce inexpensive tide tables covering the coastline within their district. Sampling should begin just before the time of low water, starting on the lower shore and moving up well in advance of the incoming tide. Special care is required when working in estuaries, where tidal currents can be very fast. Always seek local knowledge before working on sand- and mud-flats.

The type of shore being studied will determine the method of sampling, and survey methods are described in Baker & Wolff (1987). Many animals are immediately visible on rocky shores, but careful searching of crevices and rocky overhangs will reveal many more specimens. The moist conditions under a dense cover of seaweed offer refuge for a diversity of plant and animal species, while the surface of the weed is often colonized by hydroids, bryozoans and tubiculous polychaete worms. The holdfasts of the large laminarian seaweeds provide shelter for a surprising variety of small invertebrates, and boulders and large rocks can be lifted (*but always returned to their original position*), to reveal brittle-stars, crustaceans, polychaete worms, chitons and sea-anemones. Although one cannot emphasize too strongly the need for conservation on the shore, it can be very rewarding to take back to the laboratory small amounts of scrapings of, for example, acorn barnacles, encrusting sponges and the holdfasts of laminarians for examination in seawater under a microscope. The empty shells of barnacles provide a niche for small gastropods, bivalves and crustaceans, while sea-spiders and crustaceans are often found associated with sponges.

On sandy shores, where the fauna lives buried in the sediment, sampling has to be carried out by digging. The surface few centimetres of sediment are removed and wet sieved through a brass sieve, generally of mesh diameter of 0.5 mm. In this way the larger animals, known as the macrofauna, are retained for examination. Smaller animals, members of the meiofauna, pass through the sieve and are not included in this book. They are adapted to living in the tiny spaces between the sand grains and different sampling techniques are required for their capture.

Most macrofaunal species live in the surface 50 mm or so of sediment but there are exceptions. For example, the lugworm lives deep in the sediment, betraying its presence by worm casts on the surface. Animals living in mud often construct permanent burrows and the practised eye can detect the openings of these on the surface. Such specimens can be dug for selectively, but it must be borne in mind that much damage can be done to sand- and mud-flats by indiscriminate and reckless digging.

## Nomenclature and classification

All organisms described in this text have been given their scientific name. The system of scientific nomenclature in use today stems from the work of the Swedish naturalist Linnaeus, and the publication of the 10th edition of his *Systema Naturae* in 1758. This marked the beginning of the binomial system of nomenclature by which plants and animals are given two names. The first (the generic name) is the name of the genus to which the organism belongs and begins with a capital letter. The second is the species or specific name, which begins with a small letter. The name of the person (or persons) who assigned the specific name follows the name of the organism. This person is known as the author. If, since the original use of the specific name, that species has been transferred to a different genus, the name of the author is in parentheses. Two names are often written after the scientific name of a plant. This means that subsequent to its original description by the first author (whose name is in parentheses), the plant has been transferred to a new genus by the second named author. It should be noted that in most texts the names of the authors of plants are given in abbreviated form. The internationally accepted system of nomenclature using the scientific name followed by the author ensures that there is no ambiguity regarding the species under consideration. In some cases, the scientific names have undergone revision and here the former, sometimes more familiar, scientific name has also been included in the text. Where appropriate, the common name or names are given.

Organisms are grouped according to their similarities and affinities and are thus arranged in a system of classification. Closely related species are grouped into genera, and genera into families. Families are grouped into orders; orders into classes and related classes are grouped into phyla. Morphological and anatomical features are used by taxonomists when considering classification and in some cases the creation of subphyla, subclasses and suborders has proved useful.

## REFERENCES
### The seashore

Ballantine, W. J. (1961). A biologically-defined exposure scale for the comparative description of rocky shores. *Field Studies*, 1, 1–19.

Barnes, R. S. K. (1984). *Estuarine biology*, 2nd edn. London: Arnold.

Barnes, R. S. K. (1994). *The brackish-water fauna of northwestern Europe*. Cambridge: Cambridge University Press.

Barnes, R. S. K. & Hughes, R. N. (1988). *An introduction to marine ecology*, 2nd edn. Oxford: Blackwell Scientific.

Boaden, P. J. S. & Seed, R. (1985). *An introduction to coastal ecology*. Glasgow: Blackie.

Lewis, J. R. (1964). *The ecology of rocky shores*. London: English Universities Press.

McLusky, D. S. (1981). *The estuarine ecosystem*. Glasgow: Blackie.

Moore, P. G. & Seed, R. (eds.) (1985). *The ecology of rocky coasts*. London: Hodder & Stoughton.

Yonge, C. M. (1949). *The sea shore*. London: Collins.

### Collection of specimens

*Admiralty Tide Tables*. Vol. 1. *European waters including Mediterranean Sea*. Published by the Hydrographer of the Navy. Available from agents for the sale of Admiralty Charts.

Baker, J. M. & Wolff, W. J. (eds.) (1987). *Biological surveys of estuaries and coasts*. Estuarine & Brackish-Water Sciences Association Handbook. Cambridge: Cambridge University Press.

Lincoln, R. J. & Sheals, J. G. (1979). *Invertebrate animals. Collection and preservation*. Cambridge: British Museum (Natural History) and Cambridge University Press.

Nichols, D. (ed.) (1983). *Safety in biological fieldwork. Guidance notes for codes of practice*, 2nd revised edn. London: Institute of Biology.

### Nomenclature and classification

Jeffrey, C. (1977). *Biological nomenclature*, 2nd edn. London: Arnold.

# Design and layout of the book

This book has been designed both as a guide to the identification of the common and widespread organisms of the shore and as a biological text giving information on their biology and ecology. It is intended for use by students and teachers, and also by naturalists and others who might have little or no formal scientific training.

In the following pages, the different groups of organisms are arranged in phylogenetic sequence. Readers who are unable initially to assign an organism to its phylum or group are referred to the illustrated key on p. 10. When the phylum or group to which an organism belongs is known, reference can be made directly to the appropriate chapter where the reader will find an outline classification of the group or phylum in which the classes, subclasses, orders, etc. included in the text are shown in bold type. A brief statement on the morphology and biology of the group is also included. For most groups a simple dichotomous key is provided. In some cases this is to the level of families and leads to a full statement of family characteristics, followed where appropriate by a key to species. In other cases the phylum or group has been keyed directly to species. Morphological features are the basis on which identification is made but in many cases the type of substratum on which the organism is found, position on the shore and distribution are important characteristics.

For each species there is a statement of diagnostic features, a drawing and notes on its biology. The names in brackets following the species name and authority are the synonyms by which the species might be known in older texts. Common names are included as appropriate. Scientific terminology used in the description of species has been kept to a minimum and is explained in the introductory sections at the beginning of each chapter. For ease of reference a glossary is given at the end of the book. A scale has not been included on the drawings but the size of each species is given in the diagnostic features.

The inclusion of family and species keys followed by a full statement of diagnostic features for each species has led to some repetition, but this enables those readers more familiar with the identification of shore organisms, and using the book as a reference text, to turn directly to an appropriate page reference where their provisional identification of an organism can be confirmed and where they will find detail on the general biology of the species in question.

*The keys cover only those organisms included in this text and readers will undoubtedly come across species not included here. For this reason, every effort has been made to prevent false identification. Attention is drawn to cases where confusion might arise. Where identification to species is beyond the scope of this book, identification has been made to genus. When an organism has been identified to a family, it should be checked against the family characteristics before proceeding further. When identification to species has been made, this should be checked carefully against the diagnostic features, the drawing and the habitat characteristics given for that species. If a specimen cannot be identified from the information given in this book, reference should be made to the specialist keys listed at the end of the section.*

# Illustrated guide to the plants and animals of the shore

The following guide has been designed to help those readers who are unable initially to assign an organism to its correct phylum or class. The guide is divided into nine groups, each recognized by the general characters given, for example, encrusting, worm-like, etc. Each of these groups is broken down into subdivisions which are described and illustrated, and for each subdivision a page reference is given where further information can be found. In a few cases the guide will indicate that an organism belongs to one of two or more subdivisions, and for each a page reference is given. To design a guide covering a broad range of plants and animals for use by readers with widely differing background knowledge is difficult as its success will to some extent be affected by subjective assessment of the characters listed. Representative sketches of each group of organisms have been included as a further aid to identification.

## Group 1. Plants and plant-like animals

Plant-like in form, generally attached at one end. Includes obvious plants and plant-like animals.

1 Green, brown or red in colour. Form varied; branched, unbranched, thread-like, tubular, membranous, leaf-like, strap-like, globular. Delicate, tough, leathery or calcareous. A few millimetres to several metres in length. Mainly on rocky shores, also in estuaries. Widespread at all levels of the shore.

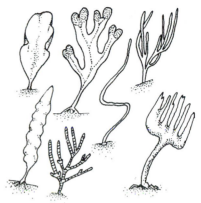

Seaweeds (Algae) (p. 29)

2 Upright or pendulous, often in tufts, normally brittle and dry. Brownish-black or greyish-green in colour. Small, usually no more than 10 mm in height. On rocky shores, extending from the terrestrial fringe to the middle shore.

Lichens (Lichenes) (p. 74)

3 Green, with flowers, seeds and roots. Typically growing in estuaries.

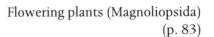

Flowering plants (Magnoliopsida) (p. 83)

4  Tubular vase or urn-like in shape, may be
   branched. With single, conspicuous opening at
   free end. Spongy. Attached to rocks, stones,
   seaweeds. Lower shore.

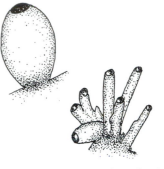

Sponges (Porifera) (p. 88)

5  Branching growths; delicate, sometimes tufted.
   Whitish, light brown or buff in colour. Tentacles
   retractile or non-retractile. Usually no more than
   50 mm in height. Attached to stones, shells,
   seaweed, etc. Middle shore and below.

Sea-firs or hydroids
(Cnidaria: Hydrozoa) (p. 97)
or Sea-mats (Bryozoa) (p. 418)

6  Delicate inverted bell/trumpet shape, attached at
   base by stalk to seaweeds and stones on the
   lower shore. Eight groups of short tentacles
   around margin.

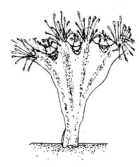

Stalked jellyfish (Cnidaria: Scyphozoa)
(p. 124)

7 Soft bodied, attached at base. Many tentacles at free end, best seen when animal is undisturbed under water. Colour varied, red, green, purple, white, orange. Middle and lower shore.

Sea-anemones (Cnidaria: Anthozoa) (p. 125)

8 Fleshy lobes, sometimes finger-like. When undisturbed, surface covered with many small, pinnate tentacles. Colour varied, white, yellow, orange, pink. Attached to rocks, stones and shells. Lower shore. **Do not confuse with sea-mats under 10 below.**

Dead men's fingers (Cnidaria: Anthozoa) (p. 126)

9 Soft bodied with many long tentacles; withdraws into hard, ridged case when disturbed. Attached to rocks, stones, shells. Lower shore (see also Group 6.1, p. 23).

Stony corals (Cnidaria: Anthozoa) (p. 139)

10  Irregular, lobed gelatinous mass, may be
    branched. Surface smooth with large number of
    tiny openings through which retractile tentacles
    project. Colour varied, yellowish-brown, brown,
    grey. Lower shore. Often washed ashore after
    storms. **Do not confuse with dead men's fingers
    under 8 above.**

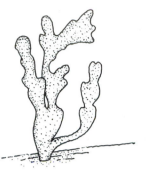

Sea-mats (Bryozoa) (p. 418)

11  Flattened fronds. Brittle and dry. Buff in colour.
    May be mistaken for dried seaweed. On the
    strandline.

Sea-mats (Bryozoa) (p. 418)

12  Flower-like with five pairs of feathery arms.
    Attached to rocks and walls of crevices on the
    lower shore and sublittoral.

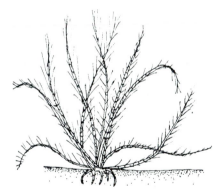

Feather-stars (Echinodermata:
Crinoidea) (p. 433)

**Group 2. Encrusting, without a hard shell, shell plates or a hard tube**
Organisms firmly attached to rock, seaweed, shells, etc. May be difficult to remove without damage.

1  Red, pink or purplish, irregular, hard patches on rock surfaces and on sides of rock pools. Sometimes as a thin, paint-like film. Sometimes thick and folded. Middle and lower shore. **Do not confuse with lichens under 2 below.**

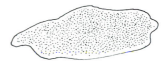

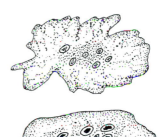

Encrusting red seaweeds (Algae) (p. 54)

2  Irregular patches on rocks and stones; sometimes as a very thin, paint-like film. Tough, brittle texture. Often with conspicuous, saucer-shaped structures on surface. Colour varied, black, yellow, orange, white. Extending from the terrestrial fringe to the middle shore.
**Do not confuse with encrusting red seaweeds under 1 above.**

Lichens (Lichenes) (p. 74)

3  Irregular, spongy patches on rock, especially under overhangs and in crevices, and among seaweed holdfasts. Sometimes with raised, volcano-like openings on surface. Colour varied, yellow, green, red. Lower shore.

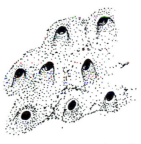

Sponges (Porifera) (p. 88)

4  Irregular patches on rocks, stones, shell and
   especially seaweeds. Made up of a large number
   of regular compartments, sometimes with 'hairy'
   appearance. Each compartment houses a tiny
   individual which has retractile tentacles. Usually
   brownish in colour. Lower shore.

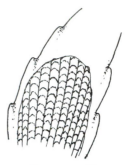

Sea-mats (Bryozoa) (p. 418)

5  Irregular, gelatinous patches on rocks, seaweed
   holdfasts, etc. Surface may be marked with
   flower-like or wavy line patterns. Brightly
   coloured, blue, yellow, brown, orange. Lower
   shore.

Sea-squirts (Chordata: Ascidiacea)
(p. 467)

**Group 3. Attached at base, lacking tentacles, gelatinous. Not plant-like***

1  Cylindrical or oval-shaped body attached at base
   or along one side. Usually gelatinous but some
   may have adhering sand and shell particles. Two
   openings at or close to free end. Occur singly or
   in groups attached to a variety of substrata on the
   lower shore. May be up to 120 mm in height.

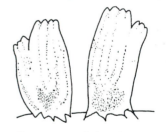

Sea-squirts (Chordata: Ascidiacea)
(p. 467)

* NB. When emersed and contracted the tentacles of some sea-anemones are not obvious and as a result
  they might be wrongly assigned to this group (see Group 1.7, p. 13).

2 Rounded or club-shaped body with one to several conspicuous openings on surface. May be covered with sand or mud. Attached to a variety of substrata on the lower shore. Up to about 50 mm in height.

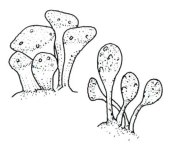

Sea-squirts (Chordata: Ascidiacea)
(p. 467)

### Group 4. Worm-like animals, varying in shape from long and rounded to flat and leaf-like

A large assemblage of animals varying widely in size but all 'worm-like'. Found under stones and rocks, burrowing into sediments and in tubes made of various materials.

1 Body long and slender, non-segmented. Ring of slender, unbranched tentacles at one end. Pinkish-brown in colour. In sand, mud or rock crevices. May be in soft, felt-like tube. Lower shore.

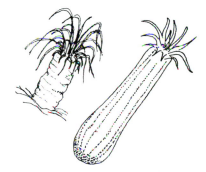

*Cerianthus* and burrowing sea-anemones (Cnidaria: Anthozoa)
(p. 96)

2 Body thin and leaf-like, non-segmented. Moves over substratum with gliding motion. Small, generally less than 30 mm in length. Eye spots and tentacles often visible. Under stones, seaweed, etc. Widely distributed on shore.

Flatworms (Platyhelminthes) (p. 145)

3   Long, ribbon-like, non-segmented. Body often
    slimy, varying in length from a few millimetres to
    several metres. Colour varied, yellow, black,
    brown. In crevices, under stones and seaweeds,
    buried in mud and sand. Lower shore.

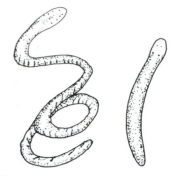

Ribbon or proboscis worms (Nemertea)
(p. 150)

4   Body flattened, leech-like, with posterior ventral
    sucker. On the gills of bivalve molluscs such as
    the cockle.

*Malacobdella* Nemertean (Nemertea)
(p. 156)

5   Cylindrical, superficially segmented body with
    anterior proboscis covered in spines and posterior
    'tassle-like' appendage. Pinkish-brown in colour.
    Burrowing in mud and muddy-gravel on the
    lower shore.

Priapulans (Priapula) (p. 158)

6   Segmented body, with lateral bristles, often borne on conspicuous paddle-shaped outgrowths. Dorsal surface sometimes covered with dense bristles or scales. Anterior end often with eyes and tentacles. Body shape varies from broad and oval to long and slender. Found under stones and rocks, and burrowing in mud and sand. Some live in tubes constructed from mucus, mud, sand and shell fragments. Widespread on shore.

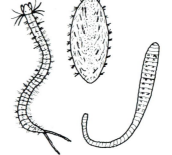

Bristle worms (Annelida: Polychaeta)
(p. 160)

7   Body long, unsegmented. Anterior end covered with small, whitish shell plates. Animal secretes a calcareous tube. Burrows into submerged timbers such as piles, piers, wooden ships, etc.

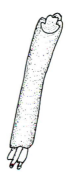

Shipworms (Mollusca: Bivalvia) (p. 326)

8   Body unsegmented and cylindrical. Divided into narrow anterior region and somewhat wider posterior region. Lobes or tentacles at anterior end. Yellow, brown, grey in colour. In burrows in sand or gravel, or in empty shells. Lower shore.

Sipunculans (Sipuncula) (p. 413)

9  Body unsegmented. Divided into narrow anterior region with ventral groove and cylindrical, wider posterior region. Colour varied, yellow, grey, blue, pink. In burrows in mud and sand, and in rock crevices. Lower shore.

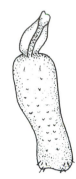

Echiurans (Echiura) (p. 416)

10  Body unsegmented, narrow, cylindrical. Bears horseshoe- or crescent- or oval-shaped group of tentacles at anterior end. Greyish, yellowish in colour. Lives in a chitinous tube, either encrusting, or burrowing into rocks, shells, sand and gravel. Lower shore.

Phoronids (Phorona) (p. 430)

11  Body elongate, unsegmented. Branched tentacles at anterior end. May be tacky to the touch. Pinkish-brown in colour. In sand and muddy-sand, and in crevices. Lower shore.

Sea-cucumbers (Echinodermata: Holothuroidea) (p. 458)

12  Body unsegmented. Divided into three regions, a conical anterior region, a short cylindrical collar-like region and a long, narrow posterior region. In U-shaped burrows in sand and muddy-sand; under rocks and stones. Lower shore.

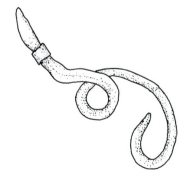

Acorn worms (Hemichordata) (p. 464)

**Group 5. Animals with jointed appendages; mobile**
Includes a wide range of animals in the Arthropoda (p. 329).

1 Animals with four pairs of legs. Spider-like. Body elongate and divided into sections. Among stones, seaweeds, sea-firs, sea-anemones, etc. Middle and lower shore.

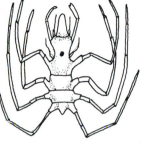

Sea-spiders (Pycnogona) (p. 330)

2 Small, delicate, often transparent shrimp-like animals. Legs with two branches, lacking pincers. Free-swimming, often in swarms in areas of reduced salinity. Do not confuse with prawns and shrimps under 5 below.

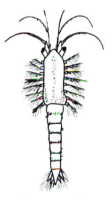

Opossum shrimps or mysids (Crustacea: Mysidacea) (p. 344)

3 Body shape resembles woodlouse. Seven (rarely five) pairs of similar legs. Wide variety of habitats; in crevices, under stones and seaweed, burrowing in sand, mud, old timbers. Widely distributed.

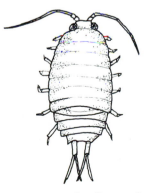

Sea-lice and sea-slaters (Crustacea: Isopoda) (p. 351)

4  Body slender, laterally compressed. Seven (rarely five) pairs of legs, not all alike. Under stones and among seaweed; when disturbed often seen to move on one side. Also found burrowing in sand, mud, gravel. On the undersurface of jellyfishes. Widely distributed.

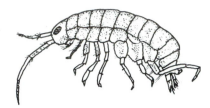

Includes sand-hoppers
(Crustacea: Amphipoda) (p. 362)

5  Shrimp-like, often transparent. One or more pairs of legs have pincer tips. Usually free-swimming in rock pools, some burrow into sand. Middle and lower shore. **Do not confuse with mysids under 2 above.**

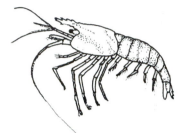

Prawns and shrimps
(Crustacea: Decapoda) (p. 383)

6  Body elongate, strongly built, with powerful pincers. Posterior region extended (lobsters) or curled underneath (squat lobsters). Some (hermit crabs) live in empty shells. Among rocks, stones, in crevices, etc. Lower shore.

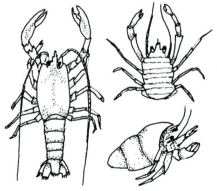

Lobsters, squat lobsters, hermit crabs
(Crustacea: Decapoda) (p. 382)

7 Body rounded, or triangular; strongly built with powerful pincers. Legs may be short or very long and spider-like. Posterior region tightly folded under main body. Among seaweed, rocks, stones, crevices and in sand and mud. Widely distributed.

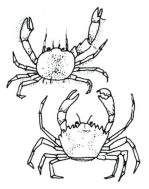

Porcelain crabs, true crabs
(Crustacea: Decapoda) (p. 382)

8 Body divided into three regions, head, middle region with three pairs of legs and a posterior region without legs. Found floating in the surface film of rock pools, and in crevices. Upper shore. Up to about 10 mm body length, but some 2–3 mm.

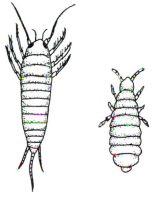

Includes the bristle tails (Hexapoda)
(p. 409)

**Group 6. Animals with a spiny skin, hard outer skin, hard shell, shell plates or hard tube; without jointed appendages (or jointed appendages not obvious)**

1 Soft body with many long tentacles; withdraws into hard, ridged case when disturbed. Attached to rocks, stones, shells. Lower shore (see also Group 1.9, p. 13).

Stony corals (Cnidaria: Anthozoa)
(p. 139)

2  Tube coiled, sinuous or almost straight; attached
   to rocks and seaweed. When undisturbed, many
   tentacles seen to project from anterior end. Tube
   whitish in colour. Lower shore.

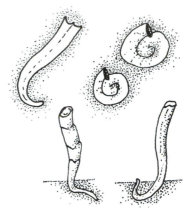

Tubiculous worms
(Annelida: Polychaeta) (p. 160)

3  Oval-shaped animal with eight overlapping shell
   plates. Large muscular foot on undersurface.
   Colour varied, dull red, olive, brownish. Under
   stones, in crevices. Middle and lower shore.

Coat-of-mail shells or chitons
(Mollusca: Polyplacophora) (p. 214)

4  Body housed in single shell; spiral, cone-shaped
   or flattened. On rocks, stones, in crevices, among
   seaweed; also on sand and mud. Widespread on
   shore.

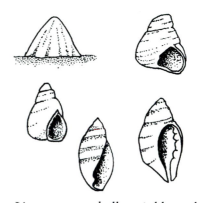

Limpets, top-shells, winkles, whelks
(Mollusca: Prosobranchia) (p. 219)
**but** see *Acteon tornatilis* and *Retusa*
*obtusa* (Mollusca: Opisthobranchia)
(p. 261) **and** (Mollusca: Pulmonata) (p. 276)

5 Body housed in shell of two parts hinged
together. Attached to rock, stones, also in gravel,
sand and mud. Widespread on shore, often in
dense beds.

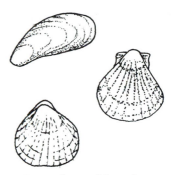

Mussels, cockles, clams, scallops
(Mollusca: Bivalvia) (p. 277)

6 Soft brown stalk with 'head' bearing number of
hard, whitish-grey plates. When under water
jointed appendages protrude from between
plates. Often washed ashore on floating
driftwood or spongy float.

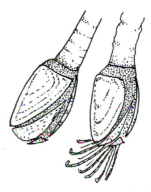

Stalked barnacles
(Arthropoda: Cirripedia) (p. 335)

7 Shell plates form cone attached to rocks and
shells. Whitish in colour. When under water the
apex of the cone opens and jointed appendages
protrude from between plates. At all shore levels
and in dense aggregations on the upper shore.

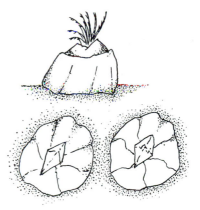

Acorn barnacles
(Arthropoda: Cirripedia) (p. 336)

8   Body star-shaped, five or more arms which are
    not sharply marked off from the central area.
    Surface rough, spiny. Lower shore. On rocks, also
    sand and mud.

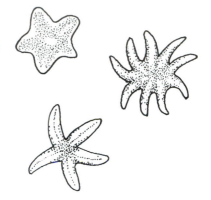

Starfishes (Echinodermata: Asteroidea)
(p. 435)

9   Body star-shaped, usually five slender arms
    which are sharply marked off from central area.
    Surface rough, spiny. Lower shore. In crevices,
    under rocks; also on sand and mud.

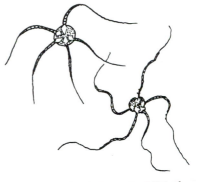

Brittle-stars
(Echinodermata: Ophiuroidea) (p. 444)

10  Body gobular, heart-shaped, oval or sometimes
    rather flattened, rigid and covered with spines.
    Lower shore. On rocks, also in sand and mud.

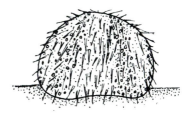

Sea-urchins
(Echinodermata: Echinoidea) (p. 450)

### Group 7. Soft and slug-like or leathery and cucumber-shaped; creeping

1 Body soft and slug-like, head with tentacles. Surface may be covered with finger-like or branched outgrowths. Often very brightly coloured. In rock pools, among seaweed; sometimes on saltmarshes. Lower shore.

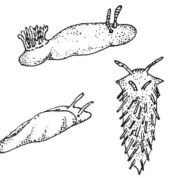

Sea-slugs (Mollusca: Opisthobranchia) (p. 261) **but** see *Lamellaria perspicua* (Mollusca: Prosobranchia) (p. 252)

2 Body cucumber- or sausage-shaped; skin may be rough and leathery. Ring of branched tentacles at anterior end. Brown, black in colour. On rocks, mud and sand. Lower shore.

Sea-cucumbers (Echinodermata: Holothuroidea) (p. 458)

### Group 8. Soft and gelatinous, swimming or floating

Animals normally found floating or swimming in the open sea but often stranded in rock pools and along the strandline.

1 Floating in surface film with sail or gas-filled float projecting above surface. Tentacles, sometimes very long, on undersurface.

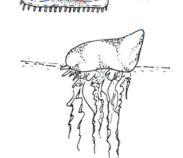

Jack Sail by-the-Wind, Portuguese Man o'War (Cnidaria: Hydrozoa) (p. 117)

2 Body saucer- or bell-shaped with arms and
tentacles trailing from undersurface. Just below
surface of sea.

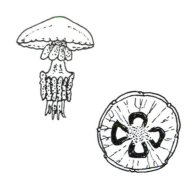

Jellyfishes (Cnidaria: Scyphozoa) (p. 119)

3 Globular or thimble-shaped body. May have pair
of long, trailing tentacles. Just below surface of
sea.

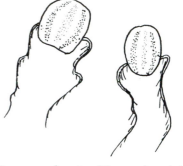

Sea-gooseberries (Ctenophora) (p. 142)

**Group 9. Fishes**

1 Body varies in shape from eel-like to that of
typical fish with conspicuous fins, gill covers,
head and tail. In rock pools and under stones.
Also on sandy shores and in estuaries. Lower
shore.

Shore fishes (Chordata: Osteichthyes)
(p. 480)

# Algae

Seaweeds are one of the most conspicuous features of rocky shores, and in areas where there is little wave action they may be present in such high densities that they dominate the shore. In common with higher plants, they are capable of photosynthesis and are thus restricted to the shore and shallow and surface waters where there is adequate light. Seaweeds are macroscopic marine algae. Most are firmly anchored to the substratum by a holdfast, although unattached forms are frequently found in calm waters.

Seaweeds are divided into three classes, commonly known as the green (Chlorophyceae), brown (Fucophyceae = Phaeophyceae) and red (Rhodophyceae) algae.

**Class Chlorophyceae** Green algae; usually green in colour, may be bleached white. Chlorophylls *a* and *b* are the main pigments and starch is the main storage product. In this respect they resemble the higher plants.

**Class Fucophyceae (Phaeophyceae)** Brown algae; colour varies from olive-green to brownish-black. Chlorophylls *a* and *c* are present but the green colour is masked by xanthophyll pigments, which give the brown or olive-green colour. The main storage product is a polysaccharide known as laminarin.

**Class Rhodophyceae** Red algae; colour varies from pink to reddish-brown, to purple. Chlorophyll *a* is present but is masked by the phycobilin pigments phycoerythrin (red) and phycocyanin (blue-green); the former being the dominant pigment in the majority of red algae. The main storage product is floridean starch, which chemically is closely related to glycogen.

The life histories of seaweeds are complex and varied, and may involve vegetative propagation, sexual and asexual reproduction and an alternation of generations involving haploid and diploid phases. Only a general outline can be given here. Alternation of generations usually involves a diploid sporophyte which produces haploid spores, which on germination give rise to haploid gametophytes producing male and female gametes. Fusion of gametes gives rise to a new sporophyte. Where sporophyte and gametophyte are morphologically similar they are known as isomorphic; when different as heteromorphic. The common green seaweed *Ulva lactuca* (p. 36) is an example of an isomorphic life history where sporophyte and gametophyte are, to all practical purposes, indistinguishable on the shore (Fig. 5). 

The laminarians are good examples of the heteromorphic life history. The conspicu-

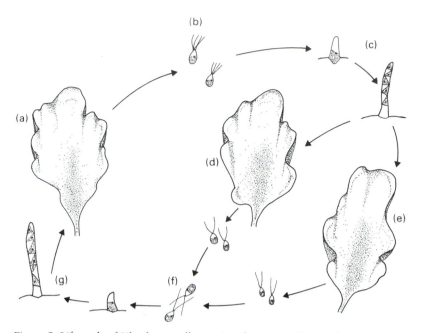

Figure 5  Life-cycle of *Ulva lactuca* illustrating the isomorphic condition (not to scale). (a) Diploid sporophyte, (b) haploid spores, (c) germination of spores, (d), (e) haploid gametophytes, (f) fusion of gametes, (g) germination to produce new sporophyte (after Boney, 1966).

ous plant seen on the shore is the diploid sporophyte. This produces haploid spores which, on germination, give rise to microscopic male and female plants, the gameto-phytes. These produce male and female gametes, and after fertilization the sporophyte develops as a filamentous outgrowth from the female gametophyte (Fig. 6). In some of the most common brown algae, such as *Fucus* (p. 50) and *Pelvetia* (p. 51), however, there is a single morphological form. This can be likened to the sporophyte of the laminarians but, unlike the latter, these species do not undergo alternation of generations. Instead, the gametes are borne in reproductive bodies known as receptacles, seen as conspicuous, swollen areas on the plant and containing many flask-shaped invaginations opening by tiny pores onto the surface. These are the conceptacles: cells lining the wall of the con-ceptacle develop into male and female gametes and, when ripe, these are discharged in a drop of mucus, usually as the thallus dries out during low water. When the tide comes in the gametes are mixed and fertilization occurs. Germination of the zygote and sub-sequent growth produce the macroscopic plant (Fig. 7).

    In the red algae some species have two morphological forms, while others have three, only two of which are free-living, and to add to the complexity these may be isomorphic or heteromorphic. As a generalization, the macroscopic sporophyte produces spores known as tetraspores, because they are produced in groups of four, and on germination

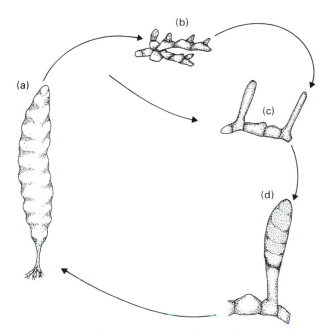

Figure 6  Life-cycle of *Laminaria saccharina* illustrating the heteromorphic condition (not
to scale). (a) Diploid sporophyte, (b) male gametophyte, (c) female gametophyte,
(d) new sporophyte developing on female gametophyte after fertilization by male
gamete (partly after Boney, 1966).

these produce separate male and female gametophytes. The male gametophyte produces
gametes and these fuse with the female reproductive organ which develops on the female
gametophyte. This fusion gives rise to a carposporophyte which produces carpospores
and is often described as being parasitic on the female gametophyte. Germination of the
carpospores and subsequent growth restore the sporophyte generation. In view of this
complexity, it is not surprising that there are examples of different stages in the life
history of a single species which are morphologically distinct and have in the past been
classified as different species.

Zonation of seaweeds is well documented and distinct bands or zones of different
species, such as the fucoids and the laminarians, can be recognized on the shore. It has
also been noted that as a rough generalization, green seaweeds are most common on the
upper shore, the browns on the middle and lower shore and the reds on the lower shore.
This distribution is often accounted for in terms of the relationship between the
efficiency of photosynthesis of the different groups of seaweeds and the changing spec-
tral composition of light with increasing depth of water during high tide. Red light,
which penetrates only the surface layers of water, stimulates rapid photosynthesis in the
green seaweeds, while the additional pigments of the brown and red seaweeds enable
them to utilize the blue and green light, which penetrates deeper. It is now believed,

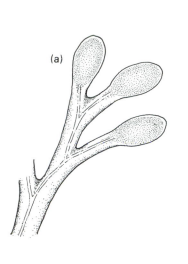
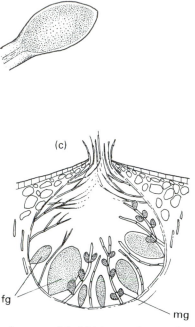

Figure 7  Stages in the development of *Fucus* sp. (not to scale). (a) Mature plant with
         receptacle (b) containing many conceptacles (c). fg, Female gametes; mg, male
         gametes (partly after Boney, 1966).

however, that this interesting hypothesis of chromatic adaptation cannot be applied as
a general rule to the distribution of seaweeds on the shore. The lower limit to which
seaweeds extend must, of course, be related to the penetration of light, but it may be a
function of the quantity rather than the quality of light. It is also worth noting that red
seaweeds are often found quite high on the shore in shaded crevices. Furthermore, there
are interesting examples in which the lower limit of distribution of a seaweed is controlled by competetive interactions between the seaweed and associated fauna. For
example, the lower limit of *Laminaria hyperborea* (p. 44) in some situations is controlled
by the grazing activity of sea-urchins. Although the upper limit of many seaweeds is
determined by physical factors associated with tidal emersion, the lower limits are
believed to be determined by an interaction of factors, including competition with other
species of seaweed and with animals.

Seaweeds are poorly represented on sedimentary shores where there is generally a
lack of solid substrata for attachment, but thrive on rocky shores where their distribution
and density are useful indicators of the degree of exposure of the shore to wave action.
Sheltered rocky shores are dominated by seaweeds, notably *Fucus* (p. 50) and *Ascophyllum* (p. 47), while in exposed areas seaweed cover is much reduced. In some species,
such as *Fucus spiralis* (p. 49) and *Fucus vesiculosus* (p. 50), the morphology of the plant

varies in response to exposure to wave action. The density of seaweeds on the shore has a marked effect on the ecology of the area. Many animals seek refuge in the moist conditions under seaweeds, and the holdfasts of the laminarians support a rich diversity of fauna. The seaweed frond provides a substratum on which invertebrates can settle and interesting algal/faunal associations have developed. It is known, however, that some seaweeds produce antibiotics which deter settlement of invertebrate larvae on at least part of the thallus; some periodically shed the outer layer of cells, so removing attached organisms. Seaweeds and their spores are grazed extensively by herbivorous invertebrates but some minimize the effects of grazing through the possession of tough, sometimes calcareous, cell walls, rapid growth and even the production of deterrent chemicals such as sulphuric acid and polyphenols. Harvesting of seaweeds for human consumption and agricultural and industrial uses is a long-established practice in many countries, and the large-scale culture of commercially important seaweeds is now widely practised, particularly in the Far East.

On some shores there is an abundance of different species of seaweeds and although some 70 species are described in the following pages, this represents no more than the very commonest. This is particularly so of the red seaweeds. In many cases identification is made difficult by variation in form and colour. Morphological characters important in identification are shown in Fig. 8. The main part of the seaweed is known as the frond, and the region of attachment is called the holdfast, which, in some species, is an important taxonomic feature. In some groups, notably the laminarians, there is a conspicuous stalk or stipe between holdfast and frond; holdfast and frond together are referred to as the thallus.

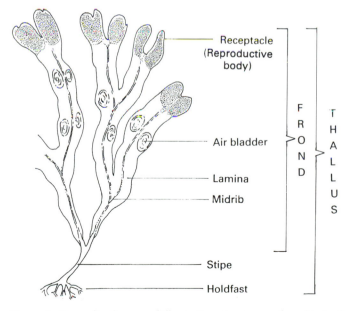

Figure 8  Generalized seaweed illustrating structures referred to in the text.

### Key to the algae

The separation of seaweeds into green, brown and red is an important first step in identification, but it must be stressed that the distinction on the basis of colour is not always clear in the field. Brown seaweeds vary in colour from olive-green to brown and brownish-black, while red seaweeds vary from pink, reddish-brown to purple and almost black. It must also be borne in mind that in some cases colour changes rapidly after death and care must be taken to examine fresh material.

Green algae – usually definite bright green in colour, but note that some specimens may show patches which are bleached white ............................... **Chlorophyceae** (p. 34)

Brown algae – olive-green to brownish-black in colour ................... **Fucophyceae** (p. 39)

Red algae – pink to reddish-brown in colour ................................ **Rhodophyceae** (p. 54)

## Class Chlorophyceae – green seaweeds
### Key to species of green seaweeds

1  Frond solid, spongy in texture. Dichotomously branched, up to 250 mm in height; cylindrical in section. Occasionally a prostrate, spongy mass 5–10 mm thick. Light green to dark green in colour ............................................................ *Codium* spp. (p. 38)

   Frond not so ............................................................................................................ 2

2  Frond thin, flat membranous; with or without a stipe ........................................ 3

   Frond not so ............................................................................................................ 4

3  Frond resembles a lettuce leaf; with wavy margin. With or without short stipe. Generally bright green; up to 300 mm or more in length ..................... *Ulva lactuca* (p. 36)

   Frond wedge-shaped; margins curled; arising from distinct stipe half as long as frond. Up to 10 mm in length. Dark green in colour ............................ *Prasiola stipitata* (p. 34)

4  Frond tubular; elongate, often inflated; sometimes branching. Green to bright green in colour, but often bleached white ............................... *Enteromorpha* spp. (p. 35)

   Frond filamentous, unbranched or branching ................................................... 5

5  (a) Frond unbranched; filaments of a single string of oblong cells which may be visible to naked eye. Attached by a large basal cell .............................. *Chaetomorpha* spp. (p. 37)

   *or* (b) Frond much branched; tufted. Light green to dark green in colour. Characteristically coarse to the touch ......................................... *Cladophora* spp. (p. 37)

   *or* (c) Frond branching to give side branches each of which bears smaller branches arranged as in a feather; glossy; limp to the touch ...................... *Bryopsis plumosa* (p. 37)

**Prasiola stipitata** Suhr ex Jessen    (Fig. 9a)

*Frond thin, membranous, wedge-shaped to oval, arising from distinct stipe half as long as length of frond; margins of frond curled. Dark green in colour. Up to 10 mm in length.*

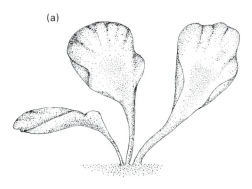

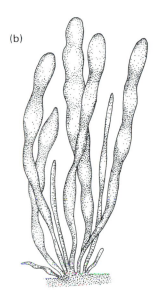

Figure 9  (a) *Prasiola stipitata*. (b) *Enteromorpha intestinalis*.

*P. stipitata* is widely distributed in north-west Europe and around Britain on rocks and stones on the upper shore; it extends into estuaries.

**Enteromorpha** spp.    (Fig. 9b)

*Frond elongate, tubular; sometimes branched; inflated or flattened and spirally twisted; sometimes constricted at intervals. Branching or disc-like holdfast. Green to bright green in colour, often bleached white. Up to 700 mm in length. Small specimens with flattened fronds may be confused with* Ulva lactuca *(below).*

The genus *Enteromorpha* is widely distributed in north-west Europe and some seven species have been recorded from Britain. Of these, **Enteromorpha intestinalis** (Linnaeus) Link is probably the commonest and most widespread. The separation of the species within the genus is difficult and depends on cell detail. The situation is further complicated by the fact that the morphology of a single species can vary in response to environmental conditions, and whether the plant is branched or not is not a reliable species characteristic. *Enteromorpha intestinalis* is found at all levels of the shore and, being tolerant of very low salinity, is particularly abundant in areas where freshwater drainage runs across the beach and in pools on the upper shore, the bright green colour making it very conspicuous. It is also common in estuaries and on saltmarshes. In sheltered situations the fronds often become detached and survive as large, floating masses. In some areas, particularly where there is an abundant supply of nutrients, growth may be prolific and the decaying plants lead to pollution problems.

*Ulva lactuca* Linnaeus    Sea-lettuce (Fig. 10a)

*Frond thin (composed of 2 cell layers), flattened, membranous; margins wavy; with or with-out short, solid stipe; attached by disc-like holdfast. Green to dark green in colour, margins often white. Usually about 100 mm in length but may reach 300 mm or more.*

(a)                                                    (b)

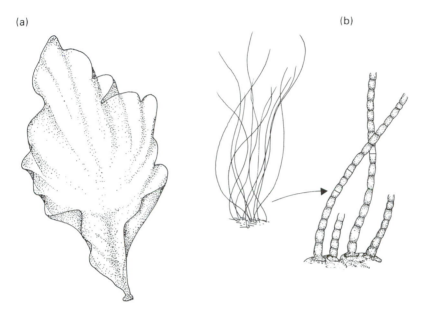

Figure 10   (a) *Ulva lactuca.* (b) *Chaetomorpha* sp.

*U. lactuca* is cosmopolitan in distribution and common over all except the upper shore on rocks and in rock pools. It is also commonly found on mud-flats attached to stones, and shells. In sheltered waters the thallus often attains a large size, sometimes becoming detached, the plants surviving as large, floating masses. The life-cycle is illustrated in Fig. 5.

A second species of *Ulva, Ulva rigida* C. Agardh, is also found on rocky shores all around Britain, especially in pools on the lower shore and into the sublittoral. The frond is dark green in colour and similar to that of *U. lactuca* in being thin and flattened with a wavy margin, but has a distinct stipe and feels somewhat stiff.

*U. lactuca* may be confused with **Monostroma grevillei** (Thuret) Whittrock, which is found in rock pools on the lower shore. The frond is up to 100 mm in length, more delicate than that of *Ulva* and is usually funnel-shaped. Microscopic examination shows that it is composed of a single cell layer.

*Chaetomorpha* spp.    (Fig. 10b)

*Frond fine, filamentous, unbranched; made up of single string of oblong cells, which are sometimes visible to the naked eye. Attached by large basal cell. Green in colour. Frond up to 300 mm in length but usually much smaller.*

There are several species of *Chaetomorpha*, some of which are common and widespread in north-west Europe. They are found attached to seaweeds and rocks, or sometimes as unattached, interwoven masses of filaments.

*Cladophora* spp.    (Fig. 11a)

*Frond filamentous, usually much branched, often tufted; attached by a branched or disc-like holdfast. Light to dark green in colour; up to 300 mm in length.*

(a)                                                                         (b)

Figure 11   (a) *Cladophora rupestris.* (b) *Bryopsis plumosa.*

There are a number of species of the genus *Cladophora* in Britain and north-west Europe, but they are difficult to separate. They occur at all shore levels, often in rock pools; some are tolerant of reduced salinity and extend into estuaries. In sheltered areas they occur as detached, floating masses. **Cladophora rupestris** (Linnaeus) Kuetzing is common all around Britain, occurring from the upper to lower shore in rock pools and under other seaweeds such as *Ascophyllum* (p. 47) and *Fucus* (p. 50). The frond is densely tufted, dark green to bluish-green in colour and characteristically coarse to the touch. It is dioecious.

*Bryopsis plumosa* (Hudson) C. Agardh    (Fig. 11b)

*Frond filamentous, branching to give side branches which each bear smaller branches arising in two opposite rows, the small branches becoming shorter distally, such that each side branch resembles a feather. Light to dark green in colour; glossy, limp to the touch. Up to 100 mm in length.*

B. *plumosa* is widely distributed in north-west Europe and around Britain, where it can be found on the sides of deep pools and under rocky overhangs on the middle and lower shore, and into the sublittoral. It is dioecious.

*Codium* spp.    (Fig. 12)

*Frond cylindrical, formed of many interwoven branching filaments; dichotomously branched; solid, spongy, felt-like texture with many colourless hairs which are best seen when the frond is held under water; attached by disc-like holdfast formed of many fine threads. Dark green or light green in colour. Length depending on species, but generally up to 250 mm.*

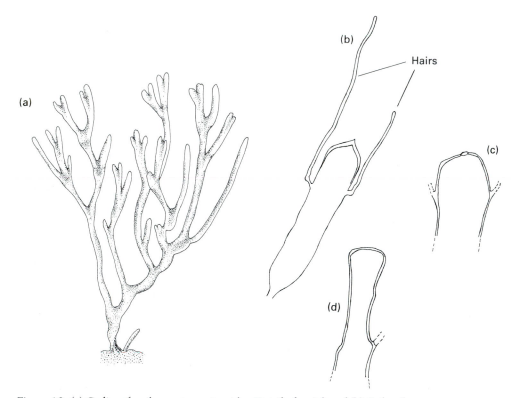

Figure 12  (a) *Codium fragile* ssp. *tomentosoides*. Detail of utricles of (b) *C. fragile* ssp.
              *tomentosoides*, (c) *C. fragile* ssp. *atlanticum*, and (d) *C. tomentosum*. ((c) and (d)
              after Burrows, 1991.)

*Codium* spp. include the two subspecies of **Codium fragile** (Suringar) Hariot, and **Codium tomentosum** Stackhouse. They are unmistakable green seaweeds widely distributed in north-west Europe growing on rock surfaces and in rock pools, their distribution depending on the species. *Codium fragile* (Suringar) Hariot subspecies *atlanticum* (Cotton) Silva is a northern form and extends as far south as Anglesey in the west and Northumberland in the east. *Codium fragile* (Suringar) Hariot subspecies *tomentoso-ides* (Van Goor) Silva is a southern form extending north to south-west Scotland, but is apparently absent from parts of the Irish Sea. *Codium tomentosum* occurs mainly in the south-west of Britain. It is now becoming rare and both pollution and competition from *C. fragile* have been put forward as possible reasons. Separation of the species is based on the shape of the apex of internal microscopic structures called utricles, which project onto the external surface of the frond. They are about 1 mm in length and clearly visible if the filaments of the frond are gently teased apart. In *C. fragile* the utricles terminate in a short point (Fig. 12b, c), while in *C. tomentosum* they are bluntly rounded (Fig. 12d).

Interestingly, there is a not uncommon species of *Codium*, *Codium adhaerens* C. A. Agardh, which grows as an encrusting spongy, felt-like mat of 5–10 mm thick on rock surfaces. It is widely distributed, reaching its northern limit in Scotland.

## Class Fucophyceae – brown seaweeds
### Key to species of brown seaweeds

1  Frond with midrib .................................................................................................... 2
   Frond without midrib ............................................................................................... 7
2  Frond not branching; long, strap-like with wavy margin; clusters of leaf-like
   reproductive bodies on either side of stipe .................................... *Alaria esculenta* (p. 46)
   Frond branching; reproductive bodies on frond, not on stipe ......................... *Fucus* spp. 3
3  Margin of frond serrated ................................................... *Fucus serratus* (p. 49)
   Margin of frond not serrated .................................................................................. 4
4  Frond with paired spherical air bladders on either side of midrib ................................
   ...................................................................... *Fucus vesiculosus* (p. 50)
   Frond without spherical air bladders .................................................................... 5
5  Frond fan-shaped, often inflated on either side of midrib; reproductive bodies at ends of
   branches, narrow pointed. In brackish water and where freshwater streams run onto
   beach ......................................................... *Fucus ceranoides* (p. 48)
   Frond and habitat not as above ............................................................................. 6
6  Frond usually twisted; reproductive bodies at ends of branches, almost round,
   surrounded by narrow rim of frond, the sterile margin. Upper shore    *Fucus spiralis* (p. 49)
   Frond not twisted. Reproductive bodies without sterile margin; often forked and
   pointed. Found on rocky shores, generally with a high degree of exposure to wave
   action ............................................... *Fucus vesiculosus* forma *linearis* (p. 50)
7  Frond large, may be well over a metre in length; strap-like or wide and deeply divided
   (digitate); not branched. Stipe and holdfast well developed. On the lower shore and
   below; the kelps ..................................................................................................... 8
   Frond not so .......................................................................................................... 11

8  Frond long, undivided, strap-like; surface wrinkled; margins wavy. Often in rock pools on the middle and lower shore .............................................. *Laminaria saccharina* (p. 45)

Frond divided (digitate); surface smooth ......................................................................... 9

9  Stipe broad and flat with frilled margin; holdfast large, bulbous, warty .........................
...................................................................................... *Saccorhiza polyschides* (p. 45)

Stipe and holdfast not so ........................................................................................ 10

10  Stipe narrow, smooth, flexible; oval in section. Frond wide, deeply divided (digitate) ....
...................................................................................... *Laminaria digitata* (p. 43)

Stipe narrow, rough, stiff; rounded in section; often with red seaweed attached. Frond wide, deeply divided .................................................. *Laminaria hyperborea* (p. 44)

11  Frond globular or lobed and irregular; often hollow; attached to other seaweeds and shells ......................................................................................................................... 12

Frond not so ......................................................................................................... 13

12  Frond globular, lobed, irregular; solid when young, hollow later; thick walled and shiny; growing on seaweeds, especially *Corallina* ................... *Leathesia difformis* (p. 42)

Frond globular, thin walled and hollow; smooth and papery, not shiny; growing on seaweeds and shells ......................................................... *Colpomenia peregrina* (p. 42)

13  Frond fine, filamentous and much branched, interwoven, slimy, *or* frond cord-like, unbranched, tangled and slimy ......................................................................................... 14

Front not so ......................................................................................................... 15

14  Frond fine, filamentous and much branched; slimy, usually growing on other seaweeds; microscopic to 300 mm in length .................................. *Ectocarpus* spp. (p. 41)

Frond cord-like, unbranched; slimy, covered with colourless hairs. Up to 8 m in length
.............................................................................................. *Chorda filum* (p. 42)

15  Frond button-like with very short stipe; with long, strap-shaped 'thong' up to 2 m in length, growing from centre of button. *Note*: the button-like fronds are without the thong-like outgrowth in the first year .................................. *Himanthalia elongata* (p. 51)

Frond not button-like ............................................................................................. 16

16  Frond with large egg-shaped air bladders at intervals along it; rounded reproductive bodies on short stalks arising from frond. Up to 2 m or more in length; widespread on middle shore ................................................................ *Ascophyllum nodosum* (p. 47)

Frond not so ......................................................................................................... 17

17  Frond branching, bushy, stiff with whorls of tiny branches *or* frond branching, bushy, spiny and nodular with small air bladders. Up to 450 mm in length ............................. 18

Frond branching; not bushy or spiny; may be flattened, cylindrical or channelled, with or without air bladders ............................................................................................... 19

18  Frond stiff, branches with whorls of tiny branches. Dirty brown in colour; up to 250 mm in length ....................................................... *Cladostephus spongiosus* (p. 46)

Frond spiny, rough texture. Reproductive bodies with spines and nodules, at ends of branches; small air bladders below reproductive bodies. Iridescent when under water. Reminiscent of heather plant. Up to 450 mm in length ... *Cystoseira tamariscifolia* (p. 53)

19  Frond delicate, flat; tips of branches rounded, usually divided into two lobes.
Reproductive bodies as fine specks over surface of frond .......... *Dictyota dichotoma* (p. 47)
Frond flat, channelled or cylindrical, with or without air bladders; reproductive bodies
at ends of branches ......................................................................................................... 20

20  Frond curled to form a marked channel; without air bladders; reproductive bodies at
ends of branches. Upper shore. Black and brittle when dry ..... *Pelvetia canaliculata* (p. 51)
Frond flattened or cylindrical; with or without air bladders. Not curled to form a
marked channel. In rock pools on the middle and lower shore ....................................... 21

21  Frond flattened, with pod-shaped air bladders divided by transverse septa.
Reproductive bodies pod-shaped, on short stalks at ends of branches. Main stem
zig-zag ......................................................................... *Halidrys siliquosa* (p. 53)
Frond cylindrical; elongate reproductive bodies at ends of branches; rounded air
bladders may be present below the reproductive bodies .......... *Bifurcaria bifurcata* (p. 52)

*Ectocarpus* spp.   (Fig. 13)

*Frond fine, filamentous, much branched; grows as mass of interwoven filaments; often slimy
in texture. Swollen reproductive bodies at ends of branches. Olive, brown, yellow in colour.
Species range in length from microscopic to about 300 mm.*

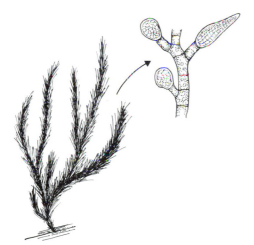

Figure 13  *Ectocarpus* sp. with detail of reproductive bodies (after Newton, 1931).

The genus *Ectocarpus* is widely distributed and common in north-west Europe. It is
found on all but the highest levels of the shore and usually grows on other seaweeds,
including *Laminaria* (below), *Himanthalia* (p. 51) and *Chorda* (below), and on rocks. In
some situations it is abundant. The different species of *Ectocarpus* are difficult to separate and require detailed microscopic examination.

### *Leathesia difformis* (Linnaeus) Areschoug    (Fig. 14a)

*Frond globular, or lobed and irregular; gelatinous; solid and rounded in young specimens, becoming hollow and lobed; thick walled and shiny (cf.* Colpomenia, *below). Yellow-brown in colour and up to 50 mm across.*

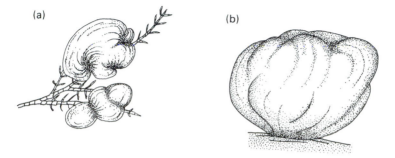

Figure 14   (a) *Leathesia difformis on Corallina.* (b) *Colpomenia peregrina.*

*L. difformis* is widely distributed in north-west Europe growing on other seaweeds, especially *Corallina* (p. 61), on the middle and lower shore. The plants make their appearance in spring and grow rapidly to reach maturity during summer. *Leathesia* is an annual.

### *Colpomenia peregrina* (Sauvageau) Hamel    Oyster thief (Fig. 14b)

*Frond globular, thin walled (cf.* Leathesia, *above) and hollow; when mature dotted with conspicuous brown specks. Smooth but not shiny; papery. Green, olive colour; up to 90 mm across.*

*C. peregrina* is widely distributed in north-west Europe and in Britain is most common in the south-west. It grows attached to seaweeds and shells on the lower shore in sheltered situations. The hollow frond sometimes becomes filled with air during low water and the plant floats on the next tide and in this way has been known to carry away an attached mollusc. *Colpomenia* was first recorded in Britain in the early 1900s, possibly as an introduced species.

### *Chorda filum* (Linnaeus) Stackhouse    Sea-lace, Dead men's ropes, Mermaid's tresses (Fig. 15a)

*Frond cord-like, unbranched, round in section and hollow; densely covered with short, colourless hairs during summer; rising from a tiny, disc-like holdfast. Slimy texture; brown to dark brown in colour. Up to 8 m in length.*

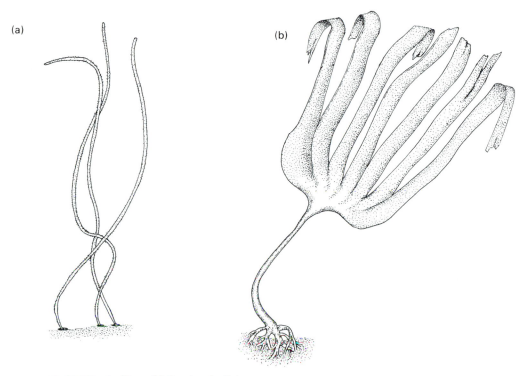

Figure 15 (a) *Chorda filum*. (b) *Laminaria digitata*.

*C. filum* is widely distributed and common in north-west Europe and in some areas is abundant. It is most frequently found in sheltered bays attached to stones and shells on the lower shore, often with the holdfast buried in sand and gravel. Sublittorally it is found to depths of about 20 m; occasionally it occurs higher on the shore in rock pools but is generally intolerant of emersion. The plants often grow together in a tangled mass of entwined fronds, sometimes supporting rich populations of epiphytes and epifauna. *Chorda* tolerates reduced salinity and is found in estuaries. It is a summer annual, disappearing during the winter months. It is often washed ashore in large quantities.

***Laminaria digitata*** (Hudson) Lamouroux    Oarweed, Tangle (Fig. 15b)

*Stipe smooth, flexible, oval in section; expanding into a wide, digitate frond. Without midrib. Attached by branched, slightly dome-shaped holdfast. Glossy, golden-brown in colour. Up to 2 m in length.*

*L. digitata* is widely distributed in north-west Europe and is a conspicuous and characteristic species at low water on most rocky shores. The blue-rayed limpet (*Helcion pellucidum*, p. 227) is often found on the stipe and frond where it can do considerable damage, weakening the stipe. The holdfast offers a refuge for a wide variety of fauna. New growth takes place at the base of the frond where it joins the stipe and, although more or less continuous, is fastest in spring. Older parts of the frond are progressively worn away. It is a perennial.

### *Laminaria hyperborea* (Gunnerus) Foslie    (Fig. 16a)

*Stipe stiff, rough texture, rounded in section; narrowing towards frond, which is expanded and deeply divided into strap-like sections. Without midrib. Holdfast large, much branched and cone-like. Glossy, golden-brown in colour. Up to 3.5 m in length.*

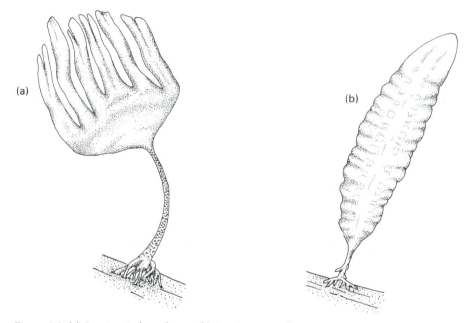

Figure 16  (a) *Laminaria hyperborea*. (b) *Laminaria saccharina*.

*L. hyperborea* is widely distributed in north-west Europe where it is common on the extreme lower shore, often forming a distinct zone below *L. digitata*. It extends into the sublittoral and where conditions are favourable forms dense forests. The rough stipe is often heavily clothed with a wide range of epiphytes such as *Palmaria* (p. 64), *Lomentaria* (p. 66) and *Delesseria* (p. 67), and is often weakened by the browsing activities of the blue-rayed limpet, *Helcion pellucidum* (p. 227). *L. hyperborea* is a perennial. The frond almost ceases growth in summer, new growth commencing the following spring at the base of the old frond which is suddenly shed. The age of individual plants during the first few years of life can, with experience, be determined by counting growth lines laid down in the stipe and revealed by sectioning. Longevity is up to about 13 years in sheltered situations but is reduced on exposed shores where the older plants are susceptible to being torn away by wave action. *L. hyperborea* reproduces over winter.

*Laminaria saccharina* (Linnaeus) Lamouroux   Sea-belt, Sugar kelp (Fig. 16b)

*Stipe smooth, flexible, broadening to a long, undivided frond with wrinkled surface and wavy margins. Without midrib. Branching holdfast. Yellowish-brown in colour. Up to 4 m in length.*

*L. saccharina* is widely distributed in north-west Europe and is often abundant on sheltered rocky shores, where large plants can be found. It occurs on the lower shore and in the sublittoral while small, isolated plants are found in rock pools on the middle shore. The specific name refers to a sweet-tasting, whitish powder which forms on the dried frond. *L. saccharina* is a perennial and although new growth occurs more or less continuously at the base of the frond, growth is most rapid in the first half of the year. Seasonal growth of the stipe results in alternate rings of light and dark tissue being laid down during periods of fast and slow growth, respectively. These rings can be seen in sectioned material and give an indication of the age of the plant. Length of life is up to three years. *L. saccharina* reproduces over winter. The bryozoan *Celleporella hyalina* (p. 427) is often found in the depressions on the frond.

*Saccorhiza polyschides* (Lightfoot) Batters   Furbelows (Fig. 17a)

*Stipe broad, flat, twisted at base and with conspicuously frilled margin; broadens into a wide frond often deeply divided into ribbon-like sections. Without midrib. Holdfast disc-like at first but grows into large, bulbous structure with warty appearance. Golden-brown in colour. Up to 4 m in length.*

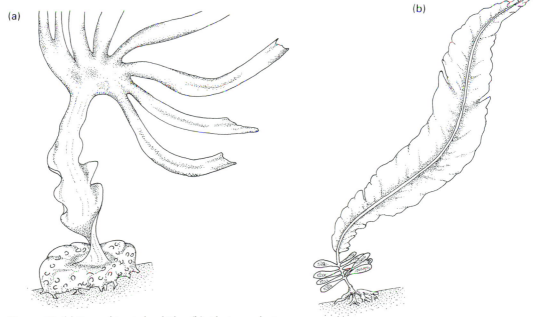

Figure 17   (a) *Saccorhiza polyschides*. (b) *Alaria esculenta*.

*S. polychides* is widely distributed in north-west Europe and is often common at extreme low water and in the shallow sublittoral, particularly on sheltered shores. The morphology of the frond shows considerable variation depending on the nature of the habitat. Plants in areas of weak current flow have broad, undivided fronds while those in areas of strong current have long fronds deeply divided into many sections. Plants growing in areas without current flow but subjected to wave action have short fronds divided into relatively few sections. *Saccorhiza* is a very fast growing annual and a growth rate of more than two meters in two months has been recorded. Reproduction occurs during autumn and winter after which the fronds decay and are shed leaving the holdfasts attached to the rock. Eventually these become detached, many becoming washed up on the strandline.

*Alaria esculenta* (Linnaeus) Greville    Dabberlocks (Fig. 17b)

*Frond elongate, with wavy margins; distinct midrib, which is a continuation of the short stipe. Clusters of leaf-life reproductive bodies, each up to 100 mm in length, on either side of stipe of mature plant. Branching holdfast. Yellowish to olive-green in colour. Up to 1 m in length.*

*A. esculenta* is widely distributed in north-west Europe and in Britain occurs most frequently in the north. It is found on the lower regions of exposed shores, particularly on vertical rock faces. The fronds are often torn by wave action and sometimes amount to little more than midrib. *A. esculenta* is a perennial, new growth being produced at the base of the frond.

*Cladostephus spongiosus* (Hudson) C. A. Agardh    (Fig. 18a)

*Frond stiff; branched, irregular or dichotomous; branches with whorls of tiny branches. Disc-like holdfast. Characteristic dirty brown colour. Up to 250 mm in length.*

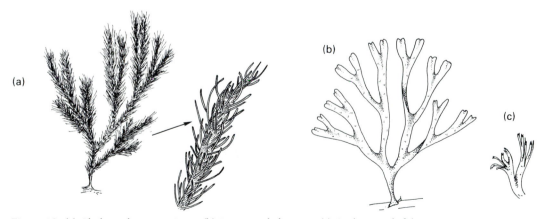

Figure 18  (a) *Cladostephus spongiosus.* (b) *Dictyota dichotoma.* (c) *Cutleria multifida,* tips of frond.

*C. spongiosus* is widely distributed in north-west Europe on rocks and in pools on the middle and lower shore. It sometimes supports dense tufts of the red seaweed *Jania rubens* (p. 61).

**Dictyota dichotoma** (Hudson) Lamouroux    (Fig. 18b)

*Frond delicate (3 cells thick), flat, no midrib; dichotomously branched, rising from disc-shaped holdfast. Tips of branches rounded, usually divided into two lobes. Reproductive bodies seen as fine specks on surface of frond. Yellow-brown, olive in colour, lighter towards tips; iridescent when under water. Up to about 150 mm in length.*

*D. dichotoma* is widely distributed in north-west Europe and is sometimes abundant on south and west coasts of Britain in pools on the lower shore and in the sublittoral. It is much less common in the north and east. Gametes are released fortnightly on spring tides.

**Cutleria multifida** (Smith) Greville is a widely distributed, sublittoral seaweed which is often washed up on the shore. It bears some resemblance to *Dictyota* (above) but is up to 400 mm in length and the tips of the branches are divided several times to give a fringe-like effect (Fig. 18c). Like *Dictyota*, the reproductive bodies are seen as small dots over the surface of the frond.

**Ascophyllum nodosum** (Linnaeus) Le Jolis    Knotted wrack (Fig. 19a)

*Frond narrow, without midrib. Large, swollen, egg-shaped air bladders at intervals along middle of frond. Reproductive bodies rounded, on short stalks rising from margins of frond. Dichotomously branched. Olive-green in colour; reproductive bodies golden-yellow. Up to 2 m or more in length.*

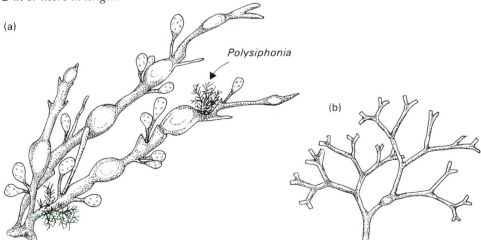

Figure 19   (a) *Ascophyllum nodosum* with the epiphyte, *Polysiphonia*. (b) *Ascophyllum nodosum* var. *mackaii*.

*A. nodosum* is widely distributed in north-west Europe. It is abundant on the middle reaches of sheltered shores and wherever a suitable substratum is present it extends into estuaries. On more exposed shores scattered plants are found but the fronds are broken, torn and only a few centimetres in length; it is absent from the most exposed shores. *A. nodosum* is dioecious. Gametes are discharged in spring after which the reproductive bodies and their stalks are shed. After one or two years' growth, each branch produces a single air bladder per year, providing a means of estimating the age of undamaged plants. A length of life of about 12 years has been suggested. *Ascophyllum* often bears tufts of the filamentous, epiphytic red alga *Polysiphonia lanosa* (p. 72) and occasionally epiphytic *Ectocarpus* or *Ectocarpus*-like seaweeds (see p. 41).

An unattached form of *Ascophyllum*, **Ascophyllum nodosum** variety **mackaii** (Fig. 19b), is widely distributed in north-west Europe. It is found on very sheltered shores, in sea lochs and is sometimes common on the west coasts of Ireland and Scotland. The frond is flat or rounded with extensive dichotomous branching and bears few, if any, small air bladders. The plants are slow growing and often reach a length of 400 mm. They drift in large, spherical masses in sheltered waters.

### *Fucus ceranoides* Linnaeus    (Fig. 20a)

*Frond thin with smooth margin; prominent midrib; without air bladders but frond on either side of midrib may be inflated. Dichotomous branching, repeated frequently to give fan-shaped frond. Reproductive bodies narrow, pointed, at ends of branches. Pale olive-green in colour. Up to 600 mm in length, but usually much smaller.*

(a)                                                      (b)

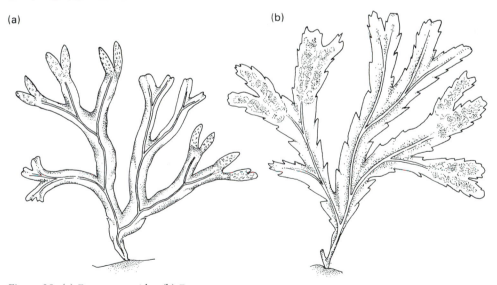

Figure 20  (a) *Fucus ceranoides.* (b) *Fucus serratus.*

*F. ceranoides* is widely distributed in north-west Europe but is common only in brackish water, and is often abundant where freshwater streams run onto the beach. It is attached to rocks and stones on the middle and lower shore and is one of the characteristic sea-weeds of estuaries. It is usually dioecious but some plants are hermaphroditic. Gametes are released on daytime high tides at about the time of the new and full moon.

**Fucus serratus** Linnaeus    Toothed wrack, Serrated wrack (Fig. 20b)

*Frond with serrated margin and conspicuous midrib; without air bladders. Dichotomously branched. Reproductive bodies at ends of fronds and surrounded by narrow rim of frond, the sterile margin; regions of frond bearing reproductive bodies flattened. Surface of frond with many tiny 'hairs'. Olive-brown in colour. Up to 1 m or more in length.*

*F. serratus* is widely distributed and often abundant in north-west Europe but absent from the most exposed shores. It is found on the lower shore, often in a distinct zone below *Ascophyllum nodosum* (above) and *Fucus vesiculosus* (below). It is dioecious and has a life-span of up to about three years. In sheltered waters the fronds often support a rich epifauna, including sponges, hydroids, bryozoans and the polychaete *Spirorbis spirorbis* (p. 208), and it has been shown that fronds with heavy encrustations of bryozoans have a reduced rate of photosynthesis. The flat periwinkles *Littorina mariae* and *L. obtusata* are also found on *F. serratus* together with their characteristic oval- or kidney-shaped, whitish egg masses (p. 240).

**Fucus spiralis** Linnaeus    Spiral wrack (Fig. 21a)

*Frond with smooth margin; prominent midrib; without air bladders; often twisted. Dichotomously branched. Reproductive bodies at ends of branches; each almost round in outline, surrounded by narrow rim of frond, the sterile margin. Olive-brown in colour. Up to 400 mm in length but usually much smaller.*

(a)                                              (b)

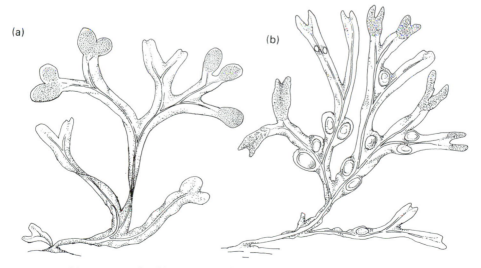

Figure 21  (a) *Fucus spiralis.* (b) *Fucus vesiculosus.*

*F. spiralis* is widely distributed and common on sheltered rocky shores. It is found just below *Pelvetia* (p. 51) and often forms a distinct zone between *Pelvetia* and the two common brown algae of the middle shore, *Fucus vesiculosus* (below) and *Ascophyllum nodosum* (above). It is usually absent from the most exposed shores; in such situations the typical plant is sometimes replaced by a dwarf form, **Fucus spiralis** forma **nanus**. *Fucus spiralis* is hermaphroditic and lives for up to about four years.

*Fucus vesiculosus* Linnaeus    Bladder wrack (Fig. 21b)

*Frond with prominent midrib and almost spherical air bladders. Air bladders usually in pairs but may be absent in very small plants. Margin of frond smooth. Dichotomously branched. Reproductive bodies at tips of branches; often forked and pointed. Olive-brown in colour. Up to 1 m or more in length.*

F. vesiculosus is widely distributed in north-west Europe and in sheltered situations is abundant. It is found on the middle shore, along with *Ascophyllum nodosum* (above), in a zone above *Fucus serratus* (above). It is tolerant of wide degrees of exposure to wave action, and the morphology of the plant varies in response to environmental conditions. On sheltered shores the frond has many air bladders, the number decreasing in exposed situations. Where exposure to wave action is high, plants without bladders may be common and these are known as **F. vesiculosus** forma **linearis**. They are no more than about 200 mm in length. The presence of air bladders, which in sheltered waters buoy up the fronds and help keep them in the illuminated surface waters, would be disadvantageous on wave-swept beaches, increasing the likelihood of the plant being torn away. On exposed shores the frond is much torn and the basal regions consist of little more than midrib. The plants are dioecious and live for up to about three years.

The genus *Fucus* includes some of the best-known brown seaweeds of temperate waters. Those described here generally possess a number of distinctive, morphological features and occupy definite zones on the shore, but the morphology of the frond varies in response to environmental conditions, leading in some cases to difficulty in identification. This is clearly illustrated by *Fucus vesiculosus* forma *linearis* and *Fucus spiralis* forma *nanus*. Difficulty in identification also arises as a result of interbreeding between species to give hybrids that are believed to be common on some shores. Unattached fucoids are found on saltmarshes and very sheltered shores. Also occurring in these habitats is **Fucus muscoides** (A. Cotton) J. Feldmann et Magne. This takes the form of a dense mossy growth no more than four centimetres in height. It may encompass several species.

**Dictyopteris membranacea** (Stackhouse) Batters has a prominent midrib and has some similarity to a fucoid seaweed. The frond is, however, thinner and more delicate and is dotted with numerous clusters of tiny hairs. It is up to 300 mm in length. Specimens just collected have an unpleasant, pungent smell. *Dictyopteris* is a southern species extending northwards to south-west Britain; it is essentially sublittoral although sometimes found on the lower shore, and is very local in distribution.

**Pelvetia canaliculata** (Linnaeus) Decaisne & Thuret    Channelled wrack (Fig. 22a)

*Frond curled to form marked channel. Dichotomously branched; reproductive bodies at ends of branches. Without air bladders and midrib. Dark olive-green colour, becoming black and brittle as the frond dries out. Up to 150 mm in length.*

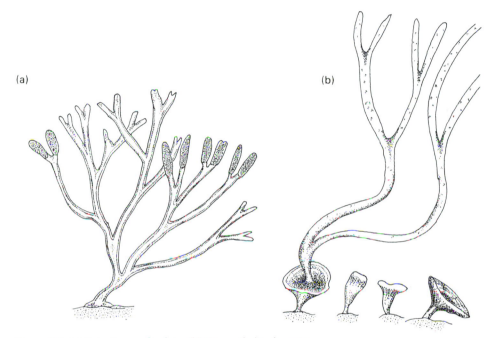

(a)

(b)

Figure 22  (a) *Pelvetia canaliculata.* (b) *Himanthalia elongata.*

*P. canaliculata* is common on rocky shores in north-west Europe and in some situations forms a distinct band on the upper shore. It is one of the highest zoned seaweeds and it has been estimated that some plants spend as much as 90% of their time out of water, surviving water loss of up to 65%. Reproduction occurs during summer and autumn when the reproductive bodies become orange coloured and form a recognizable band of colour on the upper shore. *Pelvetia* is hermaphroditic and lives for up to five years.

**Himanthalia elongata** (Linnaeus) S. F. Grey    Thong weed (Fig. 22b)

*Frond button-like, up to about 30 mm across, with very short stipe. Strap-shaped reproductive bodies grow from centre of button; dotted with brown spots when ripe; dichotomously branched. Olive-green in colour. Up to 2 m in length.*

*H. elongata* is widely distributed in north-west Europe and may be locally abundant. It is found on the lower shore on wave-swept beaches and often forms a distinct zone just above the laminarians. The young fronds are conical or club-shaped and broaden as they grow to form a characteristic button shape with a concave upper surface. This is the vegetative thallus from which the reproductive bodies, sometimes known as straps, grow in the following year. These reach maturity after about one year and after the gametes are shed, generally in summer, the plants die. *Himanthalia* is dioecious and lives for two to three years. The 'buttons', one or two centimetres across, are sometimes very numerous on the shore and often have a range of epiphytes and epifauna on the undersurface, the upper surface apparently having some property which prevents this, presumably a deterrent against settling of spores and larvae. The reproductive straps support a rich flora and fauna.

### *Bifurcaria bifurcata* Ross     (Fig. 23a)

*Frond cylindrical; unbranched near base then branching alternately or dichotomously. Rounded air bladders sometimes present. Elongate reproductive bodies at ends of branches. Yellow-olive colour, much darker when dry. Up to about 500 mm in length.*

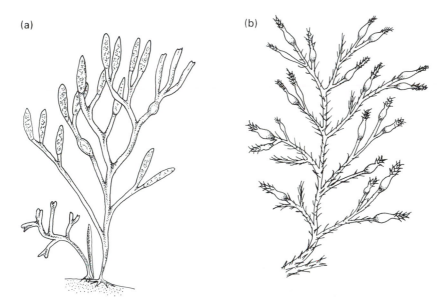

(a)                                                          (b)

Figure 23  (a) *Bifurcaria bifurcata.* (b) *Cystoseira tamariscifolia.*

*B. bifurcata* is a southern species found on the Atlantic coast of France and extending to the south and west coasts of England and the west coast of Ireland, where it is sometimes common. It is restricted to rock pools on the middle and lower shore, particularly on exposed beaches. *Bifurcaria* is perennial and hermaphroditic.

*Cystoseira tamariscifolia* (Hudson) Papenfuss    Rainbow bladder weed (Fig. 23b)

*Frond cylindrical, much branched, branches not swollen at base; bearing many small, spine-like structures giving rough texture and bushy appearance, reminiscent of a heather plant. Air bladders small, usually occur just below reproductive bodies. Reproductive bodies with nodules and spines, borne on ends of branches. Olive-green in colour, almost black when dry; blue-green iridescence when under water. Up to 450 mm in length.*

C. *tamariscifolia* is a southern species found on the Atlantic coast of France and south-west Britain and, although it has been recorded from the Hebrides, is rare in the north. It is found in rock pools on the lower shore, and is often common among shell and gravel deposits on sheltered shores. It extends into the sublittoral. C. *tamariscifolia* is hermaphroditic and perennial.

Other species of *Cystoseira* are found on south-west coasts, but C. *tamariscifolia* is readily distinguished by the iridescence seen when the plant is under water.

*Halidrys siliquosa* (Linnaeus) Lyngbye    (Fig. 24)

*Frond flattened, with alternate branches. Main stem of zig-zag appearance. Ends of some branches with pod-shaped air bladders, divided by transverse septa. Reproductive bodies similar in appearance to air bladders but without the septa and borne on short stalks at ends of branches. Young plants olive-green colour, older plants rich brown and leathery. Up to 2 m in length but usually much smaller.*

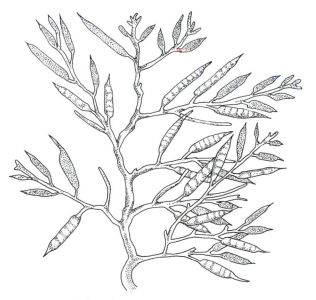

Figure 24  *Halidrys siliquosa.*

*H. siliquosa* is widely distributed in north-west Europe on the middle and lower shore in rock pools, and in the sublittoral. It often supports a range of epifauna, including hydroids, bryozoans and ascidians. *Halidrys* is hermaphroditic and perennial.

## Class Rhodophyceae — red seaweeds
### Key to species of red seaweeds

1  Thallus encrusting on surface of rock, stones, shell ........................................................ 2
   Thallus not encrusting ................................................................ 3

2  Thallus hard, calcareous, usually thin but may be thrown into folds and bumps; encrusting rock, shells; in rock pools. Pink-red, purple, yellow in colour ...................... ................................................................ *Lithophyllum incrustans* (p. 62)
   Thallus not calcareous, very thin; encrusting stones and rocks. Rose-red to dark red in colour ................................................................ *Hildenbrandia rubra* (p. 65)

3  Frond upright, much branched; made up of calcified segments; coarse to the touch; purplish, reddish, pinkish in colour ................................................................ 4
   Frond not made up of calcified segments ................................................................ 5

4  Branching mostly pinnate; branches stiff, characteristically growing in rock pools; up to 120 mm in length. Purplish, reddish, pink, often bleached white ........................... ................................................................ *Corallina officinalis* (p. 61)
   Branching dichotomous; small up to 25 mm in length; growing on other seaweeds, especially *Cladostephus* ................................................................ *Jania rubens* (p. 61)

5  Frond low growing, creeping, moss-like or carpet-like in appearance; covering rock surface ................................................................ 6
   Frond not so ................................................................ 7

6  Frond of very fine, sparingly branched filaments; bound with sand to form a spongy carpet over rocks; often extensive ........................... *Audouinella floridula* (p. 56)
   Frond much branched, constricted at intervals into different-sized segments; moss-like in appearance; purple or black in colour ........................... *Catenella caespitosa* (p. 57)

7  Frond with well-developed midrib ................................................................ 8
   Frond without midrib ................................................................ 11

8  Frond without lateral veins. Ends of fronds pointed; branches arise from the midrib ..... ................................................................ *Hypoglossum hypoglossoides* (p. 68)
   Frond with lateral veins which are conspicuous, or fine and difficult to see; frond narrow or leaf-like ................................................................ 9

9  Frond narrow; lateral veins very fine, seen with the aid of hand lens. Branches arise from margin of frond (cf. *Hypoglossum*). Irregular dichotomous branching ................... ................................................................ *Membranoptera alata* (p. 68)
   Frond leaf-like, veins conspicuous ................................................................ 10

10  Margin of frond wavy, not serrated; with the appearance of a beech leaf; bright red in colour ................................................................ *Delesseria sanguinea* (p. 67)

Margin of frond divided and serrated; with the appearance of an oak leaf; crimson-brown in colour ................................................................ *Phycodrys rubens* (p. 69)

11 Frond broad, flat, not branching but may be divided; tough and opaque *or* very thin and membranous ........................................................................................ 12

Frond not so ........................................................................................................... 16

12 Frond flattened, mostly tough, may be leathery; undivided, or divided into broad segments. Margin with or without outgrowths ............................................... 13

Frond flattened, very thin, membranous ............................................................. 15

13 Frond widening from branching holdfast; undivided or irregularly divided; margin with many small (5 mm) outgrowths forming fringe around edge ............................................. ................................................................................ *Calliblepharis ciliata* (p. 59)

Frond not so ........................................................................................................ 14

14 Frond broad, widening from small disc-like holdfast and branching into broad segments; no stipe. Margin of frond often with some small outgrowths. Common epiphyte on stipes of laminarians ........................................ *Palmaria palmata* (p. 64)

Fond broadening from small, disc-like holdfast, on short, cylindrical stipe; undivided or split into a number of broad segments. Margin without outgrowths. Found on stones and rocks .................................................................... *Dilsea carnosa* (p. 63)

15 Frond with irregular folded margin; central disc-like holdfast; dark purple, reddish in colour ...................................................................... *Porphyra umbilicalis* (p. 72)

Frond divided only slightly or deeply into strap-like sections; disc-like holdfast; surface of frond spotted; clear rose-pink to brownish-red colour ... *Nitophyllum punctatum* (p. 69)

16    (a) Frond fine, tough and wiry in a tangled mass; often partly buried in sand ............ ................................................................................ *Ahnfeltia plicata* (p. 58)

*or* (b) Frond narrow, tubular, inflated and twisted; dark brown-red in colour; in rock pools .................................................................... *Dumontia contorta* (p. 64)

*or* (c) Frond very shiny, constricted at intervals into bead-like segments usually bright red in colour. On rocks, in rock pools .......................... *Lomentaria articulata* (p. 66)

*or* (d) Frond shows alternate light and dark bands. When viewed under microscope seen to be made up of a row of large cells – monosiphonous. Usually with pincer-like tips. On rocks, stones, *Fucus, Laminaria* .............. *Ceramium* spp. (p. 67)

*or* (e) When viewed under microscope frond seen to be made up of columns of cells – polysiphonous (Fig. 40b,c). On rock, stones, *Ascophyllum* ................................... ................................................................................ *Polysiphonia* spp. (p. 71)

*or* (f) Frond not fitting any of the above ....................................................... 17

17 Frond flat, channelled or cylindrical, not constricted into segments; dichotomous branching ............................................................................................................. 18

Frond flat or cylindrical, not constricted into segments; alternate or opposite branching ............................................................................................................. 21

18 Frond cylindrical ................................................................................................. 19

Frond flat or channelled ...................................................................................... 20

19  Frond rising from much-branched holdfast; elongate reproductive bodies at ends of branches ....................................................................../ *Furcellaria lumbricalis* (p. 56)
Frond rising from disc-like holdfast; reproductive bodies as oval swellings along sides of branches .................................................................... *Polyides rotundus* (p. 57)

20  Frond flat, axils of branches rounded; dark red-purple in colour, edges often bleached green-yellow; tips of frond iridescent under water ...................... *Chondrus crispus* (p. 60)
Frond channelled, margins thickened, papilla-like reproductive bodies on surface of frond; dark reddish-brown, purple in colour .......................... *Mastocarpus stellatus* (p. 60)

21  Frond flattish, with alternate branching, very distinctive; small branches in alternate groups, each small branch bearing tiny branches on one side only; so resembling teeth on a comb; clear rose to dark red in colour ...................... *Plocamium cartilagineum* (p. 58)
Branching not so ............................................................................................ 22

22  Frond flattened, branching alternate in one plane only; tips of branches bluntly rounded; holdfast of intertwined creeping stolons; reddish-brown in colour often bleached to yellowish-green .................................................. *Laurencia pinnatifida* (p. 71)
Frond cylindrical; branching alternate or opposite; holdfast disc-like ............................. 23

23  Branching opposite or alternate, many smaller branches; branches narrow at base and tip; bright pink-red in colour .................................................. *Lomentaria clavellosa* (p. 66)
Branching mostly alternate; branches become shorter towards apex; ends of branches bluntly rounded; dark purple to yellow in colour ...................... *Laurencia hybrida* (p. 70)

**Audouinella floridula** (Dillwyn) Woelkerling    (*Rhodochorton floridulum*) (not illustrated)

*Frond of fine, sparingly branched filaments up to about 30 mm in length; some upright, some creeping. Brownish-red in colour. Filaments bound with sand to form a spongy, carpet-like covering over rocks; often extensive.*

There are many species of *Audouinella* and although microscopic examination of the fronds is generally required for species separation, *A. floridula* can be recognized by its habit of binding sand particles into a spongy, turf-like mass. It is a southern species extending to all parts of Britain, where it is widely distributed on the lower shore, often covering large areas of rock. It is perennial. The polychaete *Fabricia stellaris* (p. 202) often occurs among *Audouinella*.

**Furcellaria lumbricalis** (Hudson) Lamouroux    (*Furcellaria fastigiata*) (Fig. 25a)

*Frond cylindrical, rising from much-branched holdfast (cf. Polyides, below); dichotomous branching; elongate reproductive bodies at ends of branches. Red-brown to brownish-black in colour (blackish-brown in transmitted light), glossy; sometimes bleached green. Up to 300 mm in length.*

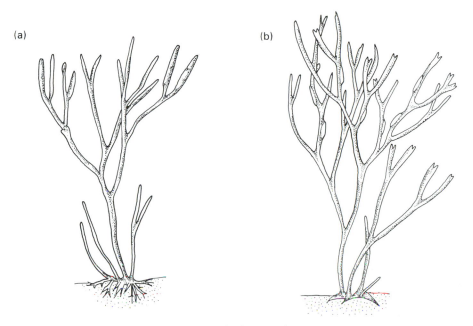

Figure 25  (a) *Furcellaria lumbricalis*. (b) *Polyides rotundus*.

*F. lumbricalis* is widely distributed and common in north-west Europe on rocks and in pools on the lower shore, often in situations where the holdfast is buried in coarse, sandy deposits. Unattached, floating plants are found in sheltered localities. It is perennial and dioecious.

**Polyides rotundus** (Hudson) Greville    (Fig. 25b)

*Frond cylindrical, rising from disc-like holdfast (cf. Furcellaria, above); dichotomous branching. Reproductive bodies seen as oval-shaped swellings along sides of branches. Dark red to black in colour (red in transmitted light). Up to 200 mm in length.*

*P. rotundus* is widely distributed in north-west Europe on rocks and in pools on the lower shore, in similar situations to *Furcellaria lumbricalis* (above). It is perennial and dioecious.

**Catenella caespitosa** (Withering) L. Irvine    (*Catenella repens*) (Fig. 26a)

*Frond irregular and much branched, creeping; constricted at intervals into different-sized segments; holdfast of tangled fibres. Moss-like in appearance. Dull purple or black in colour. Up to 20 mm in height. Not to be confused with the lichen* Lichina pygmaea *(p. 81).*

Figure 26  (a) *Catenella caespitosa*. (b) *Plocamium cartilagineum*. (c) *Ahnfeltia plicata*.

*C. caespitosa* is widely distributed and common in north-west Europe on rocks on the upper shore, and is sometimes abundant in sheltered situations, forming dense, matted growths. It is perennial and hermaphroditic.

**Plocamium cartilagineum** (Linnaeus) P. Dixon    (*Plocamium coccineum*) Cockscomb (Fig. 26b)

*Frond compressed or nearly flat, cartilaginous, much branched; branching alternate, very characteristic; larger branches have groups of smaller branches, each group arising alternately; each smaller branch with tiny branches arising from one side only, so resembling the teeth on a comb. Disc-like holdfast. Clear rose to dark red in colour, often bleached. Up to 150 mm or more in length.*

*P. cartilagineum* is widely distributed in north-west Europe on the lower shore and below, often growing epiphytically on other seaweeds, especially on the stipe of *Laminaria hyperborea* (p. 44). It is often washed up after storms. It is perennial and dioecious.

**Ahnfeltia plicata** (Hudson) Fries    (Fig. 26c)

*Frond very fine, tough, wiry; branching irregular or dichotomous. Ends of branches blunt. Disc-like holdfast. Purple when moist, almost black when dry. Growing in tangled masses; up to 150 mm in length.*

A. *plicata* is widely distributed and common in north-west Europe in rock pools on the middle and lower shore, often partly buried in sand. It is perennial and dioecious.

**Calliblepharis ciliata** (Hudson) Kützing    (Fig. 27a)

*Frond tough, cartilaginous; flattened, widening to about 70 mm across from a narrow, cylindrical stipe. Undivided or irregularly divided into strap-like sections. Margin of frond with many outgrowths, 5 mm or so in length, and giving the impression of a fringe around the margin; outgrowths sometimes also arise from surface of frond. Conspicuous branching holdfast. Dark red, purplish-red in colour, dull. Up to 300 mm in length.*

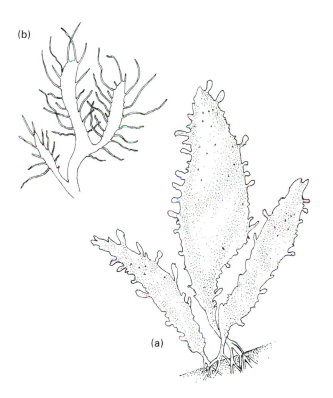

Figure 27   (a) *Calliblepharis ciliata*. (b) *Calliblepharis jubata*.

C. *ciliata* is a conspicuous red seaweed reaching its northern limit in south and west Britain, becoming rare in Scotland. It is essentially sublittoral but also occurs in pools on the lower shore. It is dioecious and annual, the plants reach maturity in winter, and are often washed ashore in the early part of the year.

**Calliblepharis jubata** (Goodenough & Woodward) Kützing (*Calliblepharis lanceolata*) is similar to *C. ciliata* but has a much more divided frond, up to 300 mm in length, and the margins are fringed with numerous long (up to 30 mm), narrow outgrowths (Fig. 27b) which often become entwined with one another and with other algae. *C. jubata* is found in pools on the lower shore attached to rock and laminarian stipes.

### *Chondrus crispus* Stackhouse    (Fig. 28a,b)

*Frond flat, widening from narrow, unbranched stipe; dichotomous branching, axils of branches rounded. Disc-like holdfast. Dark red, purple in colour, edges of frond often bleached green-yellow. Tips of frond iridescent under water. Up to about 200 mm in length.*

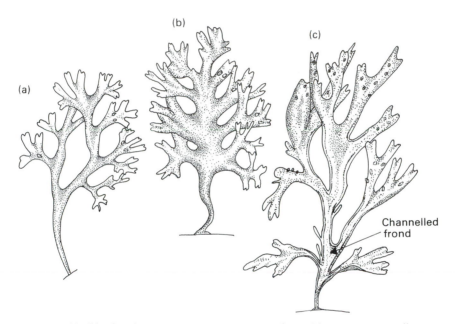

Figure 28  (a), (b) *Chondrus crispus*, note variations in form. (c) *Mastocarpus stellatus*.

*C. crispus* is widely distributed in north-west Europe and often abundant on rocks and in pools on the middle and lower shore. It tolerates some reduction in salinity and is found in estuaries. Morphologically it is a highly variable species and the differences in the appearance of plants collected from shores of differing exposure, and even from different levels on the same shore, can lead to confusion in identification. *Chondrus* is easily confused with *Mastocarpus stellatus* (below). Together with the latter species it is harvested commercially as Irish moss or Carragheen, a substance used in the medical and food industries. It is perennial and dioecious, living for up to six years.

### *Mastocarpus stellatus* (Stackhouse) Guiry    (*Gigartina stellata*) (Fig. 28c)

*Frond channelled, particularly at base, with thickened margin; widening from narrow stipe; dichotomous branching. Disc-like holdfast. Mature plants with conspicuous, papilla-like reproductive bodies on surface of frond. Dark reddish-brown, purple in colour, may be bleached. Up to about 170 mm in length.*

M. *stellatus* is widely distributed and common in north-west Europe on rocky shores, particularly those exposed to wave action, where it grows among barnacles and mussels. It is found on the lower shore and in rock pools; in less exposed conditions it is often abundant under fucoids. Like *Chondrus* (above), with which it is easily confused, it is variable in form. The channelled frond and the papilla-like reproductive bodies of *Mastocarpus* are important features in separating the species. It is a perennial, living for several years, and is probably dioecious.

### *Corallina officinalis* Linnaeus    (Fig. 29)

*Frond made up of calcified segments. Coarse to the touch. Branching mostly pinnate; branches stiff. Encrusting holdfast. Colour varied, purple, red, pink and shades of yellow; extremities often bleached white. Up to 120 mm in length.*

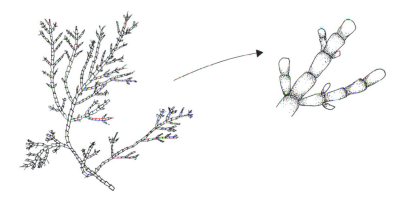

Figure 29  *Corallina officinalis*, with detail of frond.

C. *officinalis* is widely distributed and often abundant in north-west Europe. It is one of the characteristic algae of rock pools on the middle and lower shore and provides a habitat for a range of invertebrates including *Spirorbis corallinae* (p. 208), *Janua pagenstecheri* (p. 210) and *Mytilus edulis* (p. 283). It is dioecious.

### *Jania rubens* (Linnaeus) Lamouroux    (Fig. 30)

*Frond made up of calcified segments. Dichotomous branching; branches fine. Encrusting, sometimes branching holdfast. Reproductive bodies as rounded swellings at fork of dichotomous divisions. Colour pink, red. Up to 25 mm in length.*

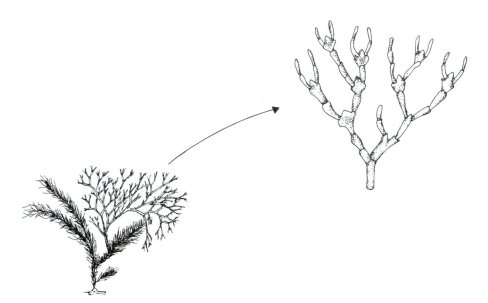

Figure 30  *Jania rubens* var. *rubens* on *Cladostephus spongiosus* (detail of frond after Irvine & Chamberlain, 1994).

*J. rubens* is widely distributed in north-west Europe. It grows in dense tufts on other seaweeds on the middle and lower shore, especially on *Cladostephus spongiosus* (p. 46) on which two varieties of *Jania* are known to occur in Britain. These are **Jania rubens** var. **rubens** and **Jania rubens** var. **corniculata**, which are separated on the morphology of the calcified segments.

Two other species with fronds made up of calcified segments, **Corallina elongata** Ellis & Solander and **Haliptilon squamatum** (Linnaeus) Johansen, L. Irvine & Webster, occur in southern and western Britain. They resemble *C. officinalis* and *J. rubens*, and while separation of the species relies on a number of characters, the stiffness of the fronds and widespread occurrence of *C. officinalis* and the relatively small size of *J. rubens* (up to 25 mm) are important indicators. Reference to Irvine & Chamberlain (1994) is recommended for further detail of these species and the varieties of *Jania*.

**Lithophyllum incrustans** Philippi    (Fig. 31)

*Thallus in the form of a hard, smooth, calcareous patch; chalky texture when dry; encrusting rock surfaces, shells and sometimes seaweed. Usually very thin, less than a millimetre in height, thicker at edges; but may be several millimetres in height and thrown into folds and bumps. Pink, red, purple, grey-yellow in colour.*

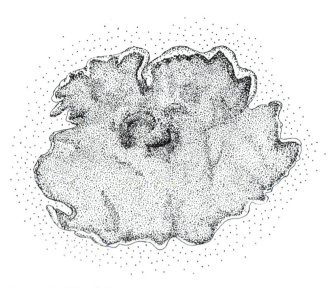

Figure 31  *Lithophyllum incrustans*.

*L. incrustans* is widely distributed in north-west Europe. It is commonly found on exposed shores on the south and west coasts of Britain and extends along the English Channel to the Isle of Wight. It occurs in rock pools on the middle and lower shore, often covering large areas of rock, and extends into the sublittoral. On death, the white, calcareous skeleton of the plant remains attached to the rock for some time, often giving a white fringe to rock pools. *Lithophyllum* is long-lived, reaching 12 or 13 years of age. Spores are released mainly during the winter months. On the west coast of Wales, the plants reproduce when about two years old.

Other encrusting red algae are commonly found in north-west Europe and identification to species can be difficult. Reference to Hiscock (1986) and Irvine & Chamberlain (1994) is recommended.

**Dilsea carnosa** (Schmidel) O. Kuntze    (Fig. 32)

*Frond flat, thick; tough and leathery; broadening from a short, cylindrical stipe; small, disc-like holdfast. The entire frond of small specimens is pear-shaped but in larger specimens the margin is split and the frond divided into broad segments; margin without outgrowths. Dark red in colour. Up to about 1 mm thick, 500 mm in length and 200 mm or so broad; smaller specimens more common.*

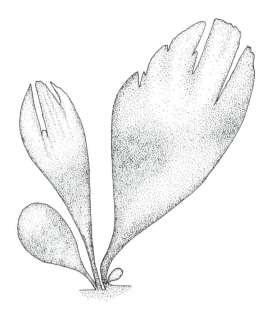

Figure 32  *Dilsea carnosa.*

*D. carnosa* is widely distributed and common on the lower shore attached to stones and rocks, and in the shallow sublittoral to depths of about 25 m. It is a perennial species, the young plants generally being seen on the shore in the autumn and reaching maturity the following spring.

**Dumontia contorta** (S. G. Gmelin) Ruprecht    (*Dumontia incrassata*) (Fig. 33a)

*Frond tubular, narrow and rounded at base and at origin of branches, becoming wider and twisted; inflated in older specimens. Alternate or irregular branching. Disc-like holdfast. Dark brownish-red in colour, sometimes bleached. Up to 230 mm in length.*

*D. contorta* is widely distributed and common on rocky shores in north-west Europe. It is found on rocks and, particularly, in rock pools on the middle and lower shore. The plants are small in winter and grow rapidly in spring and summer before dying back in the autumn. It is dioecious.

**Palmaria palmata** (Linnaeus) O. Kuntze    (*Rhodymenia palmata*) Dulse (Fig. 33b)

*Frond broad flat, very varied in form, generally widening from base and dividing into broad segments. Membranous, tough, opaque. Margins of frond often with some small out-growths. Disc-like holdfast. No obvious stipe. Purplish-red in colour. Up to 500 mm in length, about 80 mm across, but usually much smaller.*

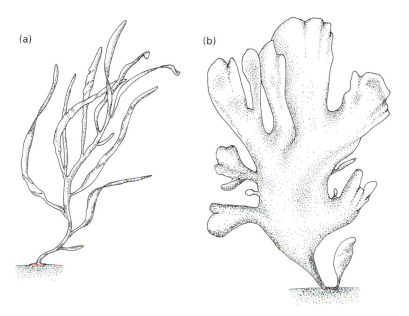

Figure 33  (a) *Dumontia contorta*. (b)*Palmaria palmata*.

*P. palmata* is widely distributed and very common on rocky shores in north-west Europe. It is found on the lower shore attached to rocks or other seaweeds, particularly the laminarians, often being abundant on the stipe of *Laminaria hyperborea* (p. 44).

In the south and west of Britain, a somewhat similar species, **Rhodymenia pseudopalmata** (Lamouroux) P. C. Silva (*Rhodymenia palmetta*) is often found growing on the stipe of *L. hyperborea* (p. 44). The pinkish-red frond arises by a short stipe from a disc-like holdfast and is up to about 50 mm in length. The frond is membranous, tough, opaque and much divided to become fan-shaped in outline. Unlike *P. palmata*, the margin of the frond does not have outgrowths.

**Hildenbrandia rubra** (Sommerfelt) Meneghini    (not illustrated)

*Thallus very thin, encrusting on stones and rock, not calcareous; may cover large areas of rock surface. Reproductive bodies in pits on surface of thallus. Rose-red to dark red in colour; dull when dry.*

*H. rubra* is widely distributed on rocky shores in north-west Europe and is common over most of the shore, particularly in shaded, moist situations where it forms conspicuous patches up to about 150 mm across.

The very similar **Hildenbrandia crouanii** J. Agardh is found in the same sort of locations as, and often with, *H. rubra*. When the two species occur together, colour can be useful in distinguishing them; *H. crouanii* is brownish-red compared with the rose-red to dark red of *H. rubra*.

### *Lomentaria articulata* (Hudson) Lyngbye    (Fig. 34a)

*Frond constricted at intervals into conspicuous bead-like segments, hollow; branches occur at constrictions. Disc-like holdfast. Dark brown to red in colour, very shiny; bleaching to pink and orange. Up to 100 mm in length, occasionally much longer.*

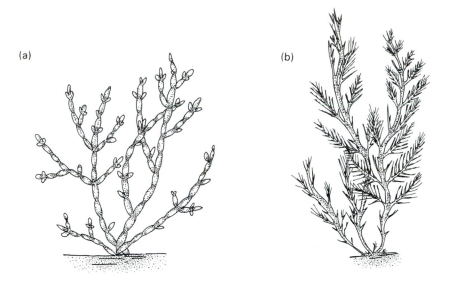

(a)                                             (b)

Figure 34  (a) *Lomentaria articulata*. (b) *Lomentaria clavellosa*.

*L. articulata* is widely distributed and common in north-west Europe. It is found on rocks and in pools on the middle and lower shore, sometimes occurring in profusion. It has been recorded growing on *Laminaria* spp. (p. 43) and is perennial and dioecious.

### *Lomentaria clavellosa* (Turner) Gaillon    (Fig. 34b)

*Frond cylindrical, not constricted into segments; main axis hollow, narrow at base. Branches opposite or alternate, with many smaller branches. Branches narrow at base and tip. Disc-like holdfast. Bright pinkish-red colour. Up to 400 mm in length.*

*L. clavellosa* is widely distributed in north-west Europe on the lower shore on rocks and in pools. The extent and nature of branching varies with exposure to wave action. It is dioecious and probably annual.

**Ceramium** spp.   (Fig. 35)

*Frond filamentous, made up of row of large cells (monosiphonous) overlain by smaller cells at intervals to give a banding effect. Irregular, dichotomous branching; branches narrower near tips which are usually pincer-like. Reddish-brown, red or purple in colour, usually with alternate light and dark bands. Up to 300 mm in length.*

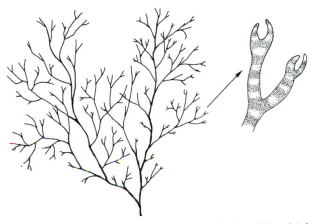

Figure 35  *Ceramium nodulosum*, with detail of tip of frond (after Newton, 1931).

Some 15 species of *Ceramium* have been described from the British Isles: they are difficult to separate. The widely distributed 'Ceramium rubrum' is now believed to be an aggregate of four species in British waters and two of these, **Ceramium nodulosum** (Lightfoot) Ducluzeau and **Ceramium pallidum** (Nägeli ex Kützing) Maggs & Hommersand are widespread and common. Both are found on a range of seaweeds such as *Fucus* spp., *Mastocarpus* (p. 60) and *Laminaria*, particularly *Laminaria hyperborea* (p. 44). They are also found on rocks and stones and in rock pools on the middle and lower shore.

**Delesseria sanguinea** (Hudson) Lamouroux    Sea-beech (Fig. 36a)

*Frond leaf-life; prominent midrib with lateral veins; margin wavy, not serrated, similar to beech leaf in appearance. Disc-like holdfast. Stipe tough, branched. Bright red in colour. Up to 250 mm in length.*

D. sanguinea is widely distributed in north-west Europe in deep pools on the lower shore. It is a conspicuous plant which thrives best in shade and is sometimes common. The leafy fronds are best seen in spring and summer; reproductive bodies are borne on the stipe after the fronds have been cast in winter. It is perennial, probably living five or six years.

*Hypoglossum hypoglossoides* (Stackhouse) F. Collins & Hervey (*Hypoglossum woodwardii*) (Fig. 36b)

*Frond narrow, leafy, usually up to 2 mm wide; with prominent midrib, without lateral veins. Often much branched, the branches arising from the midrib. Ends of fronds pointed. Rose-pink colour. Up to 200 mm in length.*

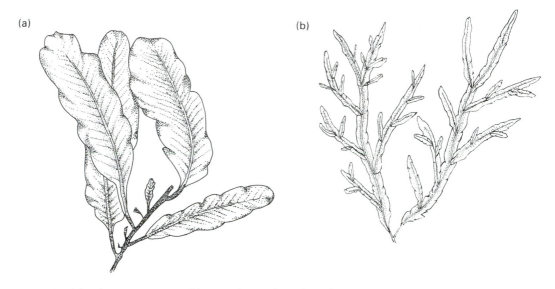

(a)                                                                                    (b)

Figure 36  (a) *Delesseria sanguinea*. (b) *Hypoglossum hypoglossoides*.

*H. hypoglossoides* is widely distributed and common. It is found on rocks and stones, and on algae, particularly the stipes of *Laminaria hyperborea*. It has been recorded in pools on the lower shore and sublittorally to depths of 30 m. It is dioecious.

*Hypoglossum* is similar in appearance to **Apoglossum ruscifolium** (Turner) J. Agardh, which is found in the same type of habitat but is not as common as *Hypoglossum*. The two are distinguished by the presence of microscopic lateral veins on the frond of *Apoglossum*, which reaches a length of 100 mm and is bright red in colour.

*Membranoptera alata* (Hudson) Stackhouse    (Fig. 37a)

*Frond flat, narrow, with midrib and many fine, lateral veins seen clearly under a hand lens. Branches arise from margin of frond. Irregular, dichotomous branching; branches narrower towards tip. Red or reddish-brown in colour. Up to 200 mm in length.*

M. *alata* is widely distributed and common in north-west Europe in pools on the middle and lower shore, frequently growing on other seaweeds, particularly laminarians. The lamina is often damaged and torn during winter. It is perennial.

## *Nitophyllum punctatum* (Stackhouse) Greville    (Fig. 37b)

*Frond flat, very thin (one cell thick over most of frond); delicate, without veins. Short stipe or stipe absent; disc-like holdfast. Frond divided only slightly or deeply divided into strap-like sections. Surface of frond with many elongate reproductive bodies giving spotted appearance. Clear rose-pink to brownish-red in colour. Up to 500 mm in length.*

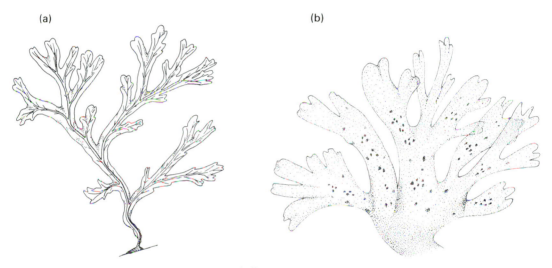

Figure 37   (a) *Membranoptera alata.* (b) *Nitophyllum punctatum.*

N. *punctatum* is widely distributed in north-west Europe, and in Britain is commonest on the west coast. It occurs on the lower shore and below attached to rocks, stones, shells and other seaweeds such as *Laminaria hyperborea* (p. 44). The shape of the frond varies considerably and this is believed to be related to habitat. It is dioecious and an annual.

## *Phycodrys rubens* (Linnaeus) Batters    (Fig. 38)

*Frond leaf-life, with midrib and lateral veins; margin divided and serrated, giving appearance of oak leaf. Disc-like holdfast. Stipe tough, elongated. Crimson-brown in colour. Up to 150 mm in length.*

Figure 38  *Phycodrys rubens*.

*P. rubens* is widely distributed in north-west Europe and is common in pools on the lower shore and sublittorally. It is a perennial, frequently found growing on the stipes of laminarians. As in *Delesseria* (p. 67) much of the lamina is lost in winter.

**Laurencia hybrida** (A. P. de Candolle) Lenormand ex Duby    (Fig. 39b)

*Frond cylindrical, with alternate, occasionally opposite, branches; tough, cartilaginous. Branches become shorter towards apex; ends of branches bluntly rounded; disc-like holdfast. Dark purple to yellowish in colour. Up to 150 mm in length.*

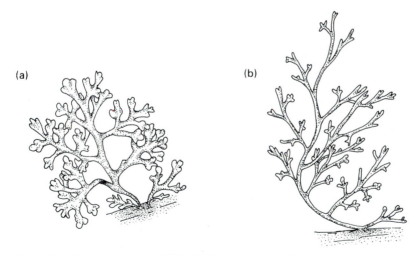

Figure 39  (a) *Laurencia pinnatifida*. (b) *Laurencia hybrida*.

*L. hybrida* is a southern species occurring all around Britain. It is found on the middle and lower shore on shells, rocks and in rock pools. It is dioecious and annual or perennial.

### *Laurencia pinnatifida* (Hudson) Lamouroux    Pepper dulse (Fig. 39a)

*Frond flattened, branches alternate, decreasing in length to apex; tough, cartilaginous. Branching in one plane only. Tips of branches bluntly rounded. Holdfast of intertwined creeping stolons. Reddish-brown in colour bleaching to yellow-green. Up to 80 mm in length.*

*L. pinnatifida* is widely distributed all around Britain and extends south to the Mediterranean. It is found on the middle and lower shore and shows considerable variation in form and this, to some extent, is dependent on the position on the shore. Plants at the higher shore levels are often small, rarely more than 30 mm in length, and are yellow-green as a result of prolonged exposure to light. Large areas of rock are often covered by a dense growth of these small plants and it is believed that many fail to reach maturity. The species is perennial and dioecious.

### *Polysiphonia* spp.    (Fig. 40a,b,c)

*Frond filamentous, polysiphonous (central cell is surrounded by a number of long, siphonlike cells). Generally dark reddish-brown colour. Some species up to 350 mm in length but most are much smaller. On seaweeds, rocks and stones.*

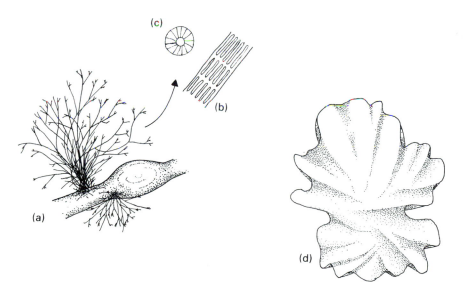

Figure 40   (a) *Polysiphonia lanosa* on *Ascophyllum nodosum*, with (b) diagrammatic representation of frond when viewed under a microscope, and (c) when seen in transverse section, to show polysiphonous nature. (d) *Porphyra umbilicalis*.

The genus *Polysiphonia* is widely distributed and common and some 19 species have been described from British waters. One of the characters used to identify the species is the number and arrangement of cells, or siphons, seen in transverse section to surround a central siphon. *Polysiphonia* spp. are found on the middle and lower shore on rocks, stones, the shells of snails and limpets, and are epiphytic on a wide range of seaweeds, including fucoids, *Corallina*, the stipe of laminarians and *Chorda*, and also on *Zostera*. *Polysiphonia lanosa* (Linnaeus) Tandy (Fig. 40a) grows as conspicuous, dark brown-red filamentous tufts on *Ascophyllum nodosum* (p. 47), less frequently on *Fucus* spp. It has 12–24 siphons around a central siphon. Root-like fibres penetrate the tissues of *Ascophyllum*. *P. lanosa* is rarely found on stones. The species is dioecious.

*Porphyra umbilicalis* (Linnaeus) J. G. Agardh     Purple laver (Fig. 40d)

*Frond membranous (1 cell thick), irregular folded margin; widening from narrow base; delicate, gelatinous texture. Small, central disc-like holdfast. Dark purple, red in colour, blackens when dry. Up to 200 mm in length.*

*P. umbilicalis* is widely distributed and common on rocky shores and extends into the sublittoral. Several other species of *Porphyra* occur around Britain and these can be separated from *P. umbilicalis* by their basal holdfast. One of these species, *Porphyra purpurea* (Roth) C. Agardh, is harvested in South Wales and sold as laver bread.

REFERENCES

Boney, A. D. (1966). *A biology of marine algae*. London: Hutchinson Educational.

Burrows, E. M. (1991). *Seaweeds of the British Isles*. Vol. 2. *Chlorophyta*. London: British Museum (Natural History).

Dickinson, C. I. (1963). *British seaweeds*. The Kew Series. London: Eyre & Spottiswoode.

Dixon, P. S. & Irvine, L. M. (1977). *Seaweeds of the British Isles*. Vol. 1. *Rhodophyta*. Part 1. *Introduction, Nemaliales, Gigartinales*. London: British Museum (Natural History).

Fletcher, R. L. (1987). *Seaweeds of the British Isles*. Vol. 3. *Fucophyceae (Phaeophyceae)*. Part 1. London: British Museum (Natural History).

Hiscock, S. (1979). A field key to the British brown seaweeds (Phaeophyta). *Field Studies*, 5, 1–44.

Hiscock, S. (1986). *A field key to the British red seaweeds (Rhodophyta)*. Field Studies Council. Occasional Publication No. 13.

Irvine, L. M. (1983). *Seaweeds of the British Isles*. Vol. 1. *Rhodophyta*. Part 2A. *Cryptonemiales* (sensu stricto), *Palmariales, Rhodymeniales*. London: British Museum (Natural History).

Irvine, L. M. & Chamberlain, Y. M. (1994). *Seaweeds of the British Isles*. Vol. 1. *Rhodophyta*. Part 2B. *Corallinales, Hildenbrandiales*. London: HMSO.

Jones, W. E. (1962). A key to the genera of the British seaweeds. *Field Studies*, 1, 1–32.

Maggs, C. A. & Hommersand, M. H. (1993). *Seaweeds of the British Isles*. Vol. 1. *Rhodophyta*. Part 3A. *Ceramiales*. London: HMSO.

Newton, L. (1931). *A handbook of the British seaweeds.* London: British Museum (Natural History).

Parke, M. & Dixon, P. S. (1976). Check-list of the British marine algae – third revision. *Journal of the Marine Biological Association of the United Kingdom,* **56**, 527–94.

South, G. R. & Tittley, I. (1986). *A checklist and distributional index of the benthic marine algae of the North Atlantic Ocean.* St. Andrews & London: Huntsman Marine Laboratory & British Museum (Natural History).

# Lichenes

Although the majority of lichens are terrestrial, some grow in areas subject to salt spray and are a conspicuous feature of the upper reaches of many rocky shores. Some species are able to withstand daily submersion by the tide and are found in the eulittoral zone, where they compete for space with other organisms, for example, barnacles and sea-weeds. Lichens are complex plants comprised of fungal and algal components existing in what is believed to be a mutually beneficial association known as symbiosis. The exact nature of the relationship is still debated, but photosynthesis of the algal cells provides the fungus with organic nutrients, while the fungus provides the alga with support and protection.

The main part of the lichen, known as the thallus, is typically seen in one of three forms; crustose, foliose and fruticose. The flattened, crust-like thallus of crustose lichens is firmly attached to the substratum and very difficult to remove without damage to the lichen. Foliose lichens have a horizontal, leaf-like thallus usually loosely attached to the substratum and often divided into lobes. In the fruticose lichens the thallus stands upright or hangs downwards from the base.

Reproduction of lichens is poorly understood but it is known to vary within the group. Reproductive bodies, known as apothecia, are seen on the surface of some lichens as saucer-shaped structures (Fig. 43). In other species, flask-shaped reproductive bodies known as perithecia are embedded in the thallus and open onto the surface via a pore. The reproductive bodies contain fungal spores and their structure together with that of the spores is important in taxonomy. The spores are dispersed by wind and after germi-nation must combine with suitable algal cells in order to re-establish the partnership, a process known as lichenization. Vegetative reproduction has also been described in lichens. Pieces of the thallus become detached and are dispersed by wind, insects and birds and grow into new lichens.

Lichens grow best on stable surfaces and do not flourish on friable slates and shales or in areas where there is atmospheric pollution. They are affected by the chemical nature of the substratum, different species growing on acidic rocks compared with calcareous rocks. Lichens are generally regarded as slow-growing and long-lived, and while some show almost no growth, others attain growth rates of 10–30 mm per year. Longevity in excess of 50 years has been recorded for some species. Lichens found intertidally and at the littoral-fringe—terrestrial boundary often form distinct zones according to their ability to withstand submersion in seawater and salt spray. On rocky shores, the most striking of these is a zone of the black lichens *Verrucaria maura* (p. 76) and *Lichina* (p. 80), above which is a zone of the orange lichens, *Caloplaca marina* (p. 79) and *Xanthoria parietina* (p. 80) together with the grey-green coloured *Ramalina siliquosa* (p. 78).

### Key to species of lichens

Although some lichens are difficult to identify, those common on the shore can generally be recognized on the basis of external characters seen under a good quality hand lens. For more advanced studies, examination under a microscope is essential and chemical testing may be necessary.

1  **Thallus crustose** – crust-like, firmly attached to the substratum; difficult to remove ........ 2
   **Thallus foliose** – horizontal leaf-life thallus often divided into lobes, usually only
   loosely attached to the substratum ................................................................................ 8
   **Thallus fruticose** – thallus upright or hangs downward from the base; much branched
   and shrubby or strap-like ............................................................................................ 9
2  Thallus black, smooth, crossed by a meshwork of fine cracks; covers large areas;
   resembles an oil stain on rock surface; at upper limit of littoral fringe often forming
   distinct band ..................................................................... *Verrucaria maura* (p. 76)
   Thallus not resembling a black oil stain ...................................................................... 3
3  Thallus dark green, smooth, almost gelatinous in irregular patches up to 300 mm
   across, but not a distinct band; littoral fringe to mid-shore ....... *Verrucaria mucosa* (p. 76)
   Thallus not dark green ................................................................................................ 4
4  Thallus orange ........................................................................................................... 5
   Thallus as small black dots embedded in barnacle or mollusc shells, particularly
   barnacles, and calcareous rocks *or* thallus grey, encrusting ....................................... 6
5  Thallus rusty-orange, granular; just above littoral fringe ............ *Caloplaca marina* (p. 79)
   Thallus shiny, bright orange colour; just above littoral fringe    *Caloplaca thallincola* (p. 79)
6  Thallus as small black dots embedded in shells of barnacles and molluscs and
   calcareous rocks; as blackish patches on harder substrata    *Pyrenocollema halodytes* (p. 81)
   Thallus grey, encrusting ............................................................................................. 7
7  (a) Thallus thick, cracked warty surface; apothecia with black centres and pale irregular
   margins; above littoral fringe ....................................................... *Tephromela atra* (p. 77)
   *or*
   (b) Similar to above, but apothecia with smooth margins ... *Lecanora gangaleoides* (p. 77)
   *or*
   (c) Thallus thick, warty, edged with white; apothecia with pinkish-grey centres and
   thick margins; above littoral fringe ......................................... *Ochrolechia parella* (p. 78)
8  Thallus bright orange, lobed, leafy; loosely attached to substratum; just above littoral
   fringe ............................................................................... *Xanthoria parietina* (p. 80)
   Thallus dark green, spongy when wet, brown and stiff when dry; thick with branching
   overlapping lobes; just above littoral fringe ......................... *Anaptychia runcinata* (p. 79)

9  Thallus light grey-green, strap-shaped; upright or hanging downwards; upper limit of littoral fringe and above ........................................................ *Ramalina siliquosa* (p. 78)

Thallus dark brown, black; upright branching lobes; shrubby; may be reminiscent of small dried seaweed ........................................................................................... 10

10  Thallus up to 10 mm in height; flattened, branching lobes; grows in fairly open tufts, often covering large areas; from lower littoral fringe to mid-shore. Reminiscent of small dried seaweed ......................................................................*Lichina pygmaea* (p. 81)

Thallus up to 5 mm in height; rounded, much-branched lobes, grows in dense tufts; not covering extensive areas; upper littoral fringe and above ......... *Lichina confinis* (p. 80)

**Verrucaria maura** Wahlenberg    (Fig. 41)

*Thallus crustose, smooth; adheres closely to substratum and is difficult to remove. Surface of thallus crossed by meshwork of fine cracks. Black in colour. Perithecia visible as black spots.*

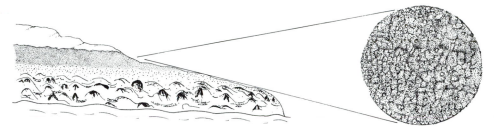

Figure 41  *Verrucaria maura*, seen as a characteristic band at the upper limit of the littoral fringe. Detail inset.

*V. maura* is a very common lichen, widely distributed in north-west Europe. Where conditions are favourable it forms a broad, black band on the upper reaches of the littoral fringe, contrasting sharply with the grey-white barnacles below and the bright orange-yellow *Xanthoria* (p. 80) and *Caloplaca* (p. 79) above. It is often described as being similar in appearance to an oil stain on the rocks. The upper limit of *Verrucaria* forms the upper limit of the littoral fringe. In the sheltered sea lochs of the west coast of Scotland, it is one of the most characteristic organisms of the upper shore.

**Verrucaria mucosa** Wahlenberg    (Fig. 42)

*Thallus crustose, smooth, almost gelatinous; in irregular patches up to 300 mm across; adheres closely to surface of substratum. Green, olive in colour, becoming darker in bright sunlight. Perithecia visible as black spots.*

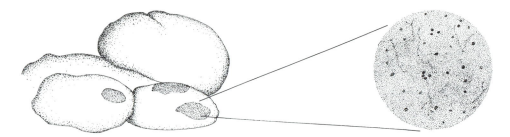

Figure 42  *Verrucaria mucosa*, on boulders on the middle shore. Detail inset.

*V. mucosa* is widely distributed and often abundant on the west coast of Britain on rocks and boulders. On some shores large patches are very conspicuous because of the bright green colour. It extends from the littoral fringe to the middle shore, but does not form a distinct band so characteristic of *V. maura* (p. 76).

**Tephromela atra** (Hudson) Hafellner ex Kalb    (*Lecanora atra*) (Fig. 43a)

*Thallus crustose, thick with cracked, often warty surface; grey in colour. Apothecia with black centres and pale, irregular margins. When apothecium is cut open it is seen to be purple-brown in colour.*

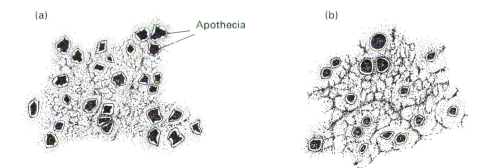

Figure 43  (a) *Tephromela atra*. (b) *Lecanora gangaleoides*.

*T. atra* is a common terrestrial lichen of north-west Europe, found on stones and walls. It extends into the lower reaches of the terrestrial fringe where it withstands salt spray.

**Lecanora gangaleoides** Nylander    (Fig. 43b)

*Thallus crustose, thick with cracked, often warty surface; grey or greenish-grey in colour. Apothecia with black centres and pale, smooth margins. When apothecium is cut open it is seen to be green-brown in colour.*

*L. gangaleoides* is common on rocks close to the sea. It is similar in appearance to *T. atra* and is often confused with that species. The tendency to a greenish-grey colour of the thallus in *L. gangaleoides* is a useful character in the separation of the species, as is the colour difference seen when the apothecia are cut open.

### *Ochrolechia parella* (Linnaeus) Massalongo    (Fig. 44a)

*Thallus crustose, thick and warty; grey in colour, edged with white. Apothecia with pinkish-grey centres and thick margins; with powdery surface.*

(a)

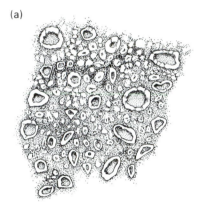

(b)

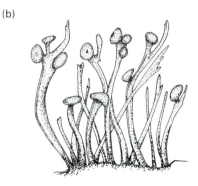

Figure 44   (a) *Ochrolechia parella.* (b) *Ramalina siliquosa.*

*O. parella* is a common terrestrial lichen of north-west Europe, found on stones, walls and trees. It extends into the lower reaches of the terrestrial fringe where it is sometimes abundant on rocks exposed to salt spray.

### *Ramalina siliquosa* (Hudson) A. L. Smith    Sea-ivory (Fig. 44b)

*Thallus fruticose, upright or hanging downwards and may be sparsely branched; strap-shaped; up to 100 mm in length. Light grey-green colour. Apothecia usually at distal ends of branches, white or pale brown in colour.*

*R. siliquosa* is widely distributed in north-west Europe. It is found in patches at the upper limit of the littoral fringe and above, and in some situations forms a distinct band above *Xanthoria* (p. 80) and *Caloplaca* (below). It often grows with the moss *Grimmia maritima* (below), and in parts of Britain is grazed by sheep.

*Anaptychia runcinata* (Withering) Laundon    (*Anaptychia fusca*) (Fig. 45a)

*Thallus foliose, thick, with branching, overlapping lobes. Dark green and spongy when wet; brown and stiff when dry. Apothecia with black centres and brown, wavy margins.*

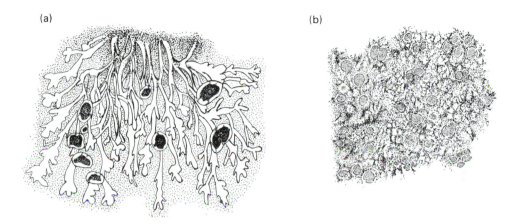

Figure 45   (a) *Anaptychia runcinata.* (b) *Caloplaca marina.*

A. *runcinata* is widely distributed in north-west Europe, growing on rocks at the littoral-fringe—terrestrial boundary.

*Caloplaca marina* (Weddell) Zahlbruckner ex Du Rietz    (Fig. 45b)

*Thallus crustose, with many small granules, particularly in centre of thallus, together with apothecia. Thallus and apothecia rusty-orange colour. May be confused with* Xanthoria parietina *(below), but thallus flatter and not leafy.*

C. *marina* is widely distributed and common in north-west Europe. Along with *Xanthoria parietina* it often forms a broad, yellow-orange band just above the littoral fringe.

*Caloplaca thallincola* (Weddell) Du Rietz    (not illustrated)

*Thallus crustose, with apothecia in the centre of the thallus. Thallus and apothecia shiny, bright orange colour.*

C. *thallincola* is widely distributed in north-west Europe and on western coasts of Britain. It often occurs with C. *marina* and is easily confused with that species, but is brighter orange in colour.

*Xanthoria parietina* (Linnaeus) Th. Fries    (Fig. 46)

*Thallus foliose, lobed; broad and leafy. Loosely attached to substratum. In patches up to about 100 mm across. Bright orange colour changing to greenish-yellow in shaded, humid conditions. Apothecia deep orange in colour. May be confused with* Caloplaca marina *(p. 79) but thallus leafy.*

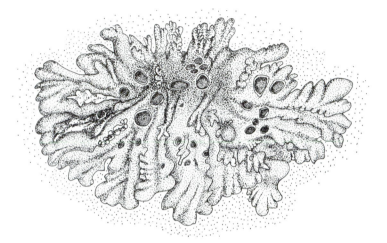

Figure 46  *Xanthoria parietina.*

*X. parietina* is widely distributed in north-west Europe, growing in profusion on rocks, walls, trees and roofs, especially in areas enriched by nitrogenous compounds. It is also abundant on rocks on the high shore and along with *Caloplaca marina* often forms a distinct orange-yellow band just above the littoral fringe. In some places it is covered by the highest spring tides.

The maritime moss, **Grimmia maritima** Turner, often occurs with *Xanthoria parietina* and *Ramalina siliquosa*, especially on the west coast of Britain, but rarely on calcareous rocks. It is dark, brownish-green in colour and grows as small, compact clumps up to 40 mm in height.

**Lichina confinis** (O. F. Müller) C. A. Agardh    (not illustrated)

*Thallus fruticose, formed of upright, rounded, much-branched lobes. Up to 5 mm in height. Dark brown, black. Rounded apothecia at ends of branches. Grows in dense tufts; does not cover extensive areas of rock.*

*L. confinis* is widely distributed and common in north-west Europe. It occurs higher on the shore than *L. pygmaea* (below) and is found on the upper littoral fringe and above, often with *Caloplaca marina* (above). It is most abundant on sheltered shores in sunny positions.

### *Lichina pygmaea* (Lightfoot) C. A. Agardh    (Fig. 47)

*Thallus fruticose, formed of upright, flattened, branching lobes. Up to 10 mm in height. Dark brown, black; brittle when dry. Rounded apothecia at ends of branches. Grows in fairly open tufts, often covering large areas. Not to be confused with the red seaweed* Catenella caespitosa *(p. 57).*

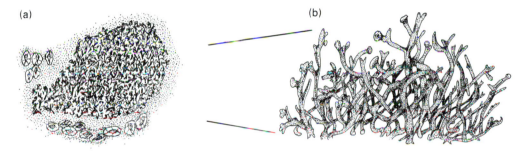

Figure 47  *Lichina pygmaea*, small patch on rocks in barnacle zone (a), with detail of frond (b).

*L. pygmaea* is widely distributed and common in north-west Europe. It has been recorded from the lower littoral fringe to about the middle shore, where it is regularly covered by the tide and tolerates exposure to wave action. It is usually found in association with barnacles, and in favourable situations such as steep slopes in open sunny positions on exposed shores it can cover several square metres of rock surface. The lichen harbours large numbers of invertebrates, particularly the isopod *Campecopea hirsuta* (p. 359), which feeds on *Lichina*, and the bivalve *Lasaea adansoni* (p. 296).

### *Pyrenocollema halodytes* (Nylander) R. C. Harris    (*Arthopyrenia halodytes*) (Fig. 48)

*Thallus crustose. On hard rocks seen as small, blackish-brown patches; on calcareous rock and shells of barnacles and molluscs is embedded in the substratum and appears as small black dots.*

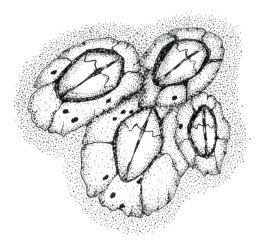

Figure 48 *Pyrenocollema halodytes*, seen as black dots on the acorn barnacle, *Chthamalus*.

*P. halodytes* has a worldwide distribution. It occurs on the upper shore on rocks and on the shells of a number of common intertidal invertebrates, such as barnacles, mussels and limpets.

REFERENCES

Alvin, K. L. (1977). *The Observer's book of lichens*. London: Frederick Warne.

Dobson, F. S. (1992). *Lichens. An illustrated guide to the British and Irish species*. Slough: Richmond Publishing.

Duncan, U. K. (1970). *Introduction to British lichens*. Arbroath: T. Buncle.

Fletcher, A. (1975). Key for the identification of British marine and maritime lichens. I. Siliceous rocky shore species. *Lichenologist*, 7, 1–52.

Fletcher, A. (1975). Key for the identification of British marine and maritime lichens. II. Calcareous and terricolous species. *Lichenologist*, 7, 73–115.

Purvis, O. W., Coppins, B. J., Hawksworth, D. L., James, P. W. & Moore, D. M. (1992). *The lichen flora of Great Britain and Ireland*. London: Natural History Museum Publications.

# Magnoliopsida (Angiospermae)

The Magnoliopsida, commonly known as the angiosperms, are flowering plants and include the trees, shrubs, herbs and grasses so conspicuous in the terrestrial environment. They are characterized by seeds which develop and ripen inside an ovary. While the vast majority are terrestrial, some live in freshwater ponds, lakes and rivers, but few are adapted to the marine environment. Of these, some are tolerant of periodic submersion in seawater and are often common on the shore, while others are tolerant of salt spray and are characteristic of the littoral-fringe–terrestrial boundary. The angiosperms included here tolerate daily submersion by the tide, and together with seaweeds such as *Enteromorpha* (p. 35), are important colonizing and stabilizing species of sand and mud deposits. Almost invariably, the presence of these plants slows down water movement, resulting in the deposition of fine particles and an increase in the level of the beach leading to a plant succession typical of saltmarshes and dominated by angiosperms such as thrift (*Armeria maritima* (Miller) Willd.), sea aster (*Aster tripolium* L.) and seablite (*Suaeda maritima* (L.) Dumort.). These latter species are not included in this text and for information on saltmarsh vegetation the reader is referred to Ranwell (1972) and Long & Mason (1983). The three genera described here are readily identified and common on sheltered shores and estuaries.

*Salicornia europaea* Linnaeus    Glasswort, Marsh samphire, Pickle-weed (Fig. 49c)

*Plant branched, fleshy and shiny; leaves opposite; stem cylindrical. Bright green in colour changing to reddish colours in autumn. Up to 150 mm in height. Tiny flowers in axils of leaves. Dies back in winter.*

There are several species of *Salicornia* in north-west Europe and of these *Salicornia europaea* is widely distributed. It is one of the first angiosperms to colonize mud-flats but is not able to colonize soft sands. The plants withstand regular submersion by the tide and are often found along with the alga *Enteromorpha* (p. 35) and isolated plants of *Spartina* (p. 85). The seeds of *Salicornia* are an important component in the diet of wildfowl such as the teal, and the plants themselves are occasionally harvested and marketed for human consumption. *Salicornia europaea* is an annual.

(a)

(b)

(c)

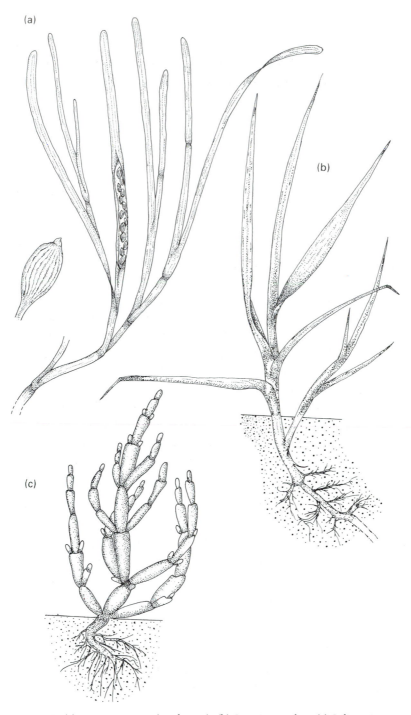

Figure 49  (a) *Zostera marina* (seed inset). (b) *Spartina anglica.* (c) *Salicornia europaea.*

*Zostera marina* Linnaeus    Eel-grass, Sea-grass, Grass-wrack (Fig. 49a)

*Long, narrow, flattened leaves with rounded tips; grass-like; creeping rhizome. Leaves 5–10 mm wide, up to 1 m in height, but usually much less. Dark green in colour. Flowers borne on one side of much-branched stem. Seeds ribbed.*

In the early 1930s much of the *Zostera*, particularly the sublittoral populations, on the Atlantic coasts of North America and Europe was destroyed by a wasting disease seen as dark necrotic patches on the leaves. Various factors including fungal infection, bacterial infection, weakening of the plants due to lack of sunshine and increased water tempera-ture due to unusually warm summers and mild winters have been suggested as possible causes of the disease, which has been seen as recurrent outbreaks in some localities. In 1988 a marine slime mould, *Labyrinthula* sp., was identified as causing the symptoms of the disease. The slime mould appears to be less active at salinities below 10‰, and this may account for the survival of some intertidal populations. *Zostera* recovered to some extent and by the 1950s had returned to many areas. *Z. marina* is now widely distributed in north-west Europe but occurs only sporadically around Britain in sheltered bays and estuaries growing on mud, sand and gravel from the lower shore into the shal-low sublittoral where it often covers extensive areas. The plants flower in summer and ripe shoots become detached and are pollinated in the water by rafts of floating pollen. The ripe seeds sink to the bottom and germinate. *Zostera* beds are important habitats for many species, including hydroids, bryozoans, crustaceans, polychaete worms, gastropod molluscs and fishes, and are grazed extensively by wildfowl such as Brent geese and widgeon. The plants have in the past been used as packing material for mattresses and cushions, and were used in the building of sea dykes in The Netherlands during the last century, but do not appear to have been used commercially since the 1930s.

A second species of *Zostera*, **Zostera noltii** Hornemann, is also widely distributed in north-west Europe. It grows in sheltered bays and estuaries in mud and sand, usually between mean high and mean low water of neap tides and is generally found higher on the shore than *Z. marina*, rarely extending into the sublittoral. The leaves are up to 200 mm in height, much narrower (1 mm in width) than those of *Z. marina* and have divided tips. The flowering stems are unbranched and the seeds smooth. *Z. noltii* is less susceptible to the slime mould *Labyrinthula* than is *Z. marina*.

*Spartina anglica* C. E. Hubbard    Cord-grass, Rice-grass (Fig. 49b)

*Tall, tough grass with stiff, pointed leaves. Large, creeping rhizome. Greyish-green in colour. Usually up to about 1 m in height. Yellowish flowers in spikes during summer. Dies back in winter.*

There are many species of *Spartina* colonizing the saltmarshes of north-west Europe and the history and development of *S. anglica* is of considerable interest. It is now believed that a North American species, *Spartina alterniflora* Loiseleur, was accidently introduced by shipping into the area around Southampton Water in the early 1800s. It was first recorded there in 1829. As a result of a cross between this species and the endemic British species, *Spartina maritima* (Curtis) Fernald, *Spartina* × *townsendii* H. & J. Groves was formed and this was first recorded in 1870 at Hythe in Southampton Water. *S.* × *townsendii* is a sterile hybrid and spreads to new areas by fragments and plants carried by currents. As a result of a natural doubling in the number of chromosomes of *S.* × *townsendii*, a new vigorous plant, *S. anglica*, was formed. It was first recorded in 1892 on the south coast of England at Lymington and spread rapidly by seedlings colonizing large areas of mud-flats, and is now widely distributed in north-west Europe.

*S. anglica*, like *Salicornia* (p. 83) is a pioneer colonizer of bare mud. Where the two species grow together, *Salicornia* occurs lower on the shore, but in areas where the sediment is very unstable, *Salicornia* cannot become established and the deeper rooting *Spartina* is the main colonizer. The deep roots are extremely effective in binding mud and stabilizing the sediment, while the shallower, more delicate roots are mainly absorptive. The plants slow down water movement and cause fine particles to be deposited with a consequent accretion and increase in the level of the sediment. *Spartina* withstands long periods of tidal submersion. Characteristically, the initial colonization of bare mud results in the development of 'islands' of *Spartina* plants which spread vegetatively and by seedlings, and dense swards soon become established. The decaying detritus associated with such areas, much of which is derived from the dying back of the grass at the end of the growing season, forms an important component of estuarine food webs. Large populations of the periwinkle, *Littorina littorea* (p. 238), the mud snail, *Hydrobia ulvae* (p. 246), and juveniles of the shore crab, *Carcinus maenas* (p. 401), are often found in association with *Spartina*. Shoals of small fishes, including the sand goby, *Pomatoschistus minutus* (p. 511), occur in the creeks and drainage channels among *Spartina*. On the west coast of Wales, estuaries colonized by the grass are important nursery grounds for young bass and mullet.

The spread of *Spartina* in north-west Europe was hastened by planting of the grass to prevent sediment erosion and as a means of reclaiming land for grazing. In 1991 it was claimed that *S. anglica* covered about 10 000 hectares of intertidal saltmarsh. This has resulted in the loss of the natural mud-flat habitat and its characteristic fauna of polychaete worms, bivalve molluscs and crustaceans which are so important in the diet of wildfowl. Such has been the rate of spread of *Spartina* in some areas that the loss of these important feeding grounds is giving cause for concern and control of the spread of the species by the application of herbicides is being investigated. Interestingly, die-back of dense stands of *Spartina* has been recorded on the south coast of England. This is attributed to rotting of the rhizome and is a phenomenon which may in some way be age-related as older populations have suffered most. *S. anglica* also suffers from infection by the ergot fungus, *Claviceps purpurea* (Fr.) Tulasne, which prevents the seed from setting.

## REFERENCES

Den Hartog, C. (1970). *The sea-grasses of the world.* Amsterdam: North-Holland Publishing Company.

Gray, A. J. & Benham, P. E. M. (1990). Spartina anglica – *a research review.* Institute of Terrestrial Ecology research publication no. 2. London: HMSO.

Hubbard, C. E. (1984). *Grasses*, 3rd edn. Harmondsworth: Penguin.

Long, S. P. & Mason, C. F. (1983). *Saltmarsh ecology.* Glasgow: Blackie.

Ranwell, D. S. (1972). *Ecology of salt marshes and sand dunes.* London: Chapman & Hall.

Stace, C. (1991). *New flora of the British Isles.* Cambridge: Cambridge University Press.

# Porifera

Sponges are frequently found on the lower shore in gullies and under rocky overhangs. Here they avoid desiccation and display a range of colours and growth forms giving them a superficial resemblance to plants. Indeed, it was not until the eighteenth century that they were classified as animals and as late as 1825 before the issue was put beyond doubt by the research of Dr R. E. Grant of Edinburgh.

Sponges are sessile, some forming encrusting, irregular growths covering large areas of rock, while others are vase-like and attached to the substratum at the base. They exhibit the cellular grade of organization, in which the cells do not form tissues and organs so characteristic of higher animals, and as a result they are often referred to as primitive animals. Despite this simplicity, they exhibit a range of complexity in body structure and have colonized both freshwater and marine habitats, extending from the shore to the greatest depths.

The basic design of a simple sponge is that in which the body wall is perforated by many tiny pores, ostia, through which water enters the sponge, and a larger opening, the osculum, through which water leaves (Fig. 50c). The cavity of the sponge is lined by flagellated cells known as choanocytes; these maintain a flow of water through the body and filter out fine suspended food particles. The body wall is often greatly folded, and in some species is supported by a skeletal network of calcareous or siliceous spicules which are of various shapes and sizes and important in taxonomy. Others have a skeleton of protein fibres known as spongin, while some have a combination of siliceous spicules and spongin fibres.

Regeneration of sponges is well documented, particularly through the now famous experiment in which a sponge is forced through a fine mesh and within a short time the dissociated cells aggregate to form many new sponges. Regeneration from fragments is an important form of asexual reproduction, others being the production of internal or external buds and the release of clusters of cells known as gemmules which develop into a new sponge.

Sponges also reproduce sexually. Most are hermaphroditic but cross-fertilization generally takes place. Sperm are discharged through the osculum and enter a neighbouring individual in the inhalent current. In some cases, fertilized eggs are discharged into the sea where they develop into ciliated larvae, known as planulae, but in the majority, the larva develops within the sponge and is then released. Larval life varies from a few hours to a few weeks and the adult lives for one to several years, some species showing regression and fragmentation during the winter months. Sponges often have a rich associated fauna of, for example, amphipod crustaceans and are preyed on by a variety of animals including sea-slugs and echinoderms.

Sponges are divided into four classes, which are distinguished by taxonomists according to the nature of the spicules.

## Phylum PORIFERA

**Class Calcarea**   Sponges with calcium carbonate spicules. Restricted to firm substrata. Most species found in shallow water.

Class Hexactinellida   Sponges with siliceous spicules mainly of six rays. Many colonize soft substrata. Found in deep water. Commonly known as the glass sponges.

**Class Demospongiae**   Sponges with siliceous spicules, spongin fibres or both. Found on a wide range of substrata from rock to soft mud and some bore into calcareous material. Found in shallow and deep water; includes the majority of the common shore species.

Class Sclerospongiae   Sponges with skeleton of siliceous spicules, spongin fibres and calcium carbonate. Few species. Found on coral reefs.

### Key to species of sponges

Accurate identification of sponges is difficult and sometimes impossible without dissection and microscopic examination of the skeleton and spicules. A small number of common species is included here and reliable identification of these can generally be made by reference to the external features described. However, it should be noted that some of these species show considerable morphological variation from one habitat to another in response to factors such as water depth and speed of current flow, and this can lead to difficulty in identification.

1 (a) Flattened, vase-shaped or tubular. **Purse sponges** ......................................................... 2
*or*
(b) Encrusting, may be in thin sheets or substantial cushion-like growths several centimetres thick. Sometimes lobed. **Encrusting sponges** .................................................. 4
*or*
(c) Limestone rocks and shells, especially oyster shells, often riddled with small round holes about 2 mm in diameter. Yellowish-orange projections may be seen at entrance to holes. **Boring sponges** ................................................................. *Cliona celata* (p. 95)
2 Delicate, tubular; upright, often branching tubes arise from encrusting network.

Osculum at end of each branch. Up to 20 mm in height. On seaweeds. Grey-white in colour .................................................................................. *Leucosolenia* spp. (p. 91)
Vase-shaped, not branching ..................................................................................... 3

3   Flattened, vase-shaped; up to 20 mm in height. Conspicuous terminal osculum, occasionally more than one. Often hanging from rocks and seaweeds .................................
.......................................................................... *Grantia compressa* (p. 90)
More cylindrical; conspicuous terminal osculum surrounded by stiff spicules. Rough, 'hairy' outer surface. Usually to about 30 mm in height. On rocks and seaweeds ...........
.......................................................................... *Scypha ciliata* (p. 91)

4   Large, thick fleshy sponge, up to 50 mm thick. May be in the form of rounded lobes. Surface channelled or smooth. Oscula conspicuous ......................................... 5
Not as above ..................................................................................... 6

5   Usually rounded in shape; up to 200 mm across. Surface smooth. Oscula few in number but conspicuous ............................................... *Suberites ficus* (p. 92)
Thick, irregular lobes; surface marked with channels, slimy. Oscula conspicuous, unevenly distributed, often aligned along ridges ...................... *Myxilla incrustans* (p. 94)

6   Flat, thin, usually about 3 mm thick; surface smooth. Oscula well defined, with small collar. Blood-red, orange-red in colour ............................... *Ophlitaspongia seriata* (p. 94)
Not as above ..................................................................................... 7

7   Surface relatively smooth. Oscula well developed, often raised above surface of sponge and likened to 'miniature volcanoes'. In some situations may be up to 20 mm thick. Usually olive-green colour. When broken has distinctive smell of seaweed .......................
.......................................................................... *Halichondria panicea* (p. 93)
Surface rough, often grooved. Relatively thin encrustations or papillate. Scattered, small oscula, not always obvious. Orange-red colour ........... *Hymeniacidon perleve* (p.94)

Class Calcarea
*Grantia compressa* (Fabricius)   (*Scypha compressa*), Purse sponge (Fig. 50b)

*Vase-shaped, sometimes rather elongate. Conspicuous terminal osculum; occasionally more than one. Up to 20 mm in height, sometimes larger. Thin and flat. Colour varied, white, grey, yellow.*

G. compressa is widely distributed in north-west Europe and around Britain in sheltered crevices and under boulders on the lower shore where it occurs singly or in groups attached to a variety of substrata, often hanging from the roofs and walls of crevices. It is an annual species. Overwintering specimens die after releasing larvae in spring and summer, and sponges settling in spring reproduce in late summer and autumn and survive the winter to reproduce in the following spring. Specimens settling later in the year are believed to reproduce only once, in the spring.

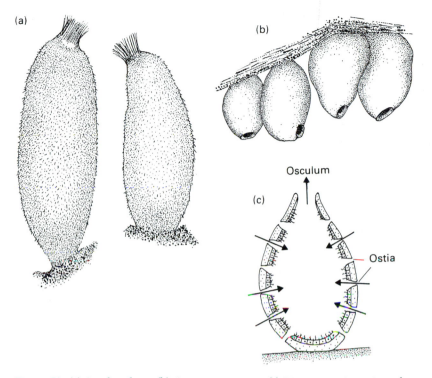

Figure 50  (a) *Scypha ciliata.* (b) *Grantia compressa.* (c) Diagrammatic section of sponge to
show direction of flow of water (arrows).

**Scypha ciliata** (Fabricius)    (*Sycon ciliatum, Sycon coronatum*) (Fig. 50a)

*Body upright, vase-shaped or tubular; more cylindrical than G. compressa (above). Height
about 30 mm but much larger specimens have been recorded. Conspicuous terminal osculum fringed with stiff spicules. Rough, 'hairy' outer surface, sometimes smooth. Yellow-grey
colour.*

S. ciliata is widely distributed in north-west Europe and around Britain where it is
common on the lower shore. It is attached singly or in groups to stones and seaweeds,
the largest specimens being found in sheltered situations. It is an annual. Overwintering
specimens release larvae in the spring and some of these develop rapidly to reproduce in
the autumn of the same year, before overwintering.

**Leucosolenia** spp.    (Fig. 51a)

*Encrusting network from which upright, often branching tubes arise. Oscula conspicuous,
terminal. Surface rough. Up to about 20 mm in height, but usually much smaller, grey-
white in colour.*

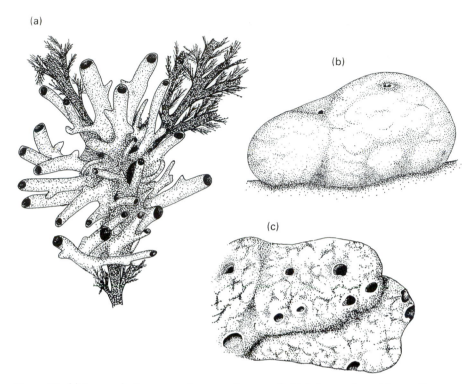

Figure 51  (a) *Leucosolenia* sp. growing on red seaweed. (b) *Suberites ficus*. (c) *Myxilla incrustans*.

The genus *Leucosolenia* is widely distributed in north-west Europe and around Britain; the species are difficult to identify. The sponges are attached to seaweed and stones on the lower shore and in the sublittoral, sometimes with the purse sponge, *Grantia compressa* (above). *Leucosolenia* is believed to be an annual, breeding twice, once in autumn and again the following summer.

**Class Demospongiae**
**Suberites ficus** (Linnaeus)  (*Suberites domuncula*), Sulpher sponge, Sea-orange (Fig. 51b)

*Large, thick, fleshy sponge with smooth surface. Usually rounded in shape and up to 200 mm across. Oscula conspicuous, few in number. Orange, yellow or brownish in colour.*

*S. ficus* is widely distributed in north-west Europe attached to a variety of hard substrata on the lower shore and in the sublittoral. It is often found on shells inhabited by hermit crabs where it completely encloses the shell. It has recently been suggested that *S. ficus* is a complex of three species separable by enzyme electrophoresis.

*Halichondria panicea* (Pallas)    Breadcrumb sponge (Fig. 52a)

*Encrusting, from relatively thin sheets to thick encrustations. Relatively smooth surface. Oscula well developed, sometimes raised well above surface of sponge. Large encrustations may be 20 mm thick. Very variable in form. Characteristically olive-green in colour but shades of yellow common. When broken has a distinctive smell of seaweed.*

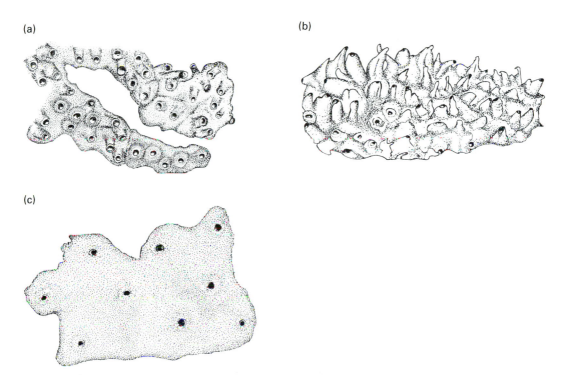

(a)

(b)

(c)

Figure 52  (a) *Halichondria panicea.* (b) *Hymeniacidon perleve.* (c) *Ophlitaspongia seriata.*

*H. panicea* is widely distributed in north-west Europe and around Britain and is sometimes abundant. It is one of the most common sponges on the lower shore where it is found in gullies, under rocky overhangs and among the holdfasts of seaweeds. It covers extensive areas of rock and often overgrows other encrusting invertebrates. The form of the sponge varies, particularly in relation to current flow, and in turbulent areas it is much thicker than in sheltered waters, the faster current presumably enhancing the flow of water through the sponge. The typical olive-green colour of the species is due to the presence of green algal symbionts in the tissue; specimens from deeper water lack the symbionts and are yellow in colour. In thick specimens, only a thin outer layer is green, the deeper tissues being yellow. Large patches of *Halichondria* often have a diverse associated fauna and there is evidence to suggest that some amphipods, e.g. *Caprella* (p. 381), are attracted to chemicals given off by the sponge. The life-span of *Halichondria* is

believed to be about three years and in some localities breeding has been recorded annually.

*Halichondria bowerbanki* Burton is similar in appearance to *H. panicea* and like *H. panicea* it is very variable in form. The two species are easily confused but useful guidelines
for separation are that *H. bowerbanki* is mostly sublittoral and rarely found on the shore,
is more tolerant of muddy and silty conditions such as in harbours, is creamish in colour,
not green, and when broken does not have the distinctive smell of seaweed that is so
characteristic of *H. panicea*.

### *Myxilla incrustans* (Johnston)    (Fig. 51c)

*Thick, encrusting, up to 200 mm across and 50 mm thick. Oscula well developed, unevenly
distributed and often aligned along ridges. Surface marked with channels, slimy. Yellow to
orange in colour.*

*M. incrustans* is widely distributed in north-west Europe, encrusting rocks and boulders
on the lower shore and in the sublittoral. Specimens with embryos have been recorded
during August and September.

### *Hymeniacidon perleve* (Montagu)    (*Hymeniacidon sanguinea*) (Fig. 52b)

*Encrusting, relatively thin or papillate intertidally, sublittoral specimens thicker and
cushion-like; sometimes buried in mud. Surface rough and often grooved. Small, scattered
oscula, randomly arranged, not always obvious. Orange-red to deep red in colour; colour,
form and texture vary widely.*

*H. perleve* is widely distributed in north-west Europe and around Britain, and like *Halichondria* (p. 93) is one of the commonest intertidal sponges. The largest encrustations
are found on the lower shore and cushion-like growths have been recorded from the
shallow sublittoral. The species extends to the middle shore where it is found as small
fragments. It is found on a variety of substrata including rocks, seaweed holdfasts and
also on muddy gravel, where only the oscula can be seen on the surface of the sediment.
Growth in the summer months is often followed by a reduction in size and by fragmentation of the sponge in winter. Off the south coast of England embryos have been
recorded in the sponge from July to October, with larval release presumably taking place
in late summer. Longevity is believed to be three or more years.

### *Ophlitaspongia seriata* (Grant)    (Fig. 52c)

*Encrusting; flat and thin, usually about 3 mm thick; growing in patches 50–100 mm
across. Surface smooth. Oscula well defined, mainly regular in distribution; with small
collar. Blood-red, orange-red in colour.*

*O. seriata* is found on west and south-west coasts of Britain and may be locally common. It is found on the lower shore and sublittorally to depths of one or two metres. Embryos are found in the sponge during summer and larval life is about two days.

## *Cliona celata* Grant    Boring sponge (not illustrated)

Limestone rocks and stones on the lower shore, and the empty shells of bivalve molluscs such as the oyster, are often riddled with small, round holes about 2 mm in diameter. These have probably been made by *C. celata*, a boring sponge sometimes seen as yellowish-orange projections at the entrance to the holes. The extent of the sponge can be seen if the rock or shell is broken open. Boring is started by the larva and is brought about by secretions, possibly including carbonic anhydrase, from special cells which eat into the rock. A small chip of the substratum is surrounded by the cell and eventually undercut and removed in the exhalant current, the sponge gradually penetrating deeper.

REFERENCES

Ackers, R. G., Moss, D., Picton, B. E. & Stone, S. M. K. (1985). *Sponges of the British Isles (Sponge IV)*. A colour guide and working document. Ross on Wye: Marine Conservation Society.

Burton, M. (1963). *A revision of the classification of the calcareous sponges*. London: British Museum (Natural History).

Hiscock, K., Stone, S. M. K. & George, J. D. (1984). The marine fauna of Lundy Porifera (Sponges): a preliminary study. *Report of the Lundy Field Society*, 34, 16–35.

Juniper, A. J. & Steele, R. D. (1969). Intertidal sponges of the Portsmouth area. *Journal of Natural History*, 3, 153–63.

# Cnidaria

The phylum Cnidaria, also known as the Coelenterata, includes the hydroids or sea-firs, the sea-anemones, jellyfishes and corals. These apparently diverse forms have in common a basic radial symmetry and a relatively simple body structure with a body wall made up of an outer epidermis and an inner gastrodermis, separated by a jelly-like mesogloea. The body cavity has a single opening to the exterior called the mouth and this is usually surrounded by tentacles, which are without cilia, but have special cells known as cnidocytes (nematocytes). These contain capsule-like structures, the cnidae, which are characteristic of the phylum. The cnidae contain a thread which is discharged on stimulation; in some (the nematocysts) the thread penetrates the prey and injects a toxin; in others (the spirocysts) the thread is adhesive. A few cnidarians are suspension feeders but most are carnivorous and when the tentacles make contact with the prey, it is immobilized and passed to the mouth. The majority of cnidarians are dioecious and both asexual reproduction by budding and sexual reproduction are common.

A striking feature of cnidarians is the high degree of polymorphism often seen in a single species. Two basic types exist; these are a sessile polyp (e.g. a hydroid or sea-anemone) and a free-swimming medusa (e.g. a jellyfish). The presence or absence of polyp or medusa in the life-cycle forms the basis of the classification of the phylum. The classification used here follows Barnes, Calow & Olive (1993).

## Phylum CNIDARIA

**Class Hydrozoa**   Usually both polyp and medusoid stages in the life-cycle, the polyp generally being the most obvious. Mainly marine; includes the hydroids or sea-firs, floating colonial forms such as the Portuguese Man o'War and a few freshwater species such as *Hydra*.

**Class Scyphozoa**   The medusa is the most conspicuous stage in the life-cycle. Includes the free-swimming jellyfishes and some bottom-dwelling species. Marine.

**Class Cubozoa**   Tall, box-shaped medusae; largely tropical and subtropical and noted for their dangerous sting. Marine.

**Class Anthozoa**   Without medusae.

**Subclass Ceriantipatharia**    Solitary, elongate anthozoans inhabiting secreted mucous tubes buried in soft sediments, and the black or thorny corals of deep water. Marine.

**Subclass Alcyonaria** (Octocorallia)    Colonial; polyps generally small, with eight pinnate tentacles. Includes the soft corals, sea-fans and sea-pens. Marine.

**Subclass Zoantharia** (Hexacorallia)    Colonial or solitary; polyps usually with many smooth tentacles. In temperate waters best known through the sea-anemones. Also includes the stony corals. Marine.

## Class Hydrozoa

Hydrozoans exist as either polyp or medusa and the majority of species has both forms in the life-cycle. Most are colonial animals, a notable exception being the freshwater hydras which are solitary polyps without a medusa. Colonies are formed of groups of polyps arising from a common stolon linking them together and anchoring them to the substratum. The polyps are connected to one another by thread-like continuations of the body wall containing extensions of the body cavity. These may be in the form of horizontal stolons or upright stems and are supported by a stiff, protein—chitinous sheath, the perisarc. Hydrozoan colonies generally show division of labour, with some polyps (the hydranths or gastrozooids) specialized for feeding, some (the gonozooids or blastostyles) for reproduction and others (the nematophores and dactylozooids) for defence. A few hydrozoans form floating colonies and are found in the open sea (p. 117).

Life-cycles are complex and varied. In species such as *Obelia* (p. 111), the reproductive polyps bud off free-swimming medusae and at certain times of the year these are abundant in the plankton of inshore waters. The medusae reproduce sexually to produce free-swimming larvae known as planulae and, after a pelagic existence of a few hours to a few days, they settle on the bottom to establish new colonies (Fig. 53). In many hydrozoans the medusa is not free-swimming but stays attached to the hydroid where it reproduces sexually to give a free-swimming planula larva. This eventually settles on the bottom to begin a new colony. In a few species such as *Tubularia* (p. 100), there is no planula; an actinula larva is released. The actinula moves about on its tentacles and eventually attaches to the substratum to establish a new colony (Fig. 54).

The majority of hydroids occurs sublittorally, but many are seen as delicate, plant-like growths at low water where they sometimes occur in profusion attached to the walls of gullies and overhangs, and attached to stones, shells and seaweed. Sublittoral species are often washed ashore during storms. Hydroids are preyed on by a variety of invertebrates, including sea-slugs, sea-spiders and ghost shrimps.

### Key to the groups of hydrozoans

The hydrozoans included here can conveniently be divided into three groups. The first two of these, the athecate and thecate hydroids, are the typical hydroids of the shore

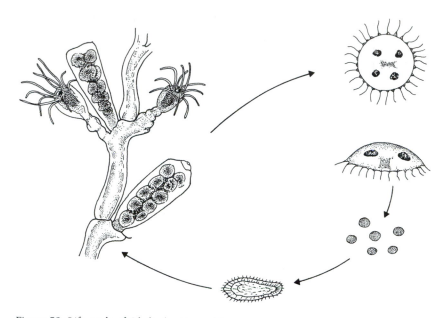

Figure 53  Life-cycle of *Obelia* (not to scale).

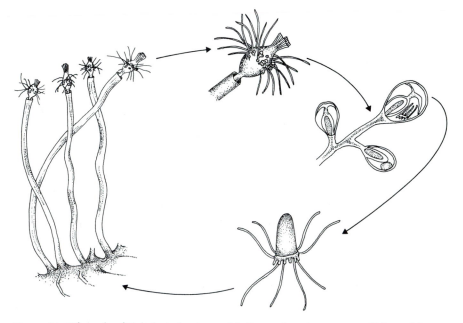

Figure 54  Life-cycle of *Tubularia* (not to scale). For explanation see text. (Adapted from
        Hincks, 1868.)

and shallow sublittoral and are usually of small size, attached to rocks, shells and seaweeds. The remaining group comprises the colonial hydrozoans characteristically found floating in open waters; they show a high degree of polymorphism.

1 **Athecate hydroids** – the perisarc stops short of the hydranth, gonozooids and nematophores and forms a protective covering only for the interconnecting strands. The hydranth cannot be withdrawn (Fig. 55a).
2 **Thecate hydroids** – the perisarc extends around the hydranth, forming a protective cup, the hydrotheca, into which the hydranth can be partially or wholly withdrawn, and around the gonozooid and nematophore to form a gonotheca and a nematotheca, respectively (Figs. 60, 70).
3 **Floating colonial hydrozoans** – species such as the Portuguese Man o'War and Jack Sail by-the-Wind, occasionally stranded on the shore.

Athecate hydroids

### Key to families of athecate hydroids

1 Large, solitary, extensible hydranth covered with large number of small, blunt tentacles. Attached by a basal region with root-like processes ....... **Candelabridae** (p. 102)
Hydroid not so ......................................................................................................... 2
2 Hydranth with knobbed tentacles, scattered or in several rings ............ **Corynidae** (p. 101)
Hydranth with long narrow tentacles ...................................................................... 3
3 Tentacles scattered over hydranth ......................................................... **Clavidae** (p. 102)
Tentacles in one or more rings on hydranth ............................................................ 4
4 Hydranths arise directly from an encrusting growth, especially on shells occupied by hermit crabs. Several polyp types and small spines present .......... **Hydractiniidae** (p. 103)
Hydranths borne on stems ...................................................................................... 5
5 Tentacles arranged in 2 distinct rings; mouth on conical projection ... **Tubulariidae** (p. 99)
Tentacles arranged in 1 ring; mouth on conical projection .......... **Bougainvilliidae** (p. 105)

Family Tubulariidae
Hydranths on upright stems; with two rings of long, narrow tentacles, one ring near the mouth and the other near the base of the hydranth. Mouth on conical projection. Gonozooids usually in clusters on branched stalks between the two rings of tentacles.

Stems rarely branching, with very few, if any, annulations; up to 150 mm in height. Tentacles around mouth more numerous than those near base of hydranth ........................ ............................................................................................ *Tubularia indivisa* (p. 100)
Stems branched, often extensively; annulated at regular intervals; up to 35 mm in height. Hydranth with up to about 20 tentacles in each ring ..... *Tubularia larynx* (p. 100)

### *Tubularia indivisa* Linnaeus    (Fig. 55a)

*Stems rarely branching, very few, if any, annulations; up to 150 mm in height; brown. Stems narrow at base and often twisted; rising from a stolon of tangled fibres. Hydranths pink to dark red with a ring of short, white tentacles around mouth more numerous than the ring of long, white, mostly drooping tentacles close to base. Gonozooids in clusters on branched stalks between the 2 rings of tentacles.*

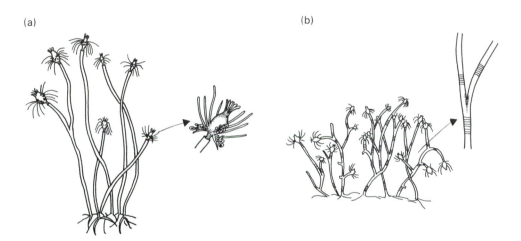

(a)                                                                              (b)

Figure 55   (a) *Tubularia indivisa*, note athecate condition (after Hincks, 1868). (b)
              *Tubularia larynx* (after Hincks, 1868).

*T. indivisa* is widely distributed in north-west Europe and around Britain, attached to rocks, stones and shells at low water of spring tides, and extends into the sublittoral where it may be abundant. It is often found at the mouths of estuaries. Although breeding occurs throughout the year in some localities, peaks of recruitment have been recorded in spring and summer in Britain. There is no free-swimming medusa; a creeping actinula larva is produced. Longevity of the hydroid is believed to be about a year.

### *Tubularia larynx* Ellis & Solander    (Fig. 55b)

*Stems branched, often extensively; annulated at regular intervals; yellow-brown colour. Annulations sometimes difficult to see in living material. Generally up to 35 mm in height; sometimes more. Hydranths small, pale red; up to about 20 tentacles in each ring. Gonozooids as in* T. indivisa.

*T. larynx* is similar in both habit and distribution to *T. indivisa*. The base of the colonies often accumulate mud in which small *Mytilus* (p. 283), *Jassa* (p. 380) and other organisms can be found. In Britain, breeding has been recorded from spring to autumn and gonozooids have been recorded on individuals as early as seven days after settlement of the larva. Maturity can be reached in three weeks. There is no free-swimming medusa; an actinula larva is released from the hydroid.

Family Corynidae
Hydranths on upright stems; with knobbed tentacles scattered or in several rings. Gono-zooids scattered over hydranth or below the tentacles.

Stems and branches annulated throughout length. Up to 25 mm in height ................... ........................................................................................... *Coryne pusilla* (p. 101)
Much branched. Stems and branches without annulations, except at base of stem and above origin of branches. Up to 100 mm in height ......................... *Sarsia eximia* (p. 102)

*Coryne pusilla* Gaertner    (Fig. 56a)

*Irregularly branched, brown stems rising from thread-like stolon; about 25 mm in height. Branches annulated throughout length. Hydranths slender, reddish, with 30 or more knobbed tentacles, either scattered or in several ill-defined rings. Spherical gonozooids scattered over hydranth.*

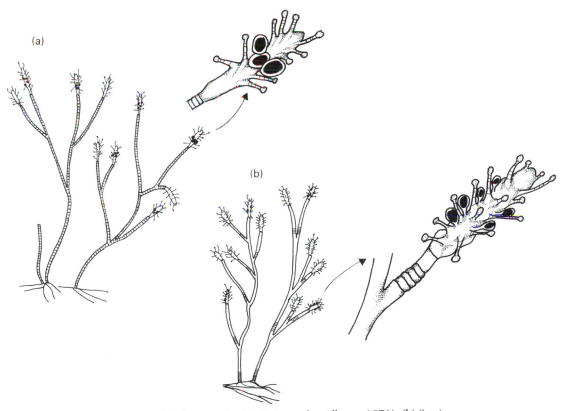

Figure 56  (a) *Coryne pusilla* (after Hincks, 1868; inset after Allman, 1871). (b) *Sarsia eximia* (after Hincks, 1868; inset after Allman, 1871).

*C. pusilla* is widespread in north-west Europe and around Britain, often attached to sea-weeds in rock pools on the lower shore. Breeding has been recorded on the east coast of England from May to September. There is no free-swimming medusa; a planula larva is liberated.

**Sarsia eximia** (Allman)    (Fig. 56b)

*Much branched, mainly from one side of stem; 50–100 mm in height. Yellow-brown colour. Annulations present at base of stem and above origin of branches. Reddish hydranths; 20–30 knobbed tentacles scattered over hydranth. Egg-shaped gonozooids borne on stalks rising from bases of tentacles.*

*S. eximia* is widely distributed in north-west Europe and around Britain where it is found on a number of seaweeds, including *Laminaria* (p. 43) and *Corallina* (p. 61), and on the sides of rock pools on the lower shore. Breeding has been recorded in spring and summer, and off the north-west coast of England medusae have been recorded in the plankton from April to July.

Family Candelabridae
Hydranth large, solitary, extensible, on root-like processes; covered with many small blunt tentacles.

**Candelabrum phrygium** (Fabricius)    (*Myriothela phrygia, Arum cocksi*) (Fig. 57a)

*Hydranth solitary, cylindrical, with pointed apex; extensible, 40–100 mm in height. Apart from apical region, hydranth covered with large number of small blunt tentacles with reddish-brown tips. Attached to stones and rocks by a brown, basal region with root-like processes. Rounded, pink gonozooids borne on projections from base of hydranth.*

*C. phrygium* is widely distributed in north-west Europe and around Britain where it is commonest in the south and west. It is found under stones and rocks on the lower shore and into the sublittoral. On the south coast of England breeding has been recorded throughout the year. During development the fertilized eggs are held by specialized tentacles known as claspers. There is no free-swimming medusa; actinula larvae are released from the gonozooids and crawl about for several days before settling.

Family Clavidae
Hydranths arising directly from stolon or on upright stems; many scattered long, narrow tentacles.

*Clava multicornis* (Forskål)    (Fig. 57b)

*Pink, white or reddish hydranths rising directly from thread-like stolon; up to 25 mm in height. Hydranths with 30–40 tentacles. Spherical gonozooids below tentacles.*

(a)                                        (b)

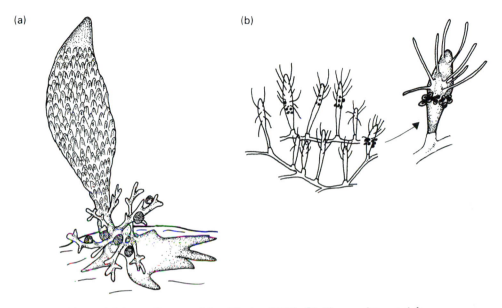

Figure 57   (a) *Candelabrum phrygium* (after Hincks, 1868). (b) *Clava multicornis* (after Allman, 1871).

C. *multicornis* is widely distributed in north-west Europe and around Britain. It is commonly found on *Fucus vesiculosus* (p. 50), *F. serratus* (p. 49) and *Ascophyllum nodosum* (p. 47) on the middle and lower shore, and also on shells and stones. The form of the colony ranges from a diffuse network of stolons with scattered hydranths, to a dense mat of stolons with many hydranths, and although previously recorded as two separate species, C. *multicornis* and *Clava squamata* (Müller), they are now believed to belong to a single species, C. *multicornis*. The exact form of the colony is believed to be governed by environmental factors such as the nature of the substratum, water movement, food supply and tidal exposure. Reproduction takes place from spring to autumn. There is no free-swimming medusa; crawling planula larvae are released from the gonozooids during daylight hours. They are believed to seek out fronds of seaweeds such as *Ascophyllum*, where attachment and metamorphosis occur within a few days of their release.

Family Hydractiniidae
Hydroid encrusting, especially on gastropod shells occupied by hermit crabs; exhibits polymorphism; spines present among polyps. Hydranths arise directly from encrusting growth; one or two rings of slender tentacles.

*Hydractinia echinata* (Fleming)    (Fig. 58)

*Encrusting growth on prosobranch shells inhabited by hermit crabs; also occasionally recorded on other substrata such as rock, mussel shells, seaweed and barnacles. Large numbers of pink, white and brown polyps, some up to 10–15 mm in height. Four distinct types of polyp present – hydranths with 2 adjacent rings of slender tentacles; oval gonozooids; much-coiled spiral polyps and thin, extensible, tentacular polyps. Small serrated spines also present.*

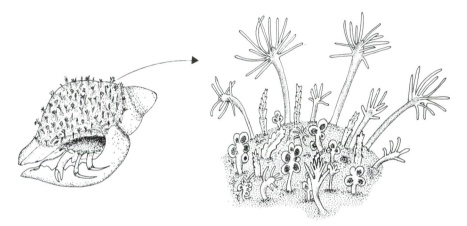

Figure 58  *Hydractinia echinata.*

*H. echinata* is widely distributed in north-west Europe and around Britain where it is found on the lower shore extending well into the sublittoral. The spiral and tentacular polyps are believed to be associated with defence, the spiral polyps predominating on the edge of the colony towards the mouth of the shell. Interestingly, they disappear in a few days if the hermit crab is removed and reappear within a matter of days if the crab is returned. It appears that both the presence of the crab and the current systems it produces are essential for development of the spiral polyps. In Britain, breeding has been recorded throughout the year with peak activity in summer. There is no free-swimming medusa. Eggs and sperm are shed into the water and crawling planulae develop, settling on shells occupied by hermit crabs. Apparently, the planulae detect moving shells to which they become attached by nematocysts. They crawl over the surface of the shell where metamorphosis is induced by the presence of bacteria. Asexual division leads to the development of a new colony.

Other hydractinids encrust living gastropod shells and empty shells inhabited by hermit crabs, and these belong to the genus *Podocoryne*. They are predominantly sublittoral and can be separated from *Hydractinia* by the presence of a single ring of tentacles on the hydranth.

Family Bougainvilliidae

Hydranths usually on upright stems; with a single ring of long, narrow tentacles; mouth on a conical projection. In addition to being found on rocks and stones, some members of the family are found on shells occupied by hermit crabs. They are readily separated from the Hydractiniidae (above).

*Bougainvillia ramosa* (van Beneden)    (Fig. 59)

*Main stem relatively thick, much branched; branches alternate. Height of 50 mm or more; yellow-brown colour. Hydranths pink to red, with single whorl of about 20 whitish tentacles. Stalked gonozooids just below hydranths.*

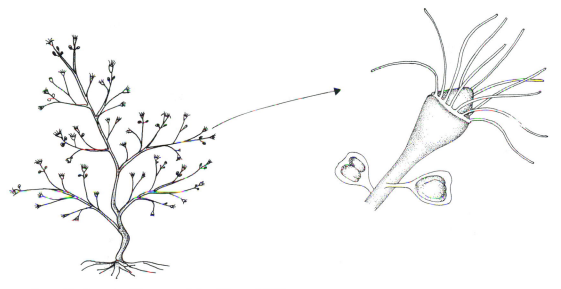

Figure 59  *Bougainvillia ramosa* (after Allman, 1871).

*B. ramosa* is widely distributed in north-west Europe and in Britain is probably commonest on southern shores. It is found on the lower shore attached to rock, shells and seaweed, and reproduces during spring and summer. At Plymouth, medusae have been recorded in the plankton from April to October. Other species of *Bougainvillia* are found in north-west Europe but *B. ramosa* is probably the most widespread and common around Britain.

Thecate hydroids

### Key to families of thecate hydroids

1 Hydrothecae bell-shaped, on stalks. Hydrothecae open, not closed by operculum; aperture rounded. Hydroid upright or creeping ........................ **Campanulariidae** (p. 106)
  Hydrothecae not on stalks, arise directly from the stem; aperture oval. Hydroid upright . 2
2 Hydrothecae on both sides of stem. Hydroid without nematophores ...................................
  ................................................................................... **Sertulariidae** (p. 111)
  Hydrothecae on one side of stem only. Hydroid with nematophores ...................................
  ................................................................................. **Plumulariidae** (p. 113)

Family Campanulariidae
Hydrothecae generally bell-shaped; on stalks; hydranth with a single ring of long, narrow tentacles; mouth on a trumpet-shaped projection. Aperture of hydrotheca open, not closed by an operculum; aperture rounded. Hydroid upright or creeping.

1 Hydrothecae on long stalks; arranged in whorls at regular intervals along stem. Gonothecae flask-shaped, apex elongate and narrow ..... *Rhizocaulus verticillatus* (p. 107)
  Hydroid not so ....................................................................................... 2
2 Stem distinctly zigzag in appearance and 'jointed'; thickened below each joint, especially on one side; occasionally branched .......................... *Obelia geniculata* (p. 110)
  Stem not so ........................................................................................ 3
  **NB** The following four species may be difficult to identify with certainty. This key should be used in conjunction with the details given for each species. For further information reference to Cornelius (1990) and Hayward & Ryland (1990) is recommended.
3 Rim of hydrotheca smooth or nearly so ............................................................. 4
  Rim of hydrotheca toothed (may be worn smooth by abrasion) ...................................... 5
4 Hydroid small, delicate (up to about 30 mm in height). Hydrothecae cup-shaped, rims smooth. Gonothecae large, apex blunt ................................... *Laomedea flexuosa* (p. 108)
  Hydroid small to large (up to 300 mm in height). Hydrothecae bell-shaped, rim smooth to toothed. Gonothecae slender with raised apex ......... *Obelia dichotoma* (p. 109)
5 Hydroid with slender sinuous stem, dark brown to black, much-branched branches, rising alternately. Up to 350 mm in height. Gonothecae with raised apex ...........................
  ................................................................................. *Obelia longissima* (p. 109)
  Hydroid with stem slightly curved between branches, pale yellow-brown colour, up to 30 mm in height. Gonothecae with up to 4 attached medusae protruding from aperture ........................................................ *Gonothyraea loveni* (p. 107)

***Rhizocaulus verticillatus*** (Linnaeus)    (*Campanularia verticillata*) (Fig. 60)

*Irregularly branched, thick stems up to about 60 mm in height. Hydrothecae bell-shaped, about 12 cusps on rim; on relatively long stalks sometimes annulated and arranged in whorls at regular intervals along stem. Gonothecae smooth, flask-shaped, apex elongate and narrow; borne on very short stalks.*

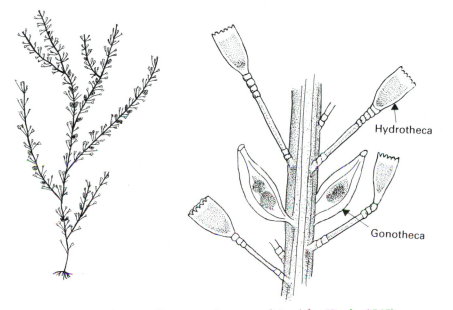

Figure 60  *Rhizocaulus verticillatus*, note thecate condition (after Hincks, 1868).

R. *verticillatus* has been recorded in a wide range of localities from north-west France to Norway. It is essentially a sublittoral species extending to depths of 200 m and more, attached to stones and shells, particularly in sandy areas. Reproduction occurs in spring and summer. There is no free-swimming medusa; a planula is released.

***Gonothyraea loveni*** (Allman)    (Fig. 61)

*Stems irregularly branched; annulated above origins of branches; up to about 30 mm in height; slightly curved between branches; pale yellow-brown colour. Hydrothecae alternate, elongate and rather slender; rim notched, with about 14 cusps to give castellated effect but may be worn smooth by abrasion; on short, annulated stalks. Gonothecae on short, annulated stalks. Up to 4 attached medusae, containing male or female gonads, protrude from aperture of gonothecae.*

Figure 61  *Gonothyraea loveni* (after Hincks, 1868).

*G. loveni* is widely distributed in north-west Europe and around Britain in intertidal pools on seaweeds and stones, extending to depths of about 200 m. It tolerates salinities down to about 12‰. Breeding apparently occurs during most of the year. There is no free-swimming medusa; planula larvae are released.

**Laomedea flexuosa** Alder    (*Campanularia flexuosa*) (Fig. 62)

*Stems about 25–30 mm in height, rising from branching stolon; may be irregularly branched; annulated above origins of branches. Hydrothecae cup-shaped, alternate, on annulated stalks; rims smooth. Gonothecae large, elongate, smooth; apex blunt; on short, annulated stalks.*

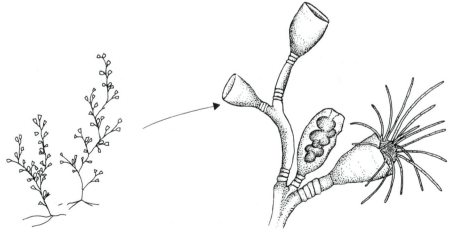

Figure 62  *Laomedea flexuosa* (after Hincks, 1868).

*L. flexuosa* is widely distributed in north-west Europe and is one of the commonest inter-tidal hydroids in Britain, extending from the middle shore into the sublittoral. It is found on seaweeds, rocks and stones. Breeding has been recorded in spring and summer from a number of localities. There is no free-swimming medusa; planula larvae are released.

### *Obelia dichotoma* (Linnaeus)   (Fig. 63)

*Slender, light brown stem, almost straight, irregularly branched. Annulated above origins of branches; up to about 40 mm in height but sometimes up to 300 mm. Hydrothecae bell-shaped with smooth to toothed rim; rising alternately on annulated stalks. Gonothecae in axils on annulated stalks; smooth, slender with raised apex.*

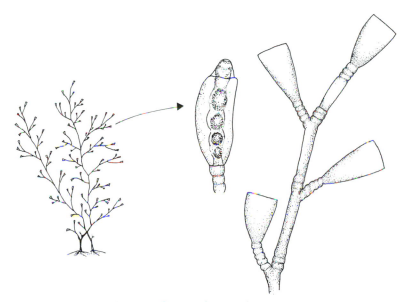

Figure 63  *Obelia dichotoma* (after Hincks, 1868).

*O. dichotoma* is widely distributed in north-west Europe and around Britain. It occurs on the lower shore and into the sublittoral attached to rocks, stones, shells and other hydroids.

### *Obelia longissima* (Pallas)   (Fig. 64)

*Slender, sinuous stem, with much-branched branches rising alternately. Stem dark brown, black; annulated above origins of branches; up to 350 mm in height. Branches sinuous with annulations above every side branch. Hydrothecae bell-shaped with 11–17 cusps on slightly toothed rim; on annulated stalks. Gonothecae in axils on annulated stalks; smooth, slender with raised apex.*

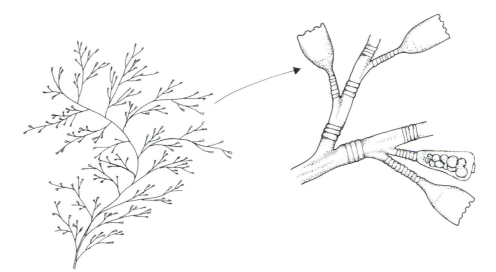

Figure 64  *Obelia longissima* (after Hincks, 1868).

O. *longissima* is widely distributed in north-west Europe and around Britain. It occurs sublittorally, usually in deep water attached to stones, shells and seaweed, often being washed ashore in large quantities.

**Obelia geniculata** (Linnaeus)    (Fig. 65)

*Stem only occasionally branched; distinctly zigzag in appearance and 'jointed'; thickened below each joint, especially on one side; up to 40 mm in height. Hydrothecae bell-shaped with smooth rim; alternate, 1 per joint on short, annulated stalks. Gonothecae slender, with narrow opening; rising from axils on short, annulated stalks.*

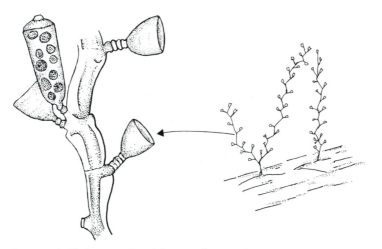

Figure 65  *Obelia geniculata* (after Hincks, 1868).

*O. geniculata* has a worldwide distribution. It occurs usually on brown seaweed, less frequently on stones, and extends into the sublittoral.

*Obelia* spp. release free-swimming medusae but these cannot be identified to species. They have been recorded in the plankton throughout the year.

Family Sertulariidae

Hydrothecae not on stalks; arise directly from both sides of the stem; aperture oval. Hydranth able to retract completely into hydrotheca; with a single ring of long, narrow tentacles; mouth on a conical projection. Hydrothecae usually closed by a flat operculum. Hydroid upright; without nematophores. A large family of hydroids, many species of which are sublittoral. Only three common and widespread species are described here.

1  Hydrothecae alternate. Gonothecae elongate, bell-shaped with a tooth on each side of aperture ................................................................... *Sertularia cupressina* (p. 112)
   Hydrothecae opposite or nearly so. Gonothecae large, pear-shaped ................................. 2
2  Hydroid with long, very slender stems alternately branched. Hydrothecae with outer edge prolonged into a spine, sometimes with 2 short lateral spines ......................................
   ................................................................... *Amphisbetia operculata* (p. 111)
   Hydroid with straight stems, usually unbranched; if branched, then irregularly branched. Hydrothecae curve outwards ................................... *Dynamena pumila* (p. 112)

*Amphisbetia operculata* (Linnaeus)   (*Sertularia operculata*) Sea-hair (Fig. 66)

*Long, very slender stems alternately branched; up to 100 mm in height; creamish colour. Hydrothecae very small; arranged in opposite pairs. Aperture of cup faces towards stem; outer edge with long spine, sometimes with short, lateral spine on either side. Gonothecae large, pear-shaped or elongate.*

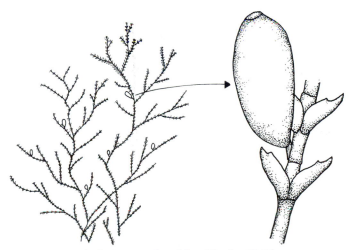

Figure 66 *Amphisbetia operculata* (after Hincks, 1868).

*A. operculata* apparently reaches its northern limit at about Shetland but has not been reliably recorded from Germany, Denmark or Scandinavia. It is found on the lower shore and in rock pools, attached to seaweeds, especially the stipe of *Laminaria digitata* (p. 43). It extends into the sublittoral to depths of about 70 m. Reproduction apparently takes place during summer. There is no free-swimming medusa.

**Dynamena pumila** (Linnaeus)    (*Sertularia pumila*) (Fig. 67)

*Straight, pale brown stems rising from stolon. Stems may be unbranched or sparsely branched and little more than 10 mm in height, or irregularly branched and up to 50 mm in height. Hydrothecae small, tubular, curving outwards; in opposite pairs on main stem and branches. Gonothecae large, pear-shaped.*

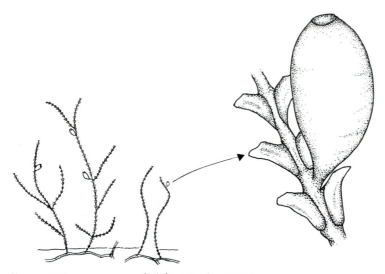

Figure 67 *Dynamena pumila* (after Hincks, 1868).

*D. pumila* occurs throughout north-west Europe and all around Britain. It is found mainly on the middle and lower shore attached to rocks and particularly fucoid seaweeds. Specimens on the lower shore are often more branched than those on the middle shore and the distribution on rocks and seaweeds may be related to the degree of exposure of the shore to wave action. *Dynamena* also occurs in brackish water. Breeding has been recorded in spring and summer from a wide range of localities. There is no free-swimming medusa.

**Sertularia cupressina** Linnaeus    Whiteweed (Fig. 68)

*Stout, more or less straight stem with alternate branches rising almost spirally; up to 350 mm or more in height, tapering gradually towards apex. Stem dark, branches brown in colour. Hydrothecae alternate, narrowing towards upper part which curves outwards only slightly; margin with 2 teeth usually of equal size. Gonothecae elongate, bell-shaped, most often with a single tooth on either side of aperture.*

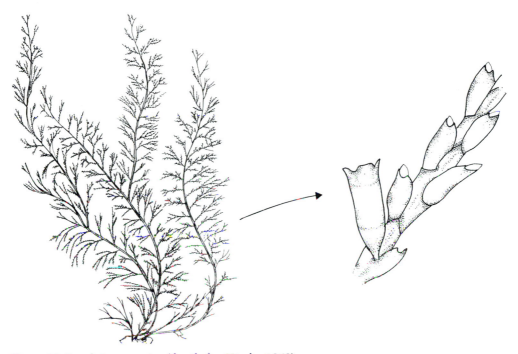

Figure 68  *Sertularia cupressina* (detail after Hincks, 1868).

*S. cupressina* is widely distributed in north-west Europe and occurs all around Britain. It is often found in deep water but also on the very low shore on firm sand and in shallow pools attached to stone and shells. On the south coast of England breeding has been recorded from November to January. There is no free-swimming medusa; planula larvae are released and have a free-swimming life of 2–3 days, possibly longer. When pressed and dyed whiteweed has some commerical value and is used for a variety of decorative purposes. A small industry existed round the Thames estuary, where the weed was collected from the sea bed by raking from small boats. Commercial exploitation has also been carried out in Germany and The Netherlands.

Family Plumulariidae
Hydrothecae not on stalks, arise directly from stem on one side only. Hydranth with single ring of long, narrow tentacles; mouth on a conical projection. Hydroid with nematophores. Hydroid upright. A large family, many species of which are sublittoral. Four common and widespread species are described here.

1  Hydrothecae with bluntly toothed rim; stalked gonothecae arise from main stem in
   place of branch, protected by basket-like structures known as corbulae ...............................
   ........................................................................................ *Aglaophenia pluma* (p. 116)
   Hydrothecae with smooth rim; gonothecae not in corbulae ............................................ 2
2  Hydroid large, up to 120 mm or more; white to yellowish in colour. Gonothecae
   roughly pear-shaped with spiny projections at top; on short stalks along both sides of
   main stem ........................................................ *Kirchenpaueria pinnata* (p. 114)
   Hydroids usually small, up to about 30 mm in height. Gonothecae not as above ............. 3
3  Hydrothecae on every other joint of branches. Gonothecae smooth, flask-shaped, in
   axils between main stem and branches, often in a continuous row along one side of
   main stem ................................................................ *Plumularia setacea* (p. 115)
   Hydrothecae separated from each other by 2, rarely 3, joints and so appear widely
   spaced on branches. Gonothecae large, on short stalks; with transverse grooves
   (female), smooth (male) .................................................. *Ventromma halecioides* (p. 115)

*Kirchenpaueria pinnata* (Linnaeus)    (*Plumularia pinnata*) (Fig. 69)

*White to yellowish, delicate stems; fairly straight, rising in groups from stolon; up to
120 mm or more in height. Side branches alternate along main stem; hydrothecae on upper
side of branches only, with smooth rim. Gonothecae on short stalks along both sides of main
stem; roughly pear-shaped, with spiny projections at top.*

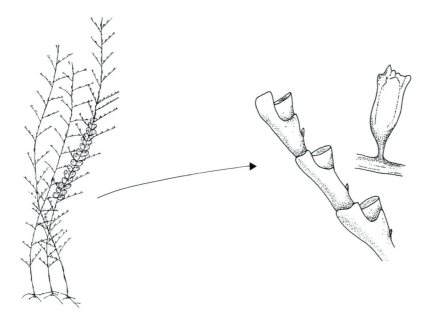

Figure 69  *Kirchenpaueria pinnata* (after Hincks, 1868).

*K. pinnata* is widely distributed in north-west Europe and around Britain in pools on the lower shore attached to stones, shells and seaweed. It extends into the sublittoral and is one of the invertebrates frequently recorded on the carapace of the spider crab, *Hyas araneus* (p. 407). Breeding has been recorded throughout summer; there is no free-swimming medusa.

### *Plumularia setacea* (Linnaeus)   (Fig. 70)

*Delicate stem with alternate branches; up to about 30 mm in height; pale yellow-brown colour. Branches with alternating long and short joints; hydranths on upper side of long joints. Hydrothecae small with smooth rim. Gonothecae smooth, flask-shaped, in axils, often forming a continuous row along one side of main stem.*

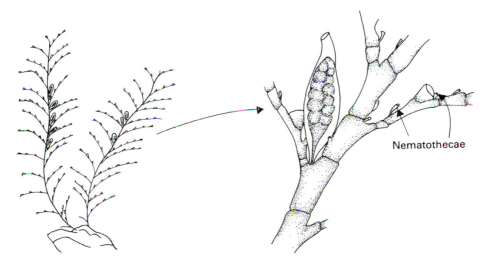

Figure 70  *Plumularia setacea* (detail after Hincks, 1868).

*P. setacea* is a southern species infrequently recorded north of Britain. It is usually sublittoral and grows attached to rock, hydroids, sponges, ascidians, molluscan shells and the tubes of polychaete worms. On the south coast of England breeding has been recorded from January to June. There is no free-swimming medusa; planula larvae are released. Longevity is believed to be up to one year.

### *Ventromma halecioides* (Alder)   (*Plumularia halecioides*) (Fig. 71)

*Delicate, irregularly branched stems up to 30 mm in height; branches arise alternately. Hydrothecae restricted to one side of branch; widely spaced and separated from each other by 2, rarely 3, joints. Hydrothecae with smooth rim. Gonothecae large, more or less egg-shaped, on short stalks; female gonothecae with transverse grooves; male gonothecae smooth.*

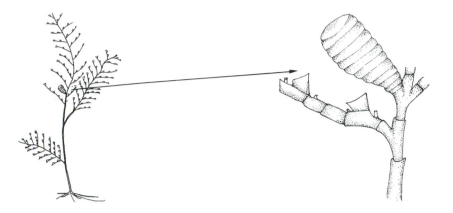

Figure 71 *Ventromma halecioides*, showing female gonotheca (after Hincks, 1868).

*V. halecioides* is widely distributed around Britain on stones, seaweed and hydroids on the lower shore. There is no free-swimming medusa.

*Aglaophenia pluma* (Linnaeus)*    (Fig. 72)

*Delicate, dark brown stems rising from a creeping stolon; up to 100 mm in height. Branches alternate; originating from front of stem, not the edge; may be lighter in colour than main stem. Hydrothecae with bluntly toothed rim of 9 cusps; arranged on one side of branch. Conspicuous nematotheca on front of hydrotheca. Stalked gonothecae protected by white, ribbed basket-like structures known as corbulae which arise from main stem in place of a branch.*

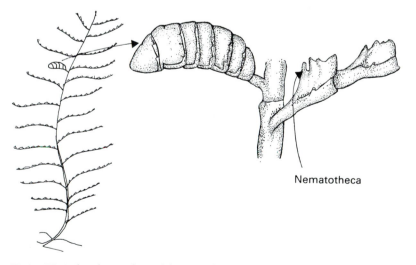

Nematotheca

Figure 72 *Aglaophenia pluma* (after Hincks, 1868).

* Placed by some authors in the family Aglaopheniidae.

*A. pluma* appears to have a patchy distribution around Britain, being most common in the south and west extending eastwards to the Isle of Wight. It is found on the lower shore attached to stones and shells but usually grows on the brown seaweed *Halidrys siliquosa* (p. 53). In south-west Britain breeding has been recorded in spring and summer. There is no free-swimming medusa.

### Floating colonial hydrozoans

In addition to the sea-firs, the class Hydrozoa also includes floating colonial animals, each colony being derived from a single larva. They are well known through descriptions of the much-feared Portuguese Man o'War (p. 118). These species have a high degree of polymorphism and some have a gas-filled float which supports the colony in the surface waters. They are oceanic and essentially tropical and semi-tropical; the two species described here are occasionally driven into British waters by prolonged south-westerly winds, and may be stranded on the shore in large numbers.

*Velella velella* (Linnaeus)    Jack Sail by-the-Wind (Fig. 73)

*Oval disc, with deep blue to blue-violet margins; 80–100 mm in length. Erect sail set diagonally on disc. Central mouth on underside surrounded by gonozooids. Ring of fairly short tentacles round edge of disc.*

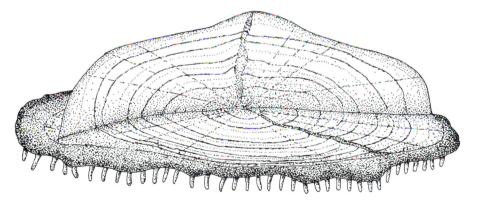

Figure 73  *Velella velella.*

*V. velella* is the only species of the genus and is classified with the athecate hydroids (p. 99). It is tropical and semi-tropical; however, in some years it occurs in shoals off south-west Britain and large numbers may be stranded on the shore. The disc contains a float and the stiff sail stands erect. The sail is set along either diagonal of the float and this line of insertion determines whether the individual drifts to the right or left of the wind direction. *Velella* is eaten by gastropod molluscs of the genus *Janthina*, warm water species often washed ashore with *Velella*. *Janthina*, in contrast to the snails of the shore, drifts in the surface waters on a bubble float secreted by the snail. *Velella* feeds on young fishes, crustaceans and other pelagic organisms caught by stinging cells on the tentacles. Free-swimming medusae are produced in large numbers and these are believed to sink into deep water. They develop ripe gonads and die after discharging the gametes. A planktonic larva develops into a new colony.

*Physalia physalis* (Linnaeus)    Portuguese Man o'War (Fig. 74)

*Large, gas-filled float may be 300 mm in length, 100 mm wide; sky-blue or light green tinged with pink. Large crest along length of float. Many feeding, reproductive and defensive polyps beneath float, the defensive polyps often many metres in length.*

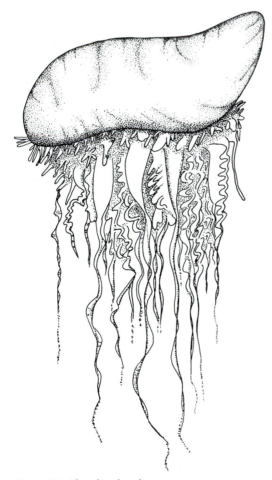

Figure 74  *Physalia physalis.*

*P. physalis* is the only species of the genus and belongs to a group known as the siphon-ophores. It is tropical and semi-tropical in distribution but occasionally found off south-west Britain, sometimes in large shoals. Specimens stranded on the shore are often damaged but they are easily recognized and contact of the tentacles with bare skin should be avoided because of their dangerous sting. The conspicuous float contains a gas, similar in composition to air but with up to 8% carbon monoxide. The float sits on the surface of the water and the crest acts as a sail. As in *Velella* (above), the position of the sail on the float determines whether the animal drifts to the right or left of wind direction. The long trailing tentacles are equipped with stinging cells which stun pelagic fishes and other prey coming into contact with them. *Physalia* is dioecious and fertilization external, leading to the development of a planula-like larva.

### Class Scyphozoa

The scyphozoans include the well-known jellyfishes. They are typically pelagic and exist for the greater part of their life as medusae. Most are dioecious and in the species described here, sperm are liberated through the mouth, which is on the undersurface, and drawn into a female, where internal fertilization takes place. A free-swimming plan-ula larva is liberated and in most species this settles as a fixed polyp known as a scyphis-toma. In the common jellyfish, *Aurelia* (p. 122), the scyphistoma reaches a height of about 14 mm. It eventually produces tiny medusae by a process of transverse fission. These medusae, also known as ephyrae, are no more than a few millimetres across. They become detached from the scyphistoma, feed and grow, and after a period of time which may be as long as two years, they develop into sexually mature jellyfishes (Fig. 75).

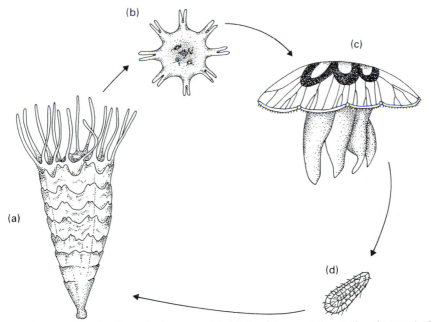

Figure 75  Life-cycle of *Aurelia* (not to scale). (a) Scyphistoma (after Bullough, 1958), (b) ephyra larva, (c) mature jellyfish, (d) planula larva.

Jellyfishes are familiar inhabitants of the inshore waters of Britain and Europe, often occurring in shoals in late summer. Six species are found in the inshore waters around Britain one of which, *Pelagia noctiluca* (Forskål), is infrequently recorded; it is best described as oceanic. The remaining five species are described here. They are propelled through the water by gentle pulsations of the bell and almost inevitably many are driven ashore by the wind and become stranded and quickly perish.

Key to species of scyphozoans

1 Thick, dome-shaped. Margin of bell divided into large number of lobes; without tentacles. Up to 900 mm across ............................................ *Rhizostoma octopus* (p. 124
  Saucer-shaped. Margin of bell with tentacles ................................................................ 2
2 Margin of bell with 16 lobes and 8 groups of tentacles, each group with 40–150 tentacles ..
  ........................................................................................................................................ 3
  Margin of bell not as above ................................................................................................ 4
3 Margin of bell with 8 groups of tentacles, each with 70–150 tentacles ..........................
  ............................................................................ *Cyanea capillata* (p. 121)
  Margin of bell with 8 groups of tentacles, each with 40–60 tentacles ..........................
  ............................................................................ *Cyanea lamarkii* (p. 122)
4 Margin of bell with 8 small notches and numerous short tentacles. Four violet-coloured semi-circular or horseshoe-shaped patches on upper surface .......... *Aurelia aurita* (p. 122)
  Margin of bell with 32 semi-circular lobes, each spotted with brown; 24 tentacles in 8 groups of 3. Upper surface of bell with 16 V-shaped brown markings radiating from centre ................................................................ *Chrysaora hysoscella* (p. 120)

*Chrysaora hysoscella* (Linnaeus)    Compass jellyfish (Fig. 76)

*Saucer-shaped; up to 300 mm diameter. Margin with 32 semi-circular lobes, each spotted with brown; 24 tentacles in 8 groups of 3, a sense organ situated between each group. Upper surface with 16 V-shaped, brown markings radiating from central brown spot. Varying degrees of brown pigment. Four arms surrounding mouth, longer than diameter of jellyfish.*

*C. hysoscella* has been recorded in inshore waters all around Britain but is most commonly found in the south and west where it is generally recorded from May to September. It is a protandrous hermaphrodite. Fertilization is internal and planulae, with a free-swimming life of a few days, are liberated in summer and autumn. Ephyrae are produced from the scyphistomae the following spring and grow rapidly to produce large, mature individuals by the summer. Longevity is about one year. *Chrysaora* is omnivorous, feeding on the wide range of zooplankton coming into contact with the highly extensible, marginal tentacles from where it is transferred to the arms surrounding the mouth.

Figure 76  *Chrysaora hysoscella* (after Russell, 1970).

### *Cyanea capillata* (Linnaeus)    (Fig. 77)

*Saucer-shaped, up to 500 mm diameter, occasionally much larger; yellow, brown or reddish in colour. Margin with 16 lobes and 8 groups of tentacles set a little distance in from margin; each group with 70–150 tentacles. Four large arms surrounding mouth, shorter than tentacles and much folded.*

*C. capillata* is a northern species extending to the Arctic Circle. Commonly recorded in the North Sea and the Irish Sea, it is less frequently found off south-west Britain, and is rare in the English Channel. The life-cycle is similar to that of *Aurelia* (below); it feeds on plankton and small fishes which make contact with the tentacles. It has a powerful sting.

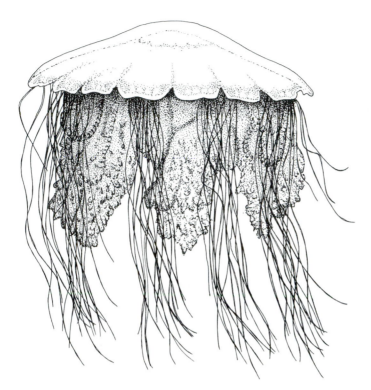

Figure 77   *Cyanea capillata* (after Russell, 1970).

**Cyanea lamarkii** Péron & Lesueur    (not illustrated)

*Similar to Cyanea capillata but with 40–60 tentacles in each marginal group. Up to 150 mm in diameter but specimens as large as 300 mm have been recorded. Pale yellow to blue in colour.*

C. *lamarkii* is recorded all around Britain and has a more southerly distribution than C. *capillata*, extending from the Bay of Biscay to Iceland and the Faroes. The life-cycle is believed to be similar to that of C. *capillata*. The sting is much less powerful.

**Aurelia aurita** (Linnaeus)    Common jellyfish, Moon jellyfish (Fig. 78a, b)

*Saucer-shaped; almost colourless apart from 4 violet semi-circular or horseshoe-shaped gonads. Up to 250 mm across, sometimes more. Margin with numerous short tentacles, smooth except for 8 small notches marking position of sense organs. Four thick arms surrounding mouth.*

Figure 78  (a), (b) *Aurelia aurita*. (c) *Rhizostoma octopus*. (d) *Haliclystus auricula*.
((d) Modified from a number of sources.)

*A. aurita* is widely distributed in the northern hemisphere and is essentially an inshore species. It is by far the most common jellyfish on British coasts and in some years is so abundant during summer and autumn that it is a nuisance to fishermen and bathers. It withstands salinities down to about 6‰ and extends into estuaries. The sexes are separate and size at onset of sexual maturity varies but is generally about 50 to 60 mm diameter. Fertilization is internal with sperm being drawn in through the mouth. When the embryos leave the ovaries they become lodged in brood chambers on the arms surrounding the mouth where they remain until they reach the planula stage. Planulae are liberated in the autumn and after a brief free-swimming existence settle on hard substrata such as rock surfaces, seaweeds and barnacles. They develop into benthic scyphistomae which produce ephyrae during winter and spring. These feed in the plankton by capturing prey with the aid of stinging cells on the tentacles and grow quickly to develop into fully grown jellyfishes in about three months. After spawning, the adults disappear from inshore waters during September and October. Most are believed to die but some survive the winter in deeper water. Adult *Aurelia* are plankton feeders, trapping a wide range of organisms in sheets of mucus transported across both upper and lower surfaces of the body by ciliary currents. The mucus collects round the margins of the bell and is removed by the arms and passed to the mouth. Both the marginal tentacles and the arms bear nematocysts which stun larger prey; *Aurelia* is known to capture small fishes.

### *Rhizostoma octopus* (Linnaeus)    (Fig. 78c)

*Thick, dome-shaped, up to 900 mm diameter; colour varied, from whitish to pale shades of green, blue, pink and brown. Margin without tentacles, finely divided into large number of lobes slightly darker in colour. Four pairs of very large arms on undersurface, each with many lobes bearing thousands of very small mouth openings. Four large pits on undersurface in which the amphipod* Hyperia galba *(p. 380) is often found.*

*R. octopus* has been recorded as far north as southern Norway but is mainly found off south and west coasts of Britain, being rare off the east coast of Scotland and north-east England. Although it occurs throughout the year, it is most common in summer and autumn and is abundant in some years. The sexes are separate and the possibility of sexual dimorphism in colour has been suggested, generally the marginal lobes and gonads of males being bluish, those of the females reddish-brown. Fertilization is internal and the life-cycle is similar to that of *Aurelia* (above). It is believed that some specimens overwinter, probably in deep water, and that these are the large specimens recorded in the first half of the following year. It feeds on microscopic planktonic organisms, drawing in currents of water through the mouth openings.

In addition to the jellyfishes described above, the scyphozoans include the Order Stauromedusae, the stalked jellyfishes. They are not pelagic but live attached to stones and seaweeds by a stalk with the mouth and bell uppermost. They are usually no more than about 50 mm in height and 30 mm across. They are not generally common on the shore, but one species, *Haliclystus auricula* (Rathke) (Fig. 78d) is not uncommon in south-west Britain. It is up to 50 mm in height, green to reddish in colour and the margin of the bell has eight lobes, each with a group of short tentacles. It lives on the lower shore attached to laminarians and other seaweeds, and on *Zostera*. Longevity is about one year.

### Class Anthozoa

The anthozoans are the largest group of cnidarians and include the corals, sea-fans, sea-pens and sea-anemones. They exist only as polyps; there is apparently no medusa in the life-cycle. The anthozoan polyp is generally much bigger and more complex than the hydrozoan polyp and both sexual and asexual reproduction are common. Asexual reproduction occurs by budding and also by longitudinal or transverse fission, or by basal laceration. During basal laceration, small pieces of the anemone break away from the basal region of the polyp and develop into new anemones. In sexual reproduction, both internal and external fertilization are known to occur. When fertilization is internal, the eggs are often retained inside the parent and released as planula larvae or as tiny anemones. Such is the complexity of reproduction in sea-anemones that in some species the method differs from one population to another.

Cnidae (see p. 96) are found in all anthozoans, being used in defence, food capture and in some cases they aid in attachment to the substratum. Some anemones possess structures known as acontia. These are long and thread-like, and attached to internal mesenteries. When the anemone contracts they are discharged through the mouth and through pores in the wall of the column, and being armed with nematocysts they presumably play a role in defence. They can be withdrawn into the anemone. The presence of discharged acontia can be an important aid to identification. Anemones have few predators, the most notable being sea-slugs, some of which feed only on a single species. The pycnogonids *Nymphon* spp. (p. 331) and *Pycnogonum littorale* (p. 332) are often found on sea-anemones but are not believed to be serious predators.

### Subclass Ceriantipatharia

The subclass comprises the black or thorny corals of deep water (not included here) and solitary, elongate anthozoans which live in secreted mucous tubes in mud and sand.

### *Cerianthus lloydii* Gosse    (Fig. 79)

*Column elongate, base rounded; up to 150 mm in length. About 70 long, tapering outer tentacles surrounding many much smaller, smooth inner tentacles; tentacles cannot be retracted into column. Column brownish in colour, outer tentacles brown-green or white often banded with darker brown; inner tentacles not banded. Secretes a soft, felt-like, mucous tube up to 400 mm in length, in which it lives.*

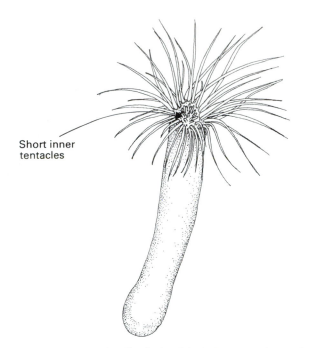

Short inner
tentacles

Figure 79  *Cerianthus lloydii* (modified after Manuel, 1988).

*C. lloydii* is widely distributed in north-west Europe and all around Britain and in some areas may be abundant. It is found on the lower shore and into the sublittoral to 100 m, buried in sand, mud and gravel, the anterior end of the tube projecting a little way above the surface. *Cerianthus* is able to move freely in the tube and can contract rapidly. There is a pelagic larva, the arachnactis, which has been recorded in the plankton from January to August.

### Subclass Alcyonaria (Octocorallia)

Colonial anthozoans in which the polyps are small and have eight pinnate tentacles. This group includes the sublittoral sea-fans and sea-pens, and the soft corals, one species of which is sufficiently common on the lower shore in north-west Europe to be included here.

### *Alcyonium digitatum* Linnaeus    Dead men's fingers (Fig. 80)

*Colonial, much-lobed thick, irregular mass; usually with stout, finger-like lobes; up to 250 mm in height and usually more than 20 mm across. Colour varied, grey, white, yellow, orange or pink. When under water and undisturbed it is covered by many small polyps, each with 8 pinnate tentacles giving a 'feathery' appearance. Polyps retractable.*

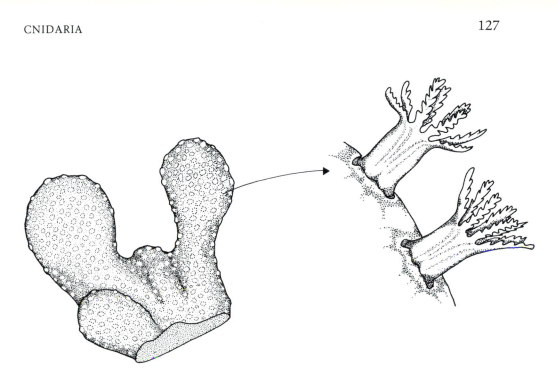

Figure 80  *Alcyonium digitatum*, with detail of polyps.

*A. digitatum* is widely distributed in north-west Europe and around Britain and, although recorded at low water of spring tides, the most substantial colonies are found sublittorally down to about 50 m, especially in areas where there is much water movement. The majority of colonies are either male or female, a few being hermaphroditic. Colonies as small as 10–20 mm in height can be sexually mature, and in the Irish Sea spawning occurs in December and January. Fertilization is external and after five to seven days a free-swimming planula larva develops and normally metamorphoses in one or two days, but has the potential for a prolonged pelagic life. Sexual maturity is reached in the second or third year. *Alcyonium* feeds on both phytoplankton and zooplankton, the polyps remaining retracted and non-feeding from about July to November, during which time the surface of the colony becomes covered with encrusting algae and hydroids. When the colony recommences feeding in December the surface film, together with the outer epithelium, is shed.

A second species of *Alcyonium*, **Alcyonium glomeratum** (Hassall), is found in deeper water but is occasionally seen on the very low shore in sheltered situations. It can be distinguished from *A. digitatum* by the more slender lobes and red-orange coloration.

### Subclass Zoantharia (Hexacorallia)

The Zoantharia includes both colonial and solitary polyps, some of which reach a considerable size. All have smooth, undivided tentacles surrounding the mouth, which is situated at the free end of the body or column. The most familiar examples in temperate waters are the sea-anemones of rocky shores, but anemones are also adapted to living in mud, sand and gravel and some are capable of burrowing into the sediment. Although commonly referred to as sessile and attached to the substratum by a basal disc, most sea-anemones are able to change position by a sliding movement and, indeed, when being attacked by predators and during aggressive encounters, some species, including the beadlet anemone, *Actinia* (p. 130), move away slowly. Sea-anemones are potentially long-lived and specimens are known to have lived for more than 65 years in aquaria. The subclass also includes the stony corals, characterized by an external calcareous skeleton into which the polyps can almost completely withdraw when disturbed; a few species are found in British waters.

### Key to species of zoantharians

(This key includes *Cerianthus* of the Subclass Ceriantipatharia.)

1  Polyp elongate, with about 70 long, slender tentacles surrounding many short tentacles. Inhabiting soft, mucous, felt-like tube buried in sand, mud or gravel ....................
...................................................................... *Cerianthus lloydii* (p. 126)
Polyp not so ...................................................................................................... 2

2  Polyp without external, calcareous skeleton; tentacles without terminal knob ................ 3
Polyp with external calcareous skeleton and knobbed tentacles *or* without a calcareous skeleton but with knobbed tentacles ................................................................... 16

3  Polyp attached to prosobranch shell occupied by hermit crab ........................................... 4
Not as above ...................................................................................................... 5

4  Polyp cloak-like, enclosing prosobranch shell occupied by hermit crab. Tentacles and mouth lie underneath shell. Usually with distinctive red or mauve spots ...........................
........................................................................ *Adamsia carciniopados* (p. 137)
Polyp firmly attached to shell by base; not enclosing shell ..... *Calliactis parasitica* (p. 137)

5  Polyp firmly attached to substratum (rocks, stones, piers, pilings, sometimes in soft sediments with base attached to small stone or piece of shell) by a basal disc. More than 48 tentacles ...................................................................................................... 6
Polyp long and narrow with rounded base, not firmly attached to substratum. Less than 36 tentacles. Found in holes and cracks in rocks or burrowing in sand and gravel .. 17

6  Column smooth without protuberances or suckers; without attached sand or gravel ....... 7
Column with protuberances or suckers, with or without attached sand, shell, gravel .... 11

7  Up to several thousand short, fine tentacles giving a 'feathery' appearance. Often white in colour .................................................................... *Metridium senile* (p. 135)
Up to about 200 longish, slender tentacles, without a 'feathery' appearance (but see *M. senile*) ...................................................................................................... 8

8  Tentacles long, slender, snake-like, almost always extended, even on contact; tips of tentacles usually purplish ........................................................ *Anemonia viridis* (p. 131)
Tentacles contract readily when anemone is disturbed .................................................... 9

9  Without conspicuous blue spots (acrorhagi) round inside of top of column. Column

very narrow and elongate when fully extended; usually orange in colour; tentacles long and slender .................................................................... *Diadumene cincta* (p. 133)

Conspicuous blue spots (acrorhagi) round inside of top of column ................................. 10

10  Column red, green or brown. Basal diameter up to 50 mm ............ *Actinia equina* (p. 130)

Column reddish covered with many green spots and resembling a strawberry in appearance. Basal diameter up to 100 mm ................................ *Actinia fragacea* (p. 131)

11  Column with 6 longitudinal rows of large white protuberances without attached shell or gravel. When contracted resembles denuded test of a sea-urchin ......................................
.................................................................... *Bunodactis verrucosa* (p. 133)

Column not so ..................................................................................... 12

12  Column with many protuberances with or without attached shell, gravel. Tentacles up to 200, short, thick. Large sea-anemones, basal diameter up to 250 mm; very firmly attached to substratum .................................................................................. 13

Column with white or greyish suckers, with or without attached sand, shell, gravel. Basal diameter about 50 mm or so. Up to 200 longish tentacles or several hundred very short tentacles .................................................................................... 14

13  Column with attached shell gravel. Colour of column varied, reddish, green; with grey protuberances. Tentacles banded red, white, green, blue. Lower shore and below .........
.................................................................... *Urticina felina* (p. 132)

Column without attached shell and gravel. Colour varied but paler than *U. felina*. Usually sublittoral .................................................................... *Urticina eques* (p. 133)

14  Column with greyish suckers near free end, usually with attached shell gravel; free end wider than base and convoluted in outline. Several hundred very short tentacles. Found in crevices filled with coarse sand and gravel into which the anemone withdraws; also in sand and mud ........................................ *Cereus pedunculatus**(p. 136)

Column with suckers, with or without attached sand, shell, gravel; free end wider or narrower than base. Up to about 200 longish tentacles .................................................. 15

15  Column with large white suckers, especially near free end, rarely with attached shell, gravel; free end wider than base. Found on rocky shores, often in holes in the rock. Rarely on sand or mud .................................................................... *Sagartia elegans**(p. 135)

Column with small suckers and attached shell, gravel; free end narrower than base. Typically found buried in sandy, muddy shores with the base attached to a small stone or piece of shell .................................................................... *Sagartia troglodytes**(p. 136)

16  Polyp with external, white calcareous skeleton; up to 80 tentacles, each ending in a white or brown knob .................................................................... *Caryophyllia smithii* (p. 139)

Polyp without external calcareous skeleton; column smooth, brilliantly coloured, often bright green; up to 100 tentacles, each ending in bright red knob ......................................
.................................................................... *Corynactis viridis* (p. 138)

*  These three species may be difficult to separate.

17  Most of column covered with thick brownish cuticle; usually 28 long slender tentacles;
    translucent, pink. Small anemone, up to 30 mm in length, 4 mm in diameter. Lives in
    holes and cracks in rocks ....................................................... *Edwardsiella carnea* (p. 138)
    Column translucent, pink, pale brown with brown marks; 12 translucent tentacles
    patterned brown and cream. May be up to 300 mm in length, 25 mm diameter.
    Burrowing in sand and gravel .................................................. *Peachia cylindrica* (p. 138)

*Actinia equina* (Linnaeus)    Beadlet anemone (Fig. 81a)

*Column smooth; basal diameter up to 50 mm, broader than column. Up to 192 tentacles
arranged in 6 circles. Column red, green or brown, sometimes with yellow, green or blue
spots randomly distributed. Usually with narrow blue line round edge of base. Conspicuous
blue spots, called acrorhagi, each containing nematocysts, occur round inside of top margin
of column. These increase in number, up to 48, with age of specimen.*

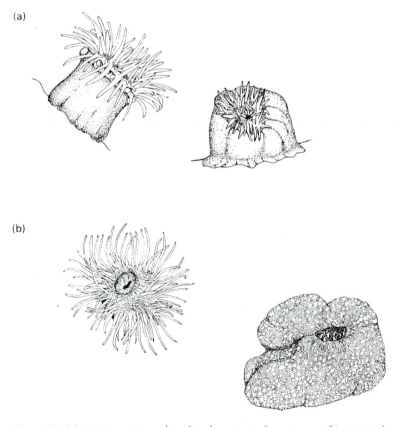

Figure 81  (a) *Actinia equina*, relaxed and contracted specimens. (b) *Actinia fragacea*,
          relaxed and contracted specimens.

*A. equina* is widely distributed and very common on rocky shores in north-west Europe. It extends from the upper shore to depths of about 20 m, and is highly adapted to the intertidal zone, tolerating high temperature and desiccation. Interestingly, intra-specific aggression has been recorded between neighbouring individuals. During these encounters, which involve tentacle contact and discharge of nematocysts, one of the animals is eventually forced to move away from the other. This behaviour pattern is perhaps related to the spacing out of individuals on the shore. The sexes of *Actinia* are separate but cannot be identified without microscopic examination of the gonad. Adults of both sexes can be found brooding young throughout the year and it is believed that the broods are produced by internal budding, although parthenogenesis has been suggested. The extent and significance of sexual reproduction are not fully understood but it has been suggested that *Actinia* reproduces sexually in the warmer months of the year. Embryos are brooded by the female for a few weeks and then discharged into the overlying seawater. No details of fertilization are known. Length of life of *Actinia* has been estimated to be about three years; many specimens are believed to live considerably longer.

A considerable amount of research, including electrophoretic studies, has recently been carried out on the colour morphs of the beadlet anemone. One of these morphs, the green form in which the column and basal disc are plain green in colour, has been shown to be a separate species, **Actinia prasina** Gosse. It is found predominantly on the lower shore, particularly under rocks and stones. Continuing research suggests that the remaining colour morphs now described as *A. equina* may in fact encompass a number of species.

**Actinia fragacea** Tugwell    Strawberry anemone (Fig. 81b)

*Column smooth and generally similar to A. equina but red to reddish-brown in colour (not green) and covered with many greenish spots, resembling a strawberry in appearance. More plump than A. equina and up to 100 mm basal diameter. Conspicuous blue, occasionally red, white or pink spots (acrorhagi), round inside of top of column.*

For many years, *A. fragacea* was considered a variety of *A. equina*. It has been recorded from north-west France and south-west Britain. Its intertidal distribution is more restricted than that of *A. equina* (above), being limited to rocks and stones on the lower shore. Apparently, it does not brood young but little is known of reproduction in this species.

**Anemonia viridis** (Forskål)    (*Anemonia sulcata*), Snakelocks anemone (Fig. 82a)

*Column smooth; basal diameter broader than column and up to 70 mm. About 200 long, snake-like tentacles; almost always extended. Two colour varieties predominate, yellow-brown, and green; cream or white specimens sometimes occur. Tips of tentacles purple.*

(a)

(b)

Figure 82  (a) *Anemonia viridis*. (b) *Urticina felina*, relaxed and contracted specimens.

*A. viridis* is a southern species recorded as far north as the west of Scotland. It is common on the south-west coast of England but absent from the east coast and the easterly half of the English Channel. It is found in rock pools on the middle shore and below, extending into the sublittoral to depths of about 20 m. *Anemonia* often occur in well-lit situations and this is probably related to the presence of photosynthesizing, symbiotic algae in the tissues. Asexual reproduction by longitudinal fission has been recorded. When groups of *A. viridis* are placed close together in aquaria, they make contact with their tentacles, in what has been described as aggressive encounters, and eventually separate from one another, resulting in spacing out. Interestingly, when the brown and green colour varieties are close together, it is apparently always the green which move away. On the shore it appears that there is some separation of the colour varieties.

*Urticina felina* (Linnaeus)    (*Tealia felina*), Dahlia anemone (Fig. 82b)

*Column with many protuberances with (in intertidal specimens) pieces of shell and gravel stuck to them making the contracted animal sometimes difficult to detect. Basal diameter up to 120 mm; may be much larger in sublittoral specimens; very firmly attached to substratum. About 160 short, thickish tentacles. When fully extended, central area around mouth is free of tentacles. Column very varied in colour, frequently reddish with green mark-*

*ings; protuberances grey; tentacles grey, with bands of red, white, green or blue. The largest British intertidal anemone.*

*U. felina* is widely distributed in north-west Europe and around Britain, where it is found from the lower shore to depths of about 100 m. It thrives in areas exposed to wave action and is often abundant in gullies and crevices, being readily identified in the contracted state by the presence of shell and gravel stuck to the column. Gametes are shed throughout the year and give rise to pelagic planula larvae.

A second species of *Urticina*, **Urticina eques** (Gosse), is widely distributed in north-west Europe and around Britain. It is essentially a sublittoral species growing up to 250 mm basal diameter, but is sometimes found on the extreme lower shore on rocks and shells. It is varied in colour, usually red or orange and generally paler than *U. felina*, but may be off-white, yellow, red, orange and lilac, often with a pale reticulate pattern on the lower part of column. Unlike the intertidal specimens of *U. felina*, it does not have bits of shell and gravel stuck to the column.

### *Bunodactis verrucosa* (Pennant)    Wartlet anemone, Gem anemone (Fig. 83a, b)

*Column pink; up to 48 longitudinal rows of protuberances, those in 6 of the rows large and white in colour. Protuberances not covered with shell or gravel. Base up to 25 mm across; broader than column; 48 banded tentacles. Closes tightly when disturbed giving a hemispherical shape resembling denuded test of* Echinus *(p. 452).*

*B. verrucosa* has a southerly distribution and apparently reaches its northern limit in the British Isles where it is most common on south and west shores, extending to the Hebrides. It occurs in rock pools on exposed shores, often well camouflaged among the red seaweed, *Corallina* (p. 61) and extends to depths of about 20 m. Young are brooded in the parent.

### *Diadumene cincta* Stephenson    (Fig. 83c)

*Column smooth, narrow and elongate when fully extended. Basal diameter up to 15 mm; column up to 60 mm in height. Tentacles long, slender, up to 200 in number. Column and tentacles usually orange in colour, occasionally fawn. Collar visible below tentacles in contracted specimens, 'disappears' in expanded specimens. Bears some resemblance to small specimens of* Metridium senile *(below) but can be distinguished by the absence of a collar below the tentacles in expanded specimens and by the absence of a white band on the tentacles.*

*D. cincta* has been recorded from the southern North Sea, the west coast of France and around Britain, where it is widely distributed. It occurs on stones, rocks and bivalve

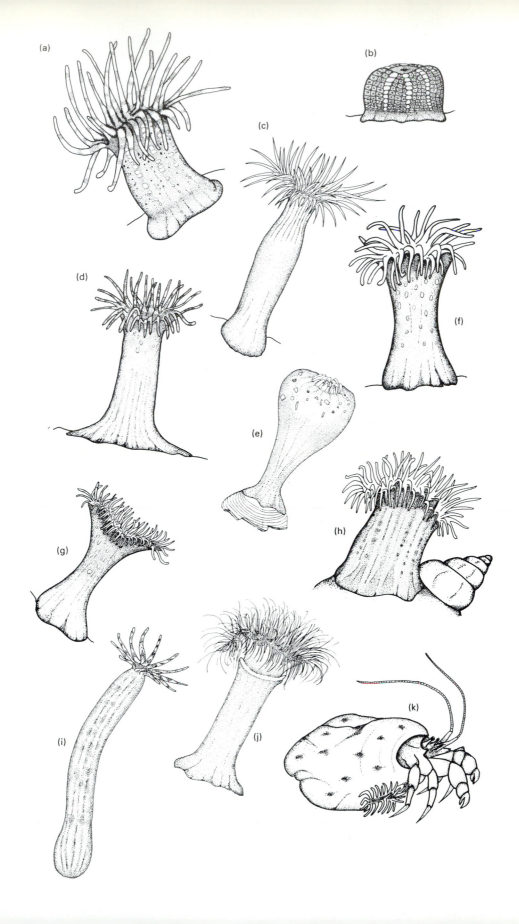

molluscs such as *Mytilus* (p. 283), from the lower shore to depths of about 20 m. When disturbed it contracts in what has been described as 'jerky' movements. Reproduction is by basal laceration.

### *Metridium senile* (Linnaeus)    Plumose anemone (Fig. 83j)

*Column smooth, narrower than base, which may be deeply lobed and up to 150 mm basal diameter, height up to 300 mm. Tentacles slender, up to several thousand in number, giving 'feathery' appearance. When fully relaxed prominent collar visible below tentacles. Colour varied, most often white or orange; may be brown, red or yellow; tentacles usually have a whitish band near the base.*

*M. senile* is widely distributed in north-west Europe and around Britain where it is commonly found in crevices, on piers, pilings, stones and under rocky overhangs and other situations where there is good water flow. It is found on the lower shore, extending sublittorally to depths of 100 m and is occasionally found in brackish water. The fine tentacles trap plankton on which the anemone feeds. Sexual reproduction with pelagic larvae, and asexual reproduction by basal laceration have been recorded, the latter often leading to a much-lobed basal disc.

Although adult specimens of *M. senile* have a number of distinctive characters and can be readily identified, the species is very varied in form. Two distinct morphological varieties are generally recognized from British waters; the variety *dianthus* as described above and a smaller variety, *pallidum*, in which the tentacles are long and much fewer in number (about 200) than in the variety *dianthus*, have a conspicuous white band near the base and do not have a 'feathery' appearance. The variety *pallidum* is up to 25 mm in basal diameter and up to 25 mm in height. Specimens intermediate between the two varieties have been described.

### *Sagartia elegans* (Dalyell)    (Fig. 83f)

*Base often indented, wider than column, but generally not as wide as free end; up to 30 mm basal diameter. Column with rounded white suckers; larger and more numerous towards free end; rarely with shell or gravel attached to them. Up to about 200 tentacles. Very varied in colour and patterning, from red, brown, orange to white; 5 colour varieties have been described. Thread-like acontia are freely discharged by the anemone when disturbed.*

---

Figure 83   (a), (b) *Bunodactis verrucosa*. (c) *Diadumene cincta* (after Stephenson, 1935). (d) *Sagartia troglodytes*. (e) *Sagartia troglodytes*, contracted specimen from sand, attached to piece of shell gravel. (f) *Sagartia elegans*. (g) *Cereus pedunculatus*. (h) *Calliactis parasitica*. (i) *Peachia cylindrica* (after Stephenson, 1928). (j) *Metridium senile*. (k) *Adamsia carciniopados*.

*S. elegans* is widespread in north-west Europe and around Britain from the middle shore to depths of about 50 m. It occurs in pools, crevices and in holes in the rock. Large numbers are often found together in crevices and overhangs. It has also been recorded from soft sediments. Reproduction is by basal laceration; sexual reproduction also occurs.

**Sagartia troglodytes** (Price)    (*Sagartia troglodytes* variety *decorata*) (Fig. 83d, e)

*Similar to S. elegans (above) but less brightly coloured. Base wider than column and generally wider than anterior region with tentacles expanded; up to 50 mm basal diameter and 120 mm in height. Column with small suckers; often with attached shell, gravel. Up to about 200 tentacles. Colour and patterning highly varied; generally dull shades of yellow, green, grey; basal part of column with pale vertical stripes; tentacles plain or banded, generally with a B-shaped mark at the base. Thread-like acontia are not freely discharged when the anemone is disturbed.*

*S. troglodytes* is widely distributed in north-west Europe and is common around Britain with the exception of the English Channel. It is found intertidally, and sublittorally to 50 m, generally in soft sediments such as sand and mud attached to stones, pebbles and pieces of shell well below the surface, but does also occur on rocky shores. The sexes are separate and sexual reproduction is believed to occur, with planktonic larvae.

Until recently, *S. troglodytes* was described as existing in two forms, variety *decorata* and variety *ornata*. It is now known that these two varieties are, in fact, two separate species, *S. troglodytes* (formerly variety *decorata*) described above, and **Sagartia ornata** (Holdsworth), formerly variety *ornata*. *S. ornata* is widespread and common intertidally but seldom occurs in soft sediments, being more typical of exposed rocky shores where it is found in cracks and crevices, and with mussels and barnacles. It has a basal diameter of up to 15 mm, is usually without attached shell and gravel, and is less varied in colour than *S. troglodytes*, being greenish or reddish-brown. The tentacles are fewer in number (up to 100) and are banded. Populations consist only of females and these brood young. It has been suggested that the broods are produced parthenogenetically.

**Cereus pedunculatus** (Pennant)    Daisy anemone (Fig. 83g)

*Base up to 40 mm diameter; wider than lower part of column. Column widening out towards free end when fully relaxed, particularly in young specimens. Up to 120 mm in height. Greyish suckers near free end, generally with attached shell, gravel. Tentacles very short and numerous, always at least 200, and more than 700 in fully grown specimens. Region from which tentacles arise convoluted in outline and wider than base. Column very varied in colour, usually buff-grey, orange, darkening towards free end, often with a purplish tinge; tentacles light brown, may be banded or mottled. Relatively wide area round mouth without tentacles. Thread-like acontia freely discharged when anemone disturbed.*

*C. pedunculatus* is a southern species extending north to Britain where it commonly occurs on south and west coasts but is rare on the east coast. It lives in rock pools, typically in inaccessible crevices, sometimes as high as the middle shore, and extends into the sublittoral to depths of about 50 m. Large numbers are often found attached to shell debris and buried in mud and sand with the tentacles exposed on the surface. *Cereus* is sometimes found at the mouths of estuaries. Young are brooded in the parent.

### *Calliactis parasitica* (Couch)    (Fig. 83h)

*Base wider than column; up to 80 mm diameter and 100 mm in height. Column cream to light brown with faint longitudinal lines; often mottled with dark brown or red. Up to 700 translucent, yellowish to orange tentacles with longitudinal reddish-brown lines. Thread-like acontia discharged from lower part of column when anemone disturbed. Usually on prosobranch shell occupied by hermit crab.*

*C. parasitica* is a southern species extending as far north as the west coast of Wales and the south of Ireland. It occurs sublittorally down to about 60 m, only rarely being found on the lower shore. The anemone usually lives on shells of the whelk, *Buccinum undatum*, occupied by the hermit crab *Pagurus bernhardus* (p. 395), although it has been recorded on other substrata. Both the anemone and the crab can survive alone. When associated with the hermit crab, some protection against predators must be gained by the crab, while the anemone must gain some advantage in food collection. The anemone establishes this commensal relationship by a series of complex manœuvres by which it eventually attaches to the shell while the crab apparently remains passive. It is believed that the anemone uses a chemical sense to recognize suitable shells. In aquaria, anemones have been seen to transfer to the shells of living *Buccinum*, but in such cases they are shaken off by the whelk. A single shell may provide a home for more than one anemone. Young are brooded in the parent.

### *Adamsia carciniopados* (Otto)    (*Adamsia palliata*), Cloak anemone (Fig. 83k)

*Base and column enclose prosobranch shell occupied by hermit crab in such a way that mouth and tentacles come to lie under crab. Base up to 700 mm diameter. Column white to fawn; usually with red or mauve spots. Tentacles white. Thread-like acontia, pinkish-white in colour, discharged when disturbed.*

*A. carciniopados* is common around the coasts of Britain and north-west Europe, particularly in the sublittoral to depths of about 200 m. It is found at low water on sandy shores and is readily identified by its coloration and the distinctive way it folds itself round the prosobranch shell. Almost without exception around Britain, it lives in association with the hermit crab *Pagurus prideaux* (p. 396), without which the anemone apparently does not reach sexual maturity. As the crab grows a chitinous membrane is secreted from the base of the anemone and this protects the crab and overcomes the need for it to move to a larger shell as it grows.


**Peachia cylindrica** (Reid)    (*Peachia hastata*) (Fig. 83i)

*A burrowing form. Varied in shape, often long and slender; base not adhesive, rounded. May be up to 300 mm in length, and 25 mm diameter. Small lobed structure (the conchula) projects from one side of mouth. Column translucent pink or pale brown flecked with brownish marks. Twelve translucent tentacles, conspicuously patterned with brown and cream. In buried specimens the tentacles can extend to become very long.*

*P. cylindrica* is widely distributed, but not abundant, in north-west Europe and around Britain, extending from the extreme lower shore to depths of about 100 m. It is well known for its characteristic habit of burrowing in sand or gravel where it remains free and not attached to the substratum. Its life-cycle is unusual in that the planulae are parasitic on a number of species of planktonic medusae. They have been recorded in the plankton from about April to August.


**Edwardsiella carnea** (Gosse)    (Fig. 84a)

*Column narrow, elongate; base not adhesive, rounded. Up to 30 mm in length, 4 mm diameter. Up to 36 (usually 28) long, slender tentacles. Column translucent, pale pink, orange, yellow; tentacles translucent, pink. Most of column covered with thick, brownish cuticle. Contracts rapidly.*

*E. carnea* has been recorded from southern Norway to the Atlantic coast of France and occurs all around Britain, but its distribution is very localized. It extends from the lower shore to the shallow sublittoral living in cracks and holes in rocks; it does not burrow. Little is known of its life-cycle.


**Corynactis viridis** Allman    Jewel anemone (Fig. 84b)

*Column smooth; base up to 10 mm diameter, height up to about 15 mm. Up to about 100 tentacles, each ending in a small knob. Brilliantly coloured in a variety of colours, often bright green with bright red, terminal knob on each tentacle.*

*C. viridis* is a southern species, common on south and west coasts of Britain, and reaching its northern limit in the north of Scotland. It occurs on the lower shore and into the sublittoral to depths of 50 m, on rocks, in crevices and under overhangs, often in large numbers. Reproduction is by longitudinal fission.

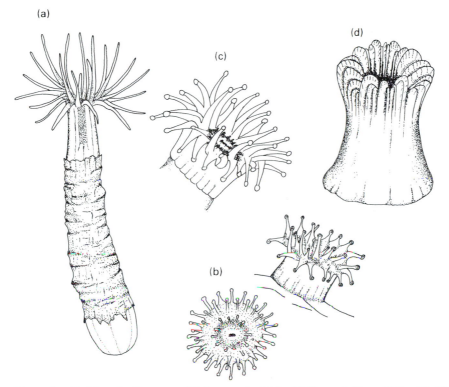

Figure 84   (a) *Edwardsiella carnea.* (b) *Corynactis viridis.* (c) *Caryophyllia smithii,* with (d)
calcareous skeleton. ((a), (c), (d) After Manuel, 1988.)

**Caryophyllia smithii** Stokes & Broderip    Devonshire cup-coral (Fig. 84c, d)

*External whitish, calcareous skeleton with many curved ridges. Up to 15 mm in height and
25 mm basal diameter. Polyp varied in colour, red, pink, orange, white, green, brown, often
with contrasting colour surrounding mouth. Up to about 80 tentacles, usually transparent,
each ending in white or brown knob. Solitary coral.*

C. *smithii* is a southern species found on south and west coasts of Britain and extending
northwards to Shetland. It occurs on the lower shore where it is relatively rare, but is
often abundant in the sublittoral to depths of 200 m, attached to rocks, stones, shells
and in crevices. A barnacle, **Pyrgoma anglicum** Sowerby, is often attached to the calcareous skeleton, particularly on corals in the south and west. Several barnacles may be
found on a single coral.

REFERENCES
**Hydrozoa**
**Allman**, G. J. (1871). *A monograph of the gymnoblastic or tubularian hydroids.* London: Ray
    Society.

Cornelius, P. F. S. (1975). The hydroid species of *Obelia* (Coelenterata, Hydrozoa: Campanulariidae), with notes on the medusa stage. *Bulletin of the British Museum (Natural History) (Zoology)*, 28, 249–93.

Cornelius, P. F. S. (1979). A revision of the species of Sertulariidae (Coelenterata: Hydroida) recorded from Britain and nearby seas. *Bulletin of the British Museum (Natural History) (Zoology)*, 34, 243–321.

Cornelius, P. F. S. (1982). Hydroids and medusae of the family Campanulariidae recorded from the eastern North Atlantic, with a world synopsis of genera. *Bulletin of the British Museum (Natural History) (Zoology)*, 42, 37–148.

Cornelius, P. F. S. (1990). European *Obelia* (Cnidaria, Hydroida): systematics and identification. *Journal of Natural History*, 24, 535–78.

Cornelius, P. F. S. (1995). *North-west European thecate hydroids and their medusae*. Parts 1 and 2. Synopses of the British fauna (New Series), no. 50. Keys and notes for identification of the species. Shrewsbury: Field Studies Council.

Edwards, C. (1966). The hydroid and the medusa *Bougainvillia principis*, and a review of the British species of *Bougainvillia*. *Journal of the Marine Biological Association of the United Kingdom*, 46, 129–52.

Hayward, P. J. & Ryland, J. S. (eds.) (1990). *The marine fauna of the British Isles and north-west Europe*, vols. 1 and 2. Oxford: Clarendon Press. (Vol. 1. Ch. 4. Cnidaria.)

Hincks, T. (1868). *A history of the British hydroid zoophytes*, vol. I, text, vol. II, plates. London: John Van Voorst.

Kirkpatrick, P. A. & Pugh, P. R. (1984). *Siphonophores and velellids*. Synopses of the British fauna (New Series), no. 29. Keys and notes for the identification of the species. Leiden: E. J. Brill/Dr W. Backhuys.

Millard, N. A. H. (1975). Monograph of the Hydroida of southern Africa. *Annals of the South African Museum*, 68, 1–513.

Russell, F. S. (1953). *The medusae of the British Isles. Anthomedusae, Leptomedusae, Limnomedusae, Trachymedusae and Narcomedusae*. Cambridge: Cambridge University Press.

Svoboda, A. & Cornelius, P. F. S. (1991). The European and Mediterranean species of *Aglaophenia* (Cnidaria: Hydrozoa). *Zoologische Verhandelingen Leiden*, 274, 1–72.


Scyphozoa

Bullough, W. S. (1958). *Practical invertebrate anatomy*. London: Macmillan.

Russell, F. S. (1970). *The medusae of the British Isles. II. Pelagic Scyphozoa with a supplement to the first volume on Hydromedusae*. Cambridge: Cambridge University Press.


Anthozoa

Manuel, R. L. (1983). *The Anthozoa of the British Isles – a colour guide*, 2nd edn. Produced for the Underwater Conservation Society by R. Earll.

Manuel, R. L. (1988). *British Anthozoa (Coelenterata: Octocorallia and Hexacorallia)*. Keys and notes for the identification of the species. Synopses of the British fauna (New Series), no. 18, revised edn. Leiden: E. J. Brill/Dr W. Backhuys.

Shaw, P. W., Beardmore, J. A. & Ryland, J. S. (1987). *Sagartia troglodytes* (Anthozoa: Actiniaria) consists of two species. *Marine Ecology Progress Series*, **41**, 21–8.

Stephenson, T. A. (1928–35). *The British sea anemones*, vols. I and II. London: Ray Society.

# Ctenophora

The ctenophores are often referred to as sea-gooseberries or comb-jellies. A few are able to creep on solid surfaces, some of these being ectoparasitic, but the majority are pelagic, and at certain times of the year may be so abundant in inshore waters that they are trapped in rock pools by the ebbing tide. Body shape varies from ribbon-like, laterally flattened to spherical, and characteristically the surface is traversed by up to eight bands of cilia known as the comb-rows. Each comb-row consists of a series of plates bearing fused cilia or ctenes. The cilia beat in waves and the animal usually moves through the water with the mouth forward. During darkness luminescence is a characteristic feature. In addition to comb-rows, some species have a pair of long, branched retractile tentacles bearing adhesive cells, the colloblasts, which are used in food capture. Ctenophores are carnivorous. Those lacking tentacles swallow the prey whole, or in some cases pieces of prey are cut away with specialized cilia. They do not possess nematocysts; although these have been recorded in one species, they are now believed to originate from the cnidarian medusae on which the ctenophore feeds. Most ctenophores are hermaphroditic and although eggs are brooded in some, the majority release gametes into the sea. Fertilization is external and leads to a free-swimming, spherical larva known as a cydippid.

The phylum is divided into two classes based on the presence or absence of tentacles.


## Phylum CTENOPHORA

**Class Tentaculata**   Ctenophores with tentacles.

**Class Nuda**   Ctenophores without tentacles.

**Class Tentaculata**

***Pleurobrachia pileus*** (O. F. Müller)   Sea-gooseberry (Fig. 85a)

*Spherical body up to 30 mm in length; 8 comb-rows. A pair of branched tentacles up to 500 mm in length; completely retractile into pouches. Virtually transparent; comb-rows iridescent; some colour associated with gut.*

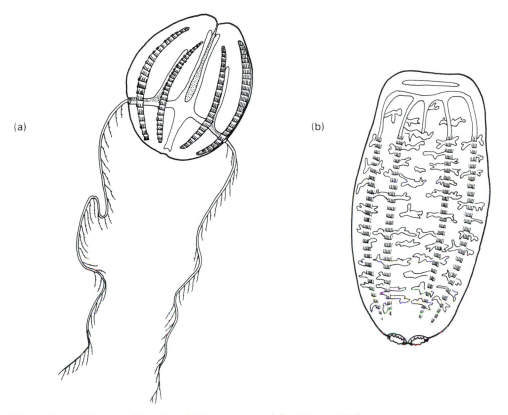

(a)                                          (b)

Figure 85  (a) *Pleurobrachia pileus*. (b) *Beroë cucumis* (after Greve, 1975).

*P. pileus* is widely distributed and almost cosmopolitan. It is typically found in the plankton of inshore waters, sometimes occurring in swarms. Around Britain it is most abundant in summer and late autumn with breeding taking place from spring to autumn. *Pleurobrachia* is hermaphroditic; eggs and sperm are shed through the mouth and fertilization is external. There is a free-swimming cydippid larva and some of the animals spawned in spring reproduce in the autumn of the same year. Most animals die after breeding. *Pleurobrachia* is a voracious carnivore and when present in high densities is a major competitor with pelagic and larval fishes for zooplankton. It feeds mainly on copepod crustaceans. The long, branched tentacles form an effective net and, once trapped, prey is held on the adhesive cells. *Pleurobrachia* is preyed on by fishes, medusae such as *Aurelia* (p. 122) and other ctenophores, particularly *Beroë* (below).

**Class Nuda**

*Beroë* spp.    (Fig. 85b)

*Body somewhat oval-shaped, 50 mm or more in length; 8 comb-rows; no tentacles. Mouth large. Body transparent or pale pink.*

There are believed to be two species of *Beroë* in north-west Europe, **Beroë cucumis** Fabricius and **Beroë gracilis** Künne. They are difficult to distinguish (see Greve, 1975) and apparently the young of the species cannot be separated. *B. cucumis* is widely distributed while *B. gracilis* is more restricted in its distribution and is found in the southern North Sea and off south and west Britain. Both species are voracious carnivores. The mouth has large, specialized cilia which are used to cut up prey too large to be swallowed whole. *Beroë* feeds extensively on other ctenophores.

REFERENCE
**Greve**, **W**. (1975). Ctenophora. *Fiches d'Identification du Zooplancton*. Sheet 146.

# Platyhelminthes

The platyhelminthes or flatworms are perhaps best known through their parasitic representatives, the flukes and tapeworms, but they also include a group of free-living worms, the turbellarians, with marine, freshwater and terrestrial representatives.

## Phylum PLATYHELMINTHES

**Class Turbellaria**   Free-living flatworms found in the terrestrial, marine and freshwater environments.

**Class Monogenea**   Mainly ectoparasitic on fishes.

**Class Trematoda**   Endoparasites, mostly of vertebrates.

**Class Cestoda**   Endoparasites of vertebrates.

### Class Turbellaria

The turbellarians are the most primitive of the flatworms. The dorso-ventrally flattened body has a bilateral symmetry and the digestive system has only one opening, the mouth, on the undersurface. Eye spots and tentacles are often visible at the anterior end. The animals move over the substratum with a gliding motion which results from the beating of cilia on the undersurface of the body, but muscular contraction is also involved, particularly in the larger species. Most turbellarians are carnivorous, feeding on a wide range of invertebrates. They also scavenge on dead and decaying material and some feed on microscopic algae. Turbellarians have considerable powers of regeneration and asexual reproduction by transverse fission has been recorded. Almost all are hermaphroditic but copulation and cross-fertilization generally take place. The eggs are usually laid in gelatinous masses and hatch after one to several weeks. The majority of species do not have a pelagic larva.

Flatworms extend from the intertidal into deep water. They are found in diverse habitats including the undersurface of stones and in mud, sand and shell deposits, and among seaweed, including laminarian holdfasts. They have to be searched for carefully and should be handled with care to avoid fragmentation.

The class is divided into a number of orders, the species included here belonging to two of these:

**Order Tricladida**   The planarians. Mainly freshwater and terrestrial, with a few marine species. The gut has three branches, e.g. *Procerodes*.

**Order Polycladida**   All polyclads in north-west Europe are marine. They have a central gut with many branches, e.g. *Leptoplana, Oligocladus, Prostheceraeus*.

### Key to species of turbellarians

Turbellarians are difficult to identify and in some cases accurate identification to species is based on microscopic detail of the anatomy. This key is based on distinctive morphological characters which can be seen with the aid of a hand lens in living specimens observed under water.

1  Animal without tentacles; 4 groups of eye spots at anterior ............................................................ *Leptoplana tremellaris* (p. 147)
Animal with 2 tentacles; head with 2 eyes, or tentacles with many eye spots .................. 2
2  Tentacles short, stumpy (Fig. 86a); head with 2 eyes ........... *Procerodes littoralis* (p. 146)
Tentacles distinct, well developed (Fig. 87), each with many eye spots ........................... 3
3  Dorsal surface of body with a number of dark longitudinal lines, some of which merge in posterior region .......................................................... *Prostheceraeus vittatus* (p. 147)
Dorsal surface of body whitish, flecked with brownish-red ........................................... *Oligocladus sanguinolentus* (p. 148)

*Procerodes littoralis* (Ström)   (*Procerodes ulvae*) (Fig. 86a)

*Body up to 9 mm in length, 1–2 mm wide. Dorsal surface olive-brown with pale streak down centre of body; 3 dark stripes on head; ventral surface pale. Anterior margin of head convex. Two short, stumpy tentacles. Two eyes on head, situated in the pale areas.*

*P. littoralis* is widely distributed in north-west Europe and is common all around Britain. It is well known for its ability to tolerate large changes in salinity, even surviving in fresh water for one or two days. It lives under stones and boulders on the upper and middle shore, and is often abundant in areas where freshwater streams run across the shore. *Procerodes* becomes sexually mature when about 5 mm in length. Fertilization is internal and yellowish-brown, oval-shaped egg capsules, about 1 mm in diameter, are attached to the underside of stones. On average, each capsule contains five embryos and on the south coast of England they are commonly found on the shore during spring and summer. There is no larva in the life-cycle, the young hatch with a body length of 2 to 4 millimetres.

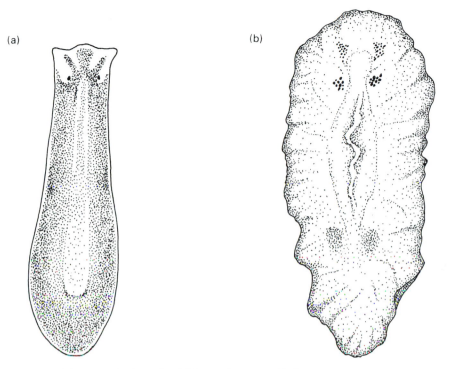

Figure 86 (a) *Procerodes littoralis.* (b) *Leptoplana tremellaris.*

### *Leptoplana tremellaris* (O. F. Müller) (Fig. 86b)

*Elongate, delicate oval body up to 25 mm in length, wider at anterior than posterior. Pale brown, buff coloured, sometimes with white patches. Margins of body frilly. Without tentacles. Four groups of black eye spots anteriorly; anterior 2 groups elongate with 20–25 eyes, posterior 2 rounded and with 6–12 larger eye spots.*

*L. tremellaris* is widely distributed in north-west Europe and found all around Britain. It lives under stones, on mussel beds, in laminarian holdfasts, among seaweed and is associated with a variety of sessile invertebrates such as sponges and ascidians. There is no larva in the life-cycle.

### *Prostheceraeus vittatus* (Montagu) (Fig. 87a)

*Body up to 50 mm in length and 25 mm wide. Dorsal surface yellowish-white with number of dark, longitudinal lines, some of which merge in posterior region. Two distinct tentacles, each with a group of darkly pigmented eye spots. Further eye spots grouped in mid-dorsal region posterior to tentacles.*

(a)                                                (b)

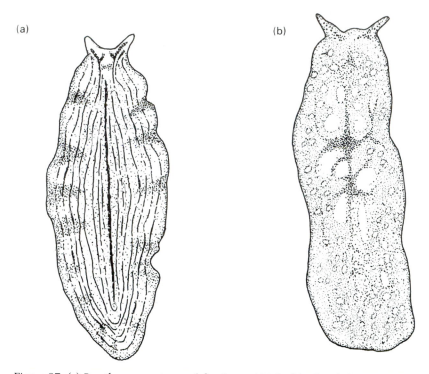

Figure 87  (a) *Prostheceraeus vittatus* (after Lang, 1884). (b) *Oligocladus sanguinolentus*
(after Lang, 1884).

*P. vittatus* is widely distributed in north-west Europe, and in Britain occurs mainly in the
south and west. It is found on the lower shore and below, under stones and seaweed and
among colonies of the sea squirt, *Ciona* (p. 474).

### *Oligocladus sanguinolentus* (Quatrefages)    (Fig. 87b)

*Body up to 22 mm in length and 13 mm wide. Dorsal surface whitish, flecked with brown-
ish-red. Two distinct tentacles each with many eye spots. Further eye spots grouped in mid-
dorsal region posterior to tentacles.*

*O. sanguinolentus* is widely distributed in north-west Europe. It occurs all around Britain
under stones and among seaweed on the lower shore and below, and is often found in
laminarian holdfasts.

REFERENCES
Ball, I. R. & Reynoldson, T. B. (1981). *British planarians Platyhelminthes: Tricladida*. Keys
    and notes for the identification of the species. Synopses of the British fauna (New
    Series), no. 19. Cambridge: Cambridge University Press. (Includes *Procerodes*.)

Hartog, C. den (1968). Marine triclads from the Plymouth area. *Journal of the Marine Biological Association of the United Kingdom*, **48**, 209–23. (Includes *Procerodes*.)

Lang, A. (1884). Die Polycladen (Seeplanarian) des Golfes von Neapel und der angrenzeden Meeresabschnitte. Eine Monographie. *Fauna und flora des Golfes von Neapel*, **11**, ix–688.

Prudhoe, S. (1982). *British polyclad turbellarians*. Keys and notes for the identification of the species. Synopses of the British fauna (New Series), no. 26. Cambridge: Cambridge University Press. (Includes *Leptoplana*, *Prostheceraeus*, *Oligocladus*.)

# Nemertea

The nemerteans are elongate, often ribbon-like worms with bilateral symmetry. They are non-segmented and have a ciliated epidermis, but unlike the flatworms have a separate mouth and anus. Dorsal to the gut, and separate from it, is a muscular, eversible proboscis used in the capture of food and in defence. It is from this structure that the animals get one of their common names, the proboscis worms. They are also known as ribbon worms. Nemerteans are carnivorous and feed on a wide variety of crustaceans, polychaete worms and molluscs, which are captured on the proboscis. In some cases the prey is ingested whole, while in others the body of the prey is torn by the proboscis, which in some species has one or more piercing stylets, and the nemertean feeds suctorially. They also scavenge on dead and decaying animals. Most nemerteans have separate sexes and fertilization is usually external. In some species the gametes are deposited in an egg sac, where fertilization and development take place. Several different types of larvae have been described, including a ciliated pelagic larva known as a pilidium. Length of life is believed to be about one year but the larger species possibly live for several years.

Nemerteans are found in crevices, under stones and seaweed, and in mud and sandy sediments. One genus lives commensally in the mantle cavity of bivalve molluscs and another is parasitic on the egg masses and gill lamellae of crabs. Most are no more than 300 mm in length, but *Lineus longissimus* reaches a length of many metres. All show considerable extension and contraction, and require careful handling if fragmentation is to be avoided.

## Phylum NEMERTEA

**Class Anopla**   Proboscis without stylets.

**Class Enopla**   Proboscis with one or more stylets, but note *Malacobdella* (p. 156).

### Key to species of nemerteans

Accurate identification of nemerteans can be difficult and may depend on detail of the internal anatomy. The species described here can generally be identified according to morphological characters such as the shape of the head, the number and arrangement of eyes (when present), and the form of slits on the head, known as the cephalic grooves or slits. These are lined by ciliated epithelium and usually open to the cerebral organs.

1 Body with posterior, ventral sucker; leech-like. Living in the mantle cavity of bivalve molluscs ....................................................................... *Malacobdella grossa* (p. 156)
Body without posterior ventral sucker. Not living in the mantle cavity of bivalve molluscs .................................................................................................................. 2

2 Body broad, very flattened, up to 150 mm in length; posterior part of body wider than anterior, truncate at posterior margin with single median cirrus ........................................
.............................................................................. *Cerebratulus fuscus* (p. 151)
Body not so .......................................................................................................... 3

3 Head with 4 eyes arranged in square or rectangular pattern (in some cases eyes may be difficult to see because of presence of pigment) ............................................................. 4
Head with more than 4 eyes, may be arranged in groups or rows; sometimes difficult to see ...................................................................................................................... 6

4 Conspicuous, large rectangular, black pigment patch on head; often obscuring anterior pair of eyes .......................................................... *Tetrastemma melanocephalum* (p. 156)
Head without rectangular black pigment patch ........................................................... 5

5 Body flattened, narrow (0.5 mm wide); colour pale pink to white. Anterior pair of eyes the larger .................................................................... *Nemertopsis flavida* (p. 155)
Body rounded, up to 1–2 mm wide; colour varied; banded, striped and spotted forms common ................................................................................ *Oerstedia dorsalis* (p. 155)

6 Eyes in 2 rows, 1 row on each side of head ................................................................ 7
Eyes not in 2 rows ................................................................................................. 8

7 Row of 10–20 dark coloured eyes on each side of head; animal of very great length, up to 10 m. Brown-black in colour with purplish tints ................ *Lineus longissimus* (p. 152)
Row of 2–8 eyes in an irregular row on each side of head; up to 80 mm body length. Reddish-brown in colour ................................................................ *Lineus ruber* (p. 152)

8 Eyes in 4 conspicuous groups. Animal pink to white in colour; 2 pinkish-red blotches on head .............................................................. *Amphiporus lactifloreus* (p. 154)
Eyes tend to be arranged in 4 groups; may be difficult to see. Dorsal surface of body with broken longitudinal brown streaks ............................... *Emplectonema neesii* (p. 154)

## Class Anopla

### *Cerebratulus fuscus* (McIntosh)   (Fig. 88)

*Body very flattened, broad, up to 150 mm in length and 5 mm wide. Pale grey, brown, pinkish, yellowish in colour with reddish markings; transversely wrinkled. Head bluntly pointed with 4–13 eyes on each side; posterior part of body wider than anterior, truncate at posterior margin with single median cirrus. One pair of horizontal cephalic slits. Characteristically fragments when handled.*

Figure 88  *Cerebratulus fuscus* (after McIntosh, 1873–4).

*C. fuscus* is widely distributed in north-west Europe and is usually found in sandy deposits on the lower shore and sublittorally to over 1000 m. It is a large, conspicuous nemertean, able to swim actively on being disturbed. The sexes are separate.

**Lineus longissimus** (Gunnerus)    Bootlace worm (Fig. 89a)

*Rounded body up to 10 m in length, 5 mm wide; larger specimens not uncommon. Brownish-black in colour, with purplish tints; anterior margin of head pale. Row of 10–20 dark coloured eyes on each side of head, may be difficult to see because of brownish-black colour of body; behind each row a horizontal, cephalic slit.*

*L. longissimus* is widely distributed in north-west Europe, and because of its great length is probably the best known species. It is the longest nemertean recorded and is widely distributed on the lower shore, lying as a coiled mass in crevices and under stones, particularly in areas where there is muddy-sand and shingle. When disturbed it produces large quantities of mucus. It feeds on a wide range of invertebrates such as polychaete worms and ascidians, some of which are consumed whole. The sexes are separate.

**Lineus ruber** (Müller)    (Fig. 89b)

*Body slightly flattened, up to 80 mm in length, 3 mm wide. Reddish-brown, paler on ventral surface. Head slightly wider than rest of body; anterior margin rounded; 2 to 8 eyes in an irregular row on each side of head; behind each row a horizontal, cephalic slit.*

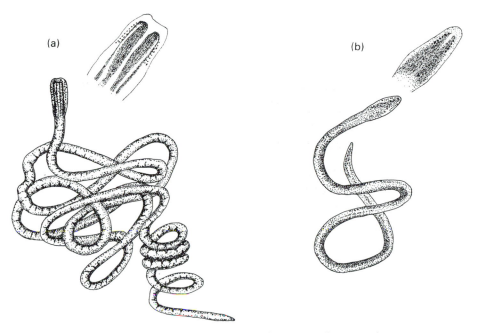

Figure 89  (a) *Lineus longissimus*. (b) *Lineus ruber* (after McIntosh, 1873–4).

*L. ruber* is common in north-west Europe and widely distributed around Britain under stones, among barnacles, mussels and seaweeds such as *Cladophora* (p. 37), *Ceramium* (p. 67) and in laminarian holdfasts. It tolerates salinities down to about 8‰ and is commonly found in estuarine sands and muds and on saltmarshes. It feeds on living and dead animals, such as polychaete worms and crustaceans captured by means of the proboscis. The sexes are separate and on the south coast of England and the Channel coast of France reproduction occurs from December/January to March. Groups of worms deposit gametes in egg sacs on the undersurface of rocks and on seaweed, where development takes place. A creeping, non-pelagic larva, known as a Desor larva, emerges. *L. ruber* breeds annually.

Two other species, **Myoisophagus sanguineus** (Rathke) (*Lineus sanguineus*) and **Lineus viridis** (Müller) are morphologically similar to *L. ruber*. Both are found intertidally under rocks and stones in muddy sediments. *M. sanguineus* is up to 200 mm in length, 3 mm in width, of reddish-brown colour, and with 4–6 eyes in a regular row on each side of the head. It is separated from *L. ruber* by its habit of contracting into a tight, spiral coil when disturbed. *L. viridis*, once considered a colour variety of *L. ruber*, is up to 80 mm in length, 3 mm width, with 2–8 eyes in an irregular row on each side of the head. As the specific name suggests, it is olive-green to greenish-black, occasionally pale green in colour.

Class Enopla

*Amphiporus lactifloreus* (Johnston)    (Fig. 90a)

*Body rounded, up to 35 mm in length, 1.5 mm wide. Pale pink to white; 2 pinkish-red blotches on head. Head rounded; eyes in 4 groups; 2 pairs transverse cephalic slits.*

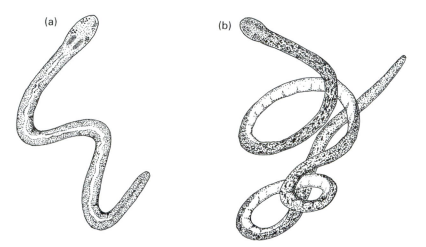

Figure 90  (a) *Amphiporus lactifloreus.* (b) *Emplectonema neesii.* (Both after McIntosh, 1873–4.)

*A. lactifloreus* is widely distributed in north-west Europe and around Britain, from the upper shore into the sublittoral to depths of about 250 m. It is often found on laminarians and other seaweeds, also in sand and gravel. The amphipod crustacean *Gammarus locusta* (p. 369) is an important food item. It is trapped on the worm's proboscis and the worm then inserts its head into the body of the prey and feeds on the soft tissues. The sexes are separate and breeding occurs during spring and early summer. Groups of mature worms are found together and gametes are deposited in egg sacs on rocks and seaweed. There is no pelagic larva.

*Emplectonema neesii* (Örsted)    (Fig. 90b)

*Up to 500 mm in length, 4 mm wide; usually smaller. Dorsal surface pale to dark brown, with broken longitudinal brown streaks; ventral surface white to pale pink. Head rounded; many eyes, tend to be arranged in 4 groups but often difficult to see. Cephalic slits inconspicuous.*

*E. neesii* is widely distributed in north-west Europe and around Britain. It is found under stones, among seaweed and in sand and gravel, from the middle shore into the sublittoral to depths of about 30 m. The sexes are separate and reproduction occurs from January to October. There is no pelagic larva.

***Nemertopsis flavida*** (McIntosh)    (Fig. 91a)

*Body flattened, very narrow; up to 40 mm in length, 0.5 mm wide, narrowing to a point at posterior end. Pale pink to white. Head rounded; 4 small black eyes arranged in a rectangle, anterior pair the larger; 2 pairs transverse cephalic slits, one slit behind each eye.*

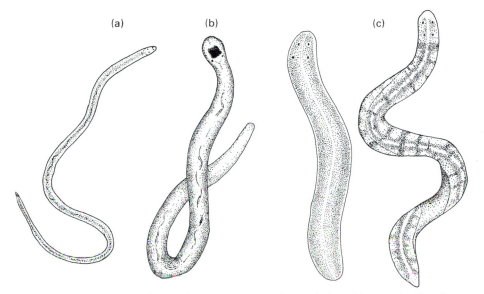

Figure 91  (a) *Nemertopsis flavida*. (b) *Tetrastemma melanocephalum*. (c) *Oerstedia dorsalis*, showing variation in body pattern (after McIntosh, 1873–4).

*N. flavida* is widely distributed around Britain in pools on the lower shore, under stones and in laminarian holdfasts. It also lives in mud, sand and gravel and extends into the sublittoral. The sexes are separate, reproduction occurs during summer and autumn. There is no pelagic larva.

***Oerstedia dorsalis*** (Abildgaard)    (Fig. 91c)

*Rounded, short body usually up to 15 mm in length, 1–2 mm wide. Colour varied, brown, red, green, yellow often with pale line along mid-dorsal surface; banded, striped and spotted forms common. Ventral surface pale. Head rounded; 4 eyes arranged in a square. Without cephalic slits.*

*O. dorsalis* is widely distributed in north-west Europe and Britain, extending from the middle shore to depths of 80 m. It is commonly found among seaweeds, in rock pools and in laminarian holdfasts where it feeds suctorially on amphipods. The sexes are separate and breeding occurs from September to November. There is no pelagic larva, young develop within a gelatinous capsule attached to the substratum. In terms of colour and patterning *O. dorsalis* is one of the most varied nemerteans and research carried out on populations on the west coast of Sweden suggests that on the basis of electrophoretic studies another species, **Oerstedia striata** Sundberg, should be recognized.

### *Tetrastemma melanocephalum* (Johnston)    (Fig. 91b)

*Body up to 60 mm in length, 2.5 mm wide; yellowish-green in colour with conspicuous, large rectangular black patch on head. Head flattened; 4 eyes arranged in a rectangle, the anterior pair often obscured by black pigment patch. Without cephalic slits.*

*T. melanocephalum* is widely distributed in north-west Europe and around Britain, where it is found on rocky shores in crevices, laminarian holdfasts and with seaweeds such as *Corallina* (p. 61). It also occurs on sand- and mud-flats and saltmarshes, where it has been recorded feeding suctorially on copepods and amphipods, especially *Corophium volutator* (p. 377) and *C. arenarium* (p. 378). It has been found in the sublittoral to depths of up to 40 m.

### *Malacobdella grossa* (Müller)    (Fig. 92)

*Body flattened, leech-like, with a posterior ventral sucker. Up to 40 mm in length and 15 mm wide. White or grey-coloured when immature; colour of gonad shows in mature specimens, males with pinkish coloration, females greenish.*

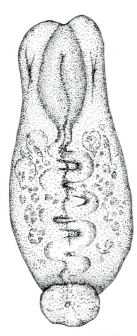

Figure 92  *Malacobdella grossa* (after Gibson, 1982).

*M. grossa* lives as an ectocommensal in the mantle cavity of bivalve molluscs. It occurs in a wide variety of species of hosts, including *Cerastoderma edule* (p. 300) and *Zirfaea crispata* (p. 325), and is widely distributed in north-west Europe. It is found between the gill lamellae of the host; usually there is only one nemertean in each host, but as many as five have been recorded. The proboscis of *Malacobdella* is not armed with piercing stylets, the animal is a suspension feeder. The pharynx is alternatively expanded and contracted and water is forced through a network of ciliated papillae, trapping food particles, which are transported to the oesophagus.

REFERENCES

Gibson, R. (1982). *British nemerteans.* Keys and notes for the identification of the species. Synopses of the British fauna (New Series), no. 24. Cambridge: Cambridge University Press.

Gibson, R. (1995). Nemertean genera and species of the world: an annotated checklist of original names and description citations, synonyms, current taxonomic status, habitats and recorded zoogeographic distribution. *Journal of Natural History*, 29, 271–562.

McIntosh, W. C. (1873–4). *A monograph of the British annelids*, vol. I, part 1. *The nemerteans.* London: Ray Society.

# Priapula

The priapulans are benthic, marine animals varying in length from a millimetre or so to about 200 mm. There are only a few species in the phylum. The body is divided into two distinct parts, an anterior proboscis, or introvert, and a posterior trunk. The proboscis is the shorter of the two regions and is retractile and covered with a large number of spines. The mouth is terminal. In some species, the posterior region of the body bears one or two much-branched processes. The trunk often has papillae on the surface, and is annulated but the body is not segmented. The animals live in sandy and muddy sediments on the shore and in the sublittoral. They are carnivorous, feeding on polychaete worms and other invertebrates captured by means of the proboscis. The sexes are separate and the gametes are discharged into the sea where fertilization occurs. The larvae live in soft sediments.

Only one species is likely to be encountered on the shore.

### *Priapulus caudatus* Lamarck    (Fig. 93)

*Body elongate, cylindrical up to 150 mm in length. Large proboscis with rows of spines. Posterior of trunk with single, much-branched process. Pinkish-brown in colour.*

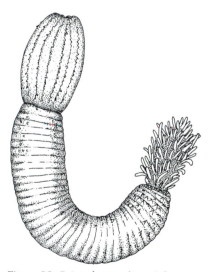

Figure 93  *Priapulus caudatus* (after M'Intosh, 1875).

*P. caudatus* is widely distributed and may be locally common in some areas of north-west Europe, burrowing in mud, muddy-gravel and muddy-sand from the middle shore into the sublittoral. It is carnivorous, feeding on polychaete worms such as *Aphrodita* (p. 166) and brittle-stars. The sexes are separate, fertilization external and breeding has been recorded in winter in specimens from the North Sea. The larvae inhabit the surface layers of sediment and are believed to feed on detritus. Development to the adult takes about two years.

REFERENCES

M'Intosh, W. C. (1875). *The marine invertebrates and fishes of St. Andrews.* Edinburgh: A. & C. Black.

Stephen, A. C. (1960). *British echiurids (Echiuroidea), sipunculids (Sipunculoida) and priapulids (Priapuloidea) with keys and notes for the identification of the species.* Synopses of the British fauna (Series 1), no. 12. London: Linnean Society.

# Annelida

The annelids include earthworms, bristle worms and leeches, and are characterized by the division of the body into a number of similar, externally visible segments. This is known as annulation, or metameric segmentation, and distinguishes them from other worm-like animals. In addition to being an important external feature, metamerism also affects the body cavity of the annelid, dividing it by septa into a number of units. In some annelids there is specialization and fusion of segments, particularly at the anterior end, and some segments have conspicuous lateral outgrowths. This is a large and diverse phylum with marine, freshwater and terrestrial representatives. It is divided into three classes.


## Phylum ANNELIDA

**Class Polychaeta**   The bristle worms, characterized by paired, lateral outgrowths known as parapodia which bear chaetae (bristles). Marine.

Class Aeolosomata   Very small, interstitial freshwater and brackish-water forms.

Class Clitellata   Without paired lateral outgrowths. Marine, freshwater and terrestrial. Includes the earthworms (subclass Oligochaeta) and the leeches (subclass Hirudinoidea).

Marine oligochaetes, known as sludge worms, are often found in sandy and muddy deposits. They are generally very small, rarely exceeding 20 mm in length and a few millimetres in width. In polluted estuaries they may be present in enormous densities, virtually to the exclusion of other macrofauna. Identification of species is difficult and beyond the scope of this book. The reader is referred to Brinkhurst (1982).


### Class Polychaeta

The polychaetes are widely distributed in the marine environment, being commonly found on the shore under stones and rocks and buried in mud and sand. In addition to species such as the ragworms and lugworms, the class also includes a range of very small worms previously assigned to a separate phylum, the Archiannelida. Many of these tiny forms are interstitial and live in sedimentary deposits. They are members of the meiofauna (p. 6) and are not included in this text. Westheide (1990) should be consulted for identification.

Typically, the segmented body of the polychaete bears paired, lateral outgrowths, the parapodia (Fig. 94b). Each parapodium is divided into a dorsal notopodium and a ventral neuropodium, very often differing in structure. The parapodia usually bear chaetae, which vary in shape and are very important in taxonomy. The anterior segments of the body are often modified and fused, and carry a number of structures associated with feeding and sensory perception (Fig. 94a, c). The anus is terminal. Within the class there is wide diversity of form; this is related to whether the worm is sedentary or active and mobile and therefore referred to as errant. Errant forms usually have well-developed

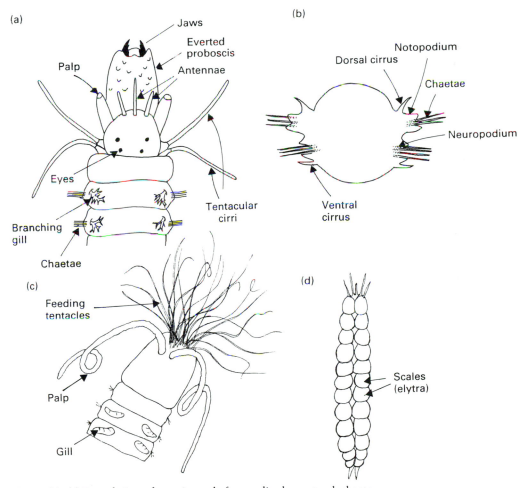

Figure 94  (a) Dorsal view of anterior end of generalized errant polychaete.
(b) Diagrammatic tranverse section through polychaete to show arrangement of parapodia. (c) Dorsal view of anterior end of generalized sedentary polychaete.
(d) Dorsal view of scale worm.

sensory structures such as eyes, antennae and palps, well-developed parapodia and an eversible proboscis (referred to in some texts as a pharynx) often armed with jaws. In the sedentary forms, eyes, antennae and palps are often reduced or absent, the parapodia are poorly developed, or very much modified, and complex food gathering structures may be present at the anterior end of the body (Fig. 94c). Some polychaetes burrow into sediments and usually live in a mucus-lined burrow, while others live in tubes made of sand grains, mud or calcium carbonate.

Polychaetes show a wide range of feeding habits including carnivorous, scavenging, deposit and suspension feeding. The sexes are usually separate and fertilization generally external. The eggs and sperm may be liberated through special ducts or the body wall of the worm may rupture, thereby releasing the gametes. An unusual form of reproduction known as epitoky occurs in some polychaetes. Highly modified individuals, or epitokes, contain the gametes and are equipped with large parapodia enabling them to swim in the surface waters where they swarm. Here synchronized release of gametes ensures cross-fertilization. In some cases it is not the whole worm that is modified for reproduction but chains of reproductive individuals are formed and these eventually break away. Many polychaetes have a free-swimming pelagic larva, the trocophore; in others the eggs are either brooded by the parent, or in the parental tube, or are laid in mucous egg sacs. Asexual reproduction by budding has been recorded in a few species.

The terminology used in the description of polychaetes is complex and there is often variation between authors. Fig. 94 shows characters used in identification, but it should be noted that many of the appendages such as tentacular cirri and antennae are often lost by rough handling of specimens. The scales (elytra) of worms belonging to the families Aphroditidae, Polynoidae and Sigalionidae may also be lost when the animal is handled, but the position of the scales (which is important in identification) can always be determined by the attachment scars which are left. The morphology of the proboscis is an important character in identification of several families; gentle pressure just behind the head is usually sufficient to cause the proboscis to be everted.

### Key to families of polychaetes

Polychaetes are often difficult to identify to species but family groupings are based on morphological features which are distinctive and should be readily identified. The family groupings of George & Hartmann-Schröder (1985) are used in this text. In some cases, an initial separation of species into the errant or sedentary way of life is a useful first step.

**Errant polychaetes** — usually with well-developed sensory structures such as eyes, antennae, palps and tentacular cirri, and an eversible proboscis and well-developed parapodia. Mobile, found under stones, seaweed, in sediment. The 10 families of errant polychaetes featured in this text are the Aphroditidae, Polynoidae, Sigalionidae, Phyllodocidae, Glyceridae, Hesionidae, Syllidae, Nereidae, Nephtyidae and Eunicidae.

**Sedentary polychaetes** — eyes, antennae and palps often reduced or absent; parapodia poorly developed or highly modified. Complex food gathering structures may be present at the anterior end of the body. Often living in tubes. The 14 families of sedentary polychaetes featured are Orbiniidae, Spionidae, Chaetopteridae, Cirratulidae, Capitellidae, Arenicolidae, Opheliidae, Oweniidae, Pectinariidae, Sabellariidae, Terebellidae, Sabellidae, Serpulidae and Spirorbidae.

1　Dorsal surface of the body covered wholly or partially by overlapping scales (elytra) (Fig. 94d) or with scales covered by a dense mat of chaetae giving the appearance of felt (NB the scales may be lost when the worm is handled, leaving attachment scars as evidence of their position) .................................................................................... 2
　　Dorsal surface of body without scales or mat of felt-like chaetae; worm inhabiting a tube or not ............................................................................................................ 3

2　(a) Short, broad body, not worm-like. Scales completely cover body and are often covered by a dense mat of felt-like chaetae. Head with single median antenna; eversible proboscis ................................................................ **Aphroditidae** (p. 166)
　　*or*
　　(b) Body long and narrow or short and broad, but worm-like; 12, 15 or 18 pairs of scales present at least on the anterior part of body, usually attached on every other segment. Head usually with 3 antennae; eversible proboscis .............. **Polynoidae** (p. 166)
　　*or*
　　(c) Body long, thin; many segmented, with many pairs of scales. Scales attached on every other segment anteriorly, and on all posterior segments. Head with 1–3 antennae; eversible proboscis ................................................................ **Sigalionidae** (p. 171)

3　Worm not inhabiting a tube ........................................................................................ 4
　　Worm inhabiting a tube which may be calcareous, parchment-like, of mucus, mud particles, sand grains, shell fragments ...................................................................... 17

4　Body regularly divided into similar segments. Conspicuous parapodia all along body. Head generally square, rounded or heart-shaped; with 1–5 antennae, often with eyes, palps and tentacular cirri. With a large, eversible proboscis ...................................... 5
　　Body regularly divided into similar segments. Parapodia not always conspicuous. Head pointed or bluntly rounded and without appendages, or with 4 very small antennae set closely together at tip *or* head or segments just posterior to head with 2 long, coiled palps or many long narrow tentacles. With or without eversible proboscis ................... 10

5　Head small, squarish, with 4 short, pointed antennae, 1 at each anterior corner, giving some resemblance to the head of a cat; no palps, no tentacular cirri; eyes not obvious. Large eversible proboscis with papillae and jaws ................................ **Nephtyidae** (p. 181)
　　Head usually with 2 or 4 eyes, 1–5 antennae and either or both palps and tentacular cirri .............................................................................................................................. 6

6　Parapodia with large expanded paddle-shaped dorsal cirri. Head usually with 2 large eyes, 4 or 5 antennae, 2–4 pairs of tentacular cirri. Without palps. Large eversible proboscis sometimes with papillae but without jaws ...................... **Phyllodocidae** (p. 171)
　　Not so ....................................................................................................................... 7

7   Parapodia divided into similar-sized notopodia and neuropodia. Head with 4 eyes, usually 2 antennae, 2 stout, knobbed palps, usually 4 pairs tentacular cirri. Large eversible proboscis with 2 large jaws and many paragnaths (Fig. 102a) .. **Nereidae** (p. 176)
Parapodia not clearly divided into similar-sized notopodia and neuropodia; dorsal cirri of parapodia slender and cylindrical, may be long and annulated ...................................... 8

8   Dorsal cirri of parapodia slender and cylindrical; not conspicuously long. Head usually with 2 eyes, 1, 3 or 5 antennae, 2 palps and 1 pair of tentacular cirri or none. Eversible proboscis with complex jaws. First 2 segments of body without parapodia and chaetae ....... ...................................................................................................... **Eunicidae** (p. 182)
Dorsal cirri of parapodia mostly very long, often annulated. Head usually with 4 eyes ..... 9

9   Head usually with 2 palps or none, if present they are 2 jointed; 2 or 3 antennae, usually 6–8 pairs of tentacular cirri. Large tubular, eversible pharynx, small jaws may be present ...................................................................................... **Hesionidae** (p. 174)
Head with 2, non-jointed palps (may be closely applied to ventral surface and not visible dorsally); 3 antennae, 2 pairs of tentacular cirri. Eversible proboscis often with 1 to many teeth ...................................................................................... **Syllidae** (p. 174)

10  Head pointed or bluntly rounded. Without appendages or with 4 very small antennae set closely together at tip ........................................................................................ 11
Head or segment just posterior to head with 2 long, often coiled palps or 2 to many long narrow tentacles ...................................................................................... 15

11  Head elongate, 4 very small antennae at tip. Body with many segments, each with 2 or 3 annulations. Very large eversible proboscis with 4 jaws .................. **Glyceridae** (p. 173)
Head without appendages ........................................................................................ 12

12  Long, many-segmented body, anterior region flattened; posterior region cylindrical. Large gills on dorsal surface of all segments except a few anterior ones. Tip of posterior region with 2 long cirri ...................................................................... **Orbiniidae** (p. 183)
Body not as above ........................................................................................ 13

13  Body solid, cylindrical with many segments, thicker anteriorly; almost all segments with 5 annulations. Large branching dorsal gills present on at least some segments ............ ...................................................................................... **Arenicolidae** (p. 191)
Body and gills not as above ........................................................................ 14

14  Body shape reminiscent of an earthworm; parapodia and chaetae inconspicuous. Gills present or absent, but not large and branching .................................. **Capitellidae** (p. 191)
Body short and cylindrical with small number of segments; usually a deep ventral groove along part or all of length. Gills present on dorsal surface of some segments ............ ...................................................................................... **Opheliidae** (p. 194)

15  Long, contractile thread-like gills present on at least some segments along sides of body. Two groups of 2 to many long, slender tentacles attached dorsally behind head ......... ...................................................................................... **Cirratulidae** (p. 189)
Without long, contractile thread-like gills along sides of body ...................................... 16

16  Head with 2 long, often coiled palps, usually with 4 eyes, sometimes 2 frontal horns. Large gills on dorsal surface of at least some segments (NB the 2 long tentacles are often lost when the worm is handled making identification difficult) ... **Spionidae** (p. 184)

Head with many long, narrow intertwined tentacles; usually 3 pairs of dorsal, branching gills just behind head ........................................................ **Terebellidae** (p. 198)

17  Worm inhabiting a calcareous tube; tube encrusting or upright; spirally coiled, straight or sinuous on seaweed, rocks, stones, shells, carapace of decapods ............................... 18

Tube not calcareous ..................................................................................................... 19

18  Tube small, encrusting, spirally coiled on seaweed, rocks, stones, shells ...........................
........................................................................................ **Spirorbidae** (p. 208)

Tube straight, sinuous, twisted, encrusting or upright; may be several entwined together, on rocks, stones, shells ....................................................... **Serpulidae** (p. 203)

19  Large, tough, parchment-like U-shaped tube. Body of worm of very distinctive appearance; divided into 2 or 3 regions with much-modified segments and parapodia; some parapodia fused to form fan-like structures ......................... **Chaetopteridae** (p. 188)

Tube and worm not as above ...................................................................................... 20

20  Tube made of sand grains and shell fragments arranged in overlapping fashion. Head of worm without appendages, but bearing a collar or frilled membrane ... **Oweniidae** (p. 194)

Tube and worm not as above ...................................................................................... 21

21  (a) Tube made of fine sand grains and shell fragments; rigid, narrow, cone-shaped, open at both ends. Worm as in Fig. 115a ....................................... **Pectinariidae** (p. 195)

*or*

(b) Tube made of coarse sand grains; attached to rock and stones, often forming large colonies firm enough to walk on. Worm as in Fig. 115c ................... **Sabellariidae** (p. 197)

*or*

(c) Tube made of sand and shell fragments; anterior end flattened and extended into distinctive finger-like projections extending above surface of sediment. Worm as in Fig. 117a ............................................................. **Terebellidae** (*Lanice conchilega*) (p. 198)

Tube and body of worm not as above ......................................................................... 22

22  Mucous or membranous tube often covered with mud or sand particles. Head of worm with crown of pinnate projections which are often brightly coloured and banded, and sometimes bear eyes .............................................................................. **Sabellidae** (p. 200)

Mucous tube; may be flimsy, or covered with mud particles, sand grains, shell fragments. Head of worm not as above ........................................................................ 23

23  Mucous tube often covered with mud particles, sand grains, shell fragments. Head of worm with 2 long, coiled palps or many long, narrow tentacles .................................... 24

Mucous tube, may be rather flimsy; head of worm not as above .................................. 25

24  Head of worm with 2 long, coiled palps ............................................. **Spionidae** (p. 184)

Head of worm with many long, narrow intertwined tentacles .......... **Terebellidae** (p. 198)

25  Parapodia divided into similar-sized notopodia and neuropodia. Head with 4 eyes, usually 2 antennae, 2 stout knobbed palps, usually 4 pairs tentacular cirri. Large eversible proboscis with 2 large jaws and many paragnaths (Fig. 102a) .. **Nereidae** (p. 176)

Parapodia not divided into similar-sized notopodia and neuropodia, with mostly very long, dorsal cirri. Head with 4 eyes, 2 non-jointed palps (may be closely applied to ventral surface and not visible dorsally); 3 antennae, 2 pairs of tentacular cirri. Eversible proboscis often with 1 to many teeth ........................................ **Syllidae** (p. 174)

**Errant polychaetes**
Family Aphroditidae
Errant; short, broad body; dorsal surface covered by large overlapping scales (elytra —
expanded dorsal cirri of the parapodia) which are often covered by a dense mat of chaetae
giving the appearance of felt. Head usually with 2 pairs of eyes, in some species on stalks;
single median antenna; 2 palps; 2 pairs tentacular cirri. Eversible proboscis with jaws.

*Aphrodita aculeata* Linnaeus    Sea-mouse (Fig. 95)

*Wide, oval body up to 200 mm in length, usually smaller. Up to 40 segments. Dorsal sur-
face rounded, ventral surface flat. Dorsal surface with 15 pairs flattened scales hidden by
mat of felt-like chaetae. Head concealed, bears 2 pairs of sessile eyes. Dorsal surface with
greyish-brown chaetae, lateral and ventral surfaces with iridescent green and gold chaetae
and stout brown chaetae. Animal of distinctive and unmistakable appearance.*

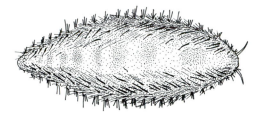

Figure 95  *Aphrodita aculeata.*

A. *aculeata* is widely distributed in north-west Europe on mud or muddy-sand at depths
of 10–170 m, but occasionally large numbers are washed ashore. It normally feeds only
when buried in the sediment and there are conflicting reports of the food taken; it is,
however, believed to be carnivorous and feeds on large polychaetes such as *Nephtys
hombergii* (p. 182), *Nereis diversicolor* (p. 178), *N. virens* (p. 178) and *Arenicola marina*
(p. 192). The sexes are separate and mature females can be recognized by the cream-
coloured eggs visible through the walls of the parapodia. Little is known of the breeding
cycle; mature animals have been recorded in spring and autumn on the south coast of
England.

Family Polynoidae – scale worms
Errant; body long and narrow or short and broad; dorsal surface wholly or partly covered
by scales (elytra – expanded dorsal cirri of the parapodia); depending on species there
are generally 12, 15 or 18 pairs. The scales are present at least on the anterior part of the
body; usually attached on every other segment. Some or all of the scales may be lost
when the animal is handled, leaving attachment scars. Head with 4 eyes; usually
3 antennae; 2 long flexible palps; 2 pairs tentacular cirri. Eversible proboscis with
concealed jaws.

1  Dorsal surface of body with 12 pairs of scales ...................................................... 2
   Dorsal surface of body with 15 pairs of scales ................................................... 3
2  Scales overlap to some extent, but leave part of anterior mid-dorsal region uncovered;
   scales without a fringe of papillae ............................................. *Lepidonotus clava* (p. 167)
   Scales overlap and completely cover dorsal surface; scales with tubercles and posterior
   fringe of papillae ............................................... *Lepidonotus squamatus* (p. 168)
3  Scales cover only the anterior half of body; 50 or so posterior segments not covered.
   Worm long and narrow ...................................................... *Polynoë scolopendrina* (p. 170)
   Scales cover the whole length of the body or leave a relatively small part (posterior 8 or
   so segments) uncovered ..................................................................................... 4
4  Posterior 8 or so segments not covered by scales ............... *Harmothoë extenuata* (p. 169)
   All of dorsal surface covered by scales ................................................................. 5
5  Scales covered with tiny tubercles; $\frac{1}{4}$ of scale margin with fine papillae. Worm
   longish, up to 65 mm in length ........................................ *Harmothoë imbricata* (p. 168)
   Scales covered with many small and a few large tubercles; more than half of scale
   margin with papillae. Worm smallish, up to 25 mm in length  . *Harmothoë impar* (p. 169)

### *Lepidonotus clava* (Montagu)    (Fig. 96a)

*Body up to 35 mm in length. Up to 26 segments. Dorsal surface rounded, ventral surface*
*flattened; 12 pairs of rounded scales, leaving part of anterior mid-dorsal region uncovered.*
*Scales patterned with shades of brown with lighter central spot.*

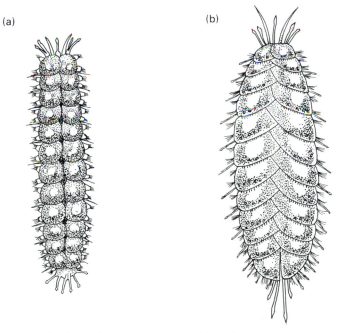

Figure 96  (a) *Lepidonotus clava* (after McIntosh, 1900). (b) *Lepidonotus squamatus* (after
McIntosh, 1900).

*L. clava* is widely distributed in north-west Europe, and in Britain is commonly found on south and west coasts where it occurs under stones on the lower shore. It is carnivorous, feeding on polychaetes, crustaceans and other invertebrates.

**Lepidonotus squamatus** (Linnaeus)    (Fig. 96b)

*Body up to 50 mm in length. Up to 26 segments. Dorsal surface rounded, ventral surface flattened; 12 pairs of overlapping scales, oval to kidney-shaped, completely covering dorsal surface; scales covered with tubercles and with posterior fringe of papillae. Colour of scales varied, brown, yellow, grey, with clear central spot.*

*L. squamatus* is widely distributed in north-west Europe and around Britain among stones and seaweed on the middle and lower shore and into the sublittoral. It is mainly carnivorous, feeding on small invertebrates, although it is also believed to feed on algae. The sexes are separate and breeding probably occurs during June and July. When disturbed the worm rolls up into a ball and relies on the scales, which are not easily dislodged, for some protection.

**Harmothoë imbricata** (Linnaeus)    (Fig. 97a)

*Body flattened; up to 65 mm in length. Up to 39 segments; 15 pairs of overlapping scales covering dorsal surface, the first pair almost circular; following pairs kidney-shaped, covered with tiny tubercles; ¼ of scale margin fringed with fine papillae. Colour of scales varied, grey, blue, brown.*

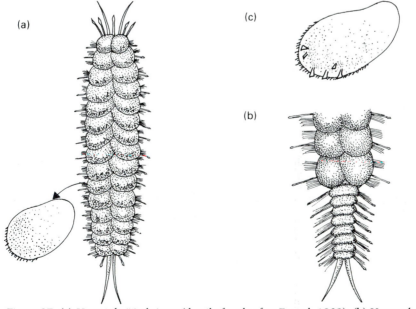

Figure 97   (a) *Harmothoë imbricata* (detail of scale after Fauvel, 1923). (b) *Harmothoë extenuata*, posterior region. (c) *Harmothoë impar*, detail of scale (after Fauvel, 1923).

*H. imbricata* is widely distributed in north-west Europe and Britain, extending from the middle and lower shore to depths of over 3000 m. It has been described as the commonest polynoid of rocky shores, where it is found under stones and in laminarian holdfasts. It is an active predator, seizing prey such as amphipods and polychaetes by sudden eversion of the proboscis. The sexes are separate and on the north-east coast of England breeding occurs during March and April, the females spawning twice during these months. The worms come together in pairs; sperm are released onto the eggs, which are held under the scales of the female where early development takes place. After two weeks, trochophore larvae are released into the plankton. Settlement of the young worms occurs during May; they breed during the following year. Length of life is probably up to four years.

### *Harmothoë extenuata* (Grube)    (*Lagisca extenuata*) (Fig. 97b)

*Body flattened; up to 70 mm in length. Up to 50 segments; 15 pairs of scales, not covering posterior 8 or so segments; first pair of scales round, others kidney-shaped; scales covered with very small tubercles and with row of larger, round tubercles near posterior margin; less than ¼ of scale margin with tiny papillae. Colour of scales varied, red, grey, brown, often with clear central spot.*

*H. extenuata* is widely distributed in north-west Europe and Britain, under stones and among laminarian holdfasts extending into the sublittoral to depths of about 550 m. It is carnivorous. The sexes are separate and breeding has been recorded from November to January. The eggs are held by the female between the parapodia and also on the ventral surface of the body until they are released as free-swimming larvae.

### *Harmothoë impar* (Johnston)    (Fig. 97c)

*Body flattened; up to 25 mm in length. Up to 40 segments; 15 pairs of overlapping scales covering dorsal surface; first pair of scales round, following pairs kidney-shaped, covered with many small and a few large tubercles; more than half of scale margin with papillae. Scales brownish in colour, often with central, yellow spot.*

*H. impar* is widely distributed in north-west Europe and Britain and is common under stones and rocks and among laminarian holdfasts from the middle shore into the sublittoral. It is carnivorous. The sexes are separate and breeding takes place during spring. The worms come together in pairs; after fertilization the eggs are held under the scales of the female until they are liberated as free-swimming larvae.

Several other species of *Harmothoë* are recorded from around Britain. These are mostly rare and/or offshore.

*Polynoë scolopendrina* Savigny    (Fig. 98a)

*Body narrow, elongate; up to 120 mm in length with up to 100 segments. Dorsal surface rounded, ventral surface flattened; 15 pairs of scales covering only the anterior half of the body, 50 or so posterior segments not covered; scales rounded, with small tubercles anteriorly. Segments in posterior half of body each with 3 tubercles on the dorsal surface, giving 3 longitudinal rows along the length of the body. Anterior region reddish in colour; scales brownish with darker central area.*

(a)                                                    (b)

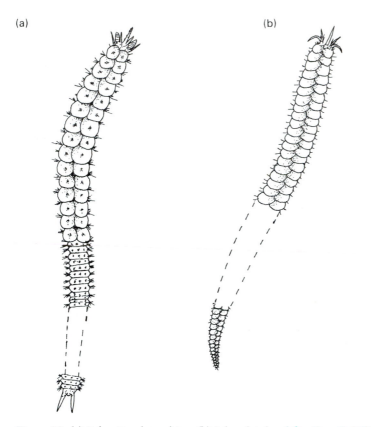

Figure 98  (a) *Polynoë scolopendrina*. (b) *Sthenelais boa* (after Fauvel, 1923).

*P. scolopendrina* is widely distributed in north-west Europe under stones and in crevices on the lower shore. It is most often commensal with other polychaetes such as *Eupolymnia nebulosa* (p. 199).

Family Sigalionidae
Errant; body long and rather flattened, many segmented. Dorsal surface of body with many pairs of scales attached on every other segment anteriorly, and on all posterior segments. Head with 1 to 3 antennae, 1 pair of palps; usually with 4 eyes; eversible proboscis with 4 jaws.

***Sthenelais boa*** (Johnston)    (Fig. 98b)

*Body up to 200 mm in length. Up to 200 segments. Head with 1 median antenna and 2 lateral antennae; 2 pairs of eyes. Dorsal surface with about 180 pairs of overlapping scales completely covering surface; first pair rounded, others kidney-shaped; outer edge of scales fringed with papillae. Colour varied, brown, yellow, orange, sometimes with darker transverse bandings.*

S. *boa* is widely distributed in north-west Europe burrowing in intertidal sands and gravel, and is commonly found under stones embedded in gravel and coarse sand.

Family Phyllodocidae – paddle worms
Errant; body long and slender with many segments. Parapodia with large, expanded paddle-shaped dorsal cirri. Head usually with 2 large eyes; 4 or 5 antennae; 2–4 pairs tentacular cirri. First 2 or 3 segments without paddle-shaped dorsal cirri. Large eversible proboscis sometimes covered with papillae but without jaws.

A large family of polychaetes found in crevices and under stones on rocky shores, in laminarian holdfasts and in a wide range of sediments from shell gravel to mud and sand. The species can be difficult to separate. Only two genera are described and illustrated here; both are widely distributed and common.

***Phyllodoce*** spp.    (*Anaitides* spp.) (Fig. 99a)

*Body up to 450 mm in length and with up to 700 segments, depending on species, the 2 described below being much smaller, up to 100 mm in length with up to 270 segments. Dorsal and ventral cirri of parapodia flattened; dorsal cirri very conspicuous, heart-shaped, kidney-shaped, oval. Head heart-shaped; 2 eyes; 4 antennae; 4 pairs longish, slender tentacular cirri. Large, eversible proboscis with papillae. Colour varied depending on species; body often yellowish-green with brown markings on each segment, but sometimes whitish with brown spots; dorsal cirri often greenish-brown, but sometimes clear with brown spots.*

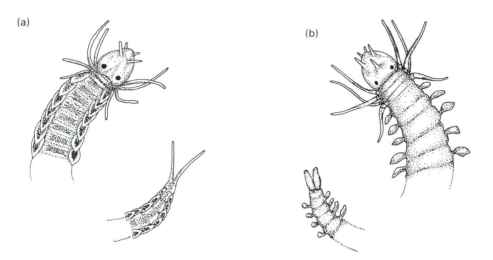

Figure 99  (a) *Phyllodoce maculata*. (b) *Eulalia viridis*.

There are several species of the genus **Phyllodoce**, two of which, **Phyllodoce maculata** (Linnaeus) and **Phyllodoce mucosa** Oersted, are widely distributed in north-west Europe and around Britain, but the species are difficult to separate and reference to Pleijel & Dales (1991) is recommended. Both occur on the lower shore and shallow sublittoral in sand, muddy-sand and under stones, and are carnivorous, feeding on small invertebrates. The sexes are separate and tangled masses consisting of several males with a single female have been observed on the surface of the sediment. Eggs and sperm are shed into mucous cocoons secreted by the female and attached to sediment and stones. The eggs are green in colour and develop into pelagic larvae.

### *Eulalia viridis* (Linnaeus)    (Fig. 99b)

*Body narrow; up to 150 mm in length. Up to 200 segments. Parapodia with flattened, paddle-shaped dorsal cirri. Head with 2 large eyes; 5 antennae; 4 pairs tentacular cirri, the longest reaching back to about segment 7. Eversible proboscis covered with small papillae. Bright green in colour, without any distinct patterning.*

*E. viridis* is widely distributed in north-west Europe and Britain from the middle shore to the shallow sublittoral in rocky crevices, among barnacles and mussels, and in laminarian holdfasts, being very conspicuous because of its bright green colour. It is a scavenger, feeding mainly on damaged barnacles and mussels. The sexes are separate and breeding has been recorded in July and August. Green gelatinous egg masses, usually found attached to seaweed, are often described as being those of *E. viridis* but this requires confirmation. There is a planktonic larva. The worms probably live for several years, breeding in the second and subsequent years.

Family Glyceridae

Errant; rounded body narrowing towards anterior and posterior. Head an elongate, pointed cone with annulations; 4 very small antennae set closely together at tip. Extremely large, eversible proboscis with many papillae and 4 jaws. Body with many segments, each with 2 or 3 annulations.

**Glycera tridactyla** Schmarda    (*Glycera convoluta*) (Fig. 100a, b)

*Body round, narrowing markedly at anterior and posterior; up to 100 mm in length. Up to 180 segments but seems more as each segment has 2 annulations. Finger-shaped, non-retractile gills present on dorsal surface of parapodia from 25th segment to posterior of body. Head elongate, pointed cone with about 8 annulations. Reddish-pink in colour. Body of worm usually thrown into spiral coils.*

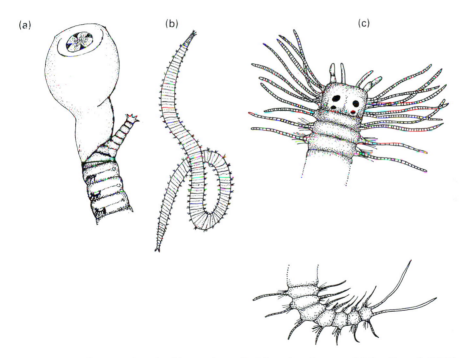

Figure 100  *Glycera tridactyla*, (a) anterior end with proboscis everted (after Fauvel, 1923), (b) whole worm. (c) *Kefersteinia cirrata*.

G. tridactyla is a southern species extending to the southern North Sea. It lives on the lower shore in mud and muddy-sand where it inhabits branching burrows. The worms are carnivorous, feeding on small invertebrates, and glands near the jaws are known to secrete a toxic substance. Breeding occurs from June to August on the south coast of England; there is a pelagic larva.

Several other species of *Glycera* occur in European waters: for further study reference to O'Connor (1987) is recommended.

Family Hesionidae
Errant; cylindrical body; parapodia with long, usually annulated, dorsal cirri; not divided into similar-sized notopodia and neuropodia. Head with 4 eyes; 2 or 3 antennae; 2 jointed palps or none; usually 6–8 pairs of tentacular cirri. Large tubular eversible proboscis, jaws present or absent.

*Kefersteinia cirrata* (Keferstein)    (Fig. 100c)

*Body up to about 75 mm in length. Up to 70 segments. Parapodia with long, annulated dorsal cirri; 2 antennae; 2 palps; 8 pairs of annulated tentacular cirri. Proboscis eversible, fringed with fine papillae. Colour varied, yellow, pink, red.*

*K. cirrata* is widely distributed in north-west Europe and around Britain on the lower shore and in the shallow sublittoral under stones, in coarse gravel, among laminarians and in mussel beds. It is carnivorous, feeding on small invertebrates. The sexes are separate and on the north-east coast of England breeding occurs during June and July. There is no epitokous stage; the worms swim for short periods at a time and release gametes.

Family Syllidae
Errant, some species inhabit mucous tubes. Small, slender body; parapodia not divided into similar-sized notopodia and neuropodia, often with well-developed dorsal cirri. Head usually with 4 eyes; 3 antennae; 2 palps, may be closely applied to ventral surface of head and not visible dorsally; 2 pairs of tentacular cirri. Antennae, tentacular and dorsal cirri sometimes distinctly annulated. Eversible proboscis often armed with 1 to many teeth.

A large family of polychaetes found in a wide variety of habitats, including laminarian holdfasts, among seaweeds, bryozoans and hydroids, and occasionally in muddy deposits. Only two of the most common genera are included here.

*Eusyllis blomstrandi* Malmgren    (Fig. 101b)

*Body up to 30 mm in length. Up to 124 segments. Anterior parapodia with long, irregularly annulated dorsal cirri; posterior parapodia with short, more or less smooth dorsal cirri; ventral cirri of parapodia oval and short. Antennae and tentacular cirri irregularly annulated. Palps long, oval, visible dorsally. Median antenna longer than lateral ones. Proboscis eversible, with single large tooth and ring of many small teeth. Yellowish-orange in colour, tips of cirri brown.*

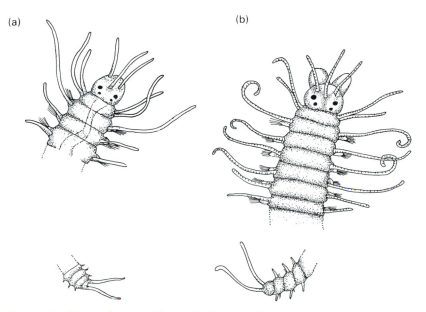

(a)                                          (b)

Figure 101  (a) *Autolytus* sp. (b) *Eusyllis blomstrandi*.

*E. blomstrandi* is widely distributed in north-west Europe and Britain, often living in mucous tubes among stones, seaweeds, laminarian holdfasts, bryozoans and especially hydroids. It is carnivorous, feeding mainly on hydroids. The sexes are separate and on the east coast of England breeding occurs during spring and summer. The epitokous stages swarm in the surface waters, where the gametes are discharged. *Eusyllis* often shines with a green luminescence.

### *Autolytus* spp.   (Fig. 101a)

*Small, thin worms; about 3–15 mm in length. Number of segments varies, about 30–60 in the commoner species. Parapodia without ventral cirri. All appendages lack annulations. Palps not always visible dorsally. Proboscis armed with ring of teeth, visible through semi-transparent body wall when withdrawn. Colour varied, depending on species; some almost colourless.*

Species of *Autolytus* are widespread in north-west Europe and Britain but are very difficult to separate. They are found on the lower shore and into the sublittoral, among hydroids, seaweeds, stones and shells where they build thin, transparent, mucous tubes. The worms are carnivorous and feed mainly on hydroids. Their reproductive biology is complex. Individuals are budded from the parent giving rise to chains of individuals, each individual being known as a stolon. These are specialized for sexual reproduction and are either male (the polybostrichus) or female (the sacconereis). They become pelagic and swarm together in the surface waters. In some species the male discharges sperm while swimming in a circle round the female. The fertilized eggs are carried by the female in a ventral egg sac until they hatch as free-swimming larvae.

Family Nereidae – ragworms
Errant, some species inhabiting mucous tubes. Parapodia conspicuous, divided into similar-sized notopodia and neuropodia. Head with 4 eyes; usually 2 antennae; 2 stout, knobbed palps; usually 4 pairs tentacular cirri. Large, eversible proboscis with 2 large jaws and a number of small teeth known as paragnaths.

The distribution of the paragnaths on the dorsal surface of the proboscis is shown for each species, and although the number of paragnaths varies, the overall pattern of distribution can be an important aid to identification and where appropriate this has been highlighted on the drawings. The proboscis of nereid worms can usually be everted by *gentle* pressure behind the head while the worm is alive. Colour and habitat are also useful aids to identification, but in some cases accurate identification can only be made following a detailed examination of the parapodia and chaetae. While the detail of individual parapodia may vary along the length of the body a useful character is the relative lengths of the dorsal cirrus and notopodial lobes in parapodia from the first 20 or so segments (excluding segments 1 and 2 which are rather atypical). Parapodia from this region of the body are illustrated for each species described here. For further detail reference to Chambers & Garwood (1992) is recommended.

1  Some of the paragnaths on the dorsal surface of the proboscis grouped to form 2 blackish crescents (Fig. 104a) .............................................. *Perinereis cultrifera* (p. 177)
   Paragnaths on the dorsal surface of the proboscis not grouped to form 2 crescents .......... 2
2  Paragnaths on the dorsal surface of the proboscis very small, arranged in short rows (Fig. 104c). Tentacular cirri long, 1 pair extending backwards to about segment 15 ......
   .................................................................................... *Platynereis dumerilii* (p. 180)
   Paragnaths and tentacular cirri not so ............................................................................ 3
3  (a) Paragnaths arranged as in Fig. 102c. Greenish-yellow, reddish in colour. In mud, muddy-sand; often in estuaries .............................................. *Nereis diversicolor* (p. 178)
   *or*
   (b) Paragnaths arranged as in Fig. 103c. Yellowish-brown in colour with white band in middle of dorsal surface. Lives as a commensal in whelk shells occupied by hermit crabs .....
   ............................................................................................... *Nereis fucata* (p. 179)
   *or*
   (c) Paragnaths arranged as in Fig. 102a. Reddish-brown, greenish-yellow in colour with an iridescent sheen. In crevices, laminarian holdfasts and under stones on the lower shore, usually in a mucous tube .............................................................. *Nereis pelagica* (p. 177)
   *or*
   (d) Paragnaths arranged as in Fig. 103a. Body large and thick; notopodia large and leaf-like. Greenish-blue in colour with iridescent sheen. In mud, muddy-sand on the lower shore ........................................................................................... *Nereis virens* (p. 178)

## Perinereis cultrifera (Grube)    (Fig. 104a, b)

*Body up to 250 mm in length. Up to 125 segments. Paragnaths, some of which are grouped to form 2 crescents, arranged as in figure. Antennae shorter than palps. Dorsal cirrus of parapodium shorter than or equal to notopodial lobes. Green with bronze tints; parapodia reddish.*

*P. cultrifera* is widely distributed in north-west Europe and around Britain on the lower shore in crevices, under stones and in laminarian holdfasts. It feeds mainly on algae and diatoms. The sexes are separate and both show epitokous changes and swarm in the surface waters, swimming in circles round each other. Breeding has been recorded in May and June in the English Channel; fertilization is external and there is a pelagic larva. Length of life is believed to be three years.

## Nereis pelagica Linnaeus    (Fig. 102a, b)

*Body cylindrical; up to 120 mm or more in length. Up to 100 segments. Paragnaths on proboscis arranged as in figure. Antennae shorter than palps. Dorsal cirrus of parapodium longer than notopodial lobes. Reddish-brown, greenish, yellow in colour with an iridescent sheen.*

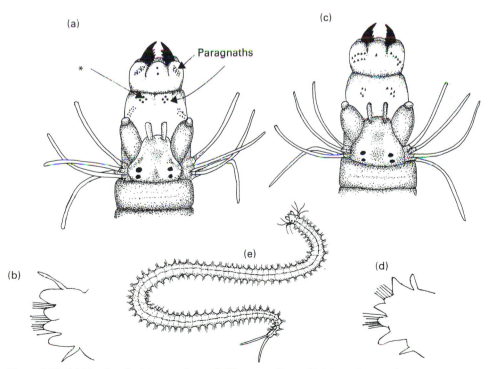

Figure 102  (a) *Nereis pelagica*, anterior end, (b) parapodium. (c) *Nereis diversicolor*, anterior end, (d) parapodium, (e) whole specimen. ((b), (d) Modified after Chambers & Garwood, 1992.)

*N. pelagica* is widely distributed in north-west Europe and Britain on the lower shore in crevices, among seaweed, in laminarian holdfasts and under stones. It extends into the sublittoral to depths of 120 m. The worms usually live in mucous tubes that are encrusted with sand particles and attached to stones and seaweed. They are omnivorous, feeding mainly on microorganisms. The sexes are separate, and on the north-east coast of England the breeding season extends from December to April. The body changes into the epitokous condition and swarming occurs. Fertilization is external but there are conflicting views on the mode of development of the larva. Both pelagic and non-pelagic larvae have been recorded and laboratory studies suggest a free-swimming life of some 18 days for the pelagic larvae. Length of life is about three years. The worms breed once then die.

**Nereis diversicolor** (O. F. Müller) (known by some authors as *Hediste diversicolor* (O. F. Müller))   Ragworm (Fig. 102c, d, e)

*Body somewhat flattened; up to 200 mm in length. Up to 120 segments. Paragnaths arranged as in figure. Antennae shorter than palps. Dorsal cirrus of parapodium shorter than notopodial lobes. Colour varied, greenish-yellow, reddish.*

*N. diversicolor* is widely distributed in north-west Europe and is one of the commonest intertidal polychaetes. It lives in a more or less permanent, often U-shaped, burrow in mud and muddy-sand, most often in estuaries and is tolerant of salinities down to 1‰. It is omnivorous, feeding on a variety of plant and animal material on the surface of the sediment and also suspension feeds on phytoplankton contained in the water drawn in to irrigate the burrow. The phytoplankton is trapped on a mucous funnel secreted by the worm at the entrance to the burrow and periodically ingested. The main predators of *N. diversicolor* are fishes and birds. The sexes are separate and many populations have a high proportion of females. Ripe males can be distinguished by their bright green colour, compared to the much darker green of the ripe females. There is no free-swimming epitokous stage in *N. diversicolor*. Breeding has been recorded in spring at the times of new and full moon. During breeding the females remain in the burrow where the eggs are released, while mature males crawl about on the sediment surface and release sperm near the burrow entrance. Using the proboscis the female carries the sperm into the burrow where fertilization occurs and development of the larvae takes place. After 10–14 days the young worms leave the burrow; they are bottom-dwelling and not planktonic. Length of life is probably about three years. The worms breed once then die.

**Nereis virens** (Sars) (known by some authors as *Neanthes virens* (Sars))   King ragworm (Fig. 103a, b)

*Body large and thick; generally 200–300 mm but may reach 900 mm in length. Up to 200 segments. Paragnaths arranged as in figure. Antennae a little shorter than palps. Notopodia large and leaf-life; dorsal cirrus small, much shorter than notopodial lobes. Greenish-blue in colour with iridescent sheen, often with purple tints.*

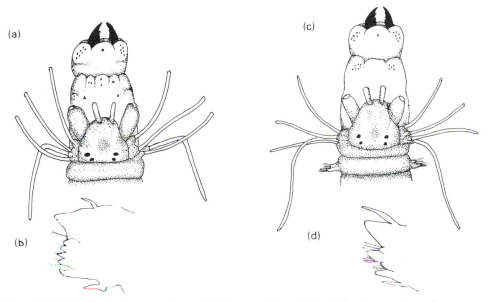

Figure 103  (a) *Nereis virens*, anterior end, (b) parapodium. (c) *Nereis fucata*, anterior end, (d) parapodium. ((b), (d) Modified after Chambers & Garwood, 1992.)

*N. virens* is a northern species extending as far south as the English Channel; its distribution is sporadic around Britain, but it may be locally abundant. It is found on the lower shore in mud and muddy-sand up to 400 mm below the surface of the sediment, as well as under stones on rocky shores and in estuaries. It is, however, less tolerant of low salinity than *N. diversicolor* (above). It is omnivorous, its diet including polychaetes such as *Nephtys hombergii* (below). The sexes are separate, and on the south-east coast of England breeding takes place during April and May. Males and females become bright green in colour; the males become epitokous and swarm at high tide. The females do not have an epitokous stage and are believed to discharge eggs at high tide while remaining in, or close to, their burrows. The larvae are more or less benthic for the first five or six days, after which they have a short pelagic phase of probably only a few hours. Length of life is believed to be two or three years. The worms breed once then die. *N. virens* has considerable economic importance as bait used by sea anglers and it is now reared in aquaria for the commercial market.

**Nereis fucata** (Savigny) (known by some authors as *Neanthes fucata* (Savigny))   (Fig. 103c, d)

*Body up to 200 mm in length. Up to 120 segments. Paragnaths arranged as in figure. Antennae slightly longer than palps. Dorsal cirrus longer than notopodial lobes. Yellowish-brown in colour, with white band in middle of dorsal surface; parapodia white.*

*N. fucata* is widely distributed in north-west Europe and Britain, occurring on the lower shore and in the sublittoral. It lives in a mucous tube as a commensal inside shells, often those of the whelk *Buccinum undatum* (p. 258), occupied by hermit crabs, especially *Pagurus bernhardus* (p. 395). It feeds on animal matter, usually scavenging on the remains of the hermit crab's prey. The sexes are separate, and on the south coast of England breeding occurs during spring. Both sexes show epitokous changes, the males becoming whitish, the females lilac or blue, while the parapodia of both sexes become pinkish-red. The worms leave the host shells and are believed to swarm in the water while discharging gametes. After a pelagic life of four to six weeks, the young worms become benthic and inhabit tubes which they construct from shell gravel and mucus. The worm extends its body from the tube and is sensitive to vibrations, eventually locating a shell occupied by a hermit crab. The worm then leaves its tube and takes up position in the host shell, where it constructs a mucous tube in which it lives. The worms breed once then die.

### *Platynereis dumerilii* (Audouin & Milne-Edwards)   (Fig. 104c, d)

*Body up to 60 mm or more in length. Up to 90 segments. Tentacular cirri long, 1 pair extend backwards to about segment 10–15. Paragnaths very small, in short rows, arranged as in figure. Antennae more or less the same length as palps. Dorsal cirrus of parapodium longer than notopodial lobes. Colour varied, iridescent, green, red, yellow.*

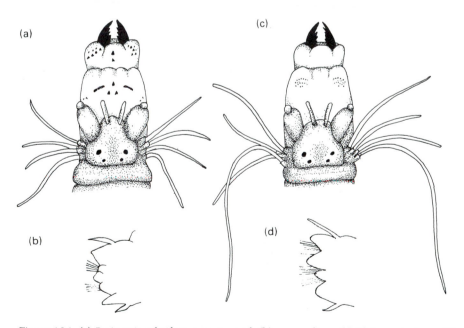

Figure 104  (a) *Perinereis cultrifera*, anterior end, (b) parapodium. (c) *Platynereis dumerilii*, anterior end, (d) parapodium. ((b), (d) Modified after Chambers & Garwood, 1992.)

*P. dumerilii* is widely distributed in north-west Europe, and around Britain is found mostly on south and west coasts. It lives in a mucous or membranous tube on the lower shore in rock crevices, among *Fucus* spp. and laminarian holdfasts. It is also found in shallow water on drifting seaweed, especially *Laminaria saccharina* (p. 45). It feeds mainly on seaweed and associated microorganisms. On the west coast of Scotland breeding occurs from April to June. The sexes are separate and both become epitokous; in some populations males have been observed discharging sperm while swimming in circles round slowly moving females. After spawning the worms die. Fertilization is external and the larvae are pelagic for a few days.

A second species of *Platynereis*, **Platynereis massiliensis** (Moquin-Tandon), also occurs in north-west Europe and is apparently morphologically indistinguishable from *P. dumerilii*, but reproduces quite differently. The animals are protandrous hermaphrodites and eggs are deposited in the parental tube where they are fertilized and hatch as non-pelagic larvae. There is no epitokous stage.

### Family Nephtyidae – catworms

Errant; more or less flattened. Head small, squarish, with 4 short, pointed antennae, 1 at each anterior corner. Large, eversible proboscis with papillae and 2 jaws. Parapodia with sickle-shaped gills between the 2 lobes; notopodia and neuropodia of similar size.

### *Nephtys* spp.    Catworms (Fig. 105)

*Length and segment number vary with species, often about 200 mm in length; up to 200 segments. Head sometimes with 2 small, visible eyes. White, pinkish, greyish, yellowish in colour, often with a pearly iridescence; gills red; chaetae yellowish.*

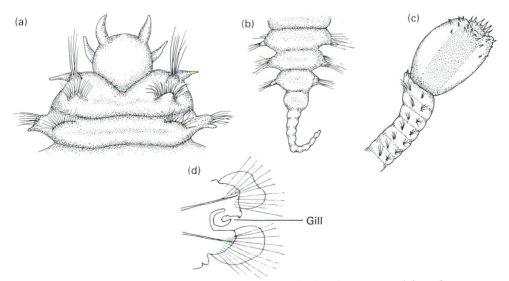

Figure 105  *Nephtys* sp., (a) anterior end, (b) posterior end, (c) proboscis everted, lateral view, (d) parapodium showing position of sickle-shaped gill.

The different species of *Nephtys* are difficult to identify, requiring detailed examination of the parapodia and chaetae. Reference to Rainer (1991) is recommended. The species are widely distributed in north-west Europe and are found intertidally and sublittorally burrowing in sand, muddy-sand, mud and gravel. They are carnivorous, feeding on molluscs, crustaceans and other polychaetes. The sexes are separate and individuals breed several times over a number of years. On the north-east coast of England, two common species, **Nephtys caeca** (Fabricius) and **Nephtys hombergii** Savigny, mature at two or three years of age and breed during April and May. The worms remain in the sediment during spawning and eggs and sperm are released onto the surface of the sediment during low tide. Fertilization occurs when the gametes are mixed by the incoming tide. The larvae are pelagic. Both species live for several years, 12 years having been recorded for *N. caeca* and six years for *N. hombergii*. The presence of growth rings on the jaws enables accurate assessments to be made of the age of individual specimens.

Family Eunicidae
Errant; cylindrical body, flattened posteriorly. Head usually with 2 eyes; 1, 3 or 5 antennae; 2 rounded palps; 1 pair of tentacular cirri or none. Parapodia not divided into similar-sized notopodia and neuropodia; dorsal cirri slender and cylindrical. First 2 segments without parapodia and chaetae. Proboscis with complex jaws. Gills usually present.

**Marphysa sanguinea** (Montagu)    Rock worm (Fig. 106)

*Body cylindrical, flattened posteriorly; up to 600 mm in length. Up to 500 segments. Five antennae; no tentacular cirri. Gills present from segments 10–40 to within 12 segments of posterior. Most gills formed of several filaments. Pink, grey, orange-brownish in colour with a marked iridescence; gills red.*

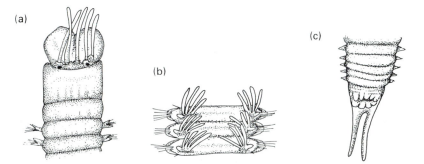

(a)

(b)

(c)

Figure 106  *Marphysa sanguinea*, (a) anterior end, (b) middle region, (c) posterior end.

*M. sanguinea* is widely distributed in north-west Europe and around Britain on the lower shore and below, living in mucus-lined burrows and galleries. It occurs in a variety of habitats such as clay, mud, gravel, rock crevices and under stones. It is probably mainly carnivorous. The sexes are separate and the eggs are deposited in masses of firm jelly. The worms are used as bait by fishermen, particularly in the Channel Islands.

### Sedentary polychaetes
Family Orbiniidae
Sedentary; burrowers in sand and mud; anterior (thoracic) region of body is flattened, posterior (abdominal) region is longer, narrower and cylindrical. Large gills present on dorsal surface of all segments except a few anterior ones. Head without appendages. Tip of posterior region with 2 long cirri.

### *Scoloplos armiger* (O. F. Müller)   (Fig. 107)

*Body up to 120 mm in length, tip of posterior region with 2 long cirri. Up to 200 or more segments. Gills present on dorsal surface from about 12th segment. Head cone-shaped, sharply pointed, with 2 eyes which may be difficult to detect. Red or reddish-orange in colour.*

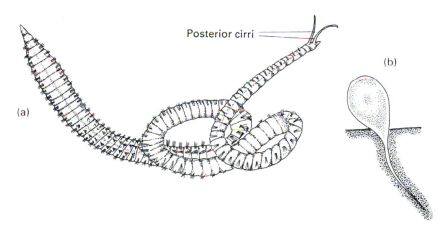

Figure 107  (a) *Scoloplos armiger*, (b) egg cocoon on surface of sand.

*S. armiger* is widely distributed in north-west Europe and around Britain on the lower shore and in the sublittoral, burrowing in muddy-sand. It is tolerant of reduced salinity and found in the lower reaches of estuaries where it feeds on organic detritus. The sexes are separate and when ripe, males are whitish in colour and the females are brownish. On the south coast of England, breeding occurs in early spring and is synchronized with spring tides. It is believed that the worms are in close association when spawning takes place and the eggs are fertilized as they leave the female. Greenish- to reddish-brown eggs are laid in pear-shaped, jelly-like cocoons, up to 15 mm in length, on the surface of the sediment. They are attached by a hollow stalk which penetrates up to 100 mm below the surface. The number of eggs per cocoon varies from about 80 to as many as 5000; they develop within the cocoon. After two to three weeks, larvae hatch at a length of about 1 mm, most of them migrating down the hollow stalk to emerge within the sediment, thereby avoiding predators. These larvae are benthic; there is no pelagic stage. Studies on *S. armiger* from eastern North Sea coasts suggest, however, that some populations have pelagic larvae which are found in the plankton in early spring and have a pelagic life of about 12 days. *Scoloplos* first breeds when two years old and lives for about four years.

### Family Spionidae

Sedentary, burrowers in sand and mud or inhabiting sand encrusted mucous tubes, or living in rock and shell. Gills usually present on dorsal surface of at least some segments. Head usually with 4 eyes; characteristically 2 long coiled palps. Tip of posterior region saucer-shaped, funnel-shaped or with several cirri.

1   Fifth segment of body enlarged, bearing stout, spine-like chaetae   *Polydora ciliata* (p. 184)
    Fifth segment not so  ................................................................................................... 2
2   Series of paired gills present in middle region of body; tip of posterior region with 4 short conical projections; worm inhabits tiny, sandy tube. About 15 mm in length. Often occurring in very high densities  ....................................... *Pygospio elegans* (p. 186)
    Series of paired gills present from 1st or 2nd segment of body to near posterior  ............... 3
3   Head pointed anteriorly and posteriorly; tip of posterior region crenulate  ....................
    ................................................................................................. *Scolelepis squamata* (p. 187)
    Head pointed posteriorly, anteriorly with 2 small frontal horns; tip of posterior region with 6–8 short cirri  ......................................................... *Malacoceros fuliginosus* (p. 187)

### *Polydora ciliata* (Johnston)   (Fig. 108a, d)

*Body up to 30 mm in length, tip of posterior region saucer-shaped. Up to 180 segments; 5th segment is enlarged and bears stout, spine-like chaetae. Gills present on dorsal surface from segment 7 to within 10 segments of posterior. Eyes often absent in adult. Yellowish-brown in colour.*

*P. ciliata* is widely distributed in north-west Europe and around Britain. It usually burrows into substrata containing calcium carbonate such as limestone, chalk and clay as well as the shells of oysters, mussels and periwinkles. It is also found in muddy sediments, wood and laminarian holdfasts. It has been suggested that burrowing is achieved by mechanical action of the chaetae, especially those of the 5th segment, but this is open to some doubt; chemical action may also be involved. The U-shaped burrow is lined by mucus and particles of sand, and the two ends extend a few millimetres above the sediment surface. As a rule *Polydora* apparently feeds on detritus, which is removed from the sediment by the two long palps. It also feeds on suspended particles in the water, and on occasions has been observed to eat dead barnacles and other dead invertebrates. The sexes are separate and breeding has been recorded in spring at a number of localities. Sperm are drawn into the burrow of the female in the respiratory current and eggs are laid in a string of capsules. A single female produces many capsules, each containing up to about 60 eggs, the individual capsules being attached by two threads to the wall of the burrow. After a week, larvae emerge and are believed to have a pelagic life of from two to six weeks before settling. Length of life is no more than one year. *Polydora* is a serious pest of oysters and mussels, but invades only the shell and does not eat the soft tissue. When infestation is heavy, the shell is weakened and this makes the mollusc more susceptible to predation by crabs. Occasionally, the inner surface of the shell is damaged by the worms and the mollusc secretes a nacreous substance to seal the perforation. The resulting blisters are an indication of attack by *Polydora*, and mussels heavily infested with the polychaete have reduced flesh content, possibly as a result of some impact on the metabolism of the mussel.

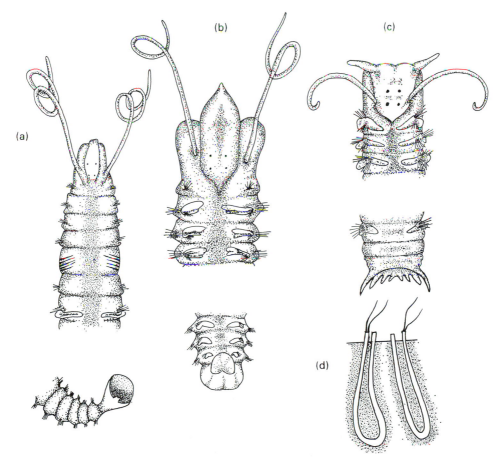

Figure 108  (a) *Polydora ciliata*. (b) *Scolelepis squamata*. (c) *Malacoceros fuliginosus*. (d) *Polydora ciliata* in U-shaped burrows.

Several other species of *Polydora* are recorded from north-west Europe. The enlarged 5th body segment is characteristic of the genus, but the species are difficult to separate. They are found in calcareous rocks, encrusting red algae and molluscan shells, and *P. ciliata* is likely to be the most widespread and common on the shore. This species, however, shows wide morphological variation and there is some evidence to suggest that it is a species complex.

**Pygospio elegans** Claparède    (Fig. 109)

*Body small, slight, up to 15 mm in length; tip of posterior region with 4 short, conical, projections, each with many papillae. Up to 60 segments. Gills present on dorsal surface, starting between segments 11 and 20; 7–9 pairs in the female, 20–28 pairs in the male; male also with 2 large gills on segment 2. Head bluntly divided anteriorly, pointed posteriorly and projecting backwards to segment 2. Eyes 2–8, irregularly placed. Yellowish-green anteriorly, brownish colour of gut visible posteriorly.*

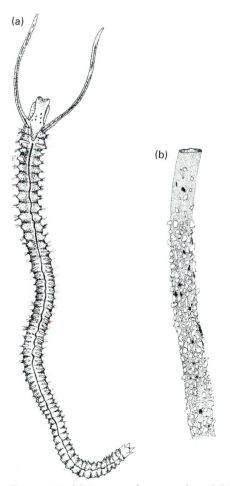

Figure 109  (a) *Pygospio elegans*, male and (b) tube.

*P. elegans* is widely distributed in north-west Europe and around Britain. It lives in thread-like chitinous tubes covered with fine particles of sand and shell fragments. The tubes are found intertidally and below in muddy sediments among rock crevices and on sandy shores and mud-flats, often in densities of 60 000 per square metre. They occur in very high densities in areas that are organically polluted and also in brackish areas, the worms being tolerant of very low salinities. *Pygospio* is a deposit feeder removing surface detritus with the paired palps. The sexes are separate and breeding has been recorded throughout most of the year. Egg capsules are deposited in rows, or egg strings, inside the tube of the female, each capsule being connected to the wall of the tube by a short stalk. Each capsule contains a number of eggs, some of which develop into pelagic larvae, the others serving as food for the developing embryos. In some populations asexual reproduction by body fragmentation has been reported. There is some evidence to suggest that *P. elegans* is a species complex.

### *Scolelepis squamata* (O. F. Müller)    (*Nerine cirratulus*) (Fig. 108b)

*Body up to 80 mm in length, tip of posterior region crenulate. Up to 200 segments. Dorsal gills present on 2nd segment with chaetae and on most segments to near posterior of body. Head pointed anteriorly and posteriorly. Bluish-green in colour, gills red.*

*S. squamata* is widely distributed in north-west Europe and around Britain. It is found on the middle and lower shore mainly in sand, or sometimes muddy-sand, in vertical burrows lined with mucus. It feeds on particles of suspended organic material using the long palps. On the south coast of England, breeding occurs from March to at least July. The sexually mature worms are not pelagic. Fertilization is external and the larva is free-swimming for some five weeks before settlement.

A second species of *Scolelepis*, *Scolelepis foliosa* (Audouin & Milne Edwards) is found on the lower shore in fairly clean sand. It is distinguished from *S. squamata* by a small median tentacle at the base of the two palps and by the absence of gills on the posterior third of the body.

### *Malacoceros fuliginosus* (Claparède)    (*Scolelepis fuliginosa*) (Fig. 108c)

*Body up to 60 mm in length; tip of posterior region with 6–8 short cirri. Up to 150 segments. Dorsal gills present from 1st segment with chaetae to near posterior end of body. Head pointed posteriorly and with 2 small frontal horns; 4 eyes. Reddish in colour.*

*M. fuliginosus* is widely distributed in north-west Europe and around Britain. It is found on the middle shore in mucus-lined tubes under stones and in sand, muddy-sand and mud. It feeds on organic matter deposited on the surface of the sediment. The sexes are separate; on the north-east coast of England, breeding takes place in spring and summer.

Fertilization is external and there is a free-swimming larva with a planktonic life of four to six weeks. Recent work has suggested that *M. fuliginosus* is a species complex.

Family Chaetopteridae
Sedentary; tube-dwelling. Tube parchment-like. Body divided into 2 or 3 regions with much-modified segments and parapodia. Head region with 2 small eyes; 2 palps; sometimes 1 or 2 pairs tentacular cirri; large terminal mouth.

### *Chaetopterus variopedatus* (Renier)    (Fig. 110)

*Body delicate, up to 250 mm in length. Up to 85 segments. Divided into 3 regions; anterior with 8–12 segments; middle with 5 segments, 1 segment with long, wing-like notopodia, 3 segments with large notopodia fused to form fan-like structures; posterior with many segments; 4th segment of body bearing conspicuous, dark-coloured chaetae. Head without tentacular cirri; palps short. Yellowish-green or whitish-yellow colour; middle region dark green. Phosphorescent. Very distinctive appearance. Tube parchment-like, U-shaped, up to 300 mm in length.*

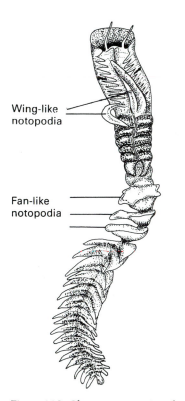

Wing-like notopodia

Fan-like notopodia

Figure 110  *Chaetopterus variopedatus.*

*C. variopedatus* is widely distributed in north-west Europe and around Britain, living in parchment-like, U-shaped tubes buried in sand. The ends of the tube are narrower than the main part and both ends are exposed above the surface of the sediment. It is normally sublittoral but is also found very low on the shore in sheltered areas. The tubes are sometimes cast up after storms. The worm remains permanently in the tube and feeds on suspended particles in the water current drawn through the tube by the action of the large fan-like notopodia of the middle region of the body. Food particles are trapped on a mucous bag suspended from the long, wing-like notopodia, and from time to time the bag is rolled up and passes along a dorsal ciliated groove to the mouth. With increase in size, the worm becomes too large for the tube but is able to tear it open with the modified chaetae on the fourth segment and add to it, thus increasing the size. The sexes are separate and fertilization is external. The larva is pelagic.

Family Cirratulidae

Sedentary; cylindrical body. Long, thread-like, contractile gills present on at least some segments along sides of body. Head sometimes with eyes, but without appendages. Two to many long feeding tentacles in 2 groups attached dorsally behind head.

Large number of tentacles in 2 groups arising behind head. Head without conspicuous eyes ............................................................................... *Cirriformia tentaculata* (p. 189)
Two groups of up to 8 tentacles arising behind head. Head with 4–8 eyes in row on each side ................................................................................ *Cirratulus cirratus* (p. 190)

*Cirriformia tentaculata* (Montagu)   (*Audouinia tentaculata*) (Fig. 111a)

*Body up to 200 mm in length. Up to 300 or more segments. Dorsal surface rounded, ventral surface flattened. Paired, long, thread-like gills present on 1st segment with chaetae and along body to near posterior. Head pointed, without eyes in adult; pigment spots present. Large number of feeding tentacles in 2 groups, arising posteriorly to the first few pairs of gills. Orange, brown or reddish in colour; gills and tentacles red.*

*C. tentaculata* has been recorded in north-west Europe from as far north as Denmark and is widely distributed around Britain, but commonest in the south. It is found on the middle and lower shore in a variety of habitats including mud, muddy-sand and gravel, and under stones where it feeds on organic matter picked up from the surface of the sediment by the tentacles, which are often seen extending over the surface of the sediment while the body of the worm remains buried. The sexes are separate and can be distinguished during the breeding season by the yellow colour of the males and the dull green colour of the females. On the south coast of England, breeding occurs during spring and summer. The gametes are discharged into the sea where external fertilization occurs. Pelagic larvae have been recorded and these settle after about 10 days. Length of life is about 15 months.

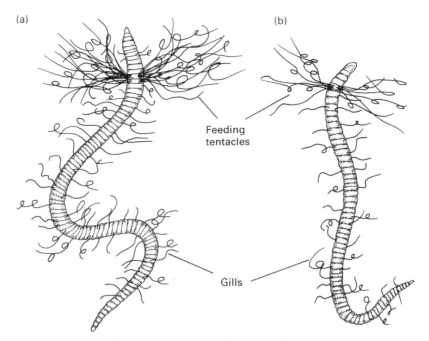

Figure 111 (a) *Cirriformia tentaculata.* (b) *Cirratulus cirratus.*

### *Cirratulus cirratus* (O. F. Müller)  (Fig. 111b)

*Body up to 100 mm or more in length. Up to 150 segments; cylindrical. Head rounded. Paired long, thread-like gills present on 1st segment with chaetae and along body to near posterior end. Head with 4–8 eyes in row on each side. Two groups of up to 8 feeding tentacles on 1st segment with chaetae. Orange, red, brown in colour; gills and tentacles red, yellow.*

C. *cirratus* is widely distributed in north-west Europe and around Britain. It occurs on the lower shore in mud and muddy-sand, under stones and in rock crevices, where it feeds on organic material on the surface of the sediment. The worms generally live in aggregations, sometimes as many as 200 individuals together. The sexes are separate and during the breeding season can be recognized by the bright yellow colour of the females and the white colour of the males. On the north-east coast of England, breeding occurs throughout the year, each female usually spawning once a year for up to four or five years. Eggs and sperm are deposited in mucous egg masses found attached to the sediment and in rock crevices. They are yellowish in colour but become covered with mud particles and are therefore difficult to detect. After about a week or so benthic larvae emerge.

A second species of *Cirratulus*, **Cirratulus filiformis** Keferstein, is found on the lower shore, usually among encrusting red algae and in crevices. It can be distinguished from

*C. cirratus* by the pointed head and lack of eyes, and in having only 1 or 2 feeding tentacles in each of the two groups.

Family Capitellidae
Sedentary; anterior (thoracic) region short and thick; posterior (abdominal) region long and narrow. Parapodia small. Gills present or absent. Chaetae inconspicuous. Head conical, eyes present or absent; without appendages. Eversible proboscis. Resembles earthworm in body shape.

### *Capitella* spp.   (Fig. 112)

*Body up to 100 mm or more in length, usually very much smaller. Up to about 100 segments. Narrowing gradually towards posterior. Gills absent. Blood-red in colour.*

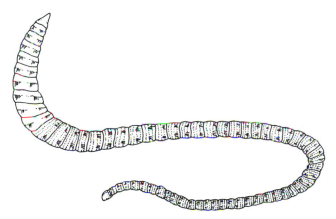

Figure 112  *Capitella* sp.

The different species of the genus *Capitella* are very similar in appearance and the reader is referred to Grassle & Grassle (1976) for their separation. The life-cycles and mode of reproduction vary from species to species; some are dioecious, others hermaphroditic and the larvae are pelagic or benthic. The most commonly described species is **Capitella capitata** (Fabricius), but it is now known that this 'species' does in fact encompass several closely related species often living in close proximity. 'C. capitata' lives in muddy-sand, muddy-grit and fine sand on the middle and lower shore. It feeds on microorganisms and detritus and sometimes occurs in enormous numbers in areas of organic enrichment such as discharges from pulp mills and sewage farms.

Family Arenicolidae
Sedentary; body cylindrical, anterior (thoracic) region thicker than posterior (abdominal) region. Almost all segments with five annulations. Branching dorsal gills always present

on at least some segments. Posterior region with or without chaetae. Head without eyes and appendages. Eversible proboscis.

Branching dorsal gills present on segments 7 to 19. Posterior region ('tail') narrow and without gills and chaetae ...................................................................................................
.......................................... *Arenicola marina* (p. 192), see also *Arenicola defodiens* (p. 194)
Branching dorsal gills on all segments posterior to the 15th. No distinct 'tail' region ....
.................................................................................... *Arenicolides ecaudata* (p. 194)

*Arenicola marina* (Linnaeus)    Blow lug, Lugworm (Fig. 113a, b, d)

*Body cylindrical, up to 200 mm in length. Anterior (thoracic) region of 19 segments, the posterior 13 of which have branching gills; narrower posterior (abdominal) region ('tail') of many segments lacking chaetae and gills. Reddish, greenish-yellow or black in colour.*

A. *marina* is widely distributed in north-west Europe and occurs all around Britain on the middle and lower shore in sand and muddy-sand. It lives in J-shaped burrows, generally about 200 mm below the surface, and is tolerant of salinities down to about 12‰. In sheltered, estuarine sediments, densities as high as 100 to 150 per square metre have been recorded. *Arenicola* feeds on organic material such as microorganisms and detritus present in the sediment. It ingests the sediment while in its burrow leaving a shallow depression on the surface. The depressions, together with the characteristic worm casts, indicate the presence of *Arenicola*. The sexes are separate, and although breeding has been recorded in spring, late summer and autumn, it occurs over a well-defined, two- to three-week period in October and November at many localities around Britain. Sperm are discharged from the burrows of the males onto the surface of the sediment, forming sperm puddles which are dispersed by the incoming tide. The female deposits eggs in the burrow where they are fertilized by sperm drawn in with the respiratory current. The larvae undergo their early development within the burrow and then move to the surface of the sediment, to be transported by tidal currents to firmer areas such as shingle and pebble, where they settle. Over the next few months they undergo development in mucous tubes attached to the substratum. At the end of this period the young worms, still in their mucous tubes, are found in the surface waters and are carried by the tide and eventually burrow into the sandy and muddy sediments where *Arenicola* is typically found. The worms become sexually mature after two or three years and spawn once a year, possibly for up to five or six years. *Arenicola* is preyed on by birds and flatfishes and is particularly prone to predation when it moves backwards in its burrow and comes close to the surface to deposit casts. Often only the 'tail' is cropped and in many cases the worm survives.

It has long been recognized that the lugworm exists in two varieties, the blow lug and the black lug. These are now recognized as two separate species, A. *marina* (above) and

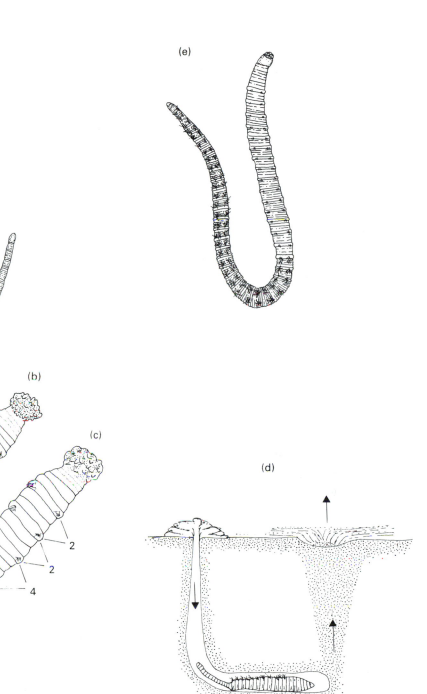

Figure 113  (a) *Arenicola marina*. (b) Anterior of *A. marina*, and (c) anterior of *Arenicola defodiens* showing the number of annuli between consecutive pairs of chaetae. (d) Diagrammatic representation of *Arenicola* in burrow, arrows indicate direction of flow of water. (e) *Arenicolides ecaudata*. ((b) and (c) After Cadman & Nelson-Smith, 1993.)

*Arenicola defodiens* (the black lug) described by Cadman & Nelson-Smith (1993). The two species are similar in general appearance but *A. defodiens* is characterized by being generally darker (usually black) in colour and reaching a greater length (270 mm). The number of annuli between the first four consecutive pairs of chaetae differs in the two species (Fig. 113b, c). *A. defodiens* occurs deeper in the sediment and does not leave conspicuous feeding depressions. The worm cast is more neatly coiled than that of *A. marina*. It also seems to favour more exposed shores and does not penetrate into estuaries. Where the two species co-exist *A. defodiens* occurs lower on the shore. So far it has been recorded mainly from South Wales.

### *Arenicolides ecaudata* Johnston    (Fig. 113e)

*Similar to* Arenicola marina *but lacks a distinct 'tail' region; external gills present on all segments posterior to the 15th. Up to 250 mm in length.*

*A. ecaudata* is widely distributed in north-west Europe but less commonly found than *A. marina* (above). It occurs on the lower shore burrowing in sand, mud, under stones and in crevices where sediment has been deposited.

### Family Opheliidae

Sedentary; short, cylindrical body with small number of segments; usually with deep ventral groove along part or all of length. Gills present on dorsal surface of some segments. Head pointed, eyes not visible; without appendages.

### *Ophelia bicornis* Savigny    (Fig. 114a)

*Body up to 45 mm in length. Up to about 32 segments. Thick anteriorly, narrowing towards posterior; ventral groove running from 10th segment with chaetae to posterior end of body; 15 pairs of gills on narrower part of body. Pinkish-purple in colour.*

*O. bicornis* is a southern species, extending as far north as the English Channel. It lives in fine, well-drained sand on the middle and lower shore but its distribution is often localized. The sexes are separate; during the breeding season the males are whitish in colour and the females dark green. The worms remain within the sediment and discharge the gametes onto the surface. The larvae are planktonic for about a week.

*O. bicornis* is perhaps the best known member of the genus. However, other species are found in sandy deposits in more northerly localities around Britain.

### Family Oweniidae

Sedentary; tube-dwelling; tube covered with sand grains and shell fragments. Cylindrical body with few segments, those of anterior region longer than posterior ones. Head bears a collar or frilled membrane; eyes present or absent.

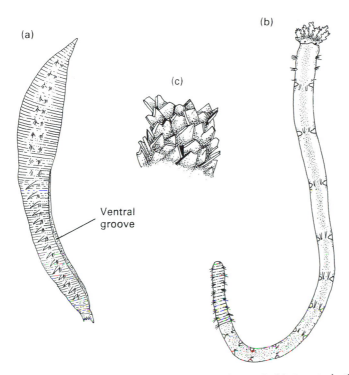

Ventral
groove

Figure 114 (a) *Ophelia bicornis* (after Fauvel, 1927). (b) *Owenia fusiformis*. (c) *Owenia fusiformis*, anterior end of tube.

### *Owenia fusiformis* delle Chiaje (Fig. 114b, c)

*Body cylindrical, up to 100 mm in length. Up to 30 segments. Anterior 3 segments short, the next 5–7 much longer; posterior segments becoming progressively shorter. Head bears much-branched membrane, with 2 eye spots at base. Greenish-yellow in colour with clear, transverse bands; head membrane reddish. Lives in tough, flexible tube covered with flat, usually whitish grains of shell debris arranged in overlapping fashion.*

*O. fusiformis* is cosmopolitan and lives buried in sand on the lower shore. The worm is able to move the tube and during feeding the anterior end is exposed above the surface of the sediment. *Owenia* feeds on suspended matter using the cilia on the head membrane, but is also able to bend over until the membrane touches the surface of the sediment, enabling it to pick up sand grains and detritus. The sexes are separate and the larvae have a planktonic life of about four weeks. On the south coast of England, breeding occurs during June and July. Length of life is four years, the worms breeding each year.

Family Pectinariidae
Sedentary; tube-dwelling; tube smooth, rigid made of sand grains and shell fragments, open at both ends. Body short, divided into 3 regions. Anterior with many stout, golden

chaetae (paleae) and 2 pairs of gills; middle with conspicuous chaetae; posterior very short, flattened. Head with large number of feeding tentacles; 2 pairs tentacular cirri.

***Lagis koreni*** Malmgren    (*Pectinaria koreni*) (Fig. 115a, b)

*Body short and wide, up to 50 mm in length. Middle region of 15 segments with chaetae. Body pinkish in colour; gills red. Tube rigid, slightly curved; made of fine sand grains and shell fragments; open at both ends, posterior end narrower than anterior.*

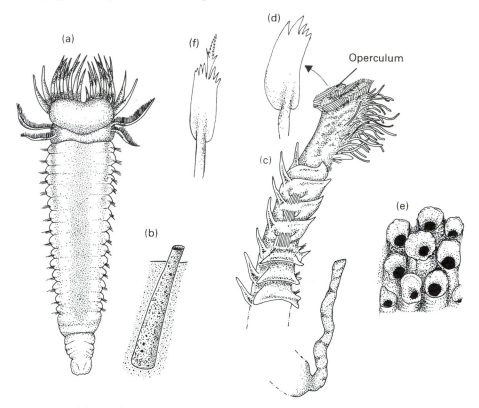

Figure 115  (a) *Lagis koreni*. (b) *Lagis koreni*, tube in sediment. (c) *Sabellaria alveolata* with (d) stout chaetae of outer ring. (e) *Sabellaria alveolata*, tubes forming part of reef. (f) Stout chaetae of outer ring of *Sabellaria spinulosa*. ((d) and (f) After Fauvel, 1927.)

*L. koreni* is widely distributed in north-west Europe and around Britain in sandy deposits from the lower shore into the sublittoral. Large numbers of empty tubes are often washed ashore after storms. The tube lies diagonally or nearly upright in the sediment with the wide end below the surface and the narrow end projecting just above it. The paleae are used for burrowing and loosening the sediment and the tentacles for sorting the sediment particles, which the animal ingests. The sexes are separate and breeding occurs during spring and summer. The larvae have a planktonic life of about a month. Length of life is thought to be about one year; the animals breed once then die.

Family Sabellariidae

Sedentary; tube-dwelling, sometimes forming large colonies; tube made of sand grains. Body short, divided into 3 regions. Anterior with rings of stout chaetae (paleae) forming an operculum which closes the tube; middle with conspicuous parapodia and gills; posterior narrow, smooth and bent under body. Head with 2 palps and large number of feeding tentacles.

Stout chaetae of the outer ring with smooth, terminal teeth (Fig. 115d). Tubes usually forming large colonies .......................................................... *Sabellaria alveolata* (p. 197)
Stout chaetae of the outer ring with terminal teeth, the median one long and barbed (Fig. 115f). Tubes usually separate ...................................... *Sabellaria spinulosa* (p. 197)

## *Sabellaria alveolata* (Linnaeus)   (Fig.115c, d, e)

*Body up to 40 mm in length. Up to 37 segments. Paleae in 3 rings, those of outer ring with smooth, unbarbed terminal teeth. Reddish-brown in colour. Tube made of sand grains and fragments of shell, inner lining membranous; may occur singly but more usually forming large colonies or reefs.*

S. *alveolata* reaches its northern limit in the North Sea and in Britain is commonest on south and west coasts. It is found on the lower shore and below on rocks close to sand and often forms large reefs firm enough to walk on. It is a suspension feeder. The sexes are separate and during the breeding season males can be distinguished by their creamy colour, while the females are purplish in colour. On the south coast of England breeding occurs in July and the larvae are pelagic, some for about six weeks, while others have a long pelagic life of at least nine months. At settlement the larvae are able to detect the tubes of adult *Sabellaria* and old colonies, often settling on them in large numbers. Length of life is usually three to four years, but some colonies have been estimated to be nine years old.

## *Sabellaria spinulosa* Leuckart   (Fig. 115f)

*Body up to 30 mm in length. Up to 40 segments. Paleae in 3 rings, those of outer ring with terminal teeth, the median one long and barbed. Reddish-brown in colour. Tubes made of sand grains; usually occur singly and do not form the large intertidal colonies so characteristic of* S. alveolata *(above).*

S. *spinulosa* is widely distributed in north-west Europe, where it is found on the lower shore and into the sublittoral, the tubes attached to rocks and stones, and in crevices between rocks. On the north-east coast of England breeding has been recorded from August to November.

Family Terebellidae

Sedentary; usually tube-dwelling; tube sometimes covered with shell fragments, mud and sand. Body long, divided into 2 regions; a thick anterior (thoracic) region, usually with 3 pairs of branching gills arising dorsally from separate segments; a narrow, posterior (abdominal) region with small neuropodia, without notopodia. Head bears large number of long, narrow sinuous feeding tentacles.

1   Twenty-four anterior segments with conspicuous chaetae. Thin mucous tube ..............
    .................................................................... *Neoamphitrite figulus* (p. 198)
    Seventeen anterior segments with conspicuous chaetae. Tube of sand, shell fragments   ...   2
2   Large, flat, lateral lobes on side of head. Worm up to 300 mm in length. Tube with anterior end flattened and extended into finger-like projections covered with sand, shell fragments ............................................... *Lanice conchilega* (p. 200)
    Without large, flat lateral lobes on side of head. Worm delicate, up to 150 mm in length. Body with many white spots. Tube delicate, without anterior, finger-like projections ....................................................... *Eupolymnia nebulosa* (p. 199)

**Neoamphitrite figulus** (Dalyell)    (*Amphitrite johnstoni*) (Fig. 116)

*Body up to 250 mm in length. Up to 100 segments. Thoracic region of 24 segments with conspicuous chaetae. Body and tentacles brown, yellow, pink; gills red.*

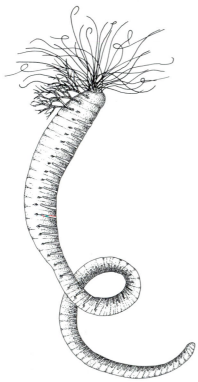

Figure 116 *Neoamphitrite figulus.*

N. *figulus* is widely distributed in north-west Europe and around Britain in mud, muddy-sand and gravel, and in laminarian holdfasts. It tolerates low salinity and is found in estuaries. It lives in a thin, mucous tube in a vertical, U-shaped burrow and feeds by picking up organic detritus and diatoms from the surface of the sediment, using the long tentacles.

**Eupolymnia nebulosa** (Montagu)    (*Polymnia nebulosa*) (Fig. 117c, d)

*Body delicate, up to 150 mm in length. Up to 100 segments. Thoracic region of 17 segments with conspicuous chaetae. Head with many eyes. Body orange, brown, with white spots, tentacles paler; gills red. Tube delicate and slimy, covered with sand and fragments of shell; without anterior projections.*

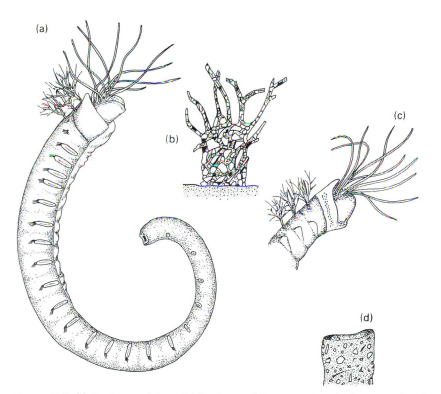

Figure 117  (a) *Lanice conchilega*. (b) *Lanice conchilega*, anterior of tube protruding from sediment. (c) *Eupolymnia nebulosa*. (d) *Eupolymnia nebulosa*, anterior end of tube. Tentacles reduced in number and length in (a) and (c).

*E. nebulosa* is widely distributed in north-west Europe and around Britain, where it is found on the lower shore and below. The tube is attached under stones and shells, and the worm feeds at night on detritus and microorganisms. The sexes are separate; on the Channel coast of France breeding occurs during May, June and July. Gametes are discharged into the sea and the larva has a short pelagic life. On the Mediterranean coast, however, the pattern of reproduction is quite different and the eggs are deposited in mucous cocoons at the entrance to the female's tube where development takes place.

Another species of *Eupolymnia*, **Eupolymnia nesidensis** (delle Chiaje), also occurs on the lower shore and below, often in laminarian holdfasts. It is smaller (up to 60 mm in length) than *E. nebulosa* and the reddish-brown body is without white spots. The tube is thin and delicate and covered with sand.

### *Lanice conchilega* (Pallas)    Sand mason (Fig. 117a, b)

*Body delicate, up to 300 mm in length. Up to 300 segments. Thoracic region of 17 segments with conspicuous chaetae; large, flat lateral lobes on segment 3. Eyes sometimes visible. Body light pink or brown in colour, gills red. Tube made of mud, sand and shell fragments; anterior end flattened and extended into finger-like projections covered with sand and shell fragments; often extends well above surface of sediment.*

*L. conchilega* is widely distributed in north-west Europe and around Britain, where it is found in sand and sandy grit on the lower shore and below, often in very large numbers. The anterior end of the tube projects above the surface of the sediment, making it very conspicuous. *Lanice* feeds on organic detritus picked up from the surface of the sediment by the long tentacles on the head. The anterior projections of the tube are set at right angles to the current flow and cause suspended organic material to be deposited behind them, where the water is relatively still. It also feeds on suspended particles trapped on mucus on the tentacles. The sexes are separate and breeding occurs during spring and summer. Gametes are discharged into the sea, where fertilizaton occurs. The larva has a long pelagic life of about two months.

### Family Sabellidae

Sedentary; tube-dwelling; tube membranous or of mucus often covered with mud or sand particles. Body divided into 2 regions; an anterior (thoracic) region of relatively few segments, with parapodia with conspicuous chaetae, and a posterior (abdominal) region, usually of many segments. Head with 2 palps and a crown of pinnate projections called radioles, which sometimes bear eyes and which project from tube when worm is feeding. A large family of polychaetes found in muddy and sandy sediments, in rock crevices, on stones and shells, in mussel beds and in laminarian holdfasts. Four common species are described here.

1  Radioles united almost to their tips by a membrane. Tube thick, mucilaginous ............. ..................................................................... ***Myxicola infundibulum*** (p. 202)
   Radioles and tube not as above ............................................................................... 2
2  Worm tiny, up to 3 mm in length with few segments. Head with 2 lobes, each bearing 3 radioles ................................................................. ***Fabricia stellaris*** (p. 202)

Worm long, may be very long (up to 250 mm in length) and many segmented. Head
with 2 lobes, each bearing more (often many more) than 3 radioles ................................. 3
3 Head with 2 lobes, each bearing 8–45 radioles, banded purple, brown, red. Tube
covered with mud particles, extends well above sediment surface .........................................
.............................................................................................. *Sabella pavonina* (p. 201)
Head with 2 lobes, each bearing 5–15 radioles, some with up to 8 eye spots. Tube
covered with mud or sand particles; free end rolls over when worm withdraws .............
............................................................................. *Pseudopotamilla reniformis* (p. 202)

**Sabella pavonina** Savigny   (*Sabella penicillus*) (Fig. 118a, b)

*Body up to 250 mm in length. Up to 600 or more segments. Thoracic region of 6–12 seg-
ments. Head with 2 lobes, each bearing 8–45 long radioles. Greyish-purple or yellowish-
orange in colour; radioles banded or patterned with purple, brown or red. Tube tough, membranous, covered with fine particles of mud at anterior end; posterior end more delicate;
may be 500 mm or more in length.*

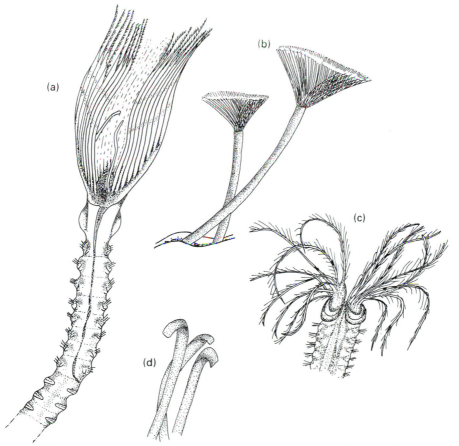

Figure 118  (a) *Sabella pavonina*. (b) *Sabella pavonina* in tubes. (c) *Pseudopotamilla
reniformis* (after McIntosh, 1922). (d) *Pseudopotamilla reniformis*, anterior end
of tubes.

*S. pavonina* is widely distributed in north-west Europe and around Britain. The tubes extend up to 100 mm above the surface of muddy and sandy sediments on the lower shore and below. They are often found in large numbers, sometimes with seaweeds, sponges and ascidians attached to them. The worms feed on suspended particles collected by cilia on the radioles. Particles are sorted on the basis of size; large particles are rejected, small ones are used for food, and medium-sized ones mixed with mucus and added to the anterior end of the tube to increase its length. The sexes are separate and breeding has been recorded in spring and summer. The larvae are pelagic.

### *Pseudopotamilla reniformis* (Bruguière)   (*Potamilla reniformis*) (Fig. 118c, d)

*Body delicate; up to 100 mm in length. Up to 300 segments. Thoracic region of 7–14 segments. Head with 2 lobes, each bearing 5–15 radioles, some with up to 8 dark eye spots; dorsally 2 pointed lobes at base of radioles. Body orange or red, often brownish anteriorly; radioles white or pink, often banded with purple or brown. Tube tough, often covered with mud or fine sand; free end flattened, rolls over when worm withdraws.*

*P. reniformis* is widely distributed in north-west Europe and around Britain, where it occurs on the lower shore and below in rock crevices and on stones and shells. It is a suspension feeder.

### *Fabricia stellaris* (Müller)   (*Fabrica sabella*) (Fig. 119a)

*Body slender, up to 3 mm in length. Up to 12 segments. Thoracic region of 8 segments. Head with 2 lobes, each bearing 3 radioles; 2 (sometimes 4) eye spots; 2 eye spots on last segment of body. Yellowish or pinkish in colour, transparent. Tube of mucus, covered with mud.*

*F. stellaris* probably has a worldwide distribution. It is found among empty barnacles, beds of mussels, the red alga *Audouinella* (p. 56), laminarian holdfasts and other situations where silt accumulates. Where conditions are suitable, densities can exceed one million per square metre. It is a suspension feeder. The sexes are separate and sperm are taken into the tubes of females in the feeding current and stored in a seminal receptacle. Breeding occurs in spring. Eggs are laid in groups within the tube of the female, each group containing one to seven eggs enclosed in a mucous cocoon. Development takes place within the parental tube and there is no pelagic larva.

### *Myxicola infundibulum* (Renier)   (Fig. 119b)

*Body up to 200 mm in length. Up to 150 segments, each with 2 annulations; 7–8 segments in thoracic region. Head with 2 lobes, each bearing 20–40 radioles united almost to their tips by a membrane; triangular projection between lobes of head on ventral surface. Body yellow or orange in colour; radioles with deep purple tips, inner surface purple or brown; palps purple. Tube thick, mucilaginous.*

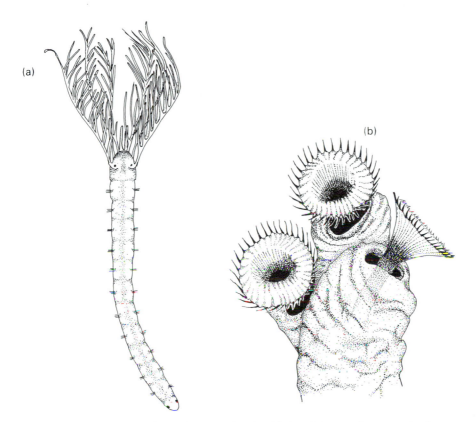

Figure 119  (a) *Fabricia stellaris.* (b) *Myxicola infundibulum* (after Stephenson, 1944).

*M. infundibulum* is widely distributed in north-west Europe and around Britain. The tubes are almost completely buried in sand and mud from the very low shore into the sublittoral. It is a suspension feeder. The sexes are separate and gametes are discharged into the sea, where fertilization occurs. The larvae have a pelagic life of one to two weeks.

Family Serpulidae
Sedentary; tube-dwelling; tube calcareous, encrusting or upright. Body divided into 2 regions; a short, anterior (thoracic) region with few segments enveloped in a membranous collar, and a posterior (abdominal) region of many segments. Head bears crown of pinnate projections called radioles which sometimes bear eyes, and which project from the tube when the worm is feeding. Palps usually absent; if present poorly developed. Usually 1 (rarely 0 or 2) of the radioles modified as an operculum which closes the tube when the worm withdraws.

1  Tube very small, thin; large numbers often entwined together. Head with 2 small
   cup-shaped opercula ................................................................ *Filograna implexa* (p. 206)
   Tube encrusting or upright. Head without operculum or with 1 operculum .................... 2
2  Head without operculum. Radioles banded red, orange and with many red eye spots.
   Tube smooth, upright, cylindrical, attached at narrow end ........ *Protula tubularia* (p. 208)
   Head with 1 operculum. Radioles banded, blue, red, white. Tube encrusting or upright .. 3
3  Stalk of operculum with winged expansions near tip (Fig. 120c,e). Tube encrusting,
   irregularly curved .................................................................................................... 4
   Stalk of operculum without winged expansions near tip. Tube encrusting or upright ....... 5
4  Tube with single, longitudinal ridge on upper surface. Operculum cone-shaped.
   Usually sublittoral ............................................................ *Pomatoceros triqueter* (p. 204)
   Tube with 3 longitudinal ridges on upper surface. Operculum cup-shaped. Lower shore
   and below ................................................................ *Pomatoceros lamarckii* (p. 206)
5  Operculum funnel-shaped, edged with rounded teeth (Fig. 120a). Tube sinuous with 3
   longitudinal ridges, or upright in older specimens and transversely ridged at intervals ..
   ................................................................................ *Serpula vermicularis* (p. 204)
   Operculum funnel-shaped with centrally 10–20 toothed spines (Fig. 121a). Tube
   encrusting or upright, often several entwined ...................... *Hydroides norvegica* (p. 206)

**Serpula vermicularis** Linnaeus    (Fig. 120a, b)

*Body up to 70 mm in length. Up to 250 segments; 7 segments in thoracic region. Head with
2 lobes, each with 30–40 radioles united at base; single operculum, funnel-shaped and edged
with rounded teeth. Varied in colour, body yellow to red; radioles red, pale pink or banded
red and pink; operculum with red and white rays. Tube calcareous, pinkish-white; young
animals with sinuous tubes usually with 3 longitudinal ridges; older animals with upright
tubes, smooth and transversely ridged at intervals.*

S. *vermicularis* is cosmopolitan in distribution. It is found at extreme low water and sublittorally to depths of 250 m, encrusting rocks, stones and especially bivalve shells. In
some areas large reefs develop from the aggregation of tubes. The worms are suspension
feeders. Breeding has been recorded in August and September. There is a pelagic larva.

**Pomatoceros triqueter** (Linnaeus)    (Fig. 120e, f)

*Body up to 25 mm in length. Up to 100 segments; 7 segments in thoracic region. Head with
2 lobes, each bearing 18–20 radioles united at base; single operculum, cone-shaped and
with winged expansions near tip of opercular stalk. Body varied in colour, yellow, red,
brown, green; radioles banded with various colours, often blue and white, or red and white.
Tube encrusting, calcareous, whitish; irregularly curved, with longitudinal ridge on upper
surface often projecting as point over anterior end.*

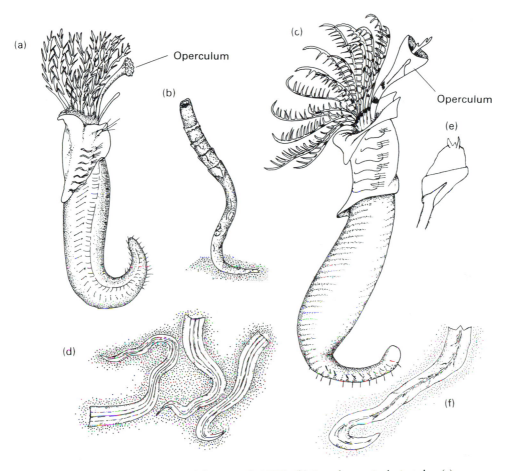

Figure 120  (a) *Serpula vermicularis* (after Fauvel, 1927). (b) *Serpula vermicularis*, tube. (c)
*Pomatoceros lamarckii*. (d) *Pomatoceros lamarckii*, tubes. (e) *Pomatoceros
triqueter*, operculum. (f) *Pomatoceros triqueter*, tube.

*P. triqueter* is widely distributed in north-west Europe and around Britain, encrusting
stones, rocks and shells, and the carapace of some species of decapods. It is predominantly sublittoral occurring to depths of about 70 m. It is a suspension feeder. The
worms are protandrous hermaphrodites and breeding probably takes place throughout
the year with a peak in spring and summer. The larvae have a pelagic life of two to three
weeks in summer but about two months in winter. The newly settled worms first secrete
a delicate semi-transparent tube and calcareous material is later added at the anterior
end. Growth is rapid and sexual maturity can be reached in about four months. Length
of life is about one and a half years and most worms die after breeding.

*Pomatoceros lamarckii* (Quatrefages)    (Fig. 120c, d)

*As* P. triqueter *but with cup-shaped operculum, and tube with 3 longitudinal ridges on the upper surface.*

P. *lamarkii* is widely distributed in north-west Europe and in some places is abundant. It is found from the lower shore to depths of about 30 m, encrusting stones, rock, shell and the carapace of some species of decapods.

There is confusion in the literature concerning the identification of the two species of *Pomatoceros* and it is now believed that some records referring to P. *triqueter* may in fact have been to P. *lamarckii*, the species more likely to be encountered intertidally.

*Hydroides norvegica* Gunnerus    (Fig. 121a, b)

*Body up to 30 mm in length. Up to 100 segments; 7 segments in thoracic region. Head with 2 lobes, each bearing 13–20 radioles united at base; single operculum, funnel-shaped, with centrally about 10–20 toothed spines. Body red in colour; radioles banded red and white. Tube calcareous, white; lying flat or upright; irregularly twisted; often many entwined together.*

H. *norvegica* has a worldwide distribution and although it does occur on the lower shore, it is mainly sublittoral on stones, shells, bryozoans, as well as causing problems as a fouling organism on ships' hulls. It is a suspension feeder. *Hydroides* is a protandrous hermaphrodite; on the south coast of England breeding has been recorded in August. There is a planktonic larva with, according to laboratory experiments, a pelagic life of eight to 21 days. The newly settled worm secretes a transparent tube and after a few days calcareous material is added at the anterior end.

*Filograna implexa* Berkeley    (Fig. 121c, d)

*Body up to 5 mm in length. Up to 35 segments; 6–9 segments in thoracic region. Anterior and posterior regions of body separated by non-segmented region without chaetae. Head with 2 lobes, each with 4 radioles; 2 opercula, very small, cup-shaped, transparent. Body greyish and transparent; radioles colourless. Tube very small, thin, calcareous and white; aperture sometimes slightly broader than tube; large numbers often entwined together and have superficial similarity to coral.*

F. *implexa* is widely distributed in north-west Europe and around Britain on the lower shore and sublittoral on pebbles, stones and shells. The sexes are separate but asexual reproduction by budding commonly occurs and the buds are most frequently seen during spring and summer.

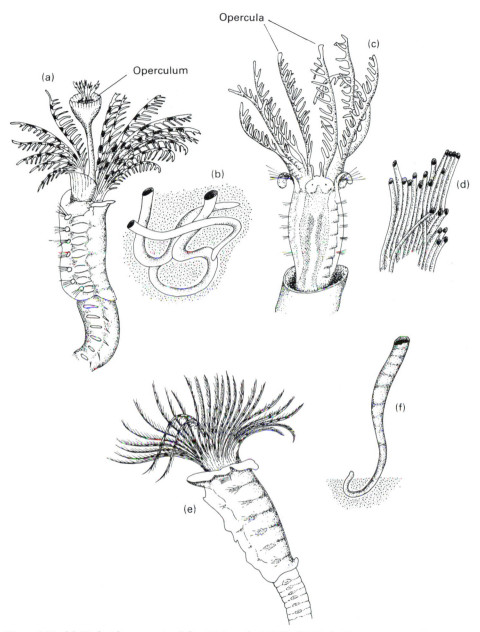

Figure 121  (a) *Hydroides norvegica* (after McIntosh, 1923). (b) *Hydroides norvegica*, tubes.
(c) *Filograna implexa* (after McIntosh, 1923). (d) *Filograna implexa*, tubes. (e)
*Protula tubularia* (after McIntosh, 1923). (f) *Protula tubularia*, tube.

*Protula tubularia* (Montagu)    (Fig. 121e, f)

*Body up to 50 mm in length. Up to 125 segments; 7 segments in thoracic region. Head with 2 lobes, each bearing 25–45 radioles arranged in spiral whorls; without operculum. Body red or orange in colour with greenish anterior region; radioles banded red and orange with many red eye spots. Tube smooth, white and cylindrical; attached at narrow end.*

P. *tubularia* is widely distributed in north-west Europe and around Britain. It occurs mainly sublittorally, attached to rocks, stones and shells; the tubes are often washed ashore.

Family Spirorbidae
Sedentary; tube-dwelling, encrusting seaweed, rocks, stones. Tube calcareous, small, spirally coiled. Body divided into 2 regions, a short, anterior (thoracic) region usually of 3 segments enveloped in a membranous collar, and a posterior, many-segmented (abdominal) region. These 2 regions separated by a region lacking chaetae. Head bears crown of pinnate projections called radioles, one of which is modified as an operculum and which closes the tube. Radioles project from tube when worm is feeding. The family Spirorbidae includes a number of genera but the two most likely to be found intertidally are *Spirorbis* and *Janua*.

Tube coiled sinistrally, i.e. tube opens clockwise; smooth or ridged ............................ .................................................................................. ***Spirorbis*** spp. (p. 208)
Tube coiled dextrally, i.e. tube opens anticlockwise; often with 3 longitudinal ridges .... .................................................................................. ***Janua pagenstecheri*** (p. 210)

*Spirorbis* spp.    (Fig. 122a, b)

*Small worms, no more than a few millimetres in length; number of segments varies with species, up to about 30. Usually 3 segments in thoracic region. Radioles few in number. Operculum flattish. Body often orange-red in colour. Embryos develop inside parental tube in egg string attached by thread to wall of tube. Tube opaque, calcareous, smooth or ridged; often white; usually coiled sinistrally, i.e. worm lies with ventral surface upwards and tube opens clockwise.*

There are several species of *Spirorbis* in north-west Europe, each of which is typically associated with a particular substratum, for example, ***Spirorbis spirorbis*** (Linnaeus) with *Fucus* spp., ***Spirorbis corallinae*** de Silva & Knight-Jones with *Corallina officinalis* (p. 61), ***Spirorbis tridentatus*** Levinsen with stones and rock crevices, ***Spirorbis inornatus*** L'Hardy & Quievreux with red algae, laminarian holdfasts and the undersurface of 'buttons' of *Himanthalia* (p. 51) and ***Spirorbis rupestris*** Gee & Knight-Jones with encrusting red algae. For more detail, reference to Knight-Jones & Knight-Jones (1977) is recommended.

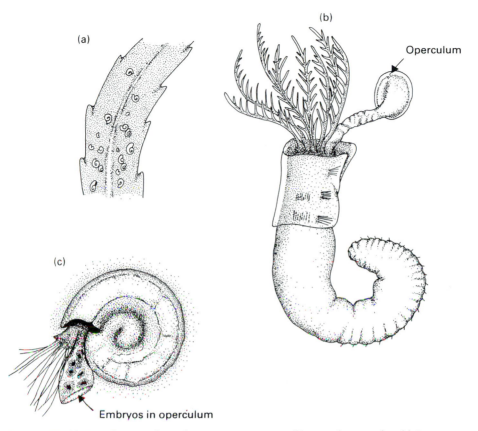

Figure 122  (a) *Spirorbis spirorbis*, tubes on *Fucus serratus*. (b) *Spirorbis spirorbis*. (c) *Janua pagenstecheri*.

One of the most widespread and common spirorbids in north-west Europe and Britain is *Spirorbis spirorbis*, formerly known as *Spirorbis borealis*. This species occurs on *Fucus*, particularly *F. serratus* (p. 49), on sheltered shores. The worms live permanently in characteristic smooth, spiral tubes, the shape being an adaptation to withstand flexure of the seaweed frond. *Spirorbis* feeds on suspended particles trapped on the radioles. It is hermaphroditic with fertilization taking place inside the tube and, although self-fertilization may occur, cross-fertilization is more usual. Sperm are released into the water and are believed to be taken into the tubes of neighbouring individuals in the feeding current and then stored in a spermatheca. Breeding occurs during summer and on the coast of south Wales has been recorded from April to November. At some localities the worms show a lunar periodicity, with the larvae being released fortnightly at neap tides; up to 12 broods a year may be produced. The eggs develop in an egg string, each string containing about 20 to 35 eggs attached to the wall of the tube. The larvae have a pelagic life of only a few hours and show a marked preference for settling on seaweeds such as *Fucus serratus* on which adult spirorbids are already present. The first tube secreted is thin and made of mucus, and it has been suggested that release of larvae at neap tides ensures that the young worms avoid the hazards of desiccation in the period before they have had time to secrete a tough, calcareous tube. The worms are gregarious and large numbers can be found on a single piece of *Fucus*. Length of life is about one and a half years.

*Janua pagenstecheri* (Quatrefages)    (Fig. 122c)

*Body up to 2 mm in length. Up to 15 segments; 3 segments in thoracic region; 8 radioles. Operculum cup-shaped with uncalcified, transparent walls closed by thin, flat plate. Embryos brooded in operculum. Two colour forms exist, a brightly coloured orange-brown form with orange radioles, and a pale form which is greyish in colour. Tube coiled dextrally, i.e. worm lies with ventral surface upwards and mouth of tube opens anticlockwise. Tube opaque, calcareous; off-white or brown in colour, often, but not always, with 3 longitudinal ridges.*

*J. pagenstecheri* is widely distributed in north-west Europe and Britain and it has been suggested that it is probably the commonest spirorbid in the world. It is found from about the middle shore to depths of 120 m on a wide range of substrata, including *Corallina officinalis* (p. 61) and other seaweeds, rocks, stones, shells and the carapace of crabs. The brightly coloured form occurs high on the shore, especially in shallow rock pools, while the pale form is found in more shaded situations and extends into the sublittoral. The worms are hermaphroditic and it is believed that both cross- and self-fertilization occur, the latter probably being more usual. On the coast of south Wales, breeding takes place from May to October but there appears to be no rhythmicity in larval release. Up to 12 broods a year are produced. About 10 to 20 embryos are brooded in the operculum and pelagic larvae are released when the wall of the operculum splits.

REFERENCES

Brinkhurst, R. O. (1982). *British and other marine and estuarine oligochaetes*. Keys and notes for the identification of the species. Synopses of the British fauna (New Series), no. 21. Cambridge: Cambridge University Press.

Cadman, P. S. & Nelson-Smith, A. (1993). A new species of lugworm: *Arenicola defodiens* sp. nov. *Journal of the Marine Biological Association of the United Kingdom*, 73, 213–23.

Chambers, S. (1985). *Polychaetes from Scottish waters*. Part 2. *Families Aphroditidae, Sigalionidae and Polyodontidae*. Edinburgh: Royal Scottish Museum Studies.

Chambers, S. J. & Garwood, P. R. (1992). *Polychaetes from Scottish waters*. Part 3. *Family Nereidae*. Edinburgh: National Museums of Scotland.

Clark, R. B. (1960). *The fauna of the Clyde Sea area. Polychaeta with keys to the British genera*. Millport: Scottish Marine Biological Association.

Day, J. H. (1967). *A monograph on the Polychaeta of Southern Africa*. Part 1. *Errantia*. Part 2. *Sedentaria*. London: British Museum (Natural History). (Includes many north-west European species.)

Fauvel, P. (1923). *Polychètes errantes*. Faune de France, vol. 5. Paris: Lechevalier.

Fauvel, P. (1927). *Polychètes sédentaires*. Addenda aux errantes, archiannélides, myzostomaires. Faune de France, vol. 16. Paris: Lechevalier.

Fitzhugh, K. (1990). A revision of the genus *Fabricia* Blainville, 1828 (Polychaeta: Sabellidae: Fabriciinae). *Sarsia*, **75**, 1–16.

Garwood, P. R. (1981). Polychaeta. Errantia. *Report of the Dove Marine Laboratory*, 3rd Series, no. 22. Cullercoats.

Garwood, P. R. (1981) Polychaeta. Sedentaria incl. Archiannelida. *Report of the Dove Marine Laboratory*, 3rd Series, no. 23. Cullercoats.

*George, J. D. & Hartmann-Schröder, G. (1985). *Polychaetes: British Amphinomida, Spintherida and Eunicida*. Keys and notes for the identification of the species. Synopses of the British fauna (New Series), no. 32. Leiden: E. J. Brill/Dr W. Backhuys.

Gidholm, L. (1967). A revision of Autolytinae (Syllidae, Polychaeta) with special reference to Scandinavian species, and with notes on external and internal morphology, reproduction and ecology. *Arkiv för Zoologi*, **19**, 157–213.

Grassle, J. P. & Grassle, J. F. (1976). Sibling species in the marine pollution indicator *Capitella* (Polychaeta). *Science*, **192**, 567–9.

Hartmann-Schröder, G. (1971). Annelida, Borstenwürmer, Polychaeta. *Die Tierwelt Deutschlands*, **58**, 1–594.

Holthe, T. (1986). *Polychaeta Terebellomorpha. Marine invertebrates of Scandinavia*, no. 7. Oslo: Norwegian University Press.

Holthe, T. (1992). Identification of Annelida Polychaeta from northern European and adjacent Arctic waters. *Gunneria*, **66**, 1–30.

Kingston, P. & Duff, A. (eds.) (1987). *Key to the polychaete annelids from the North Sea and Baltic approaches*. Edinburgh: Heriot-Watt University. (A translation of the keys in Hartmann-Schröder, 1971. Translated by M. Ingold & M. Riddle.)

Knight-Jones, P. & Knight-Jones, E. W. (1977). Taxonomy and ecology of British Spirorbidae (Polychaeta). *Journal of the Marine Biological Association of the United Kingdom*, **57**, 453–99.

Knight-Jones, P., Knight-Jones, E. W. & Kawahara, T. (1975). A review of the genus *Janua*, including Dexiospira (Polychaeta: Spirorbinae). *Zoological Journal of the Linnean Society*, **56**, 91–129.

McIntosh, W. C. (1900–23). *A monograph of the British marine annelids*, vols. 1 (part II), II–IV. London: Ray Society.

O'Connor, B. D. S. (1987). The Glyceridae (Polychaeta) of the North Atlantic and Mediterranean, with descriptions of two new species. *Journal of Natural History*, **21**, 167–89.

Pleijel, F. (1993). *Polychaeta Phyllodocidae. Marine invertebrates of Scandinavia*, no. 8. Oslo: Scandinavian University Press.

*Pleijel, F. & Dales, R. P. (1991). *Polychaetes: British phyllodocoideans, typhloscolecoideans and tomopteroideans*. Keys and notes for the identification of the species. Synopses of the British fauna (New Series), no. 45. Oegstgeest, The Netherlands: UBS/Dr W. Backhuys.

Rainer, S. F. (1991). The genus *Nephtys* (Polychaeta: Phyllodocida) of northern Europe: a review of species, including the description of *N. pulchra* sp. n. and a key to the Nephtyidae. *Helgolander Meeresuntersuchungen*, **45**, 65–96.

Stephenson, T. A. (1944). *Seashore life and pattern*. London: King Penguin Books.

Tebble, N. & Chambers, S. (1982). *Polychaetes from Scottish waters*. Part 1 *Family Polynoidae*. Edinburgh: Royal Scottish Museum Studies.

*Westheide, W. (1990). *Polychaetes: Interstitial families*. Keys and notes for the identification of the species. Synopses of the British fauna (New Series), no. 44. Oegstgeest, The Netherlands: UBS/Dr W. Backhuys.

*  Further volumes are planned in this series covering the Class Polychaeta.

# Mollusca

Molluscs are found in marine, freshwater and terrestrial environments and in terms of numbers of species are second only to the arthropods. Body form varies widely within the phylum but the basic molluscan plan shows a head, often well developed and bearing sensory structures, a muscular foot and a visceral hump. The delicate tissue covering the visceral hump is known as the mantle (or pallium). This projects beyond the edge of the visceral mass to enclose a cavity known as the mantle (pallial) cavity where gills and other structures are found. The mantle secretes a calcareous shell consisting of a number of layers. There is usually an outer proteinaceous layer known as the periostracum, beneath which are layers of calcium carbonate. In some molluscs an inner mother-of-pearl or nacreous layer is often exposed on the surface of the shell by the wearing away of the outer layers. Although described as one of the most characteristic features of the phylum, it is important to note that in some groups the shell is internal, while in others it is absent. The characteristic feeding organ is a ribbon-like, chitinous structure, the radula, which comprises rows of teeth. In many species these have a rasp-like action and remove adhering microorganisms from the rock surface. In some gastropods, the radula is modified to enable it to bore into prey. The bivalves lack a radula and are deposit or suspension feeders using the gill to filter out fine particles of food. Most molluscs have separate sexes and while some shed the gametes and have external fertilization, in others fertilization is internal following copulation. Brood protection is well known but many species have a pelagic larva.

There is no universally accepted scheme of classification for the molluscs. The following is based on a phylogenetic sequence featured in recent texts.

## Phylum MOLLUSCA

Class Chaetodermomorpha    Worm-like molluscs without a shell. No foot. Body wall with chitinous scales. Live in marine sediments; deposit feeders.

Class Neomeniomorpha    Worm-like molluscs without a shell. Foot reduced. Body wall with calcareous scales. Carnivorous on cnidarians. Marine.

Class Monoplacophora    A primitive group with a single, symmetrical shell and replicated organs. Found in the deep sea.

**Class Polyplacophora**   The chitons or coat-of-mail shells. The shell is formed of eight, transverse articulating plates. Marine.

**Class Gastropoda**   The largest class of molluscs. The shell is a single unit, often coiled but may be cone-shaped or flattened. Shell sometimes absent. Found in marine, freshwater and terrestrial environments.

**Subclass Prosobranchia**   The winkles, whelks, limpets and top-shells. Well-developed shell. Marine, with a few freshwater and terrestrial species.

**Subclass Heterobranchia**   Gastropods in which the shell is reduced, internal or lost.

**Superorder Opisthobranchia**   The sea-slugs and sea-hares. Shell usually reduced and internal, or absent, but primitive members retain external shell. Marine.

**Superorder Pulmonata**   Freshwater and land snails and slugs. Also includes a few marine representatives. Mantle cavity modified for air breathing. Shell present or absent.

**Class Bivalvia (Pelecypoda)**   Includes the mussels, oysters and cockles. Shell with two valves joined by a ligament. Marine, with a few freshwater representatives.

**Class Scaphopoda**   The tusk shells. Shell a cylindrical tube, open at both ends. Sublittoral.

**Class Cephalopoda**   Squids, cuttlefishes and octopuses. Shell usually internal and reduced. Marine. Very occasionally stranded on the shore.

Family groupings and nomenclature used throughout the phylum follow Smith & Heppell (1991).

### Class Polyplacophora

Chitons, or coat-of-mail shells, are dorsoventrally flattened molluscs which have retained some primitive features. The shell is made of eight articulating plates enabling the animal to curl up when disturbed. The shell plates are surrounded by a fleshy mantle, known as the girdle, and on the ventral surface the most conspicuous feature is a broad foot enabling chitons to creep slowly and adhere firmly to rock surfaces. Between foot and mantle is a groove, the mantle or pallial groove, in which there is a series of gills. These may be restricted to the posterior part of the groove or distributed throughout the entire groove, and their position may be important in identification. The head is generally not clearly distinguished and both eyes and tentacles are lacking but some species are believed to have specialized light-sensitive cells in the shell plates. Chitons graze on a variety of microorganisms and algae by means of the radula. Some teeth of the radula are hardened with iron and silicate compounds, enabling the animals to feed on the tougher algae, including encrusting calcareous forms. The sexes are usually separate and the gametes are discharged into the sea. Fertilization is external, the eggs developing into trochophore larvae but in a few species they are brooded in the mantle cavity. Chitons are common on most rocky shores clinging to the undersurface of rocks and stones, especially those loosely buried in coarse sand and gravel. Crabs and fishes are believed to be their main predators.

### Key to families of polyplacophorans

Of the four families of British chitons, representatives of three are widely distributed and common on the shore. They are described below using characters easily seen with the aid of a hand lens or low-power stereo-microscope. Fine detail of the shell plates, seen only when the plates are dissected out, is not used in the key.

1 Girdle with tufts (usually 18) of conspicuous bristles ............. **Acanthochitonidae** (p. 217)
  Girdle without tufts of bristles .................................................................................. 2
2 Girdle narrow, covered in scales; fringed with spines ................... **Leptochitonidae** (p. 215)
  Girdle relatively broad, with granules varying from minute to coarse, widely separated to dense; fringed with spines ..................................................... **Ischnochitonidae** (p. 216)

Family Leptochitonidae
Small, oval-shaped chitons. Shell plates finely granular. Girdle narrow, covered in scales; fringed with spines. Gills in posterior region of mantle groove.

**Leptochiton asellus** (Gmelin)   (*Lepidopleurus asellus*) (Fig. 123a)

*Oval body. Shell plates with keel, finely granular; whitish or yellow colour often marked with darker patches. Girdle narrow, covered with longitudinally ribbed scales and fringed with slender spines; 8–13 pairs of gills in posterior region of mantle groove. Body length up to 19 mm.*

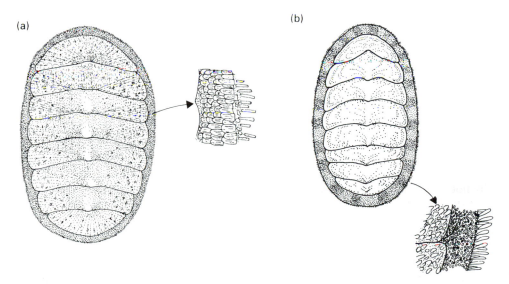

(a)

(b)

Figure 123  (a) *Leptochiton asellus*, with detail of girdle (detail after Matthews, 1953). (b) *Lepidochitona cinereus*, with detail of girdle (detail after Matthews, 1953).

*L. asellus* is widespread and common in north-west Europe. It is found all around the coast of Britain where it is common on the lower shore and into the sublittoral to depths of 250 m.

Family Ischnochitonidae
Oval to elongate-shaped chitons. Shell plates smooth and shiny or with fine granules. Girdle relatively broad, granular; fringed with spines.

1  Up to 15 pairs of gills (usually 10–15), in posterior region of mantle groove; shell plates smooth. Girdle densely covered with oval granules ............ *Tonicella rubra* (p. 217)
   With 16–26 pairs of gills ................................................................................................ 2
2  Girdle leathery, with a few well-spaced, minute granules; fringed with spatulate spines. Shell plates with minute granules but appear smooth .............................................
   ......................................................................................... *Tonicella marmorea* (p. 216)
   Girdle covered with rounded granules; fringed with large, golden spines. Shell plates granular ........................................................................ *Lepidochitona cinereus* (p. 216)

**Lepidochitona cinereus** (Linnaeus)    (Fig. 123b)

*Shell plates granular; colour varied, grey, green, blue, white; often patterned. Girdle covered with rounded granules and fringed with large golden, spines; banded colour pattern; 16–19 pairs of gills distributed throughout almost the entire mantle groove. Body length up to 28 mm.*

*L. cinereus* is widely distributed in north-west Europe and common all around Britain. It is found throughout most of the intertidal region, generally on the undersurface of stones. The sexes are separate and fertilization is external, the larvae spending only a few hours in the plankton. Winter growth checks can clearly be seen on some of the shell valves. Longevity is up to five years.

**Tonicella marmorea** (Fabricius)    (Fig. 124a)

*Shell plates covered with minute granules but appear smooth; red to dark red in colour with cream and brown areas. Girdle leathery with a few well-spaced, minute granules and fringed with spatulate spines; purple, red or green in colour, sometimes banded; 17–26 pairs of gills, usually in posterior region of mantle groove but occasionally distributed throughout entire groove. Body length up to 45 mm.*

*T. marmorea* is a northern species and in Britain is found as far south as North Wales on the west coast, and Northumberland on the east coast. It occurs on rocks and stones from the lower shore to depths of about 200 m. The sexes are separate and fertilization is external.

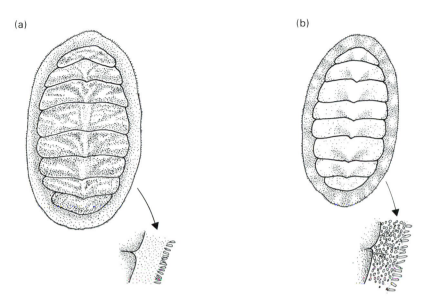

(a)                                                                                           (b)

Figure 124   (a) *Tonicella marmorea*, with detail of girdle (detail after Matthews, 1953). (b)
*Tonicella rubra*, with detail of girdle (detail after Matthews, 1953).

### *Tonicella rubra* (Linnaeus)    (Fig. 124b)

*Shell plates smooth and shiny; pinkish-red in colour with cream areas. Girdle densely
covered with oval granules and fringed with golden spines; with banded colour pattern; 10–
15 pairs of gills in posterior region of mantle groove. Body length up to 21 mm.*

*T. rubra* is widespread in north-west Europe and around Britain where it is found on
stones on the lower shore extending to 300 m. The sexes are separate and fertilization is
external.

Family Acanthochitonidae
Shell plates granular, with keel; partly covered by girdle. Girdle broad, with (usually) 18
tufts of bristles.

Shell plates with large, pear-shaped or oval granulations, irregularly spaced ..................
.............................................................................. *Acanthochitona crinitus* (p. 218)
Shell plates with tiny, rounded granulations; granulations dense, evenly spaced ...........
.............................................................................. *Acanthochitona fascicularis* (p. 218)

*Acanthochitona crinitus* (Pennant)    (*Acanthochitona discrepans*) (Fig. 125a)

*Large, pear-shaped or oval, irregularly spaced granulations on shell plates; brown-yellow colour with varied patterning; white V-shaped area in centre of plates. Girdle broad, with 18 tufts of colourless bristles; fringed with spines; 10–15 pairs of gills in posterior region of mantle groove. Body length up to 34 mm.*

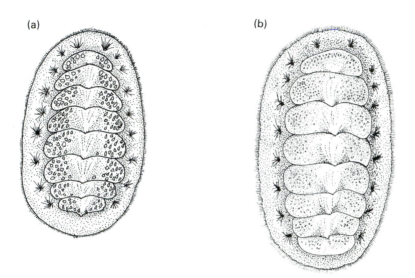

Figure 125  (a) *Acanthochitona crinitus*. (b) *Acanthochitona fascicularis*.

A. *crinitus* is widely distributed in north-west Europe, extending as far north as southern Norway. It is found all around Britain on rocky shores, extending from the lower shore to depths of 50 m. The sexes are separate and fertilization is external.

*Acanthochitona fascicularis* (Linnaeus)    (*Acanthochitona communis*) (Fig. 125b)

*Shell plates with very small, rounded granulations, closely packed and evenly spaced; varied coloration, whitish, olive, brown, often with whitish 'V' along mid-dorsal region. Girdle broad, with 18 tufts of colourless bristles; fringed with long spines; 10–15 pairs of gills in posterior region of mantle groove. Body length up to 60 mm. The largest British species of chiton.*

A. *fascicularis* is a southern species, extending to Channel coasts, the west coast of Ireland and as far north as Anglesey. It is found from the lower shore to depths of about 50 m.

## Class Gastropoda

A typical gastropod has a single, usually coiled shell, a large ventral foot and a well-developed head bearing eyes and conspicuous tentacles. In the opisthobranchs and pulmonates shell reduction and loss has occurred. A few marine gastropods are adapted to a pelagic habit but the most familiar species are benthic and include the often abundant limpets, periwinkles and top-shells belonging to the subclass Prosobranchia, and the sea-slugs and sea-hares belonging to the Superorder Opisthobranchia. During development gastropods undergo torsion, a process by which the visceral mass of the larva twists through 180° so that the mantle cavity, gills and anus come to lie just behind the head. This condition persists in the adult prosobranch and there is much controversy and debate as to its adaptive significance. In opisthobranchs and pulmonates, a process of detorsion has taken place. Torsion of the visceral mass is quite distinct from coiling of the shell which is believed to result from increase in body size during growth. While many gastropods have separate sexes, others are hermaphroditic and the class shows a variety of reproductive patterns. Fertilization may be external, with the development of a pelagic larva, or internal. When internal the eggs may be brooded in an internal brood pouch, laid in egg masses or capsules attached to the substratum, or released into the sea. Development can be either direct or pelagic. In some gastropods, for example the limpets and top-shells, the typical larva is a trochophore, similar to that of the polychaetes (p. 162), with a short free-swimming life. In others, such as the winkles, mud snails and sea-slugs, the typical larva is a veliger. This has a tiny shell and swims by means of two ciliated lobes; many veligers have a long pelagic life.

## Subclass Prosobranchia

Prosobranch molluscs are among some of the most common members of the rocky shore fauna. They are primitive members of the Class Gastropoda from which the opisthobranchs (p. 261) and the pulmonates (p. 276) are believed to have evolved. They are characterized by a conical (e.g. the limpets) or coiled shell (e.g. the periwinkles) but in one family (p. 252) the shell is internal. The external shell is to varying degrees sculptured with a surface ornamentation of ridges, lines and ribs, which are often useful in identification. The aperture of the shell is often extended into a groove, known as the siphonal canal. In life it houses a fold of mantle tissue, the siphon. Shell characteristics are shown in Fig. 126. Most prosobranchs have a tough, proteinaceous disc known as the operculum, which is sometimes impregnated with calcium. This closes the shell when the animal is contracted and offers some protection against desiccation and predation when the tide is out. The nomenclature used to describe the gastropod shell is shown in Fig. 126. Shell length as used in the families Fissurellidae, Acmaeidae and Patellidae is defined as the greatest basal diameter.

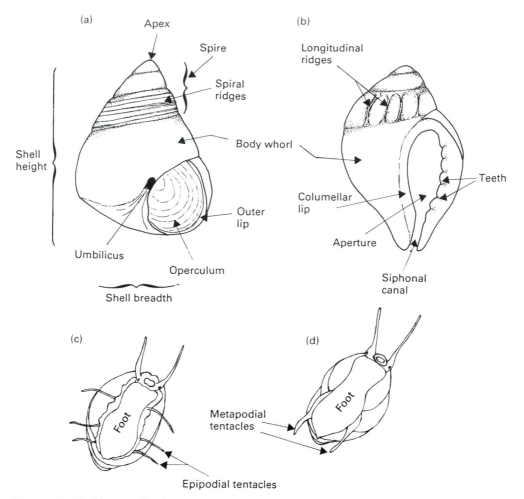

Figure 126 (a), (b) Generalized gastropod shells to illustrate features referred to in the text. (c), (d) Ventral view of animal showing (c) epipodial and (d) metapodial tentacles.

### Key to families of prosobranchs

Prosobranch molluscs are characterized by a shell made up of a single unit. Variation in shell morphology within the group allows the recognition of the different families but in a small number of cases it is convenient to key to the level of subfamily. It should be noted that some pulmonates (including *Leucophytia bidentata*, p. 276, *Ovatella myosotis*, p. 277, and *Otina ovata*, p. 277) and some opisthobranchs (including *Acteon tornatilis*, p. 263, and *Retusa obtusa*, p. 265) also possess an external shell. They should not be confused with prosobranchs.

1 Resembles sea-slug (Opisthobranchia, p. 261); shell internal and enclosed by mantle. Mantle covered with tubercles; siphon-like extension at anterior ... **Lamellariidae** (p. 252)
   External shell present ............................................................................................... 2

2 Shell a flattened spiral with conspicuous row of rounded openings ..... **Haliotidae** (p. 223)
   Shell not as above ...................................................................................................... 3

3 Shell conical (i.e. more or less straight-sided, without spiral whorls) or cap-shaped; without operculum ............................................................................................. 4
   Shell not conical, not cap-shaped; with or without operculum ............................... 8

4 Shell conical with single apical aperture or a marginal slit ............... **Fissurellidae** (p. 224)
   Shell conical or cap-shaped, without apical aperture or marginal slit ..................... 5

5 Shell with internal shelf extending across part of aperture ............. **Calyptraeidae** (p. 249)
   Shell without internal shelf extending across part of aperture ............................ 6

6 Shell conical to cap-shaped; smooth, may be transparent; oval outline, with 2–8 usually bright, blue-green rays. On stipe, frond and holdfasts of *Laminaria*, also on *Fucus*, *Himanthalia* and occasionally other seaweeds ........................... **Patellidae** (p. 226)
   Shell conical, not as above ...................................................................................... 7

7 Shell smooth; with tortoiseshell pattern or pink or brown rays. Firmly attached to rock surface, stones, shells ........................................................ **Acmaeidae** (p. 225)
   Shell strong, usually with radiating ridges. Firmly attached to rock surface ....................... ...................................................................................... **Patellidae** (p. 226)

8 Shell glossy, strongly ridged with ridges running into aperture, or smooth; body whorl very large with narrow elongate aperture; spire very small or hidden by shell. Without operculum. Typical cowrie shape ........................................... **Triviidae** (p. 251)
   Shell not so. With operculum ................................................................................. 9

9 Shell with siphonal canal which may be open or closed; shell may have large pointed expansion of outer lip .............................................................................................. 10
   Shell without siphonal canal ................................................................................ 13

10 Shell with large pointed expansion of outer lip to give appearance of pelican's foot ....... ........................................................................................ **Aporrhaiidae** (p. 248)
   Shell not so .......................................................................................................... 11

11 Shell usually large (about 100 mm shell height), stout; body whorl large; aperture large; outer lip not usually thickened; siphonal canal not closed over. Shell smooth or ridged, with spiral ridges, or spiral and longitudinal ridges. Predominantly sublittoral on soft sediment; dead shells washed ashore . **Buccinidae (Subfamily Buccininae)** (p. 258)
   Shell up to 50 mm shell height, stout, with spiral or longitudinal ridges or both. Outer lip may be thickened; sometimes with teeth in older specimens. Siphonal canal may be closed over. Often common on rocky shores ................................................................ 12

12 Siphonal canal open, short; in non-apertural or 'back view' the end of the canal is seen as a notch. Groove runs from lower end of siphonal canal to columella round 'back' of shell ......................................................... **Buccinidae (Subfamily Nassariinae)** (p. 259)
   Siphonal canal open or closed, not seen as a notch in non-apertural or 'back view' ........ .................................................................................................. **Muricidae** (p. 255)

13  Shell nacreous, especially at apex when worn, and within aperture. Umbilicus obvious
    or closed. Rounded operculum. Lateral lobes of foot with 3–6 pairs of tentacles
    (epipodial tentacles) (Fig. 126c) ........................................................ **Trochidae** (p. 230)
    Shell, operculum and foot not so ........................................................... 14

14  Shell with conspicuous umbilicus ........................................................... 15
    Shell without umbilicus or umbilicus very small ........................................... 16

15  Shell with spire or spire reduced. Umbilicus conspicuous, with groove on columellar
    region leading to it. Posterior of foot with 1 pair of tentacles (metapodial tentacles)
    (Fig. 126d) .................................... **Littorinidae (Subfamily Lacuninae)** (p. 235)
    Smooth, glossy shell with large body whorl and short spire. Umbilicus large. Foot
    large, covers head and part of shell when animal active; without metapodial tentacles ..
    ........................................................................................ **Naticidae** (p. 253)

16  Shell tall, narrow, with pronounced spire; with many (about 15–20) whorls ................. 17
    Shell not as above ........................................................................ 19

17  Shell with up to 15 whorls; ornamentation of spiral and longitudinal ridges to give
    tuberculate pattern. Aperture with short, basal extension ................... **Cerithiidae** (p. 234)
    Shell not as above ........................................................................ 18

18  Shell with up to 15 whorls; with pronounced, well-spaced, longitudinal ridges.
    Aperture rounded ........................................................ **Epitoniidae** (p. 254)
    Shell with up to 20 whorls; with spiral ridges. Aperture angular. Operculum with
    bristles around periphery ................................................ **Turritellidae** (p. 235)

19  Smooth glossy shell, of 5–6 whorls; whitish with reddish-brown streaks, with white,
    domed, calcareous operculum. Up to 9 mm height ........................ **Phasianellidae** (p. 233)
    Shell and operculum not as above ......................................................... 20

20  Shell with pronounced spire, or spire low and flattened. Body whorl accounts for 75%
    or more of shell height. Shell smooth to strongly ridged with spiral ridges. Shell height
    from 3 to 30 mm. Widespread and often abundant. Found at all levels on rocky shores;
    some species live on soft sediments and extend into estuaries ..............................
    ........................................... **Littorinidae (Subfamily Littorininae)** (p. 238)
    Shell with spire; body whorl accounts for less than 75% of shell height. Usually up to
    about 6 mm shell height, never greater than 9–10 mm shell height. Widespread and
    often abundant ........................................................................... 21

21  Shell smooth (but see *Potamopyrgus antipodarum*, p. 248); up to 6 mm in height.
    Brownish-black in colour. Found on estuarine sands and muds, often in very high
    densities. Also found in brackish and freshwater ditches. Not found on rocky shores ..........
    ........................................................................ **Hydrobiidae** (p. 245)
    Shell smooth to heavily ridged. Variously coloured, white, cream, orange, brown.
    Usually less than about 6 mm shell height; never greater than 9–10 mm. Marine,
    predominantly found on rocky shores where they may be abundant under stones and
    amongst algae ............................................................... **Rissoidae** (p. 244)

Family Haliotidae – ormers
Flattened, spiral shell with a row of rounded openings, of which five to seven are open, the rest are closed. Aperture large. Without operculum.

### *Haliotis tuberculata* Linnaeus     Ormer (Fig. 127)

*Flattened, spiral shell up to 90 mm in length; with row of rounded openings, 5–7 of which are open, remainder closed. Outer surface of shell shades of red, yellow or green; inner surface iridescent. Operculum absent.*

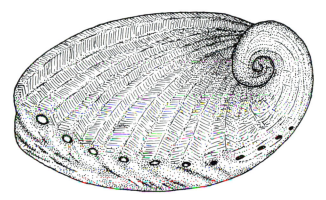

Figure 127  *Haliotis tuberculata.*

*H. tuberculata* is a southern species, extending as far north as the Channel Islands. It is found clinging to rocks, especially those covered with encrusting red seaweeds, on the lower shore and in the sublittoral to depths of about 40 m. The ormer feeds mainly on red seaweeds and the colour of the shell is believed to be related to the food eaten. The sexes are separate and fertilization external. There is a trochophore larva which settles after a brief free-swimming phase at a length of about 2 mm. In the Channel Islands breeding occurs in late summer, the animals becoming sexually mature at 40 to 50 mm shell length when they are two or three years old. They have a longevity of five to six years. Length of larval life is probably five to six days. The openings along the edge of the shell of *Haliotis* are the routes by which water leaves the mantle cavity. New openings form at the margin of the shell and can be seen as slit-like indentations which are surrounded as the shell grows. The oldest openings are closed. The ormer is greatly valued as food and is fished commercially on the French coast.

Family Fissurellidae – slit limpets, keyhole limpets
Shell conical, with radiating ribs; a single apical aperture or a marginal slit. Without operculum.

Shell with marginal slit ......................................................... *Emarginula fissura* (p. 224)
Shell with small, apical opening ................................................ *Diodora graeca* (p. 224)

**Emarginula fissura** (Linnaeus)    (*Emarginula reticulata*), Slit limpet (Fig. 128)

*Ribbed, conical shell; apex curved towards posterior; marginal slit at anterior. Up to 10 mm in length. Outer surface of shell white, grey or yellow; inner surface similar but nacreous. Short siphon-like structure, through which water leaves mantle cavity, extends through anterior marginal slit. Operculum absent.*

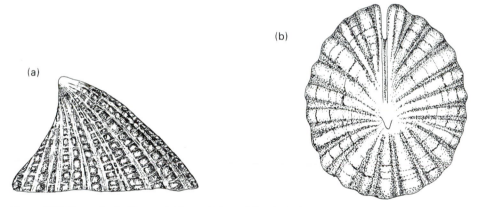

Figure 128  *Emarginula fissura*, (a) lateral view, (b) apical view.

*E. fissura* is widely distributed on rocky shores in north-west Europe but absent from the eastern English Channel and the southern North Sea. It is found on the lower shore on the undersurface of rocks and boulders and extends into the sublittoral to depths of 200 m and more. The slit limpet feeds largely on sponges but also takes some algae and detritus. The sexes are separate and fertilization is external.

**Diodora graeca** (Linnaeus)    (*Diodora apertura*), Keyhole limpet (Fig. 129)

*Shell ribbed; conical with small apical opening; up to 25 mm in length and 10 mm high. Outer surface white to grey with coloured rays; inner surface glossy white. Mantle tissue expands upwards round base of shell. Short siphon-like structure, through which water leaves mantle cavity, extends through apical opening. Operculum absent.*

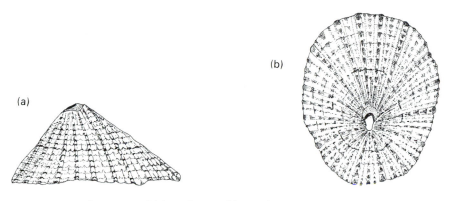

Figure 129  *Diodora graeca*, (a) lateral view, (b) apical view.

*D. graeca* is found along the west coast of Britain, extending as far north as the Faroes, but is absent from the North Sea. It occurs on the lower shore on rocky coasts extending to depths of about 250 m. It is intolerant of reduced salinity. The keyhole limpet preys extensively on the sponges *Halichondria* (p. 93) and *Hymeniacidon* (p. 94). The sexes are separate and on the south coast of England breeding extends from December to May. Sperm are believed to enter the apical aperture of the female and the fertilized eggs, yellow in colour, are laid on the undersurface of rocks and stones. There is no pelagic larva, young limpets emerge directly from the egg mass.

Family Acmaeidae – tortoiseshell limpets
Shell conical, without apical aperture or marginal slit; smooth, with tortoiseshell pattern or with pink or brown rays. Without operculum.

Outer surface of shell whitish with pink or brown rays; inner surface with reddish apical mark ...................................................................... *Tectura virginea* (p. 226)
Outer surface patterned with reddish-brown markings suggesting tortoiseshell; inner surface with dark brown apical mark ..................................... *Tectura testudinalis* (p. 225)

*Tectura testudinalis* (Müller)   (*Collisella tessulata, Acmaea testudinalis*), Tortoiseshell limpet (Fig. 130)

*Conical shell, smooth; apex anterior; up to 25 mm in length. Outer surface patterned with reddish-brown suggesting tortoiseshell; inner surface glossy, whitish, with dark brown apical mark. Operculum absent.*

(a)                                          (b)

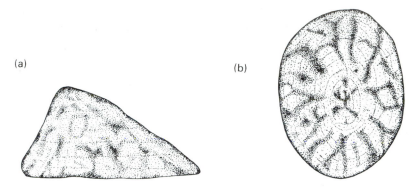

Figure 130 *Tectura testudinalis*, (a) lateral view, (b) apical view.

*T. testudinalis* is a northern species not found in the southern half of Britain. It occurs on stones and shells from the lower shore into the sublittoral among encrusting red seaweeds; these, together with detritus, are its main food. The sexes are separate and reddish-coloured eggs are laid in a mucous sheet which quickly breaks down. On the Danish coast, individuals mature after one or two years and breeding occurs in the spring. The larvae are planktonic for about four days and longevity is up to three years.

***Tectura virginea*** (Müller) (*Acmaea virginea*), White tortoiseshell limpet (not illustrated)

*Conical shell, smooth; apex anterior; up to 10 mm in length. Outer surface whitish with pink or brown rays; inner surface glossy, whitish-purple, sometimes with a reddish apical mark. Operculum absent.*

*T. virginea* is widely distributed in north-west Europe and around Britain, on stones, shells and especially among the encrusting red seaweeds on which it feeds. It extends from the lower shore to depths of about 100 m. On the west coast of France breeding has been recorded in spring. The sexes are separate, fertilization external and there is a pelagic larva.

Family Patellidae – limpets
Shell conical, strong, usually with radiating ridges, or conical to cap-shaped, thin and smooth with bright blue-green lines. Without apical aperture or marginal slit. Without operculum.

1   Shell smooth, may be transparent; with blue-green rays; attached to frond, stipe and
    holdfast of *Laminaria*, also on *Himanthalia* and *Fucus*  .......... ***Helcion pellucidum*** (p. 227)
    Shell strong, usually with radiating ridges; very firmly attached to rock surface  .............. 2

2 Tentacles around edge of mantle (Fig. 132b), transparent. Foot grey, greyish-green, yellowish. Inner surface of shell greenish-bluish nacre, may be yellowish in older shells .................................................................................. *Patella vulgata* (p. 228)

Tentacles around edge of mantle cream or white ............................................................. 3

3 Tentacles around edge of mantle, cream. Foot yellow to orange. Inner surface of shell whitish, porcelain-like ...................................................... *Patella ulyssiponensis* (p. 229)

Tentacles around edge of mantle white, set off against dark grey foot and dark rays of inner surface of shell ................................................................. *Patella depressa* (p. 230)

*Helcion pellucidum* (Linnaeus)   (*Patina pellucida*), Blue-rayed limpet (Fig. 131)

*Smooth, somewhat transparent shell; even, oval outline; apex anterior; up to 20 mm in length; 2–8 blue–green, broken rays overlying brown shell. Operculum absent.*

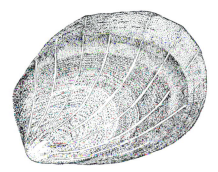

Figure 131  *Helcion pellucidum.*

*H. pellucidum* is widely distributed in north-west Europe and around Britain. It is found on rocky shores, characteristically on *Laminaria digitata* (p. 43) and *L. hyperborea* (p. 44), but also on other seaweeds such as *Fucus serratus* (p. 49) and *Himanthalia elongata* (p. 51). It extends into the sublittoral to depths of about 25 m. The sexes are separate, fertilization is external and breeding occurs mainly in winter and spring. The larvae are pelagic, settling in late winter–early spring and it has been suggested that they settle preferentially on crustose red algae on the lower shore. They then migrate to the laminarians at a shell length of 2–3 mm. It has been suggested that newly settled limpets cannot eat *Laminaria* and therefore utilize other food sources during their early life, only migrating later to *Laminaria*. Limpets are found both on the stipe and the frond of the seaweed, those on the frond being exposed to the risk of damage when the frond is shed or disintegrates in the autumn. However, there is evidence to suggest that loss of individuals in this way is minimized by a migration of many of the limpets to the basal region of the frond and the stipe. The timing of this migration is believed to be related to changes in the chemical nature of the seaweed on which the limpet is feeding. Sexual maturity is reached at a shell length of about 5 mm; most animals die after one year.

The blue-rayed limpet exists in two forms, the one described above, known as *Helcion pellucidum pellucidum*, typically found on the frond of the weed, and *Helcion pellucidum laevis* which lives in cavities in the holdfast of the laminarian. The form *laevis* has an opaque brown shell with numerous, but sometimes inconspicuous, blue rays that alternate with reddish-brown rays. When the young limpets settle they are all of the *pellucidum* form; the differences which develop subsequently are believed to be related to differences between the two habitats, the exposed frond and the more sheltered cavity in the holdfast. It is interesting to note that although most individuals die after about one year, a few survive into a second year and the majority of these are of the *laevis* form.

*Patella vulgata* Linnaeus    Common limpet (Fig. 132)

*Shell conical; apex central or slightly anterior; with radiating ridges; up to 60 mm shell length. Outer surface whitish-grey or fawn coloured; inner surface greenish-bluish nacre, may be yellowish in older shells, with a silvery apical area. Foot grey, greyish-green, yellowish; tentacles around edge of mantle, transparent. Operculum absent.*

(a)

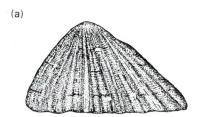

(b)

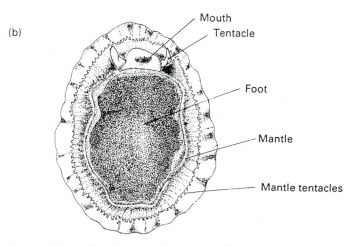

Figure 132  *Patella vulgata*, (a) lateral view, (b) ventral view.

*P. vulgata* is widely distributed in north-west Europe. Along with acorn barnacles and periwinkles, the common limpet is one of the best-known intertidal organisms. It is found on rocky shores of all degrees of exposure to wave action, from the high shore to the sublittoral, although usually not abundant on shores with a dense growth of sea-weed. It extends into the mouths of estuaries, surviving in salinities of down to about 20‰. There is considerable variation in shell shape between individuals from different shore levels, those from the high shore generally having a taller shell and smaller shell length when compared with animals from the lower shore. At high shore levels during low tide, the shell muscles of the limpet are contracted, keeping the animal firmly attached to the substratum and thereby reducing desiccation. This is believed to have the effect of pulling in the shell-secreting tissues and thus affecting shell shape.

Limpets feed on a wide range of microorganisms and algae, including *Fucus* spp., and encrusting red algae. These are removed from the rock surface by the radula, the teeth of which are hardened by iron and silicate compounds, and when feeding over relatively soft rocks such as shales, zigzag impressions are left in the rock. Feeding has been observed during high water in daylight but there is evidence to suggest that limpets are most active when emersed during the night. They undergo feeding excursions, returning to the same 'home scars' from which they started. This homing behaviour is achieved by the limpet following a mucous trail deposited on the rock surface and is controlled by an endogenous circatidal rhythm. The 'home scar' is a shallow depression in the rock formed by chemical action and abrasion between the limpet shell and the rock. The edge of the shell fits tightly into the depression and thus reduces the rate at which water is lost from the mantle tissues during emersion.

The impact achieved by limpets and other herbivorous molluscs grazing on shore algae is well known and can be demonstrated by excluding them from marked areas on the shore. Under these conditions the sporelings of green seaweeds such as *Ulva* (p. 36) and *Enteromorpha* (p. 35) grow almost unchecked, clothing the rocks in a bright green colour. The common limpet is a protandrous hermaphrodite, although a few individuals remain as males. On the east coast of England, *P. vulgata* matures as male at two years of age, and then changes to female at about four years of age. Spawning occurs mainly during October to December and is believed to be induced by rough seas and onshore winds. Fertilization is external and the trochophore larvae have a pelagic life of a few days. Settlement on the shore occurs at a shell length of about 0.2 mm and the newly settled spat are usually found in rock pools or permanently damp situations. Length of life varies considerably; where growth is fast it is generally four to five years, compared with up to 15 to 17 years where conditions lead to slow growth.

**Patella ulyssiponensis** Gmelin   (*Patella aspera, Patella athletica*), China limpet (not illustrated)

*Shell conical, flattish; apex anterior; with radiating ridges; up to 60 mm in length. Outer surface whitish-grey; inner surface whitish, porcelain-like, with cream to orange apical area. Foot yellow to orange; tentacles around edge of mantle, cream. Operculum absent.*

*P. ulyssiponensis* reaches its northern limit on the north coast of Scotland and is absent or rare on south-east shores of England from the Isle of Wight to the Humber estuary. It is found on rocks on the lower shore and in the sublittoral, also occurring in shallow rock pools on the middle shore. In exposed situations it is the commonest limpet on the lower shore. Its diet is probably similar to that of *P. vulgata* (above). Despite earlier reports that the sexes are separate throughout life, there is now evidence to suggest that it is a protandrous hermaphrodite. In south-west Ireland, maturation of the gonads begins in May and June, and spawning occurs in October, probaly induced by gales. Fertilization is external.

**Patella depressa** Pennant    (*Patella intermedia*), Black-footed limpet (not illustrated)

*Shell conical, flattened; apex anterior; with fine radiating ridges; up to 30 mm shell length. Outer surface greyish; inner surface dark with conspicuous, brown marginal rays, and creamy-yellow apical area. Foot dark grey; tentacles around edge of mantle, white. Operculum absent.*

*P. depressa* is found on the Atlantic coast of Europe, extending northwards to south-west Britain but is not found east of the Isle of Wight. The northern limit of the species is North Wales. It favours wave-beaten, rocky shores and is usually on the middle to lower shore. The sexes are separate and on the south coast of England maturation of the gonads begins in spring and spawning, which may be related to gales and heavy wave action, occurs from July to September. Fertilization is external.

Family Trochidae – top-shells
Shell with spire; nacreous within aperture and at apex of shell when worn. Aperture without siphonal canal. Umbilicus present but may be closed in some cases. Operculum rounded. Lateral lobes of foot with 3–6 pairs of tentacles (epipodial tentacles).

1  Apex of shell sharply pointed; with straight-sided whorls    *Calliostoma zizyphinum* (p. 233)
   Apex of shell not sharply pointed; sides of shell not straight-sided ................................... 2
2  Single, prominent tooth on inner margin of aperture (Fig. 134b); apex often eroded and
   silvery ................................................................ *Monodonta lineata* (p. 232)
   Shell without tooth on inner margin of aperture ................................................ 3
3  Sides of whorls angular, with steps between whorls; upper surface of whorls with
   ridges and tubercles, tubercles especially noticeable on body whorl. Umbilicus deep .....
   ........................................................................... *Gibbula magus* (p. 232)
   Sides of whorls more or less smoothly convex; upper surface of whorls without
   tubercles. Umbilicus small or large ........................................................ 4
4  Shell greyish in colour, with many narrow, red-purple lines. Umbilicus small. Apex
   sometimes eroded and silvery ................................... *Gibbula cineraria* (p. 231)
   Shell greyish-green in colour, with well-spaced, diagonal purple stripes. Umbilicus
   large. Apex often eroded and silvery ...................................... *Gibbula umbilicalis* (p. 231)

*Gibbula cineraria* (Linnaeus)    Grey top-shell (Fig. 133a)

*Shell roughly triangular in outline; up to 16 mm in height and 17 mm broad; umbilicus small. Greyish in colour with many narrow, red-purple lines. Apex sometimes eroded and silvery. Foot with 3 pairs of epipodial tentacles.*

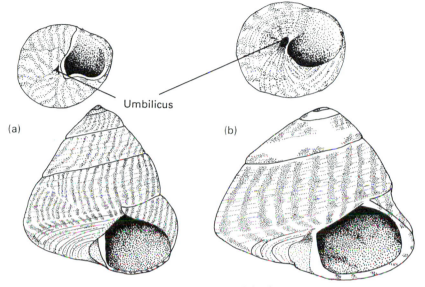

Figure 133   (a) *Gibbula cineraria*. (b) *Gibbula umbilicalis*.

*G. cineraria* is widely distributed in north-west Europe and around Britain on sheltered rocky shores. It is found from the lower shore to depths of about 130 m, feeding mainly on detritus and microalgae. The radular teeth of *G. cineraria*, and other top-shells, are not hardened with iron compounds and they are apparently unable to eat encrusting red seaweed or seaweeds with tough cell walls such as *Fucus* spp. (p. 50) which feature in the diet of *Patella* spp. (above). The sexes are separate, fertilization is external and it is believed that spawning occurs in response to increase in seawater temperature. In north-east England the main breeding period is from June to September. There is a free-swimming trochophore larva and laboratory studies suggest a pelagic life of eight or nine days.

*Gibbula umbilicalis* (da Costa)    Flat or purple top-shell (Fig. 133b)

*Shell roughly triangular in outline; up to 16 mm in height and 22 mm broad; umbilicus large. Greyish-green in colour, with well-spaced, diagonal purple stripes. Apex often eroded and silvery. Foot with 3 pairs of epipodial tentacles.*

*G. umbilicalis* occurs on the west coast of France, south-west and west coasts of Britain as far north as Orkney. It is found from the upper shore down to low water on sheltered rocky shores and, like *G. cineraria* (above), feeds on microorganisms and detritus. The sexes are separate, fertilization is external and sexual maturity is reached in the second year at a shell height of about 8–9 mm. Breeding takes place during summer and the trochophore larvae are believed to have only a short pelagic life. Length of life has been estimated to be eight or more years.

**Gibbula magus** (Linnaeus)    (Fig. 134a)

*Shell broader than tall, up to 30 mm in height and 35 mm broad; upper surface of whorls with ridges and tubercles; sides of whorls angular with steps between whorls; umbilicus deep. Whitish-yellow with reddish markings. Foot with 3 pairs of epipodial tentacles.*

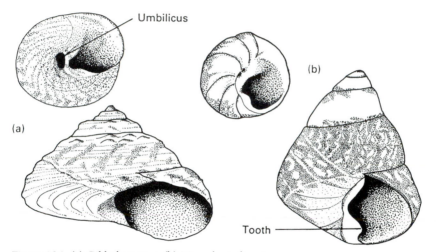

Figure 134   (a) *Gibbula magus.* (b) *Monodonta lineata.*

*G. magus* is a southern species found along the west coast of Britain but is absent from the eastern English Channel and the North Sea. It is restricted to the extreme lower shore, extending into the sublittoral to depths of about 70 m on rocky substrata, as well as being found on muddy-sand. The food consists of microorganisms and detritus. The sexes are separate, fertilization is external and breeding occurs during spring and summer. The trochophore larva probably has only a short pelagic life.

**Monodonta lineata** (da Costa)    Thick top-shell (Fig. 134b)

*Triangular shell up to 30 mm in height and 25 mm broad; umbilicus small or closed. Single prominent tooth on inner margin of aperture. Greyish-coloured shell with purplish markings; whitish and iridescent around umbilicus; apex often eroded and silvery. Foot with 3 pairs of epipodial tentacles.*

*M. lineata* is a southern species, extending as far north as Anglesey, North Wales. It occurs on south-west shores and is absent from the eastern English Channel and the

North Sea. It is found on rocks and boulders from the upper to the lower shore and, as with other top-shells, its food is predominantly microorganisms and detritus. The sexes are separate, fertilization is external and breeding occurs during summer. Length of larval life is a few days and settlement occurs on the upper shore where the juvenile snails can be found on the undersurface of stones and boulders. Longevity may be 10 years or more, and annual growth checks can clearly be seen on the shell (Fig. 134b), particularly in specimens from the northerly part of the range. With experience, these checks can be used to determine the age of individual snails.

**Calliostoma zizyphinum** (Linnaeus)    Painted top-shell (Fig. 135a)

*Conical shell with straight-sided whorls; up to 30 mm in height and 30 mm broad; umbilicus not obvious in adult. Greyish-pink colour with reddish streaks; occasionally all white. Foot with 4 or 5 pairs of epipodial tentacles.*

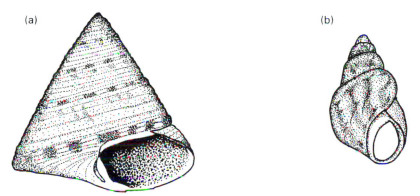

(a)                                                    (b)

Figure 135   (a) *Calliostoma zizyphinum.* (b) *Tricolia pullus.*

C. *zizyphinum* is widely distributed in north-west Europe and occurs all around the coast of Britain. It is found at extreme low water on fairly sheltered rocky shores, extending sublittorally to about 300 m. In some areas the white morph may be more abundant than the coloured form. It has been suggested that the clean, unworn appearance of the shell of *Calliostoma* is a result of a cleaning behaviour in which the foot is extended over the shell and presumably coats it with mucus, deterring settlement of larvae and algal spores. Its diet is probably microorganisms and detritus although it has been suggested that it also feeds on cnidarians. The sexes are separate; details of fertilization are not known. Breeding occurs during summer and yellowish-coloured eggs are laid in gelatinous ribbons up to 35 mm long and 3—4 mm broad. They are attached to the substratum and after about a week crawling young emerge.

Family Phasianellidae – pheasant shells
Shell small, with spire. Without umbilicus. Aperture without siphonal canal. White, opaque operculum, domed externally. Shell glossy, usually with reddish-brown streaks. Up to 9 mm shell height.

*Tricolia pullus* (Linnaeus)    Pheasant shell (Fig. 135b)

*Small glossy, whitish shell with reddish-brown streaks; up to 9 mm in height. Conspicuous, domed white operculum.*

*T. pullus* is a southern species, extending north along the west coast of Britain; it is absent from the North Sea. It occurs on rocky coasts on the lower shore and sublittorally to about 35 m, usually among red seaweed. It feeds on the weed and on microorganisms and detritus. The sexes are separate, fertilization is external and breeding occurs throughout the year but mainly in summer and autumn. The larvae are pelagic for a few days and settlement occurs predominantly on red seaweeds. Longevity is believed to be one year.

Family Cerithiidae – needle whelks
Shell tall, with 10–15 whorls, patterned with many regularly arranged tubercles. Aperture with short basal extension. Without umbilicus.

*Bittium reticulatum* (da Costa)    Needle whelk (Fig. 136b)

*Shell tall, narrow with 10–15 whorls; patterned with many regularly arranged tubercles; up to 10 mm in height, occasionally larger. Aperture with short basal extension. Brown in colour.*

(a)

(b)

Figure 136  (a) *Turritella communis.* (b) *Bittium reticulatum.*

*B. reticulatum* is widely distributed in north-west Europe, occurring in Britain mainly in the south and west. It lives on sandy and muddy shores, especially where *Zostera* (p. 85) is present, as well as on rocks and stones. Found from the lower shore sublittorally to depths of about 250 m, it feeds on a variety of microorganisms. Breeding occurs in summer. The sexes are separate but details of fertilization are not known. Whitish eggs (up to 1000) are laid in a long, coiled ribbon approximately 24 mm long and 4–5 mm broad. After a few days the eggs hatch as free-swimming veliger larvae which are common in the plankton in spring and summer. The length of planktonic life has not been determined but after settlement the snails live for one or two years.

Family Turritellidae – auger or screw shells
Shell tall, up to 20 whorls; with spiral ridges. Aperture angular, without siphonal canal. Without umbilicus. Operculum with bristles around periphery.

**Turritella communis** Risso    Auger or screw shell (Fig. 136a)

*Shell tall and narrow; up to 20 whorls, with spiral ridges; up to 40 mm in height, sometimes larger. Colour reddish-brown, yellow or white.*

*T. communis* is widely distributed in north-west Europe and a large part of Britain but is rare on the south-east coast. It is a sublittoral species and lies buried just below the surface in mud and sandy-mud with the shell aperture uppermost, drawing in a current of water which is used both for respiration and suspension feeding. It is often present in very large numbers and empty shells are frequently washed ashore, especially after storms. Breeding occurs during spring and early summer. The sexes are separate and sperm are believed to be transferred to the female in a spermatophore. Fertilization is internal. The egg capsules are rounded and attached to the sediment by a narrow stalk, each containing from six to 20 pinkish eggs. Several capsules are often deposited together in a cluster. After a week to 10 days, veliger larvae emerge, and are believed to have only a short pelagic life. The sipunculan *Phascolion strombi* (p. 414) lives in mud and sand accumulated in empty shells of *Turritella*.

Family Littorinidae
SUBFAMILY LACUNINAE – THE CHINK CELLS
Shell with prominent spire, or shell flattened and spire reduced. Aperture without siphonal canal. With or without umbilicus; the species described here have an umbilicus and a groove leading to it. Posterior of foot with 1 pair of tentacles (metapodial tentacles).

1  Shell with prominent spire, 5–6 whorls; groove leading to umbilicus smooth. Usually with broad bands of red-brown colour ............................................. *Lacuna vincta* (p. 236)

Shell with spire of 3–4 whorls; body whorl very large. Groove leading to umbilicus with longitudinal ridges ............................................................................................ 2

2 Shell flattened, spire reduced; aperture very large. Usually green colour; white specimens present in some populations ..................................... *Lacuna pallidula* (p. 237)
Shell with low spire; aperture large. Brownish in colour, sometimes with brown bands on body whorl .............................................................................. *Lacuna parva* (p. 237)

*Lacuna vincta* (Montagu)    Banded chink shell (Fig. 137a)

*Shell of 5–6 whorls; with prominent spire; up to 10 mm in height; narrow umbilicus, groove leading to it smooth. Cream or brown coloured, most specimens with red-brown bands.*

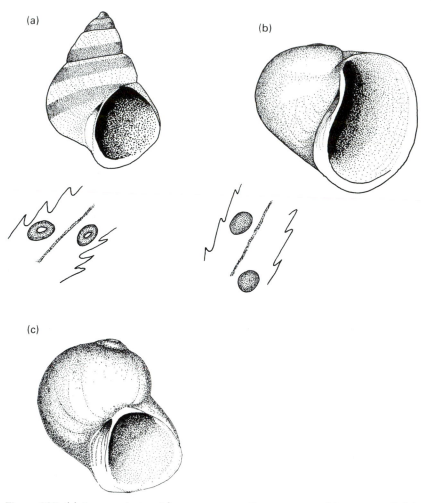

Figure 137  (a) *Lacuna vincta* with egg masses on *Fucus serratus*. (b) *Lacuna pallidula* with egg masses on *Fucus serratus*. (c) *Lacuna parva*.

*L. vincta* is widely distributed in north-west Europe and extends as far south as the Channel coast of France. It is most abundant in the sublittoral and is found all around Britain on semi-exposed to sheltered rocky shores, always with seaweed on which it feeds. *Lacuna* migrates onto the lower shore to breed, generally from January through to summer. The sexes are separate; following copulation and internal fertilization, ring-shaped egg masses, white, yellow or pale green in colour, are laid on seaweeds such as *Fucus serratus* (p. 49) and *Laminaria* (p. 43) and on *Zostera* (p. 85). A single egg mass may contain more than 1000 eggs. After two to three weeks free-swimming veliger larvae emerge, which spend two or three months in the plankton before settling. Length of life is no more than one year.

### *Lacuna parva* (da Costa)    (Fig. 137c)

*Shell of 3 whorls, with short spire and large body whorl; up to 4 mm in height. Umbilicus deep, groove leading to it with longitudinal ridges. Shell brownish in colour; body whorl sometimes banded.*

*L. parva* is widely distributed in north-west Europe and is often found along with *L. vincta*. Like other lacunids, it is found on seaweeds on which it feeds. The sexes are separate and clear, spherical egg masses are laid on seaweeds. Longevity is one year.

### *Lacuna pallidula* (da Costa)    Chink shell (Fig. 137b)

*Shell of 3–4 whorls, flattened; spire reduced, body whorl and aperture very large; umbilicus deep, with groove leading to it ridged longitudinally. Usually green in colour but white shells present in some populations. Females up to 8 mm in height, about twice as large as males.*

*L. pallidula* is widely distributed in north-west Europe and extends as far south as the west coast of France. It is found all around Britain on the lower shore on semi-exposed to sheltered rocky shores, most often on *Fucus serratus* (p. 49), on which it feeds. It occurs in the sublittoral to depths of about 70 m. The sexes are separate and, following copulation and internal fertilization, round to oval egg masses are laid on the seaweed. They are white in colour and attached firmly to the weed, each containing up to 100 or so eggs. Breeding occurs during winter and spring and after eight to 14 weeks, depending on temperature, crawling young emerge from the egg mass. Length of life is no more than one year. The egg masses of *L. pallidula* resemble those of the flat periwinkles, *Littorina obtusata* (below) and *L. mariae* (below). Egg masses of the latter species, however, are usually oval to kidney-shaped in outline, easily peeled from the surface of the seaweed (or in some cases the rock surface) and contain many more eggs than those of *Lacuna pallidula*.

SUBFAMILY LITTORININAE – PERIWINKLES

Shell with obvious spire or relatively flat (two species). The body whorl is large, accounting for 75% or more of the shell height. Shell smooth to strongly ridged with spiral ridges. Without siphonal canal; without umbilicus. Shell colour varied, black, grey, yellow, olive. A large group of common and widespread intertidal molluscs found at all levels on rocky shores. Some species live on soft sediments and extend into estuaries.

1  Shell with distinct spire  ........................................................................................  2
   Shell flattened; spire small  ...................................................................................  4
2  Transparent flap-like extension (the periostracum) projects beyond edge of outer lip of shell. Shell no more than about 9 mm in height  ................  *Melarhaphe neritoides* (p. 243)
   Shell without transparent flap-like extension projecting beyond outer lip  .......................  3
3  Shell up to 30 mm in height. Outer lip joins body whorl tangentially (Fig. 138a). Columellar lip white in small specimens but not obviously so in some larger specimens. Inner surface of outer lip of small specimens with alternate light and dark bands. Juveniles with shells less than about 7 mm in height, heavily ridged  ........................
   ...................................................................................  *Littorina littorea* (p. 238)
   Shell up to 20 mm in height, smooth to strongly ridged. Outer lip of aperture joins body whorl at about a right angle (Fig. 140d)  .  *Littorina saxatilis* species complex (p. 241)
4  Shell with flattened spire. Outer lip joins body whorl a little below apex. Large aperture. Outer lip thickened in older specimens  ......................  *Littorina obtusata* (p. 239)
   Shell has little if any spire. Outer lip joins body whorl at about level of apex (Fig. 139b). Outer lip thickened in older specimens. Aperture small  ...  *Littorina mariae* (p. 240)

***Littorina littorea*** (Linnaeus)    Edible periwinkle (Fig. 138)

*Shell with faint spiral lines on whorls but surface generally smooth; apex pointed; up to 30 mm in height. Outer lip to aperture joins body whorl tangentially. Dark grey or black colour; columella white. In small specimens inner surface of outer lip has alternate light and dark bands. Tentacles with black transverse bands.*

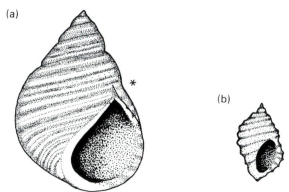

Figure 138 *Littorina littorea*, (a) adult, (b) young specimen, 4 mm shell height. Note angle at which outer lip joins body whorl (*).

*L. littorea* is widely distributed on rocky coasts in north-west Europe and all around Britain, from the upper shore into the sublittoral. It also occurs in estuaries and on mud-flats. *Littorina* grazes on microorganisms and detritus, and is also known to feed on green algae such as *Ulva lactuca* (p. 36) and *Enteromorpha* spp. (p. 35), apparently rejecting *Ascophyllum nodosum* (p. 47). The sexes are separate; on the west coast of Wales, the gonads develop during winter. Copulation takes place during early spring and after internal fertilization the eggs are shed into the sea in gelatinous capsules, usu-ally containing up to three eggs, occasionally more. Egg release is synchronized with the time of spring tides and the capsules are pelagic. After several days, free-swimming veliger larvae hatch and, after a pelagic life estimated to be as long as six weeks, they settle on the lower shore among barnacles and in crevices, usually during June and July. Generally, shells up to about 7 mm in height are heavily ridged with a crenulate margin to the outer lip and can be confused with the young of the *Littoria saxatilis* species complex (p. 241). However, they can be identified by other characteristics given above, particularly the alternate light and dark banding on the inner surface of the outer lip. Maturity is generally reached at two or three years of age and a shell height of about 12 mm. Longevity is five or more years. The edible periwinkle has some commercial value and is gathered by hand at a number of localities.

**Littorina obtusata** (Linnaeus)    (*Littorina littoralis*), Flat periwinkle (Fig. 139a)

*Shell with flattened spire; large aperture; up to 15 mm in height. Outer lip joins body whorl a little below apex of shell; thickened in older specimens. Colour varied, usually olive-green, but yellow, brown, banded and criss-cross patterned varieties also common.*

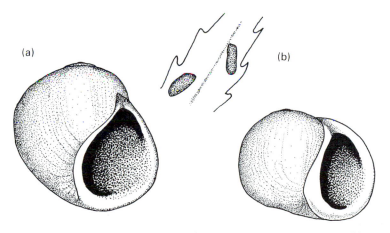

Figure 139  (a) *Littorina obtusata* with egg masses on *Fucus serratus*. (b) *Littorina mariae*.

*L. obtusata* is widely distributed in north-west Europe. It occurs all around the coast of Britain and is almost always found in association with seaweed, particularly *Ascophyl-lum nodosum* (p. 47), *Fucus vesiculosus* (p. 50) and *Fucus serratus* (p. 49). It grazes on these large algae and it has been shown that it prefers *Ascophyllum*, even though this is known to produce chemicals which deter most grazers. Shell colour is related to the degree of exposure of the shore to wave action, olive-green shells being dominant on sheltered shores and criss-cross brown shells on exposed shores. These differences are believed to be maintained by visual selection of predators such as shore fishes. The sexes are separate and fertilization is internal following copulation. Whitish egg masses, usually oval or kidney-shaped in outline and containing up to about 280 eggs, are laid on *Ascophyllum*, *Fucus vesiculosus* and *F. serratus*, and occasionally on the rock surface. They are easily confused with those of *Lacuna pallidula* (p. 237) but are more easily removed from the weed. Breeding occurs during most of the year; generally there is a peak in spring and early summer. Development within the egg mass takes up to four weeks and crawling young emerge. Sexual maturity is reached at two years of age and longevity is probably three or more years.

*Littorina mariae* Sacchi & Rastelli    Flat periwinkle (Fig. 139b)

*Shell with flattened, if any, spire; very similar to* L. obtusata *(above) but flatter; up to 10 mm in height. Outer lip joins body whorl at about level of apex; thickened in older specimens giving a relatively small aperture. Colour varied, usually yellow or brown criss-cross pattern.*

*L. mariae* has a similar distribution to that of *L. obtusata* and is found all around the coasts of Britain, usually on *Fucus serratus* (p. 49), on which it feeds by grazing epiphytes from the surface of the weed. It tolerates more exposed situations than *L. obtusata* and occurs lower on the shore, but there is some overlap in the vertical distribution of the two species. Shell colour is varied; yellow shells are dominant on sheltered shores; criss-cross brown shells on exposed shores and, as with *L. obtusata*, these colours are believed to be maintained by visual selection of predators such as shore fishes. In some populations on sheltered shores, juvenile snails up to a diameter of 3.5 mm are often white in colour and it has been suggested that white colour at this size mimics the tubeworm *Spirorbis* (p. 208), which is found attached to *Fucus serratus*. This strategy is believed to reduce predation rates by shore fishes such as the blenny. The sexes are separate and fertilization is internal; egg masses are laid on the fronds of seaweed, particularly *F. serratus*. They are very difficult to distinguish from those of *L. obtusata* (above) and are found throughout the year, predominantly in spring and summer. *L. mariae* is believed to be an annual.

Prior to 1966, the flat periwinkles were lumped together and called either *L. obtusata* or *L. littoralis*. In 1966 Sacchi and Rastelli split the group into *L. obtusata* and a new species, *L. mariae*, which is the designation used in this text and now widely accepted by shore ecologists. However, it has been suggested that *L. mariae* described by Sacchi and Rastelli was a redescription of **Littorina fabalis**, as described and illustrated by Turton in 1825.

THE *LITTORINA SAXATILIS* SPECIES COMPLEX

Members of the 'Littorina saxatilis species complex', commonly known as the rough periwinkles, are widely distributed on the shores of north-west Europe, but because they have a history of taxonomic confusion the literature is rich in specific and subspecific names. Although debate continues about the status of some members of the group, many taxonomists now accept that it includes four species, namely: Littorina nigrolineata, L. neglecta, L. arcana and L. saxatilis. This is the position adopted in this text, although there is some doubt as to whether L. saxatilis is a single species or an aggregate of species, and it has recently been suggested that L. neglecta is a 'barnacle-dwelling' ecotype of L. saxatilis. **Littorina arcana**, described in 1978 by Hannaford Ellis, cannot reliably be separated from L. saxatilis using external characters alone, and is therefore not included here. It should be noted, however, that in some situations L. arcana may be as common as L. saxatilis and has undoubtedly been confused with the latter species. Examination of the internal anatomy and reference to Hannaford Ellis (1979) are recommended.

### *Littorina nigrolineata* Gray    (Fig. 140c)

*Shell up to 20 mm in height; distinct spiral ridges on whorls separated by narrow grooves; ridges sometimes flattened. Outer lip of aperture joins body whorl at about a right angle. Background shell colour may be white, yellow or reddish-brown. The grooves are clearly pigmented (purplish-brown) in some populations, while in others the pigmentation may be faint or absent. When pigmentation absent, confusion with* L. saxatilis *may result.*

The distribution of L. nigrolineata is inadequately known. It appears to be widely distributed around Britain, occurring on the middle to upper shore of fairly exposed rocky coasts in association with L. saxatilis (below), but is much less abundant than the latter species. It browses on a wide variety of microorganisms and detritus. The sexes are separate, fertilization internal and pinkish egg masses, each containing 100–200 eggs, are laid under stones, in crevices and other damp situations. Although egg masses are found throughout the year, there is a peak in May and June. After about four to seven weeks, depending on temperature, crawling young emerge. Length of life is up to five years.

### *Littorina neglecta* Bean   (Fig. 140b)

*Shell tiny, usually no more than 3.5 mm in height; with a short spire. Outer lip of aperture joins body whorl at about a right angle. Juvenile snails pale brown or grey, most developing dark brown banding at shell height of about 1–1.5 mm; shells above 2.5 mm often tessellated.*

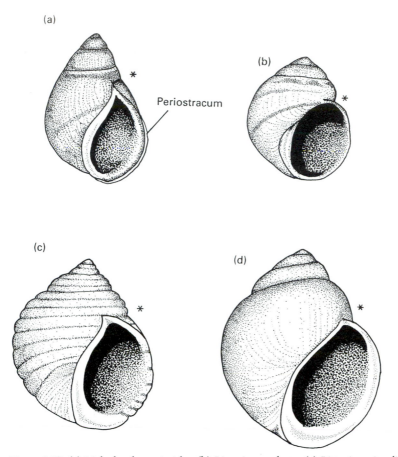

Figure 140  (a) *Melarhaphe neritoides*. (b) *Littorina neglecta*. (c) *Littorina nigrolineata*. (d) *Littorina saxatilis*. Note angle at which outer lip joins body whorl (*).

*L. neglecta* is widely distributed around Britain in exposed and semi-exposed situations among barnacles and in empty barnacle shells and is readily collected by taking scrapings of barnacles from the rock face. This habitat is also occupied by very small specimens of other members of the *L. saxatilis* group and small *L. littorea* and care must be taken to distinguish these. It has been suggested that *L. neglecta* is a 'barnacle-dwelling' ecotype of *L. saxatilis*. Work carried out on the 'neglecta shell-type' shows that the sexes are separate, fertilization internal and the embryos develop to shelled, crawling young in a brood chamber within the body of the female snail, a mode of development known as ovoviviparity. There is a peak of breeding activity in May and June and females of 2 mm shell height carry about 10 embryos at different stages of development. Snails of 3 mm shell height and above carry as many as 50 or more embryos. After about five to six weeks young snails emerge. Maturity is reached after one year and length of life is 12–18 months. The coloration and banding of the shell is believed to be cryptic in the barnacle habitat.

***Littorina saxatilis*** (Olivi)    (*Littorina rudis*), Rough periwinkle (Fig. 140d)

*Shell with pronounced spire; up to 18 mm in height, but generally smaller. Whorls with spiral ridges. Outer lip of aperture joins body whorl at about a right angle. Colour varied, yellowish-white, greenish, reddish, brown, sometimes patterned.*

*L. saxatilis* is widely distributed in north-west Europe and around Britain on the upper reaches of shores of all grades of exposure to wave action. It is found in crevices, on stones and rocks, and with seaweeds such as *Fucus spiralis* (p. 49), occurring on mud-flats and in estuaries, in salinities down to about 8‰. Variation in shell colour from one population to the next is believed to be maintained, at least in part, by visually hunting predators, the different colours being camouflaged with the colour of the rock sub-stratum. The rough periwinkle feeds on microorganisms, detritus and seaweed. The sexes are separate, fertilization internal and the embryos develop to shelled, crawling young in a brood chamber within the body of the female (ovoviviparity). Females with embryos are found throughout most of the year, with a peak of breeding activity in late spring and early summer when the brood chamber of a single female may contain more than 100 embryos at different stages of development. Longevity is up to six years.

***Melarhaphe neritoides*** (Linnaeus)    (*Littorina neritoides*), Small periwinkle (Fig. 140a)

*Shell with pointed apex; up to 9 mm in height. Outer lip of aperture joins body whorl tangen-tially. Blackish-brown in colour; transparent, flap-like extension (the periostracum), pro-jects beyond edge of outer lip. Often with pale band on base of body whorl, running into aperture.*

*M. neritoides* is widely distributed in north-west Europe and occurs around most of Bri-tain, although absent from the southern North Sea. It is found in cracks and crevices on rocky shores, high in the littoral fringe, its upper limit increasing with increasing expo-sure to wave action. In some areas it can be found many metres above the height of the highest spring tides, where it is wetted only by spray. Microorganisms, detritus and lichens are the main food, and the snails browse over the rock surface during moist con-ditions. The sexes are separate, fertilization internal and breeding takes place during winter and spring, although in some years many snails do not reproduce. Egg capsules, each containing one egg, are liberated fortnightly on spring tides. They are pelagic and after several days free-swimming veliger larvae emerge and spend about three weeks in the plankton before settling on the shore in cracks, crevices and empty barnacle shells. On the south coast of Ireland shell heights of less than 2 mm were recorded after the first year's growth. Longevity is five years and more, some of the larger specimens prob-ably being considerably older.

Family Rissoidae

Small shell with spire, usually less than about 6 mm shell height and never greater than 9–10 mm. Shell smooth to heavily ridged. Variously coloured, white, cream, orange, brown. Aperture without siphonal canal. Umbilicus may be present as a narrow slit. Predominantly found on rocky shores where they are sometimes abundant under stones and among algae, a feature which is useful in distinguishing them from the Hydrobiidae (below).

A very large family of marine snails found in littoral and sublittoral habitats. Only two species are described here. For further information the reader is referred to Graham (1988).

**Rissoa parva** (da Costa)    (Fig. 141b, c)

*Shell up to 4 mm in height; some shells smooth, others ribbed with longitudinal ridges but intermediates common; shades of brown; apex purplish; many have a brownish, coma-shaped mark on the body whorl; while useful in identification, this is absent from shells below 2 mm shell height and from many above 2 mm.*

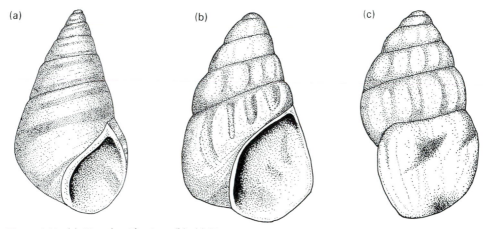

Figure 141   (a) *Cingula trifasciata*. (b), (c) *Rissoa parva*.

R. parva is widely distributed in north-west Europe and on British coasts, often in very high densities. It is found on the middle and lower shore and below, on stones, among red seaweeds such as *Corallina officinalis* (p. 61), on *Lomentaria articulata* (p. 66) and in the holdfasts of *Laminaria digitata* (p. 43). The proportion of smooth to ribbed shells varies from one locality to another. The snails feed on microorganisms and detritus. The sexes are separate, fertilization is internal and rounded egg capsules, each containing up to 50 eggs, are attached to seaweed. Veliger larvae take about three weeks to develop and probably have a long pelagic life before settling on red seaweed on the lower shore. Breeding and settlement of young apparently take place over the whole year. Length of life is up to one year, many animals living only a few months.

*Cingula trifasciata* (J. Adams)    (*Cingula cingillus*) (Fig. 141a)

*Shell up to 4 mm in height, smooth, but with spiral lines on body whorl; yellowish-brown in colour, with conspicuous, dark brown bands. Some specimens lack the dark bands and are white. Black specimens also occur.*

*C. trifasciata* is widely distributed in north-west Europe and around Britain, occurring in rock crevices and among barnacles on the upper shore along with *Melarhaphe neritoides* (p. 243), *Littorina neglecta* (p. 241), *Otina ovata* (p. 277) and *Lasaea adansoni* (p. 296). It is also found among seaweed and under stones on the lower shore, and in the sublittoral to depths of about 20 m. *Cingula* feeds on detritus. Breeding takes place in spring and summer. The sexes are separate; fertilization is internal and rounded egg capsules are deposited in crevices. Each capsule contains up to four eggs from which crawling young emerge.

Family Hydrobiidae – mud snails

Shell smooth (but see the keeled form of *Potamopyrgus antipodarum*); with spire; up to about 6 mm in height. Brownish-black in colour. Aperture without siphonal canal. With or without umbilicus but in the species described here the umbilicus is reduced to a narrow slit or closed. Found on estuarine muds and sands, often in very high densities, some species extending into brackish and fresh water. Different species of *Hydrobia* often co-exist and care is needed in their separation. Not found on rocky shores.

1   Shell whorls slightly swollen, brown-yellow colour but often black with surface deposits. Some specimens have a keel and bristles on the middle of the whorls. Tentacles with pale line down centre. Found in fresh- or brackish-water habitats ................................................................... *Potamopyrgus antipodarum* (p. 248)
    Shell whorls slightly swollen or almost straight-sided; without keel or bristles. Tentacles with or without a dark pigment patch near tip. Found in brackish to fully saline seawater ............................................................................................................................ 2

2   Shell whorls almost straight-sided; apex blunt. Tentacles pale with dark pigment patch at a distance greater than its own length from tip. On mud-flats and in estuaries, often in very high densities. During spring and summer clusters of eggs covered by sand grains are attached to many shells ................................................ *Hydrobia ulvae* (p. 246)
    Shell whorls slightly swollen. Tentacle pigmentation not as above. Usually in lagoons and ditches without direct contact with the sea ................................................................. 3

3   Tentacles pale, occasionally with a faint, terminal black streak ..........................................
    ................................................................................................... *Hydrobia ventrosa*\* (p. 247)
    Tentacles pale, with dark pigment patch at a distance less than its own length from tip ........................................................................................... *Hydrobia neglecta*\* (p. 247)

\*  Some specimens of *Hydrobia ventrosa* and *Hydrobia neglecta* may be difficult to separate. In these cases males of the two species can be separated on the shape of the penis, which is shown in Fig. 142. The penis of *Hydrobia ulvae* is included for comparison.

*Hydrobia ulvae* (Pennant)    Laver spire shell (Fig. 142a)

*Shell small, up to 6 mm in height; whorls almost straight-sided; apex blunt. Generally light brown to brownish-black in colour. Tentacles pale with dark pigment patch at a distance greater than its own length from tip.*

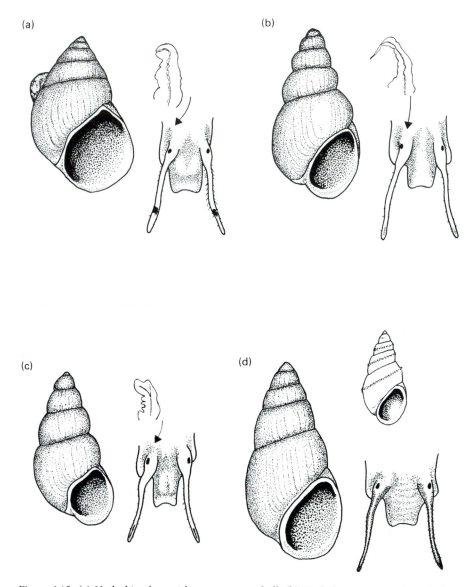

(a)

(b)

(c)

(d)

Figure 142  (a) *Hydrobia ulvae* with egg mass on shell. (b) *Hydrobia ventrosa.* (c) *Hydrobia neglecta.* (d) *Potamopyrgus antipodarum*, specimen with keel inset. Detail of pigmentation of head and tentacles shown for each species, detail of penis shown for (a), (b) and (c). (Partly based on Muus, 1963.)

*H. ulvae* is widely distributed in north-west Europe and around Britain on muddy-sand, mud-flats and in estuaries, often in densities as high as 20 000 individuals per square metre. It is most abundant on the middle and upper reaches of the shore in salinities of 10–33‰. It can, however, tolerate salinities as low as 1.5‰. When covered by the rising tide, the snail secretes a mucous raft and floats upside down in the surface film of water. The floating phase lasts for a short time during tidal cover. When the tide is out, *Hydrobia* is often seen climbing vertical objects such as the stems of *Spartina* (p. 85), with which it is often found, and from which it launches itself on the rising tide. The food of *Hydrobia* consists mainly of silt, diatoms and fungal mycelia removed from the surface of sediment particles by the action of the radula; feeding on detritus trapped in the mucous raft also takes place when the snail is floating. Breeding occurs from spring to autumn. The sexes are separate, fertilization internal and clusters of from four to eight eggs are deposited in an egg mass cemented onto the shell of a neighbouring *Hydrobia* or, less frequently, on dead shells or other firm substrata. At some localities peaks of spawning have been recorded in spring and autumn; others show a single peak. The egg mass is protected by a covering of sand grains and the eggs develop into free-swimming veliger larvae, which emerge after 20–30 days. After a pelagic life of about three weeks these metamorphose and young snails settle on the shore. Sexual maturity is reached after one or two years and the length of life is usually about two and a half years.

### *Hydrobia neglecta* Muus   (Fig. 142c)

*Shell similar to that of* H. ventrosa *(below) but whorls not as rounded; apex blunt; up to 4 mm in height. Grey, brown, black in colour. Tentacles pale, with dark pigment patch at a distance less than its own length from the tip.*

*H. neglecta* has a patchy distribution in north-west Europe, having been recorded from Denmark, the east coast of England and the west coast of Scotland and Ireland, and from a single locality in France. Apart from the record from France, it appears to be restricted to lagoons and ditches without direct contact with the sea, usually in salinities of 10–24‰. It lives on sandy and muddy sediments and on *Zostera* (p. 85), feeding on diatoms and bacteria. The sexes are separate, fertilization internal and egg capsules, similar to those of *H. ventrosa* but containing only one embryo, are laid in spring and summer on stones and empty shells. After about three weeks, crawling young emerge and the length of life is about 18 months. This species was first recognized in 1963 by a Danish biologist, Muus, and had probably been confused previously with *H. ulvae* (above) and *H. ventrosa* (below); its distribution may be more widespread than currently recorded.

### *Hydrobia ventrosa* (Montagu)   *Ventrosia ventrosa* (Fig. 142b)

*Small shell, up to 5 mm in height; whorls rounded, apex blunt. Brownish-black in colour. Great majority have pale tentacles, occasionally there is a faint, terminal black streak.*

*H. ventrosa* is widespread on sand and muddy deposits in north-west Europe, but much less so than *H. ulvae* (above), having a patchy distribution on British coasts, where it is usually restricted to sheltered lagoons and ditches that are almost cut off from the sea. It is normally found in salinities of 6–20‰. Diatoms and bacteria removed from the surface of sediment particles are the main food. Breeding occurs in spring and summer. The sexes are separate, fertilization internal and egg capsules, protected by a covering of sand grains, are laid on empty shells, stones and occasionally on the shells of neighbouring *H. ventrosa*. The capsules contain one, two or rarely three eggs and development takes about three weeks. Crawling young emerge from the capsule and length of life is about 18 months.

### *Potamopyrgus antipodarum* (Gray)    (*Potamopyrgus jenkinsi*) (Fig. 142d)

*Brownish-yellow shell, but often black due to surface deposits; whorls slightly swollen; up to 5 mm in height. Some shells have a keel and bristles on the middle of the whorls. Tentacles pale.*

*P. antipodarum* is widespread in Britain and Europe in fresh- and brackish-water habitats (more especially running water), with a salinity range from 0 to 20‰. It tends to occur in high densities and would appear to have a preference for hard water. It feeds on diatoms and detritus. Generally, all individuals in a population are females, which reproduce parthenogenetically; the young undergo development in a brood chamber and leave the parent as crawling individuals. In a few populations males have been recorded, the first in 1958 from the river Thames at Sonning. Males are present in low numbers and although they may have mature sperm, their precise role in reproduction has not been determined. Hermaphroditic animals have been recorded in very low numbers. In northwest England, breeding has been recorded all year, and longevity is about seven months. In Britain, the species was first recorded from Gravesend in 1859, since when it has spread rapidly throughout the country, although it is rare in parts of mid-Wales and Scotland.

Family Aporrhaiidae – pelican's foot
Shell tall, with 8 or so whorls. Outer lip drawn out into large pointed expansion to give appearance of pelican's foot. Without umbilicus. Aperture with siphonal canal.

### *Aporrhais pespelecani* (Linnaeus)    Pelican's foot (Fig. 143)

*Shell of distinctive appearance, up to 40 mm in height; whorls with longitudinal ridges and tubercles. Outer lip of adult expanded outwards to give appearance of pelican's foot. Greyish-white in colour with brown markings.*

Figure 143  *Aporrhais pespelecani.*

*A. pespelecani* is common in north-west Europe in mud, muddy-gravel and muddy-sand offshore to depths of about 180 m. Empty shells are often washed ashore. *Aporrhais* eats plant debris, feeding when buried in the sediment and when on the surface. Breeding has been recorded in spring. The sexes are separate, fertilization internal, with the eggs deposited either singly or in groups of two or three, on sand grains and debris. After about two weeks, free-swimming veliger larvae emerge; they are believed to have a lengthy pelagic life. The sipunculan *Phascolion strombi* (p. 414) sometimes inhabits empty shells of *Aporrhais*.

Family Calyptraeidae – slipper limpets
Shell conical or cap-shaped with internal shelf extending across part of aperture. Without operculum.

Shell oval, cap-shaped; aperture very large; occurs in chains, large specimens at base, smaller ones above ............................................................... *Crepidula fornicata* (p. 249)
Shell conical, with pointed apex; not forming chains ........... *Calyptraea chinensis* (p. 250)

*Crepidula fornicata* (Linnaeus)    Slipper limpet (Fig. 144)

*Oval, cap-shaped shell with much reduced spire; surface marked with growth lines; up to 25 mm in height, 50 mm in length. Aperture very large with shelf extending about half-way across. Orange-brown, white or yellow in colour; shelf white. No operculum. Occur in chains with large specimens at base, smaller ones above.*

(a)                                         (b)

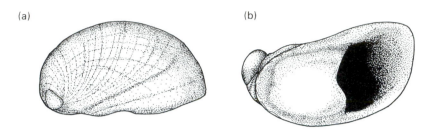

Figure 144  *Crepidula fornicata*, (a) lateral view, (b) apertural view.

*C. fornicata* is widely distributed in north-west Europe, and is found on south, south-west and south-east coasts of Britain. It is a North American species often found attached to oyster shells and is believed to have entered Britain on imported oysters, the first specimens being recorded in Essex in about 1890. High densities of *Crepidula* on oyster beds cause much damage as the oysters are smothered. Curved chains of up to 12 animals can be found on the lower shore and sublittorally to about 10 m, often attached to stones, and the shells of oysters and mussels. They feed on small particles which are drawn into the mantle cavity in a water current and trapped on a mucous filter.

*Crepidula* is a protandrous hermaphrodite; in a chain of individuals the smaller, younger animals at the top of the chain are male and the larger, older ones at the base of the chain are female. Copulation is followed by internal fertilization and the eggs are laid in stalked capsules attached to the foot or shell of the female, or to a stone or shell under the female, thereby affording them some protection. Development to the veliger stage takes three to four weeks and after a pelagic life estimated to be about five weeks, the animals settle under stones and shells. By the time they have reached a shell length of 3–5 mm, the young animals migrate and take up position on an existing chain of *Crepidula* as males. Others not locating a chain settle on stones and develop as females. Breeding occurs from April to September; most females spawn twice a year and it has been suggested that spawning is correlated with neap tides. Length of life is about eight to nine years.

### *Calyptraea chinensis* (Linnaeus)    Chinaman's hat (Fig. 145)

*Shell conical; apex sharp, sometimes coiled; surface smooth apart from near aperture where small tubercles occur; up to 5 mm in height, 15 mm in length. Shelf extending about half-way across aperture. External surface of shell yellowish-white in colour; shelf white. No operculum.*

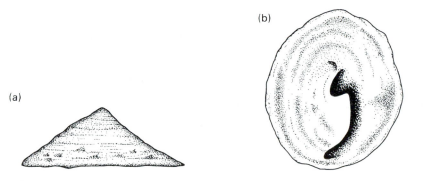

Figure 145 *Calyptraea chinensis*, (a) lateral view, (b) apertural view.

*C. chinensis* reaches its northern limit in Britain, where it is found on the west coast as far as Scotland, but is absent from the North Sea. It is a lower shore and shallow sublittoral species that lives attached to shells and stones. Small particles of food are trapped on a mucous filter. It is a protandrous hermaphrodite; the smaller, younger, more active animals are male and settle temporarily on the shells of the older, larger, more sedentary females, but do not form chains, cf. *Crepidula* (above). Copulation is followed by internal fertilization. Eggs are laid in capsules that are roughly triangular in outline and attached to a stone under the front of the shell of the female. A single capsule contains up to 25 eggs. There is no veliger larva; crawling young emerge. Breeding occurs from spring to summer and length of life is three years.

Family Triviidae – cowries
Shell glossy; smooth or strongly ridged; aperture elongate, narrow; body whorl very large, may have short spire or be typical cowrie shape. Without operculum.

Shell glossy, oval in shape, with 3 dark spots; surface conspicuously marked with ridges; aperture long, narrow; typical cowrie shape ................... *Trivia monacha* (p. 251)
Shell as above but without the 3 dark spots ...................................... *Trivia arctica* (p. 252)

Note: the shell of young specimens of both species, up to a shell height of about 5–6 mm, is smooth with a wide aperture and a short spire.

*Trivia monacha* (da Costa)    European cowrie, Spotted cowrie (Fig. 146)

*Glossy shell, oval in shape with long, narrow aperture; surface conspicuously marked with ridges; up to 13 mm aperture length. Upper surface reddish-brown with 3 brown spots; lower surface white. Mantle brightly coloured, extending to cover shell when animal is active. Operculum absent. Young shells with short spire, wide aperture and lacking the characteristic ridging of larger specimens.*

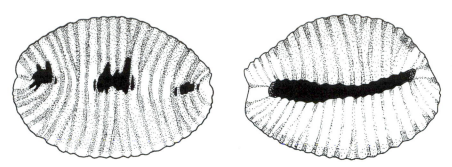

Figure 146  *Trivia monacha.*

*T. monacha* is a southern species, extending northwards to the west coast of France, and around most of Britain. It is found on the lower shore and sublittorally on rocky shores with compound ascidians such as *Botryllus schlosseri* (p. 478) and *Botrylloides leachi* (p. 479) on which it feeds. Breeding occurs during late spring and summer. The sexes are separate, fertilization internal and flask-shaped egg capsules, each with up to 800 eggs, are deposited in holes eaten out of the ascidian by the female. The tip of the egg capsule is button-shaped and can be seen protruding from the surface of the ascidian. Free-swimming larvae emerge after a few weeks and are pelagic for a month or so.

### *Trivia arctica* (Pulteney)   (not illustrated)

*Shell as described for* T. monacha *but without the 3 dark pigment spots. Up to 10 mm aperture length.*

*T. arctica* is widely distributed in north-west Europe and although predominantly found in the sublittoral it is occasionally found along with *T. monacha* on the very low shore. It feeds on ascidians and its life-cycle is similar to that of *T. monacha*.

Family Lamellariidae
In the species described here, the shell is internal and enclosed by mantle. Similar in appearance to sea-slug (Opisthobranchia, p. 261).

### *Lamellaria perspicua* (Linnaeus)   (Fig. 147)

*Shell totally enclosed by mantle; spire very small; smooth; colourless and transparent; aperture very large; up to 10 mm shell length. Animal up to 20 mm in length; colour varied, yellow, grey, lilac, sometimes with coloured flecks, white, yellow or black. Mantle covered with tubercles and thickened at edge; anteriorly forms a siphon-like extension. Tentacles smooth. May be mistaken for an opisthobranch mollusc.*

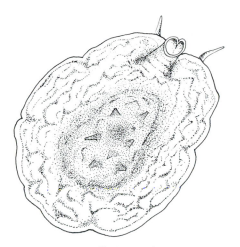

Figure 147 *Lamellaria perspicua.*

*L. perspicua* is widely distributed in north-west Europe and around Britain on the lower shore and in the sublittoral, among stones and rocks where compound ascidians such as *Botryllus schlosseri* (p. 478) occur. It feeds on ascidians, breeding in spring and summer. The sexes are separate, fertilization internal and eggs are laid in flask-shaped egg capsules deposited in holes eaten out of the ascidians by the female. Each capsule contains up to 3000 eggs. Pelagic larvae emerge after about three weeks and probably have a long pelagic life.

A second species of *Lamellaria*, **Lamellaria latens** (Müller), has a similar distribution to *L. perspicua* but can be recognized by its small size, no more than 10 mm in length, and a flatter mantle surface with fewer tubercles.

Family Naticidae – necklace shells
Smooth shell with short spire. Body whorl very large. Aperture large and without siphonal canal. The foot is large and covers head and part of shell when animal is active. The two species likely to be encountered intertidally are found on mud and sand on the very low shore; each has a distinct umbilicus.

**Polinices catena** (da Costa)    (*Lunatia catena*, *Natica catena*), Large necklace shell (Fig. 148)

*Smooth, glossy shell, up to 35 mm in height; short spire; umbilicus large and deep. Fawn to pale orange with a single row of brownish streaks or lines on upper part of whorls. Foot large; covers head and part of shell when animal is active.*

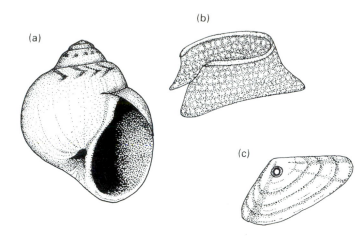

Figure 148  (a) *Polinices catena*, and (b) its egg mass. (c) Shell of bivalve mollusc bored by
*Polinices*.

*P. catena* extends as far north as Denmark and is found all around Britain, from the lower
shore to about 125 m on sandy sediments. It preys on bivalves such as *Donax* (p. 309),
entering the shell by chemical and mechanical means, the shell being softened by a chemical before being bored by the radula. Empty bivalve shells with the characteristic hole
made by *Polinices* spp. (Fig. 148c) are often washed ashore. The sexes are separate, fertilization internal and the eggs are laid in a coiled, greyish-brown mass having the appearance of a collar. The egg mass can be up to 170 mm in length. Only a small proportion
of the eggs develop, the rest serving as food for the developing embryos. There is some
doubt as to whether there is a free-living veliger larva in this species.

*Polinices polianus* (delle Chiaje) (*Lunatia alderi*) lives in deeper water than *P. catena*,
only occasionally being found on the lower shore. It is smaller than *P. catena* (up to
15 mm) and the body whorl has five rows of brown markings.

Family Epitoniidae
Shell with spire, up to 15 whorls. Pronounced, well-spaced longitudinal ridges. Aperture
rounded. Without siphonal canal.

*Epitonium clathrus* (Linnaeus)    (*Clathrus clathrus*), Wentletrap (Fig. 149)

*Shell tall, narrow; many whorls with pronounced longitudinal ridges; up to 40 mm in
height. Aperture rounded. Whitish to pale brown in colour, with blotches of pigment on
longitudinal ridges.*

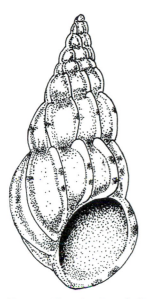

Figure 149  *Epitonium clathrus.*

*E. clathrus* is widely distributed in north-west Europe on sandy and muddy sediments. It is a sublittoral species which migrates onto the lower shore during spring and summer to spawn and is sometimes common on south and west coasts of Britain. There is little information regarding its feeding behaviour; it would appear to eat a range of prey including sea-anemones. Breeding occurs during summer. *Epitonium* changes sex each year; fertilization is internal and strings of polygonal egg capsules covered with sand and mud particles are laid on the shore. After about two weeks veliger larvae are released.

Family Muricidae —whelks and tingles
Stout shell with spire. Often heavily ridged. Aperture with siphonal canal; in some species this is closed over in older specimens. Outer lip often thick and sometimes with teeth. Common on rocky shores.

Shell with spiral ridges. Outer lip thin in young specimens, thicker in older shells, sometimes with teeth. Siphonal canal open ................................ *Nucella lapillus* (p. 255)
Shell with longitudinal and spiral ridges. Siphonal canal open in young specimens, closed in older shells ............................................................... *Ocenebra erinacea* (p. 257)

*Nucella lapillus* (Linnaeus)    Dog-whelk (Fig. 150)

*Shell with spiral ridges; usually up to 30 mm in height, may reach 60 mm. Usually whitish-grey but grey-, brown- or yellow-banded forms also occur. Siphonal canal open, short, leads from base of aperture; outer lip of aperture thick and may be toothed; thin in young shells.*

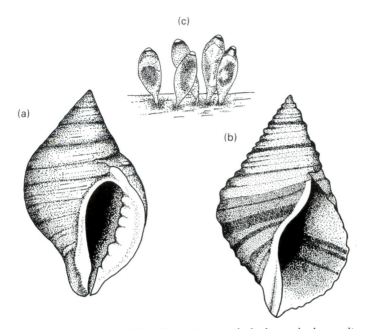

Figure 150 *Nucella lapillus.* (a) Specimen with thick, toothed outer lip, (b) specimen with thin outer lip, (c) egg capsules.

*N. lapillus* is widely distributed in north-west Europe and is common on the coasts of Britain. It occurs from the middle shore into the shallow sublittoral on rocky shores experiencing most degrees of exposure to wave action, but is not found on very sheltered shores. The shape of the shell and the relative size of the aperture vary with the exposure of the shore. In general, specimens from exposed shores have short, squat shells with a relatively large aperture, while those from more sheltered shores have taller shells with a relatively narrow aperture. These differences have been related both to the direct effect of exposure to wave action and to predation by shore crabs. A large aperture indicates the possession of a large foot, which enables the dog-whelk to cling to the rock surface, a useful adaptation in exposed conditions. A tall shell with a relatively narrow aperture has been shown to be an adaptation against predation by crabs, which are often abundant on sheltered shores. Tall narrow shells are not easily handled by the crab and the thick outer lip is difficult to crack open. It has also been shown that shells from sheltered shores are stronger than those from shores exposed to heavy wave action.

The dog-whelk is carnivorous; acorn barnacles and the mussel, *Mytilus edulis* (p. 283), are the main prey species. Adult dog-whelks force apart the opercular plates of barnacles using the snout-like proboscis, whereas bivalves (and sometimes barnacles) are bored by a combination of mechanical and chemical means. Enzymes, particularly carbonic anhydrase, are secreted to soften the shell, which can then be bored by the radula. The prey is believed to be narcotized by secretions from the dog-whelk. The whole process of boring through a shell may take several hours and whelks are often found on the shore in the act of boring into a bivalve. Cannibalism has also been recorded.

The sexes are separate, fertilization internal and vase-shaped egg capsules, a few milli-metres tall and usually yellow in colour, are attached in clusters to rock surfaces, in crevices and under overhangs. A newly deposited egg capsule may contain up to 1000 eggs but no more than a small number of these develop, the others acting as food for the embryos during the first week or so of development. After about four months crawling juveniles emerge. On the east coast of England, breeding occurs mainly in spring; in other areas it is believed to extend throughout the year. Prior to breeding, adult dog-whelks are found in non-feeding aggregations on the lower shore, where copulation takes place. Maturity is reached after about two years and individuals live for five years or more.

The dog-whelk is now well known through the development of 'imposex' in response to pollution of inshore waters by tributyltin (TBT) leached from anti-fouling paints. Imposex is recognized as the development of a penis and vas deferens in female snails and is widespread in populations of *Nucella* in north-west Europe, being particularly evident in areas close to harbours and marinas. In some cases the vas deferens overgrows and blocks the genital duct with the result that egg capsules cannot be laid and the female is sterile. Whole populations have been affected in some areas but since the intro-duction in 1986 of legislation to control the sale of TBT, levels of the pollutant in inshore waters have fallen and some recovery of the whelk populations has been detected.

**Ocenebra erinacea** (Linnaeus)    Sting winkle, Rough tingle (Fig. 151a)

*Shell with longitudinal and spiral ridges on whorls; up to 50 mm in height. Whitish in colour with orange-brown, brownish markings. Siphonal canal leads from base of aperture; open in young shells, closed over in older shells.*

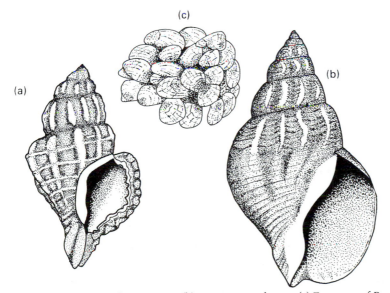

Figure 151  (a) *Ocenebra erinacea.* (b) *Buccinum undatum.* (c) Egg mass of *Buccinum undatum.*

*O. erinacea* is a southern species extending as far north as Britain where it is commonest in the west and south-west. It is found on the lower shore on sheltered rocky shores and on muddy-gravel, extending into the sublittoral to depths of 150 m. It feeds on bivalves and at times can be an important predator of oysters, but it also eats barnacles and tube-dwelling polychaete worms. The method of boring into the prey is similar to that described for *Nucella* (above). Breeding occurs in spring and summer. The sexes are separate, fertilization internal and yellowish egg capsules, not unlike those of *Nucella* (above) but more angular and with a conspicuous plug, are laid in clusters in rock crevices and on stones. They are much less common on the shore than those of *Nucella*. A single capsule may contain as many as 150 eggs (but usually far fewer), all of which normally develop into crawling, juvenile snails after about three months.

Family Buccinidae
SUBFAMILY BUCCININAE – WHELKS
Stout, usually large shell with spire; smooth, or ridged with spiral ridges, longitudinal ridges or both. Aperture oval, large; outer lip usually not thickened. Aperture with siphonal canal which is not closed over. A predominantly sublittoral group found on soft sediments.

Large shell, with both spiral and longitudinal ridges, the latter stop short of base of body whorl ................................................................... *Buccinum undatum* (p. 258)
Large shell, with many spiral ridges; no longitudinal ridges ..... *Neptunea antiqua* (p. 259)

*Buccinum undatum* Linnaeus    Whelk, Buckie (Fig. 151b, c)

*Stout shell; up to 110 mm in height. Whorls with spiral ridges and pronounced longitudinal ridges, latter stop short of base of body whorl; conspicuous ridge runs from base of siphon along columellar lip. Greyish-white, brownish-white in colour. Aperture large. Short, wide siphonal canal leads from base of aperture.*

*B. undatum* is widely distributed in north-west Europe and Britain, from extreme low water mark into the deep sublittoral, usually on soft sediments. It is one of the largest British prosobranch molluscs and is carnivorous, feeding mainly on polychaetes and bivalves. The bivalve shell is forced open by the whelk using the edge of its own shell. Carrion is also eaten and is detected from some considerable distance by olfaction. The breeding period extends from about October to May, depending on locality. The sexes are separate, fertilization internal and the eggs are deposited in capsules attached to rocks, stones, shells, often forming a substantial egg mass which can contain as many as 2000 or more egg capsules, each capsule containing up to 1000 or so eggs. Egg masses can be up to 500 mm across and contain capsules from a number of females. Only a small percentage of the eggs complete development, with the majority utilized as food by the embryos. Crawling young emerge from the capsules after several months and length of life has been estimated to be 10 years, with sexual maturity reached at a shell height of about 60 mm. Empty egg masses are often found on the strandline and are sometimes mistaken for sponges. They are known as 'sea wash balls', a name derived from the fact that the early mariners used them for washing. *Buccinum* is fished commercially using baited pots.

### *Neptunea antiqua* (Linnaeus)    (Fig. 152)

*Stout shell, up to 100 mm in height; whorls with many fine spiral ridges, without longitudinal ridges; conspicuous ridge runs from base of siphon along columellar lip. Cream to reddish in colour. Short siphonal canal leads from base of aperture.*

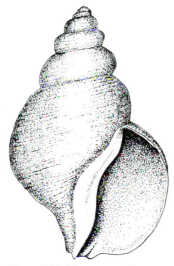

Figure 152  *Neptunea antiqua.*

*N. antiqua* is widely distributed in north-west Europe in the sublittoral, usually at depths of from 15 to 1200 m, but, as with *Buccinum*, empty shells are often found on the shore. It is carnivorous, feeding on a wide range of invertebrate fauna. It breeds in spring, when egg capsules are cemented together in masses on the shells of other whelks. Development takes about six months, the young snails emerging at a shell height of about 6 to 12 mm.

SUBFAMILY NASSARIINAE
Stout shell with spire, strongly ridged. Aperture with siphonal canal. Outer lip may be thickened, sometimes with teeth. In non-apertural or 'back' view the end of the siphonal canal is seen as a notch.

Stout shell, whorls straight-sided; arrangement of longitudinal and spiral ridges gives squared pattern ........................................................................ *Hinia reticulata* (p. 260)

Stout shell, whorls well rounded; with longitudinal and spiral ridges but without squared patterning. Conspicuous brown mark in siphonal canal    *Hinia incrassata* (p. 260)

**Hinia reticulata** (Linnaeus)    (*Nassarius reticulatus*), Netted dog-whelk (Fig. 153a)

*Brownish shell, up to 30 mm in height; criss-crossing of longitudinal and spiral ridges gives squared pattern. Whorls straight-sided. Short siphonal canal leads from base of aperture. In non-apertural or back view, the end of the canal is seen as an indentation. Mature animals with thick outer lip with several teeth.*

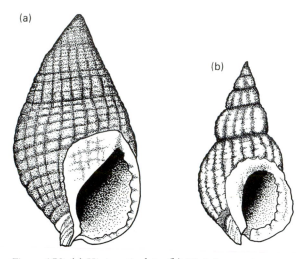

Figure 153  (a) *Hinia reticulata.* (b) *Hinia incrassata.*

*H. reticulata* is widely distributed in north-west Europe and around Britain. It is found on the lower shore and in the sublittoral down to about 15 m on sandy shores or among rocks and sand and is tolerant of salinities down to about 16‰. It feeds on dead and decaying animal matter. Breeding occurs in spring and summer. The sexes are separate, fertilization internal and flattened, vase-shaped egg capsules are laid in rows on *Zostera* (p. 85), on red seaweeds and sometimes on hydroids and stones. Each capsule contains up to 300 or more eggs, which hatch as free-swimming veliger larvae after about a month. The larvae have a long pelagic life of about two months. On the west coast of Sweden, sexual maturity has been recorded after four years and the whelks have a longevity of about 15 years.

**Hinia incrassata** (Ström)    (*Nassarius incrassatus*), Thick-lipped dog-whelk (Fig. 153b)

*Shell up to 12 mm in height, with longitudinal and spiral ridges but without squared patterning. Brownish in colour with dark bands; conspicuous brown mark in siphonal canal. Whorls well rounded. Short siphonal canal leads from base of aperture. In non-apertural or back view, the end of the canal is seen as an indentation. Mature specimens with thick outer lip and several teeth.*

*H. incrassata* is widely distributed in north-west Europe and occurs all around Britain. It is found in crevices and in silt on the lower shore, several specimens often occurring together, and extends into the sublittoral to depths of about 100 m. It feeds on dead and decaying animal matter. Breeding occurs in spring and summer. The sexes are separate, fertilization internal and flattened, vase-shaped egg capsules are laid in clusters on hydroids, bryozoans and *Zostera* (p. 85), each capsule containing 50–80 eggs. After a month or so free-swimming veliger larvae emerge; they have a long planktonic life of two months or more. Longevity is believed to be up to six years.

Subclass Heterobranchia
Superorder Opisthobranchia

The opisthobranchs are among the most colourful of marine organisms. They are marine gastropods in which the shell is usually reduced and enclosed within the body, or is absent. A few species, however, have an external shell: these are the primitive members of the group. The anterior region has up to four different types of paired tentacles. Those on either side of the mouth are the oral tentacles, while those on the dorsal surface of the head are known as rhinophores (Fig. 166a). These vary in structure and sometimes arise from a basal cup-like sheath: they are chemosensory. Posterior cephalic tentacles are sometimes present on the posterior region of the head, while propodial tentacles arise from the anterior region of the foot. Gills are often conspicuous and in some groups are in a posterior circle around the anus. Some opisthobranchs have a large number of dorsal outgrowths known as cerata (Fig. 166a) which may contain extensions of the gut. The cerata produce secretions rendering the animal unpalatable to predators, an important adaptation in animals which have lost the external shell. In some species which feed on sea-anemones, the nematocysts of the prey are arranged in the cerata where they continue to function, thus adding to the defensive role. Some opisthobranchs burrow into soft sediments; others move over the substratum on a muscular foot, often thrown into extensive lateral lobes or parapodia (Fig. 159) extending onto the dorsal surface of the animal. Although some opisthobranchs are herbivorous, for example the sea-hare, *Aplysia* (below), most are carnivorous on a wide range of invertebrates, including sponges, hydroids and bryozoans. Sea-slugs and sea-hares are hermaphroditic but copulation and cross-fertilization usually occur. The eggs are generally deposited in jelly-like egg masses, often as coiled ribbons attached to the substratum, from which free-swimming veliger larvae emerge.

### Key to families of opisthobranchs

The presence or absence of a shell is the first step in the identification of opisthobranchs. When present, as in *Acteon* (p. 263), and *Retusa* (p. 265), the animals must not be confused with prosobranch molluscs (p. 219). When an internal shell is present, it can be detected by careful handling. It is a delicate structure and if it is to be removed, this must be done with care.

The prosobranch mollusc *Lamellaria perspicua* has an internal shell and bears a superficial resemblance to the opisthobranchs with which it is easily confused. The diagnostic features of *Lamellaria* are given on p. 252.

1 External shell present .................................................................................... 2
  Shell not visible, may be internal or absent ............................................... 3
2 Operculum present .................................................... **Acteonidae** (p. 263)
  Operculum absent. Head with pair of large, posterior cephalic tentacles ..........................
  ............................................................................ **Retusidae** (p. 264)
3 Careful examination reveals presence of internal shell ..................................... 4
  Internal shell absent ................................................................................... 6
4 In dorsal view body seen to be made up of 4 lobes .............................. **Philinidae** (p. 264)
  Body not as above ...................................................................................... 5
5 Parapodia very large, flap-like. Pair of oral tentacles and pair of rhinophores ................
  ........................................................................... **Aplysiidae** (p. 268)
  Body rather flattened. Pair of oral tentacles and pair of rhinophores. Single, large gill on right side of body lying between mantle and foot .................... **Pleurobranchidae** (p. 269)
6 Body without a posterior, dorsal circle of gills and without cerata .................................. 7
  Body with a posterior dorsal circle of gills or with cerata ........................................... 9
7 With parapodia, extending along length of body .................................. **Elysiidae** (p. 266)
  Without parapodia ...................................................................................... 8
8 Three gills under posterior mantle edge on right side. Without tentacles ...........................
  ............................................................................ **Runcinidae** (p. 265)
  Without gills under posterior mantle edge on right side. With or without tentacles ......
  ......................................................................... **Limapontiidae** (p. 266)
9 Body with posterior circle of gills around anus; without cerata ..................................... 10
  Body without a posterior circle of gills; with cerata .................................................. 12
10 Circle of gills cannot be completely retracted, but individual gills can retract .................
  ......................................................................... **Onchidorididae** (p. 270)
  Circle of gills can be completely retracted ............................................................ 11
11 Dorsal surface with large, rounded tubercles of range of sizes; rough to the touch .........
  ........................................................................ **Archidorididae** (p. 273)
  Dorsal surface with very many, small, similar-sized tubercles; soft to the touch ...........
  ......................................................................... **Kentrodorididae** (p. 273)
12 Rhinophores smooth, long and slender, unbranched; arise from cup-like sheaths (Fig. 161) ....................................................................... **Dotidae** (p. 269)
  Rhinophores without cup-like sheaths ................................................................. 13
13 Rhinophores ringed. Cerata grouped on dorsal surface. ...................... **Facelinidae** (p. 274)
  Rhinophores smooth; tips pale or white. Dorsal surface with transverse rows of cerata .
  ......................................................................... **Aeolidiidae**(p. 274)

Family Acteonidae

Shell external, oval, glossy; pointed apex. Animal retracts fully into shell; operculum present.

### *Acteon tornatilis* (Linnaeus)   (Fig. 154a)

*Oval shell; up to 25 mm in height. Glossy, pinkish to brown in colour with white bands running round shell. Single tooth present on inner lip of aperture. Body retracts fully into shell. Head flattened, with 4 lobes. Operculum present. May be mistaken for a prosobranch mollusc (p. 219).*

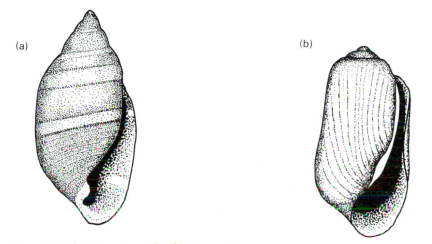

(a)          (b)

Figure 154   (a) *Acteon tornatilis.* (b) *Retusa obtusa.*

*A. tornatilis* is widely distributed in north-west Europe and, although occurring all around Britain, has a patchy distribution and may be locally abundant. It is found on the lower shore to depths of about 250 m on sheltered sandy beaches, where it burrows in the top few centimetres of sediment using its foot and flattened head. The presence of the burrows is revealed on the surface by holes 4–5 mm in diameter. Specimens are occasionally found crawling on the sediment where they leave broad furrows. *Acteon* is carnivorous, the diet including small polychaetes. It is hermaphroditic. In south-west Wales breeding has been recorded from April to August. Maturity is reached at about 10 mm shell length and following copulation club-shaped egg masses are laid on the surface of the sediment. It has been estimated that one individual can produce six or seven egg masses per year, containing as many as half a million eggs. The egg masses are about 60 mm in length and are attached at the narrow end to weed or stones on, or buried in, the sediment. Free-swimming veliger larvae emerge and these are believed to have a planktonic life of one or two months. Most individuals die after spawning at an age of about $1\frac{1}{2}$ years, but a few survive to spawn a second time and may reach three years of age. *Acteon* has many characters which resemble those of the prosobranchs. It is the most primitive British opisthobranch and one of the few species in the superorder to possess an operculum.

Family Philinidae

Shell internal, delicate; aperture large. Parapodia conspicuous and extend onto the dorsal surface but do not meet. In dorsal view body clearly divisible into 4 regions – head, visceral mass and 2 lateral parapodia.

*Philine aperta* (Linnaeus)    (Fig. 155)

*In dorsal view body seen to be made up of 4 lobes – right and left parapodial lobes, cephalic shield (head) and posterior mantle lobe over the visceral mass. Whitish to pale yellow in colour; up to 70 mm in length. Internal shell with very wide aperture; delicate, whitish in colour.*

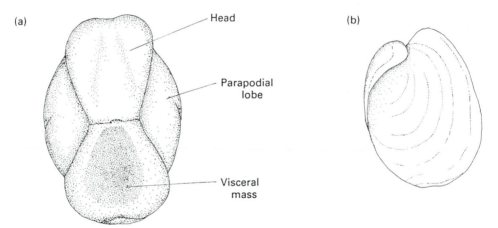

Figure 155  *Philine aperta*, (a) dorsal view, (b) shell.

*P. aperta* is widely distributed in north-west Europe. It is found on the lower shore and into the sublittoral to depths of about 500 m, burrowing in sand and muddy-sand. It preys on a wide range of invertebrates found in soft sediments, including polychaetes, small bivalves and gastropods, and foraminiferans. Acid secretion from the skin give it some protection from predators, which include fish. In Britain spawning has been recorded from spring to summer when flask-shaped egg masses are laid. Veliger larvae hatch after a few days. Longevity is believed to be three to four years.

Family Retusidae

Shell external, delicate; may have short spire or spire flattened. Aperture very elongate, extending along most or all of shell height. Animal retracts fully into shell. Head with 2 large, posterior tentacles. Posterior margin of foot not split in 2 lobes. Without operculum.

***Retusa obtusa*** (Montagu)     Pearl bubble (Fig. 154b)

*Delicate, whitish shell; up to 10 mm in height; aperture elongate but not as long as the main whorl of the shell; spire not always conspicuous. Retracts fully into shell. Operculum absent. Head with pair of conspicuous, posterior cephalic tentacles. Posterior margin of foot entire. May be mistaken for a prosobranch mollusc (p. 219).*

*R. obtusa* is widely distributed in north-west Europe and all around Britain, buried in muddy-sand and mud on the shore and sublittoral, surviving salinities down to about 20‰. It is carnivorous, feeding on *Hydrobia ulvae* (p. 246), which it ingests whole, and on foraminiferans. *Retusa* is a simultaneous hermaphrodite; copulation takes place in autumn and the eggs mature during winter and are laid in early spring. Each female produces up to four egg masses containing a total of about 50 eggs. These are found on the shells of adult snails and after three to four weeks crawling young emerge, generally from about February to May. This is followed by death of the adults, which live for a little more than one year. This is an uncommon life-cycle as most opisthobranchs have free-swimming veliger larvae.

Family Runcinidae
Shell absent. Body small, no more than 8 mm in length. Without tentacles; without parapodia. Gills under posterior mantle edge on right side.

***Runcina coronata*** (Quatrefages)     (Fig. 156)

*Small, usually up to 6 mm in length. Brownish colour, paler areas around eyes. Without tentacles; 3 gills under mantle edge on posterior right side, visible in dorsal view when speci-men relaxed. Shell absent.*

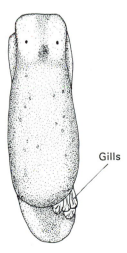

Gills

Figure 156  *Runcina coronata* (after Thompson, 1988).

*R. coronata* is found in rock pools on the middle shore with *Corallina* (p. 61), on which it is believed to browse. It extends into the shallow sublittoral and has been recorded on the Atlantic coast of France and west coast of Britain. It is hermaphroditic and egg masses, each containing up to about 30 eggs, are laid in spring. There is no pelagic phase.

Family Elysiidae
Shell absent. Parapodial lobes extend onto dorsal surface. Rhinophores and propodial tentacles present.

*Elysia viridis* (Montagu)    (Fig. 157)

*Body up to 45 mm in length, generally much smaller. Usually green, but colour varies; shiny red, blue and green spots always present. Shell absent.*

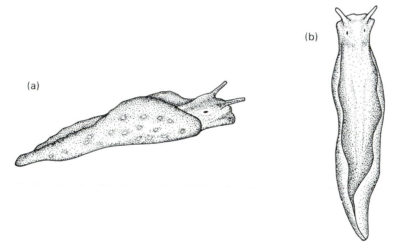

Figure 157 *Elysia viridis*, (a) lateral view, (b) dorsal view.

*E. viridis* is found throughout north-west Europe and all around Britain, especially in the south and west. It is found creeping among seaweed on the lower shore and in the shallow sublittoral. It feeds on a variety of algae including *Cladophora* (p. 37) and *Codium* (p. 38), cutting them open with the radula. Chloroplasts from the ingested algae are apparently incorporated into the digestive gland where they may continue to photo-synthesize for some months. Breeding occurs from May to October. The egg masses con-tain up to several thousand eggs, which develop into free-swimming veliger larvae after one or two weeks. Most individuals live for one year.

Family Limapontiidae
Shell absent. Body small, no more than 8 mm in length. Head with or without pair of tentacles, which may be ridge-like crests or finger-like. Without parapodia. Eyes con-spicuous, surrounded by pale areas.

(a) Head with 2 ridge-like tentacles between eyes. Found in rock pools ........................
.................................................................................... *Limapontia capitata* (p. 267)
*or*
(b) Head with 2 longish, slender tentacles. Found in rock pools ...........................................
.................................................................................... *Limapontia senestra* (p. 268)
*or*
(c) Head without ridge-like tentacles or long slender tentacles. Found on saltmarshes .........
.................................................................................... *Limapontia depressa* (p. 267)

**Limapontia capitata** (Müller)    (Fig. 158a)

*Small, usually up to 4 mm body length; brown or black in colour, area around eyes paler; 2 ridge-like tentacles between eyes. Shell absent.*

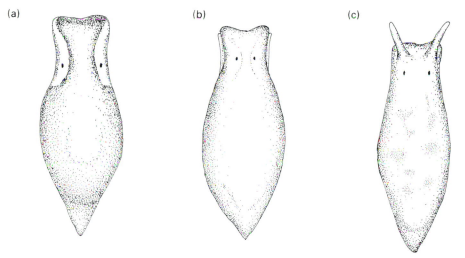

Figure 158  (a) *Limapontia capitata*. (b) *Limapontia depressa*. (c) *Limapontia senestra*.

*L. capitata* is widely distributed in north-west Europe and is found all around Britain, usually in rock pools on the middle to lower shore, where it feeds on seaweeds such as *Cladophora* (p. 37). The animals are hermaphroditic, depositing globular or sausage-shaped masses, each containing several hundred eggs, on seaweed from spring to autumn. After one or two weeks, free-swimming veliger larvae emerge. Length of life is up to one year.

**Limapontia depressa** Alder & Hancock    (Fig. 158b)

*Small, rather flattened, usually up to 6 mm body length; brown-black colour but some specimens pale colour. Head without ridge-like tentacles or long slender tentacles. Eyes surrounded by pale areas. Shell absent.*

*L. depressa* is widely distributed in north-west Europe and occurs all around Britain on saltmarshes. It feeds on algae. The animals are hermaphroditic; egg masses, each containing up to 200 or so eggs, are laid from May to September. After about four weeks, free-swimming veliger larvae emerge. Length of life is believed to be one year.

### *Limapontia senestra* (Quatrefages)    (Fig. 158c)

*Small, usually up to 6 mm body length; brown-black colour. Head with 2 longish, slender tentacles. Eyes surrounded by pale areas. Shell absent.*

*L. senestra* is widely distributed in north-west Europe and all around Britain in rock pools on the lower shore, sometimes with *L. capitata* (p. 267), feeding on algae such as *Cladophora* (p. 37) and *Enteromorpha* (p. 35). It is hermaphroditic; spawn is laid from February to September. The eggs are large (400 μm) and development direct: there is no veliger larva.

Family Aplysiidae
Shell internal, delicate. Large, flap-like parapodia. Oral tentacles and rhinophores present.

### *Aplysia punctata* Cuvier    Sea-hare (Fig. 159)

*Brownish-green body, often with black spots; up to 200 mm in length; young specimens have reddish colour. Oral tentacles and rhinophores present. Flap-like lobes, the parapodia, joined at rear. Shell internal, delicate, up to 40 mm in length. Purple secretion ejected when animal disturbed.*

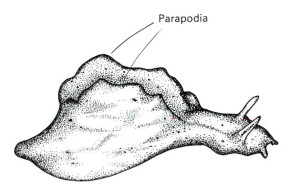

Figure 159  *Aplysia punctata.*

*A. punctata* is widely distributed in north-west Europe and around Britain where it is essentially a shallow-water species, sometimes being found on the lower shore. Contrary to earlier reports, *Aplysia* does not swim but moves over the surface of the substratum with a gliding motion, feeding mainly on seaweeds. It is hermaphroditic and breeding occurs mainly in spring. In some areas the animals move onto the lower shore to spawn. Large numbers of individuals form a chain and orange-pink egg masses, each containing many thousands of eggs, are laid among seaweed. After about two to three weeks, free-swimming veliger larvae emerge. Length of life is probably one year. *Aplysia* has been known since the time of Pliny, the Roman, and up to the Middle Ages was surrounded by a rich mythology related to the belief that it was poisonous.

Family Pleurobranchidae
Shell internal, delicate. Body rather flattened. Single, large gill evident on right side of body. Oral tentacles and rhinophores present.

***Pleurobranchus membranaceus*** (Montagu)    (Fig. 160)

*Body up to 120 mm in length; dorsal surface brownish, covered with tubercles. Single gill evident on right side. Oral tentacles and rhinophores present. Shell internal, thin, up to 50 mm in length.*

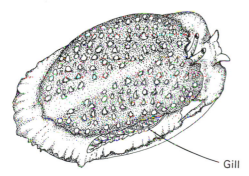

Figure 160  *Pleurobranchus membranaceus* (after Forbes & Hanley, 1853).

*P. membranaceus* is a shallow-water species found intertidally on rocky shores on the French coast and all around Britain. It feeds on ascidians such as *Ascidiella aspersa* (p. 475) and *Botryllus schlosseri* (p. 478) by boring into them using the radula. When the animal is disturbed, the body and foot secrete sulphuric acid, which acts as a deterrent to predators. *Pleurobranchus* swims by a flapping action of the lobes of the foot and is a strong swimmer. It is hermaphroditic and breeds in spring, the egg masses containing many thousands of eggs. There is a free-swimming veliger larva. The animals probably die after spawning and have a life-span of about one year.

Family Dotidae
Shell absent. Body with two rows of large, papillate cerata. Rhinophores smooth, long and slender, and unbranched; arise from cup-like sheaths.

**Doto coronata** (Gmelin)    (Fig. 161)

*Body usually up to 12 mm in length. Pale coloured with purple-red streaks and markings.*
*Up to 8 pairs of pale-coloured cerata, spotted red. Rhinophores emerge from smooth-rimmed,*
*cup-like sheaths. Shell absent.*

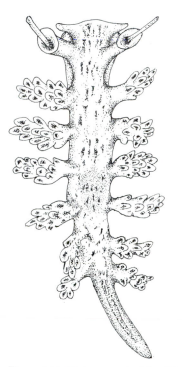

Figure 161  *Doto coronata* (after Thompson, 1988).

D. *coronata* is widely distributed in north-west Europe and has been recorded all around
Britain. It is found on the lower shore and into the sublittoral to depths of 180 m, feeding
on a wide range of hydroids such as *Obelia* (p. 109), *Sertularia* (p. 112) and *Dynamena*
(p. 112). It is hermaphroditic; coiled bands of spawn have been recorded throughout the
year, each containing many thousands of eggs, which take about 16 days to hatch as
veliger larvae. There may be up to four generations per year and the length of life is
believed to be up to nine months. D. *coronata* is a species complex.

Family Onchidorididae
Shell absent. Body flattened. Rhinophores ringed; not in cup-like sheaths. Circle of gills
around anus; gill circle cannot be completely retracted but individual gills can retract.
Mantle expansive.

Circle of up to 29 gills. Dorsal surface whitish, with brown patches and whitish tubercles ................................................................ *Onchidoris bilamellata* (p. 271)
Circle of up to 11 gills. Dorsal surface whitish with large, rounded tubercles .................
................................................................................ *Onchidoris muricata* (p. 272)

**Onchidoris bilamellata** (Linnaeus)    (*Onchidoris fusca*) (Fig. 162)

*Soft body; up to 40 mm in length. Posterior circle of up to 29 gills. Dorsal surface covered with tubercles. White patterned with brown, tubercles usually white. Rhinophores ringed. Shell absent.*

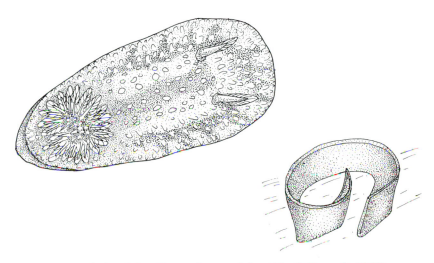

Figure 162 *Onchidoris bilamellata* and spawn (after Alder & Hancock, 1855).

O. *bilamellata* is widely distributed in north-west Europe and is common around Britain, both intertidally and in the shallow sublittoral to depths of about 20 m. It appears to feed exclusively on barnacles, especially *Semibalanus balanoides* (p. 341), forcing open the opercular plates of the barnacle and sucking out the soft parts. When disturbed, acid secretions are produced by the mantle. *Onchidoris* is hermaphroditic; on the north-east coast of England breeding occurs from December to April, but mainly in January. The spawn is in the shape of a flattened coil attached to the substratum by one edge and contains many thousands of eggs. After about five to six weeks, free-swimming veliger larvae emerge and these have a pelagic life of about three months before settling inter-tidally on the underside of boulders and rocks. The time of settlement of the young nudibranchs is believed to be closely related to the occurrence of one of their main prey items, newly settled barnacles, and it has been observed that the veligers will not meta-morphose unless living barnacles are present. As the nudibranchs grow they move to the upper rock surfaces and become more conspicuous, which led to the suggestion that they migrated upshore after settling at lower beach levels. Length of life is believed to be about one year, with some individuals spawning and dying after as little as three months on the shore.

### Onchidoris muricata (Müller)    (Fig. 163)

*Soft body; up to 14 mm in length. Posterior circlet of up to 11 gills. Dorsal surface covered with large, rounded tubercles. White or sometimes yellowish in colour. Rhinophores ringed. Shell absent.*

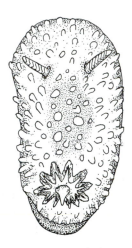

Figure 163  *Onchidoris muricata* (after Alder & Hancock, 1855).

*O. muricata* is a northern species reaching its southern limit on the Atlantic coast of France. It is widely distributed around Britain, especially in the north, on the lower shore and in the shallow sublittoral to depths of about 15 m. It feeds on encrusting bryozoans such as *Electra pilosa* (p. 425), *Membranipora membranacea* (p. 424) and *Securiflustra securifrons* (p. 427). *O. muricata* is a simultaneous hermaphrodite breeding from February to April on the east coast of England. White, gelatinous spiral ribbons, each containing thousands of eggs, are laid on rocks and stones and the larvae hatch after about two weeks. They have a pelagic life of several weeks and it appears that contact with bryozoans stimulates metamorphosis. Length of life is a year.

*O. muricata* is similar in appearance to another opisthobranch mollusc, **Adalaria proxima** (Alder & Hancock), which is sometimes common in the intertidal region and shallow sublittoral, feeding on bryozoans. The tubercles on the mantle of *Adalaria* are tapering and more elongate than the rounded tubercles of *O. muricata* and it is generally yellow in colour. However, in some cases reliable separation can only be achieved by examination of the internal anatomy. Reference to Thompson (1988) is recommended.

Family Archidorididae

Shell absent. Body flattened. Rhinophores ringed; not in cup-like sheaths except in one rare British genus. Circle of gills around anus. Gill circle can be completely retracted. Mantle expansive. Dorsal surface rough, with tubercles of range of sizes.

***Archidoris pseudoargus*** (Rapp) (*Archidoris britannica, Archidoris tuberculata*) Sea-lemon (Fig. 164)

*Large body; up to 120 mm in length. Dorsal surface covered with rounded tubercles. Colour varied, usually yellowish with green, pink or brown patches, but reddish varieties also occur. Posterior circle of up to 10 whitish gills. Rhinophores ringed. Shell absent.*

Figure 164 *Archidoris pseudoargus* and spawn.

A. *pseudoargus* is widely distributed in north-west Europe and Britain on the lower shore and in the sublittoral. It is found under stones and in crevices on rocky shores where it feeds on sponges, especially *Halichondria panicea* (p. 93). It is hermaphroditic, breeding in the spring. The egg mass is a long, convoluted whitish or yellowish ribbon containing thousands of eggs, attached by its edge to the rock surface. Free-swimming veliger larvae hatch after about four weeks and have a long pelagic life of, possibly, up to three months. Metamorphosis of the larvae appears to be stimulated by contact with the sponge *Halichondria*. Length of life has been shown to be about one year in animals from the Isle of Man, but may be longer at other localities.

Family Kentrodorididae

Shell absent. Body flattened. Rhinophores ringed; not in cup-like sheaths. Circle of gills around anus. Gill circle can be completely retracted. Mantle expansive. Dorsal surface soft, with very many small, similar-sized tubercles.

*Jorunna tomentosa* (Cuvier)    (Fig. 165)

*Body up to 55 mm in length. Small tubercles on dorsal surface. Yellowish-brown in colour with brown patches roughly in pairs along dorsal surface. Posterior circle of up to 17 whitish gills. Rhinophores ringed. Shell absent.*

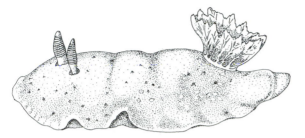

Figure 165  *Jorunna tomentosa* (after Alder & Hancock, 1855).

*J. tomentosa* is widely distributed in north-west Europe and around Britain, both inter-tidally and in the shallow sublittoral. It feeds on the sponge *Halichondria panicea* (p. 93). *Jorunna* is hermaphroditic and breeds during spring and summer when yellowish-coloured ribbons containing many thousands of eggs are deposited on the underside of stones on the lower shore. Free-swimming veliger larvae emerge after about three weeks.

Family Facelinidae
Shell absent. Body with groups of cerata. With the exception of two rare members of the family, the rhinophores are ringed. Oral tentacles long, propodial tentacles conspicuous.

*Facelina auriculata* (Müller)    (*Facelina coronata*) (Fig. 166a)

*Body up to 38 mm in length; groups of cerata on dorsal surface. Rhinophores conspicuously ringed. Body usually pinkish-white; cerata bluish, iridescent, red, brown, with white patch near tip. Shell absent.*

*F. auriculata* is widely distributed in north-west Europe and around Britain, living among stones and rocks, on the lower shore and in the sublittoral. It feeds on hydroids, especially *Tubularia* (p. 100). It is hermaphroditic and breeding occurs during summer when convoluted egg masses are laid on the undersurface of stones. There is a free-swimming veliger larva.

Family Aeolidiidae
Shell absent. Body with transverse rows of cerata. Rhinophores smooth. Oral tentacles long, propodial tentacles conspicuous.

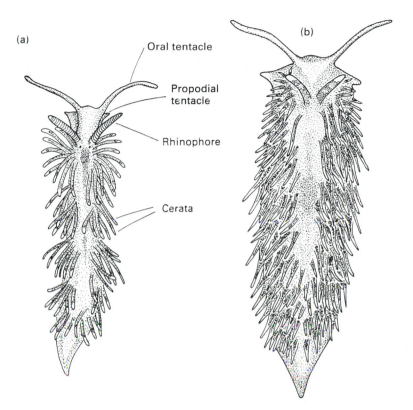

Figure 166  (a) *Facelina auriculata*. (b) *Aeolidia papillosa*.

**Aeolidia papillosa** (Linnaeus)    Common grey sea-slug (Fig. 166b)

*Body up to 120 mm in length. Many cerata on dorsal surface, separated by mid-dorsal, clear whitish area. Rhinophores smooth. Usually purplish-brown, grey in colour with white tips to the cerata and rhinophores. Shell absent.*

A. *papillosa* is widely distributed in north-west Europe and common among rocks and stones on the lower shore and shallow sublittoral. The diet includes a wide range of sea-anemones which are detected by chemosensory means. Undischarged nematocysts of the anemones are transported through the gut to the ends of the cerata where they are used by the sea-slug in defence. *Aeolidia* is hermaphroditic, breeding in spring and summer. Convoluted egg masses, usually pinkish in colour and containing thousands of eggs, are attached to stones. The larvae are planktonic. Longevity is probably about one year.

## Subclass Heterobranchia
### Superorder Pulmonata

The best-known pulmonates are the freshwater and land snails and the slugs, but the group also includes a few marine representatives. The highly vacularized mantle cavity is adapted for air breathing and opens by a single, contractile aperture on the right side of the snail. In almost all species the operculum is absent. Pulmonates are hermaphroditic but cross-fertilization normally occurs and the eggs are laid in gelatinous capsules. Except in a very few species there is no veliger larva. In many pulmonates the shell is absent; those described in this text have a shell and may be mistaken for prosobranch molluscs (p. 219).

### *Leucophytia bidentata* (Montagu)    (Fig. 167a)

*Shell thick; smooth, with pointed apex; up to 9 mm in height. Columella with 2 distinct teeth. No operulum. Creamy-white in colour. Tentacles short, flattened.*

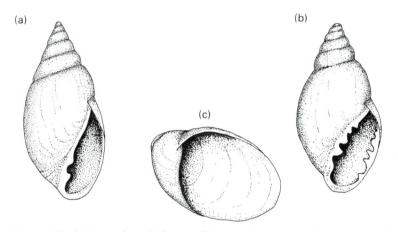

Figure 167  (a) *Leucophytia bidentata.* (b) *Ovatella myosotis.* (c) *Otina ovata.* ((a) After Macan, 1977.)

*L. bidentata* is widely distributed on south and west coasts of Britain, usually on rocky shores under stones and in crevices on the upper shore, but also in estuaries and on mud-flats. Its food is organic detritus. It is a protandrous hermaphrodite and irregularly shaped egg masses, about 4 mm in diameter and containing up to 200 eggs, are laid in crevices in the rock. On the south coast of England, egg masses are deposited during June. There is no free-swimming larva, crawling young emerge. Longevity is probably two years.

*Ovatella myosotis* (Draparnaud)    (*Phytia myosotis*) (Fig. 167b)

*Shell thin; smooth, with pointed apex; up to 9 mm in height. Columella and outer lip each with up to 3 or more teeth. No operculum. Semi-transparent, reddish-brown, yellowish in colour. Tentacles faintly annulated; with dark streaks.*

O. *myosotis* has a patchy distribution around Britain. It is found on the upper shore and above, in estuaries and on saltmarshes. It is hermaphroditic and egg masses containing up to 50 eggs are attached to stones and stems of plants. Crawling young emerge. Breeding probably occurs in spring.

*Otina ovata* (Brown)    (Fig. 167c)

*Shell with very wide aperture; about 3 mm in height. Operculum absent. Reddish-brown, purplish in colour. Shell does not enclose whole animal.*

O. *ovata* is a high shore species found on south and west coasts of Britain among empty barnacle shells, in crevices and with the lichen *Lichina pygmaea* (p. 81). It feeds on micro-organisms and detritus. On the south coast of England breeding occurs in May and June. The animals are protandrous hermaphrodites and the egg masses, each about 5 mm in diameter and containing 20–30 eggs, are laid in crevices. Crawling young emerge.

### Class Bivalvia (Pelecypoda)

As the names implies, bivalve molluscs have a shell of two valves. The valves have a dorsal hinge and are held together by an elastic ligament which is made of fibres that force them apart so as to open the shell. This action is opposed by the contraction of muscles which close the shell. These are the adductor muscles and they generally leave a clearly visible scar or scars on the inside of the shell (Fig. 168). Also visible on the inner surface along the lower edge of each valve of the shell is a line running from one adductor muscle scar to the other. This is the pallial line, marking the position where the mantle was attached to the shell. In some, but not all species, the posterior part of the pallial line is indented to give a pallial sinus showing the area occupied by the siphons (Fig. 168). In bivalves such as the common cockle (p. 300), in which the siphons are small, there is no pallial sinus; however, in those species (for example the sand gaper, p. 320) which have large siphons, the pallial sinus is well developed. For many of the bivalves decribed in this text the position of the adductor muscle scars, the pallial line and pallial sinus is shown in the figures.

    The shell ligament is internal, external or both internal and external; when external, it is often clearly visible on the outside of the shell. The internal ligament is attached to a depression known as the chondrophore. This is usually conspicuous and in a few species is developed as a spoon-shaped projection (Fig. 210b, c). The dorsal hinge line may have a number of interlocking teeth which prevent the two valves of the shell from slipping laterally. The teeth beneath the beak are known as cardinals, to either side of which are the anterior and posterior lateral teeth. The first-formed part of the shell is known as the beak and round it is a raised area, the umbo (Fig. 168). The shell may be smooth, or sculptured with grooves, lines or ribs; the ribs radiate outwards from the beaks and may bear spines. Concentric growth rings, which can be used for age determination in some species, are often clear. The outer layer of the shell, or periostracum, is sometimes thick and may be drawn out into spines, but it is often worn away, particularly in dead shells.

The body of the bivalve is laterally compressed and the gills, which can be very large, are used in food collection as well as respiration. Water circulates through the mantle cavity, passing over the gills where fine particles are filtered out. The water current enters and leaves by two posterior apertures, the inhalant and exhalant apertures, the exhalant being the more dorsal. In some bivalves, notably those living deep in sediment, these apertures are situated at the end of long siphons. In suspension-feeding bivalves the inhalant siphon is generally held upright and draws in a current of water containing food material such as plankton; the inhalant siphon of deposit feeders moves over the surface of the sediment, drawing in detritus. An anterior, laterally compressed foot is usually present as the organ of locomotion, but in those species which are not very mobile it is much reduced. Some bivalves are attached to the substratum by toughened fibres secreted by a gland in the foot and known as the byssus. The group as a whole shows adaptations to a sedentary way of life, although the scallops are able to swim by flapping the valves together. Most bivalves have separate sexes; some are hermaphroditic. Fertilization is generally external and there is usually a veliger larva in the life-cycle. This has a tiny shell of two valves and swims by means of a pair of ciliated lobes.

Bivalves are common members of the intertidal fauna. Well-known examples include mussels found on rocky shores and cockles living buried in sandy sediments. Both tend to occur in very high densities and are preyed on extensively by birds and fish. Many species occur offshore where they comprise an important part of the diet of commercial fishes and, although living specimens of such species are not likely to be found on the shore, empty shells are frequently found on the strandline. These are included in the text.

### Key to families of bivalves

The shell characteristics used in identification are shown in Fig. 168. A few groups have been keyed to the level of subfamilies for ease of use of the key.

1 Shell very small, covers only anterior end of long, worm-like body. Shell valves divided into regions with different sculpturing. Bores into submerged timbers; tunnels lined by calcareous tubes ...................................................................... **Teredinidae** (p. 325)
   Shell and body not as above ..................................................................... 2
2 Right and left valves of shell similar (equivalve); shell almost circular, beaks central.

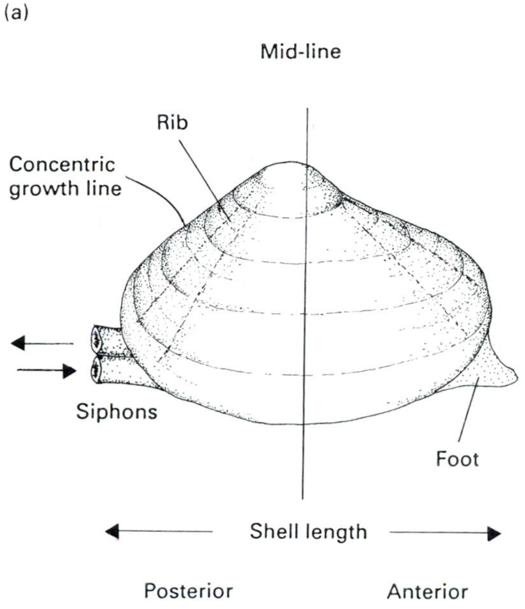

(a)

Mid-line

Rib

Concentric
growth line

Siphons

Foot

Shell length

Posterior                    Anterior

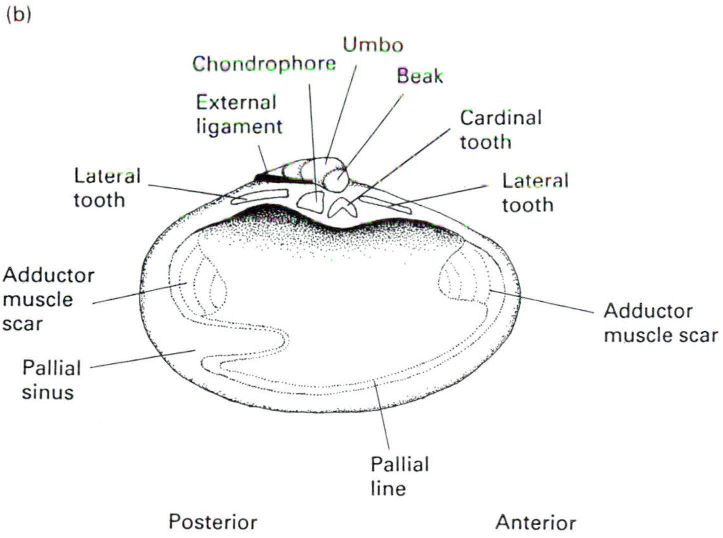

(b)

Umbo

Chondrophore

Beak

External
ligament

Cardinal
tooth

Lateral
tooth

Lateral
tooth

Adductor
muscle
scar

Adductor
muscle scar

Pallial
sinus

Pallial
line

Posterior                              Anterior

Figure 168  Generalized bivalve shell to illustrate features referred to in the text. (a) Right
valve, external features, (b) inside of left valve.

Hinge line with many small teeth alternating with sockets (taxodont arrangement) ......
.................................................................................................. **Glycymerididae** (p. 281)
Teeth on hinge line not taxodont ........................................................................ 3

3   Shell with projections ('ears') on either side of beaks (Fig. 176). Shell usually with
conspicuous radiating ribs. May be attached to substratum or free ....... **Pectinidae** (p. 289)
Shell without 'ears' on either side of beaks ...................................................... 4

4   Shell valves markedly dissimilar (inequivalve); may be thin and delicate, or thick and
rough. Valves with single adductor muscle scar; upper valve may also have additional
scars (Fig. 180). Attached to substratum .......................................................... 5
Shell not as above ................................................................................................ 6

5   Shell thin and delicate; lower valve with indented hole. Upper valve with additional
scars. Attached very firmly to rocks, stones, mollusc shells ................. **Anomiidae** (p. 294)
Shell thick, rough, with surface corrugations; shell valves without hole; without
additional scars. Attached to muddy-gravel, muddy-sand, rock ............ **Ostreidae** (p. 287)

6   Shell more or less elongate, with pronounced gape. Part of shell reflected over beaks –
the umbonal reflection (Fig. 212). Hinge line without teeth but has slender, finger-like
projection, the apophysis. Burrowing into rock, clay, peat ................. **Pholadidae** (p. 323)
Shell not as above ................................................................................................ 7

7   Shell roughly triangular; beaks very much at anterior. Hinge line without teeth but
may be crenulate; without chondrophore. Shell without gape ............... **Mytilidae** (p. 282)
Shell not as above ................................................................................................ 8

8   Hinge line without teeth. Left valve with conspicuous, spoon-shaped chondrophore
projecting from below beak (Fig. 210). Shell with posterior gape ............. **Myidae** (p. 319)
Shell not as above ................................................................................................ 9

9   Hinge line with triangular or semi-circular chondrophore (Fig. 187); cardinal teeth
present, lateral teeth may be present ................................................................ 10
Hinge line without chondrophore; cardinal and/or lateral teeth may be present ............. 11

10  Cardinal and lateral teeth present; cardinal teeth of left valve joined to give ∧-shaped
ridge in front of chondrophore .......................................................... **Mactridae** (p. 301)
Cardinal teeth present, not joined to give ∧-shaped ridge in front of chondrophore;
lateral teeth present or absent .......................................................... **Semelidae** (p. 311)

11  Hinge line of right valve with projecting cardinal tooth. Shell more or less rectangular.
Beaks close to anterior. The intertidal species in this family is found boring into soft
rocks and shell or nestling in rock crevices. .................................... **Hiatellidae** (p. 322)
Shell not as above ................................................................................................ 12

12  Shell long, narrow, much longer than broad; dorsal and ventral sides curved or
straight. Beaks at anterior, inconspicuous. Hinge line with cardinal and/or lateral
teeth. Wide gape at anterior and posterior ...................................................... 13
Shell not as above ................................................................................................ 14

13  Shell with vertical groove at anterior end. Hinge line without lateral teeth. Dorsal and
ventral sides of shell straight ............................................................. **Solenidae** (p. 304)

Shell without vertical groove at anterior end. Dorsal and ventral sides straight or curved. Hinge line with cardinal and lateral teeth ................................ **Cultellidae** (p. 305)

14 Shell elongate, rectangular; beaks inconspicuous, just anterior of mid-line. Hinge line with small, projecting cardinal teeth. Wide gape at anterior and posterior ....................
.................................................... **Psammobiidae (Subfamily Solecurtinae)** (p. 310)

Shell not as above ..................................................................................... 15

15 Pallial sinus present (but difficult to see in the Turtoniidae, p. 318) ........................... 16

Pallial sinus absent .................................................................................... 20

16 Lunule present in front of beaks (Fig. 201) ....................................................... 17

Shell without lunule ................................................................................... 18

17 Shell small, fragile; usually less than 3 mm shell length. Found in crevices, among barnacles and seaweed on rocky shores. Lunule narrow. Teeth of hinge line difficult to distinguish. Pallial line difficult to see .............................................. **Turtoniidae** (p. 318)

Shell up to 75 mm or so in length. Teeth of hinge line distinct ............. **Veneridae** (p. 313)

18 Shell wedge-shaped. Beaks towards posterior. Margin strongly crenulate or finely serrated ................................................................................ **Donacidae** (p. 309)

Shell not wedge-shaped; margin smooth ........................................................ 19

19 Shell rounded and globular or oblong; may be truncated with raised keel running to posterior margin; beaks central or slightly anterior or slightly posterior. Hinge line with cardinal teeth; lateral teeth absent ...............................................................
.................................................... **Psammobiidae (Subfamily Psammobiinae)** (p. 309)
(But note, this step includes *Macoma balthica* of the Tellinidae, p. 308.)
Thin, flattened shell, beaks usually to the posterior. Hinge line with cardinal and lateral teeth (lateral teeth conspicuous only in right valve) .................... **Tellinidae** (p. 306)

20 Rounded globular shell with conspicuous radiating ribs ...................... **Cardiidae** (p. 297)

Shell without radiating ribs ........................................................................ 21

21 Thick, heavy rounded shell, up to about 120 mm shell length. Beaks well to the anterior. Hinge line with 3 cardinal teeth in each valve. Periostracum thick and shiny .........
.................................................................................... **Arcticidae** (p. 312)

Shell small, delicate, no more than 10 mm shell length. Beaks towards the posterior. Hinge line with no more than 2 cardinal teeth in each valve .............................. 22

22 Oval shell, rather plump; up to 3 mm in length. Found in crevices, empty barnacle shells, among lichens, particularly *Lichina* (p. 81) ............................................
.................................................... **Family Kelliidae (Subfamily Lasaeinae)** (p. 296)

Similar to above, no more than 10 mm shell length; usually in association with echinoids (p. 457) ...................................................... **Montacutidae** (p. 297)

Family Glycymerididae – dog-cockles
Shell equivalve, thick rounded; beaks central. Several small teeth on either side of beaks alternating with small sockets, i.e. taxodont. Adductor muscle scars roughly same size. Pallial sinus absent.

*Glycymeris glycymeris* (Linnaeus)    Dog-cockle (Fig. 169)

*Shell almost circular; up to 65 mm in length. Beaks mid-way between anterior and posterior ends of shell. Inner surface of shell with 6–12 teeth on either side of beak; area beneath beak without teeth. Outer surface usually patterned with brown marks; inner surface white, may be dark brown. Margin strongly crenulate.*

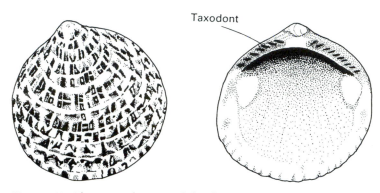

Taxodont

Figure 169  *Glycymeris glycymeris*, left valve.

*G. glycymeris* is widely distributed and common in north-west Europe and occurs all around Britain. It is found sublittorally to depths of about 70 m in all types of sediment, such as sand, gravel and mud, and dead shells are commonly washed ashore. A free-swimming veliger larva has been recorded in the plankton of the North Sea throughout the year, but mainly in late autumn.

Family Mytilidae – mussels, crenellas
Shell roughly triangular; beaks at anterior. Attached to substratum by byssus threads. Hinge line lacks teeth and chondrophore, but may be crenulate. Periostracum obvious, may be developed into spines. Adductor muscle scars usually of different size. Pallial sinus absent.

1 Shell surface smooth; but periostracum may have spines ................................................. 2
  Shell surface with radiating ribs at anterior and posterior; area between, without ribs ..... 5
2 Beaks right at anterior ........................................................................................... 3
  Beaks not quite at anterior ...................................................................................... 4
3 Beaks rounded, not pointing downwards. Edge of mantle whitish, yellow, brown .........
  ................................................................................................ ***Mytilus edulis*** (p. 283)
  Beaks pointed, point downwards. Edge of mantle dark purple or blue .........................
  ................................................................................... ***Mytilus galloprovincialis*** (p. 284)
4 Periostracum with serrated spines on the posterior part of the shell ............................
  ............................................................................................ ***Modiolus barbatus*** (p. 286)

Periostracum without spines in adult, but developed into long, smooth spines in young
animals ................................................................................ *Modiolus modiolus* (p. 285)

5 Shell noticeably tumid; posterior margin somewhat elongate; 15–18 ribs anteriorly;
20–35 ribs posteriorly. Often found living in body of ascidians ............................................
................................................................................ *Modiolarca tumida* (p. 287)

Shell not tumid; posterior margin broad and rounded; 9–12 ribs anteriorly; 30–45 ribs
posteriorly ................................................................................ *Musculus discors* (p. 286)

*Mytilus edulis* Linnaeus    Common mussel (Fig. 170)

*Up to 100 mm in length, occasionally larger. Beaks at anterior, rounded. Teeth absent from
hinge line but up to 12 crenulations near beaks. Outer surface of shell dark blue, sometimes
brownish; inner surface with pearly iridescence and dark blue edge. Edge of mantle whitish,
yellow, brown.*

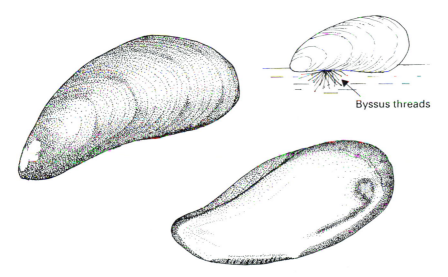

Byssus threads

Figure 170 *Mytilus edulis*, left valve, and showing attachment by byssus threads.

*M. edulis* is widespread and common in north-west Europe where it is found from the
middle shore to the shallow sublittoral. It is attached by byssus threads to stones, rocks
and piers, and on exposed shores small mussels are often abundant. It is also found on
soft sediments in estuaries and can tolerate salinities down to 4–5‰, often forming
dense beds which are fished commercially. There are fisheries at a number of localities,
including the Wash, Morecambe Bay, Menai Strait, Conway and the west coast of Scotland. At some localities mussels are farmed; banks of small overcrowded mussels are
moved to more favourable areas where growth is rapid. They are also cultivated on suspended ropes, on poles and on stakes interlaced with brushwood.

Mussels are suspension feeders. Their main predators are the dog-whelk, the shore crab, the common starfish and the oyster catcher. Despite their sessile habit and apparent vulnerability to predators, it has been shown that on dense mussel beds mussels are able to immobilize one of their predators, the dog-whelk. They do this by attaching large numbers of byssus threads to the shell of the whelk, eventually pulling it over. *Mytilus* has separate sexes, fertilization is external and around Britain there is a single peak of spawning in spring or peaks of spawning in spring and summer; the free-swimming veliger larva is planktonic for up to four weeks. Primary settlement and metamorphosis and temporary attachment by the byssus occur on hydroids and filamentous seaweeds such as *Polysiphonia* (p. 72) and *Ceramium* (p. 67). Newly settled mussels are known as early plantigrades and are between 0.25 mm and 0.55 mm in shell length. They grow quickly, detach themselves, then drift to new sites where they may again become temporarily attached to seaweeds. The velum, the main organ of propulsion in the veliger larva, is lost at metamorphosis, but when the plantigrades leave the substratum they secrete a long thread from which they are suspended in the water. They hang with the ventral margin of the shell uppermost and the drag caused by the thread reduces the rate of sinking and ensures distribution in the water currents. Drifting on threads has been recorded in young of other bivalves and likened to the 'gossamer flight' of spiders. After about four weeks the late plantigrades are between 1 and 2 mm shell length. Settlement and attachment by the byssus occur on mussel beds, but some plantigrades may remain on the primary settlement site over winter. This behaviour pattern at settlement is believed to be an adaptation to reduce competition between the newly settled and adult mussels, but direct settlement of the larvae on adult mussel beds has been recorded on the west coast of Ireland. *Mytilus* can mature during its first year and the length of life is probably four or five years; however, those living at high levels on the shore are believed to live for 17 years and more.

Careful examination of the mantle cavity of *Mytilus* may reveal the presence of the copepod **Mytilicola intestinalis** Steuer, once believed to be a damaging parasite which brought about a reduction in the mussel condition factor, with serious commerical implications. Movement of mussels from infected populations for laying in other areas was prohibited. Most recent research, however, suggests that the copepod is a commensal rather than a parasite. The pea crab, *Pinnotheres pisum* (p. 405), is also a common inhabitant of the mantle cavity of the edible mussel, and the shell of *Mytilus* is frequently bored by the polychaete worm *Polydora ciliata* (p. 184). Where heavy infestation occurs, the condition factor of the mussel is reduced and the shell is more vulnerable to predation by crabs.

**Mytilus galloprovincialis** Lamarck    Mediterranean mussel (Fig. 171)

*Shell very similar to, and easily confused with,* M. edulis. *Beaks at anterior end and pointed, pointing downwards. To the posterior of the beaks the ventral margin of the shell is slightly concave. Shell somewhat taller than that of* M. edulis. *Edge of mantle dark purple or blue.*

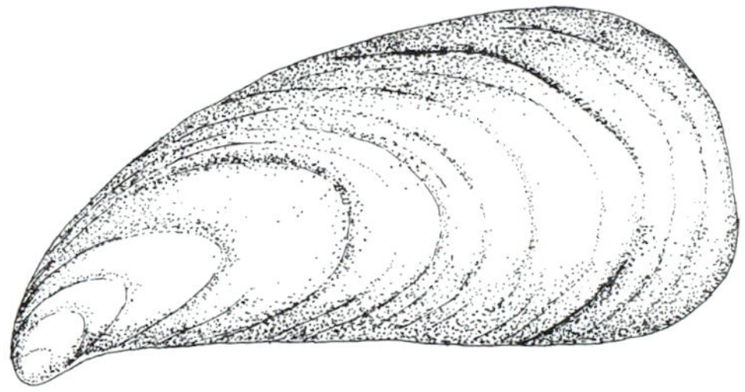

Figure 171 *Mytilus galloprovincialis*, left valve.

*M. galloprovincialis* has been recognized from the south and west coasts of Ireland, South Wales and south-west England, extending to the coast of France and the Mediterranean. Recent research suggests that it has stronger byssal attachment to the substratum than does *M. edulis* and this may be reflected in its ability to withstand exposure to wave action.

The shell morphology of *Mytilus* spp. varies in response to environmental parameters and the separation of *M. edulis* and *M. galloprovincialis* is difficult. In some populations, separation based on shell morphology alone may be impossible; in others the morphological differences given above are clear throughout all size ranges of mussel. Wherever the species are found together they hybridize, and as a result there is confusion as to their taxonomic status. It has been suggested that *M. galloprovincialis* is a subspecies of *M. edulis*, with the designation **Mytilus edulis galloprovincialis**.

**Modiolus modiolus** (Linnaeus)    Horse mussel (Fig. 172)

*Up to 200 mm in length; intertidal specimens usually smaller. Beaks not quite at anterior end of shell. Teeth absent from hinge. Outer surface of shell dark blue or purple; periostracum thick, yellow or brown and developed into long, smooth spines in young animals; inner surface of shell white. Most easily distinguished from* Mytilus edulis *(above) by the blunter appearance of the anterior end of the shell.*

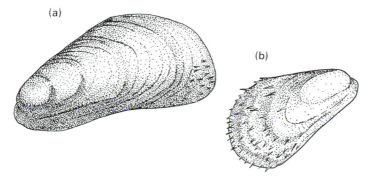

Figure 172 *Modiolus modiolus*, (a) left valve, (b) young specimen, right valve.

*M. modiolus* is widely distributed in north-west Europe and found throughout Britain in pools and crevices on the lower shore, often with *Laminaria* (p. 43). It is also found on soft sediments, where it lives with most of the shell buried in the sediment and attached by byssus threads which penetrate deep into the sediment. Essentially a sublittoral species, it extends to depths of about 60 m, often occurring in large densities. It is a suspension feeder. Unlike the common mussel, *M. modiolus* is not commercially import-ant in Britain. The sexes are separate, fertilization external and data from populations in the Irish Sea suggest that it breeds more or less continuously throughout the year in the sublittoral, while intertidal populations breed in late autumn and winter. Off the Isle of Man, *Modiolus* becomes sexually mature at three to four years of age. The horse mussel is long-lived, some specimens living for up to 48 years.

### *Modiolus barbatus* (Linnaeus)    Bearded horse mussel (Fig. 173a)

*Up to 60 mm in length, usually smaller. Beaks not quite at anterior end of shell. Hinge line without teeth. Periostracum brownish-yellow, thick and developed into serrated spines on posterior part of shell; inner surface of shell shiny, pale blue colour.*

(a)                                                    (b)

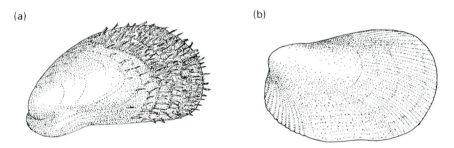

Figure 173   (a) *Modiolus barbatus*, left valve. (b) *Musculus discors*, left valve.

*M. barbatus* is widely distributed in north-west Europe, and in Britain occurs mainly on south and west coasts. It is found among rocks and stones on the lower shore, extending into the sublittoral to depths of about 100 m.

### *Musculus discors* (Linnaeus)    (Fig. 173b)

*Up to 12 mm in length. Beaks not quite at anterior end of shell. Posterior of shell broad and rounded. Shell with 2 groups of ribs; 9–12 ribs in anterior group, 30–45 in posterior group; area between the 2 groups without ribs. Outer surface yellowish, greenish, brown in colour; inner surface pearly.*

*M. discors* is widely distributed in north-west Europe and is commonly found in Britain on the middle shore and below and into the shallow sublittoral among rocks, shells and seaweed, especially *Corallina officianalis* (p. 61).

***Modiolarca tumida*** (Hanley)    (*Musculus marmoratus*) (not illustrated)

*Up to 20 mm in length; delicate, tumid shell. Beaks not quite at anterior of shell. Posterior of shell somewhat elongate. Shell with 2 groups of ribs; 15–18 broad ribs in anterior group, 20–35 finer ribs in posterior group; area between the 2 groups without ribs. Yellowish to brown in colour with darker markings; inner surface pearly.*

M. *tumida* is widely distributed in north-west Europe among stones and shells and the holdfasts of *Laminaria*. It is also found in the tests of ascidians such as *Ascidia mentula* (p. 476) and *Ascidiella aspersa* (p. 475).

Family Ostreidae – oysters
Shell rough, thick, with corrugations. Beaks not obvious. Valves dissimilar; lower (left) valve convex, 'cemented' to substratum; upper (right) valve flat. Hinge line lacks teeth in adults. Single adductor muscle scar.

Adductor muscle scar white or creamish in colour ........................... ***Ostrea edulis*** (p. 287)
Adductor muscle scar purple to reddish-brown ........................ ***Crassostrea gigas*** (p. 288)

***Ostrea edulis*** Linnaeus    Common European oyster, Flat oyster (Fig. 174)

*Thick, rounded, but often irregular shell; up to 100 mm in length, sometimes larger. Valves dissimilar, lower (left) valve convex, upper (right) valve flat. Valves sculptured with wavy concentric lines. Attached to substratum by left valve. Outer surface usually greyish-brown; inner surface pearly white. Single adductor muscle scar white or creamish in colour. Larger specimens often bored by the sponge Cliona celata (p. 95).*

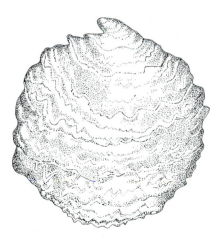

Figure 174  *Ostrea edulis*, left valve.

*O. edulis* is widely distributed in north-west Europe, and although occurring all around Britain is more common in the south. It is found sublittorally to depths of about 80 m, extending into estuaries and tolerating salinities down to about 23‰. It often occurs in large beds on muddy-sand and muddy-gravel, and also on rocks. Like the mussel *M. edulis* (above) it is commerically important and cultivated in large numbers, especially in south-east England, where Whitstable and Colchester are famous for their oysters. They are also cultivated on the French coast, especially the Bay of Arcaçhon, and in The Netherlands. During the nineteenth century oysters were very common in European waters, but overfishing led to a drastic decline. Oysters are susceptible to a number of pests and predators, notable among which are *Crepidula* (p. 249) and *Asterias* (p. 442). They are suspension feeders.

Oysters are usually male when they first become sexually mature but after spawning change to female, thereafter regularly alternating between male and female for the rest of their lives. Breeding usually takes place during the summer months and in Britain each animal spawns twice during the breeding season, once as male and once as female. Fertilization is internal, sperm entering the bivalve in the feeding current, and the embryos remain in the oyster for one or two weeks until they become veliger larvae. The eggs are attached to the surface of the gills, and as development proceeds they change colour from white through grey to black; at this time the oysters are known as 'white sick', 'grey sick' or 'black sick' and are inedible and out of season. The veligers are released and spend about two weeks in the surface waters, feeding on plankton, and by the time of metamorphosis have developed an extensible foot which enables them to move over the substratum. Once a suitable surface is found, cement is produced by the byssus gland at the base of the foot, attaching the left valve of the shell to the substratum. The larvae are attracted to surfaces already colonized by oysters. Longevity is generally 10–12 years, but records of them living 30 years are not uncommon.

**Crassostrea gigas** (Thunberg)   (*Crassostrea angulata*), Portuguese oyster, Cup oyster (Fig. 175)

*Valves dissimilar, lower (left) deeply convex, upper (right) flat. Valves sculptured with concentric ridges, radiating ribs and grooves. Single adductor muscle scar, purple to reddish-brown in colour. Margin of shell wavy. Outer surface white to pale brown in colour, with purple patches. Up to 180 mm shell length.*

*C. gigas* was introduced into Britain from America and Portugal, and is now farmed commercially at a number of localities, accounting for up to 90% of British oyster cultivation. It has a faster growth rate than the flat oyster, maturing first as males and then changing sex to female, although some animals remain as male. The species is increasingly being found outside the commercial farms, and natural spawning and recruitment have been recorded at a number of localities in England and Wales, but apparently not in Scotland.

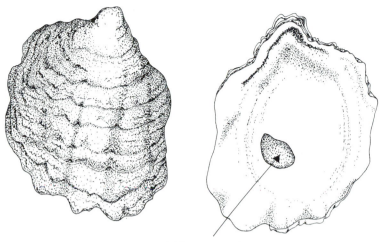

Adductor muscle scar

Figure 175 *Crassostrea gigas*, left valve.

Family Pectinidae – scallops
Shell rounded, usually with conspicuous, radiating ribs. Lies on right valve. In some species one valve is flat, the other concave. Projections ('ears') on either side of beaks. 'Ears' of equal size or anterior one the larger. May be attached by byssus or not. Hinge line lacks teeth. Single adductor muscle scar in adult.

1 Valves distorted, irregular in shape. Lower valve cemented to substratum. Upper valve with 60–70 ribs with spines ......................................... *Chlamys distorta* (adult) (p. 290)
   Valves not distorted or irregular ...................................................................................... 2
2 Lower valve convex, upper valve flat, each valve with 15–17 broad ribs. Large, up to 150 mm shell length .................................................................. *Pecten maximus* (p. 291)
   Upper and lower valves convex, at least to some extent ..................................................... 3
3 Anterior and posterior 'ears' of roughly same size. Each valve with 19–22 rounded ribs
   ........................................................................................ *Aequipecten opercularis* (p. 292)
   Anterior 'ears' 2–3 times as long as posterior ................................................................. 4
4 Shell relatively smooth or with 30 or more fine ribs, a few of which may be broad and pronounced. Colour varied, often patterned ......................... *Palliolum tigerinum* (p. 293)
   Shell with conspicuous, radiating ribs ........................................................................... 5
5 Shell with 25–35 ribs with blunt spines; spines most conspicuous towards margin of shell ......................................................................................... *Chlamys varia* (p. 290)
   Shell with 30–50 ribs, with a few spines ................... *Chlamys distorta* (juvenile) (p. 290)
   See also *Pteria hirundo* (p. 293) and *Limaria hians* (p. 293)

*Chlamys varia* (Linnaeus)    Variegated scallop (Fig. 176)

*Valves convex; up to 65 mm in length; 25–35 ribs with blunt spines; spines most conspicuous towards margin of shell. 'Ears' on either side of beaks asymmetrical and ribbed, anterior 2–3 times length of posterior. Attached by byssus or lying free. Extremely varied in colour; brown, red, yellow; often patterned. Internal surface of shell with single adductor muscle scar.*

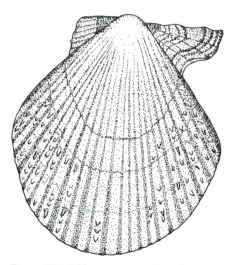

Figure 176  *Chlamys varia*, right valve.

C. *varia* is widely distributed around Britain and on the Atlantic coast of France from the lower shore to depths of about 80 m. It is fished commercially in some areas, for example the Bay of Brest. Individuals usually mature first as males then undergo sex reversal one or more times during life. Breeding occurs during spring and autumn and longevity is believed to be about six years.

*Chlamys distorta* (da Costa)    Hunchback scallop (Fig. 177a)

*Valves dissimilar, up to 35 mm in length. 'Ears' on either side of beaks asymmetrical, anterior 2–3 times length of posterior. Young shells attached by byssus, older shells have the lower (right) valve cemented to the substratum, and shell shape varies with age, being oval when young and distorted when older. Young animals have 30–50 ribs on each valve, older animals 60–70 ribs with spines on upper valve; lower valve covered with encrusting organisms. Varied in colour, white, red, brown; may be patterned. Single adductor muscle scar on inside of shell.*

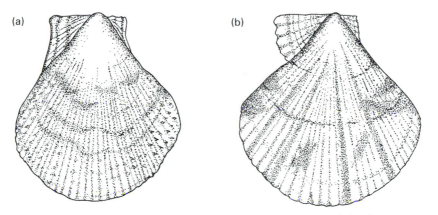

Figure 177  (a) *Chlamys distorta*, left valve. (b) *Palliolum tigerinum*, left valve.

*C. distorta* is widely distributed in north-west Europe and Britain from the extreme low shore into the sublittoral to depths of about 100 m. It is found on rocky shores attached to stones and shells. Most animals mature first as males, changing to female after they have spawned. In the Irish Sea, breeding occurs over a long period with a peak in summer and another in late autumn.

**Pecten maximus** (Linnaeus)    Great scallop, St. James' shell (Fig. 178)

*Valves dissimilar, up to 150 mm in length; lower (right) valve convex, upper (left) valve flat, each with 15–17 broad ribs. Large, equal 'ears' on either side of beaks. Lower valve white, or whitish-brown; upper valve reddish-brown. Single adductor muscle scar on inside of shell.*

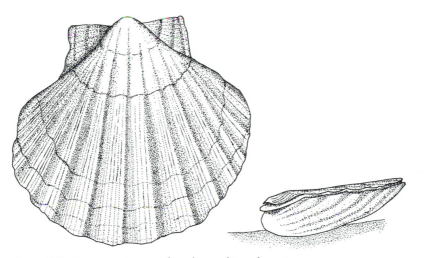

Figure 178  *Pecten maximus*, right valve, and in sediment.

*P. maximus* is widely distributed in north-west Europe. It is found all around Britain, usually sublittorally on sand and sandy-gravel to depths of about 110 m. Scallops are much prized as food and a number of commercial fisheries exist around Britain. Adult scallops are not attached to the substratum but lie in a depression in the sand. They are capable of swimming by opening and closing the valves of the shell and move through the water with the free edge of the shell leading. Scallops can also make sudden escape movements backwards if, for example, they are threatened by the starfish *Asterias* (p. 442), one of their main predators. Scallops are suspension feeders. They are hermaphroditic, the gametes of either sex being discharged first, and fertilization is external. The veliger larvae are planktonic for about three to four weeks and settle on a wide range of algae, bryozoans and hydroids. The young scallops generally remain attached by a byssus until they are anything from 4 to 13 mm in length. The byssus is then lost. They breed first when two years old, and off the Isle of Man have a long breeding period with peaks in spring and autumn. Scallops live for up to 20 years.

**Aequipecten opercularis** (Linnaeus)    (*Chlamys opercularis*), Queen scallop (Fig. 179a)

*Valves dissimilar, upper (left) convex, lower (right) only slightly convex; with 19–22 rounded ribs; up to 90 mm in length. Anterior 'ear' a little longer than posterior. Varied in colour including brown, pink, red; often patterned. Single adductor muscle scar on inside of shell.*

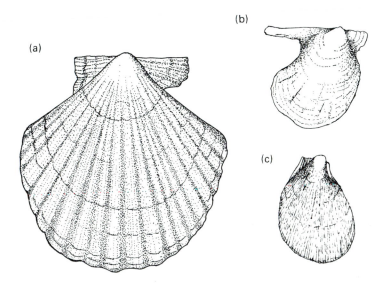

Figure 179  (a) *Aequipecten opercularis*, right valve. (b) *Pteria hirundo*, right valve. (c) *Limaria hians*, left valve.

*A. opercularis* is widely distributed in north-west Europe and occurs all around Britain. It is found sublittorally in sandy-gravel, often in high densities, only rarely occurring on the lower shore. Queen scallops are hermaphroditic and reach maturity after one year. In the Irish Sea the main breeding period is in autumn but spawning has been recorded in winter and spring. Larval life is believed to be of three to four weeks duration and newly settled young are found on bryozoans, hydroids and sometimes on *Laminaria saccharina* (p. 45). The young scallops are attached by a byssus but later become free and often swim by bringing the valves together sharply, the free edge of the shell leading. Longevity is five to six years. The queen scallop is fished commercially at a number of localities, especially around the Isle of Man.

Both valves of the shell of *A. opercularis* are often encrusted with a wide range of epifauna. Common among the epifaunal associations are sponges of the genus *Suberites* (p. 92). This relationship has been described as a 'protective–commensal mutualism' in which the sponges protect the scallop from predation by starfishes while the sponges are protected from predation by the sea-slug *Archidoris pseudoargus* (p. 273).

**Palliolum tigerinum** (Müller)   (*Chlamys tigerina*), Tiger scallop (Fig. 177b)

*Valves almost equal, small, convex, up to 25 mm in length. 'Ears' on either side of beaks asymmetrical, anterior 3 times the length of posterior. Sculpturing varied, smooth, with fine lines, or with about 30 or more ribs, a few of which may be broad and pronounced. Colour extremely varied, white, yellow, brown; may be patterned. Single adductor muscle scar on inside of shell.*

*P. tigerinum* is widely distributed in north-west Europe and is commonly found around Britain from the lower shore to depths of about 90 m among stones, gravel and sand. The sexes are separate, and in the Irish Sea breeding occurs in June. The animals spawn when they are about one year old and then die.

Two other bivalves with 'ears' on the shell, but not belonging to the family Pectinidae, are **Pteria hirundo** (Linnaeus) (Fig. 179b), the wing oyster, and **Limaria hians** (Gmelin) (Fig. 179c), the gaping file shell. *Pteria* is up to 75 mm in length, the 'ears' are very large and the shell is brown on the outer surface and pearly white on the inner surface. It is found sublittorally, mainly in southern Britain, attached by the byssus in sandy-mud and gravel.

*Limaria* is up to 25 mm shell length, the 'ears' are small and the shell gapes anteriorly and posteriorly. It is whitish to whitish-brown in colour with about 50 radiating ribs. It is found in the south and west of Britain on the lower shore and into the sublittoral on gravelly substrata, and in the holdfasts of *Laminaria* in a nest formed from byssus threads. A conspicuous feature of *Limaria* is the fringe of orange-red mantle tentacles which can be seen when the animal swims.

Family Anomiidae – saddle oysters

Shell thin, delicate, valves unequal. Attached firmly to solid substratum by a calcified byssus. Lower (right) valve with an indented hole through which byssus passes. This hole is formed by the shell growing around the byssus. Single adductor muscle scar; upper (left valve) also has byssal scars.

1  Upper valve with 3 distinct muscle scars (1 adductor, 2 byssal) on inner surface ...........
   ............................................................................................ *Anomia ephippium* (p. 294)
   Upper valve with 2 muscle scars (1 adductor, 1 byssal) on inner surface; scars separate
   or touching or fused to form a bilobed scar ....................................................................... 2
2  Upper valve with 2 muscle scars, just touching. Surface of muscle scars smooth. Shell
   small, no more than 15 mm in shell length ..................... *Heteranomia squamula* (p. 295)
   Upper valve with 2 muscle scars, separate or fused to form 1 bilobed scar. Surface of
   muscle scars furrowed. Shell up to about 40 mm in length. .............................................
   ............................................................................ *Pododesmus patelliformis* (p. 296)

*Anomia ephippium* Linnaeus    Common saddle oyster (Fig. 180)

*Shell roughly circular; delicate in appearance; valves very dissimilar, up to 60 mm in length. Lower (right) valve thin and flat, with indented hole through which calcified byssus attaches shell to substratum; upper (left) valve larger, curved, irregular in shape. Inner surface of upper valve with 3 distinct muscle scars (1 adductor, 2 byssal). Lower valve with 1 adductor muscle scar. Upper valve may be encrusted with epifauna including barnacles and tube worms. Outer surface whitish in colour; inner surface white.*

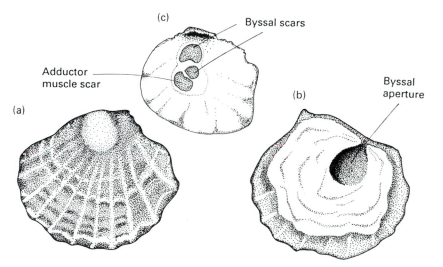

Figure 180  *Anomia ephippium*, (a) left valve, (b) right valve lying within left valve, (c) inner surface of left valve showing byssal and muscle scars.

*A. ephippium* is widely distributed in north-west Europe. It is common in Britain from the middle and lower shore to depths of about 150 m, attached to algae, stones, rocks and shells. The shape of the shell follows that of the object to which it is attached. *Anomia* is hermaphroditic and fertilization external. Breeding has been recorded during the summer months in populations in the Irish Sea; veliger larvae can be found in the plankton during this period, but especially in September. Settlement occurs after 20 to 30 days.

### *Heteranomia squamula* (Linnaeus)   (*Pododesmus squamula*) (Fig. 181a)

*Shell roughly circular; delicate. Valves dissimilar, up to 15 mm in length. Lower (right) valve thin and flat, upper (left) valve larger, convex. Lower valve with oval aperture through which calcified byssus attaches shell to substratum. Inner surface of upper valve with 2 muscle scars (1 adductor, 1 byssal) just touching; surface of scars smooth. Lower valve with 1 adductor muscle scar. Upper valve white, pinkish; lower valve translucent.*

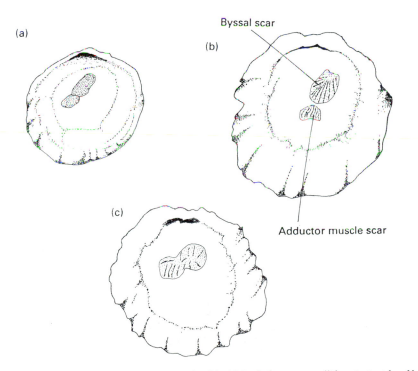

Figure 181   (a) *Heteranomia squamula*. (b), (c) *Pododesmus patelliformis*, inside of left
valves to show the arrangement of the muscle and byssal scars. Modified from
a variety of sources.

*H. squamula* is widely distributed in north-west Europe and around Britain on the lower shore and into the sublittoral. It is found on stones and seaweed.

*Pododesmus patelliformis* (Linnaeus)   (*Monia patelliformis, Monia squama*) (Fig. 181b, c)

*Shell roughly circular, delicate. Valves dissimilar, up to 40 mm in length. Lower (right) valve thin and flat; upper (left) larger, convex. Lower valve with oval to pear-shaped aperture through which calcified byssus attaches shell to substratum. Inner surface of upper valve with 2 muscle scars (1 adductor, 1 byssal), separate or fused to form 1 bilobed scar. Muscle scars with furrowed surface. Lower valve with 1 adductor muscle scar. Shell white to brownish colour, lower valve may be transparent. Shell often encrusted with barnacles, polychaete worms, etc.*

*P. patelliformis* is widely distributed in north-west Europe from the lower shore into deep water. It is found on rocks and stones and other hard substrata, such as scallop shells.

*Pododesmus patelliformis* and *Monia squama* were previously believed to be separate species showing different arrangement of the muscle scars in the upper valve of the shell. Recent observations have shown that both arrangements of scars can be seen during development in a single individual, and it is suggested that there is a single species, *Pododesmus patelliformis*.

Family Kelliidae
SUBFAMILY LASAEINAE – COIN SHELLS
Small, delicate shell; beaks towards posterior. Hinge line with poorly developed teeth. Adductor muscle scars roughly the same size. Pallial sinus absent.

*Lasaea adansoni* (Gmelin)   (*Lasaea rubra*) (Fig. 182a)

*Small, oval, rather plump shell, 2–3 mm in length. Beaks slightly posterior. Left valve with 1 cardinal tooth; right valve without cardinal teeth. Inner and outer surfaces whitish, with distinct pinkish-red tinge. Not to be confused with* Turtonia minuta *(p. 319), which can be found in a similar habitat.*

(a)                                    (b)

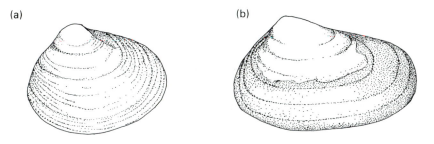

Figure 182  (a) *Lasaea adansoni*, right valve. (b) *Tellimya ferruginosa*, right valve.

*L. adansoni* is widely distributed in north-west Europe and Britain, except the south-east. It is found mainly on the upper and middle shore in crevices, empty barnacles and among tufts of the lichen *Lichina pygmaea* (p. 81), often in very large numbers on exposed rocky shores, but has also been recorded from the shingle beaches of the Fleet in Dorset on the south coast of England. It attaches by a temporary byssus, but can also crawl rapidly. *Lasaea* is a suspension feeder and highly adapted to take full advantage of the short periods of tidal cover, which it experiences on the upper shore, by having a rapid filtering rate. It is hermaphroditic and believed to be self-fertilizing but it has been suggested that the species is apomictic and reproduces without meiosis and fusion of male and female gametes. Breeding occurs during the summer months. Up to 35 embryos develop in the body on the gills, and on the west coast of Wales crawling juveniles emerge during August and September and have a shell length of about 0.5 mm. There is no pelagic larval stage. Longevity is no more than three years and the animals breed in their second year.

### Family Montacutidae

Small delicate shell; beaks towards posterior. Hinge line with poorly developed teeth. Adductor muscle scars roughly the same size. Pallial sinus absent. Usually commensal with echinoderms.

**Tellimya ferruginosa** (Montagu)     *Montacuta ferruginosa* (Fig. 182b)

*Delicate, oval shell, up to 9 mm in length. Beaks well to the posterior. Outer surface whitish, often with reddish-brown deposit; inner surface whitish-purple.*

*T. ferruginosa* is common in north-west Europe and Britain, occurring in sandy deposits. It is a commensal of the sea-potato, *Echinocardium cordatum* (p. 456), and as many as 14 or more bivalves have been recorded with a single echinoderm. Adult specimens live freely in the burrow of *Echinocardium*, while the young are attached to the spines of the echinoderm by byssus threads. The bivalve has also been recorded with the heart-urchins, *Spatangus purpureus* (p. 456), *Echinocardium pennatifidum* (p. 457) and the offshore species, **Echinocardium flavescens** (O. F. Müller). It is a suspension feeder and presumably has an enriched food supply through its association with *Echinocardium*. *Tellimya* first matures as a female, later changing to male. Breeding occurs during the summer months and the eggs are incubated to the veliger stage on the gills of the parent. When released, the veligers are believed to have a larval life of several months.

### Family Cardiidae – cockles

Rounded globular shell, with conspicuous radiating ribs. Beaks towards the anterior. Hinge line with cardinal and lateral teeth. Adductor muscle scars roughly the same size. Pallial sinus absent.

1  Grooves on inner surface of shell running from margin well in towards beak .................. 2
   Grooves on inner surface of shell running from margin to just a little way in .................... 4
2  Shell with 22–28 ribs. Anterior margin of shell crenulate, posterior margin smooth ......
   ....................................................................................... *Cerastoderma glaucum* (p. 301)
   Entire margin of shell crenulate ...................................................................... 3
3  Shell with 24–26 ribs; anterior ribs with broad scales, posterior ones with small spines
   (Fig. 184). Shell small, up to 12 mm in length ........................ *Parvicardium ovale* (p. 299)
   Shell with 18–22 ribs, each with spines joined at the base (Fig. 183). Large, stout shell,
   up to 75 mm in length ..................................................... *Acanthocardia echinata* (p. 298)
4  Right valve of shell with 2 anterior and 1 posterior lateral teeth. Shell valves with 20–
   22 ribs; tubercles on all ribs in small specimens, restricted to anterior ribs in adults ......
   ..................................................................................... *Parvicardium exiguum* (p. 299)
   Right valve of shell with 2 anterior and 2 posterior lateral teeth. Shell valves with 22–
   28 ribs, without tubercles ..................................................... *Cerastoderma edule* (p. 300)

*Acanthocardia echinata* (Linnaeus)   Prickly cockle (Fig. 183)

*Shell stout, up to 75 mm in length; 18–22 conspicuous ribs, each bearing spines joined at the base. Beaks slightly anterior. Grooves on inner surface of shell extend long way in from edge of shell. Margin crenulate. Outer surface yellow-brown in colour; inner surface white.*

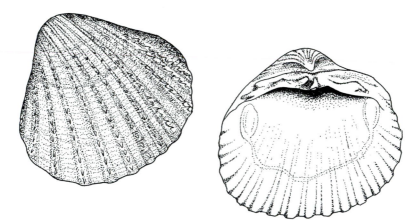

Figure 183  *Acanthocardia echinata*, left valve.

*A. echinata* is widely distributed in north-west Europe and Britain. It is a sublittoral species, occurring at depths of 3 m and below, living in the top few centimetres of sand, mud or gravel. Shells of dead specimens are often washed ashore. *Acanthocardia* is well known through its ability to 'leap', often as far as 200 mm, when disturbed by predators such as starfishes. It does this by rapidly extending the large foot.

Shells of two other species of *Acanthocardia*, **Acanthocardia aculeata** (Linnaeus) and **Acanthocardia tuberculata** (Linnaeus), are occasionally washed ashore in south and south-west Britain. *A. tuberculata* is distinguished by the fact that the spines on the ribs of the shell are not joined at the base. The spines on the ribs of *A. aculeata*, like those of *A. echinata*, are joined at the base; these two species can be separated by examination of the hinge line. In the left valve, the anterior cardinal tooth is larger than the posterior in *A. aculeata*.

### *Parvicardium ovale* (Sowerby)    (Fig. 184a, b, c)

*Small shell, up to 12 mm in length. Beaks slightly anterior. Shell with 24–26 ribs; anterior ribs with broad scales, posterior ones with small spines; ribs in between lack spines and scales. Grooves on inner surface of shell extend long way in from margin towards beak. Margin crenulate. Outer surface yellowish; inner surface white.*

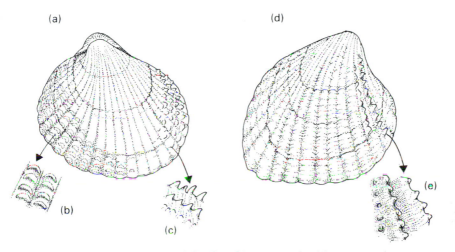

Figure 184  *Parvicardium ovale*, (a) left valve, (b) anterior ribs, (c) posterior ribs.
*Parvicardium exiguum*, (d) right valve, (e) anterior ribs. (After Tebble, 1976.)

*P. ovale* is widely distributed in north-west Europe and is common around Britain. It is a sublittoral species, occurring at depths of about 3–100 m on muddy-sand and gravel.

### *Parvicardium exiguum* (Gmelin)    Little cockle (Fig. 184d, e)

*Small shell, up to 15 mm in length. Beaks slightly anterior. Shell with 20–22 ribs; tubercles on all ribs in young specimens, restricted mostly to anterior ribs in adults. Grooves on inner surface of shell present from margin to just a little way in. Margin crenulate. Outer surface whitish-yellow to brown; inner surface whitish-green.*

*P. exiguum* is widespread in north-west Europe and around Britain in all types of sediment from the lower shore down to about 55 m. It is often found in estuaries and tolerates salinities down to about 17‰. Breeding takes place in spring and summer, with two peaks of recruitment recorded in north-west Spain and a single peak in Denmark.

***Cerastoderma edule*** (Linnaeus)    (*Cardium edule*), Common cockle (Fig. 185)

*Up to 50 mm in length, but usually less. Beaks just anterior. Shell with 22–28 ribs; concentric growth lines conspicuous. Grooves on inner surface of shell present from margin to just a little way in. Margin crenulate. Outer surface off-white, yellowish or brownish; inner surface white with brownish mark near posterior adductor muscle scar.*

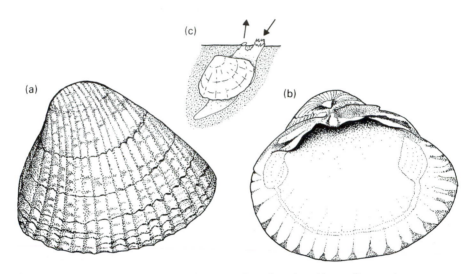

Figure 185  *Cerastoderma edule*, (a) left valve, (b) right valve, (c) in sediment. Arrows
indicate direction of flow of water.

*C. edule* is widely distributed in north-west Europe and around Britain on the middle and lower shore, sometimes sublittorally, in sand and muddy-sand. It usually lives in salinities between 15 and 35‰ and is often abundant in estuaries, as well as in sheltered bays. In areas such as Morecambe Bay, the Wash and South Wales, the cockle is fished commercially by hand-raking. Cockles are suspension feeders, found buried in the top few centimetres of sediment, where they are readily collected with a rake. They are easily dislodged by storms and during winter gales beds of cockles can be washed away. They have many predators, notable among which is the oyster catcher. The sexes are separate, fertilization external and breeding takes place during spring and summer. The free-swimming veliger larva spends some three to six weeks in the plankton and the cockle lives for up to nine years or more, but two to four years is more usual.

***Cerastoderma glaucum*** (Bruguière)  (*Cerastoderma lamarcki*, *Cardium lamarckii*, *Cardium glaucum*), Lagoon cockle (Fig. 186)

*Thinnish shell, up to 50 mm in length, usually smaller. Beaks just anterior. Shell elongate posteriorly; 22–28 ribs; concentric growth lines conspicuous. Grooves on inner surface of shell extend long way in from margin towards beak. Margin crenulate at anterior, smooth at posterior. Outer and inner surfaces of shell coloured shades of brown.*

(a)  (b)

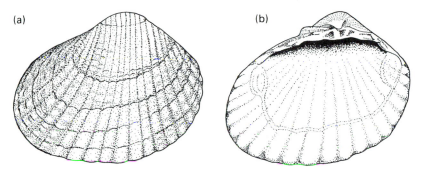

Figure 186 *Cerastoderma glaucum*, (a) right valve, (b) left valve.

*C. glaucum* is widely distributed in north-west Europe and in Britain is found mainly in the south-east. It occurs predominantly in enclosed lagoons, usually permanently submerged, and has a salinity preference of 5 to 38‰. Its distribution is believed to be restricted by the damaging effect of wave action on the newly settled spat. It is a suspension feeder. The sexes are separate, fertilization external and the veliger larvae have a pelagic life of from one to three weeks. Newly settled young attach temporarily by byssus threads to filamentous algae before becoming buried in the top few centimetres of sediment. Breeding occurs during late spring and early summer, and individuals live for about five years.

Family Mactridae – trough shells, otter shells
Triangular to oval shell; beaks near mid-line. Below each beak is a triangular depression, the chondrophore, which houses the internal ligament. Hinge line with cardinal and lateral teeth, the laterals sometimes poorly developed. Cardinal teeth of left valve joined to give a ∧-shaped ridge in front of chondrophore. Adductor muscle scars roughly the same size. Pallial sinus present.

1 Shell elongate, gaping at anterior and posterior; thick brownish-green periostracum. Large shell, up to about 125 mm length ..................................... *Lutraria lutraria* (p. 303)
  Shell triangular, not gaping; without thick periostracum. Up to about 50 mm shell length . ................................................................................................................ 2
2 Lateral teeth smooth ................................................................. *Mactra stultorum* (p. 302)
  Lateral teeth finely serrated ......................................................... *Spisula solida* (p. 302)

***Mactra stultorum*** (Linnaeus)    (*Mactra corallina*), Rayed trough shell (Fig. 187)

*Smooth, triangular shell; rather thin and fragile; up to 50 mm in length. Beaks mid-way between anterior and posterior ends of shell. Chondrophores conspicuous; ∧ -shaped ridge in front of chondrophore of left valve; lateral teeth of hinge line smooth (cf. Spisula, below). Sculpturing of many very fine, concentric lines; shell feels smooth. Outer surface whitish-brown with broad, brown or yellowish rays; inner surface purplish-white. Margin smooth.*

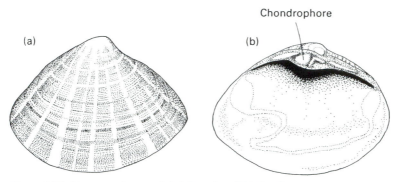

Figure 187  *Mactra stultorum*, (a) right valve, (b) left valve.

M. *stultorum* is widely distributed in north-west Europe and Britain from the very low shore to depths of about 100 m buried in clean sand. Offshore, *Mactra* oftens occurs in vast densities and empty shells are washed ashore in large numbers. It is a suspension feeder. On the south coast of England breeding occurs during spring.

***Spisula solida*** (Linnaeus)    Thick trough shell (Fig. 188)

*Thick triangular shell up to 45 mm in length. Beaks usually mid-way between anterior and posterior ends of shell. Chondrophores conspicuous; ∧ -shaped ridge in front of chondro-phore of left valve small, spans about half the depth of the hinge line. Lateral teeth of hinge finely serrated (cf. Mactra, above). Sculpturing of many concentric lines and several flat concentric ridges. Outer surface off-white; inner surface white. Margin smooth.*

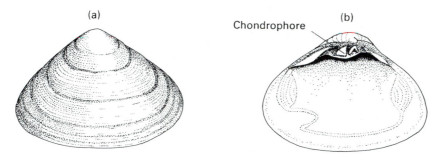

Figure 188  *Spisula solida*, (a) right valve, (b) left valve.

*S. solida* is widely distributed in north-west Europe and Britain, living in sand from extreme low water into the sublittoral where it often occurs with *Mactra* in enormous densities. It is a suspension feeder.

Two other species of *Spisula*, **Spisula elliptica** (Brown) and **Spisula subtruncata** (da Costa), commonly occur in the shallow sublittoral around Britain. Separation of the three species is difficult but *S. solida* is distinguished by the small ∧-shaped ridge formed by the fusion of the cardinal teeth of the left valve of the shell. In *S. elliptica* and *S. subtruncata* the ridge is relatively large and spans almost the complete depth of the hinge line.

**Lutraria lutraria** (Linnaeus)    Common otter shell (Fig. 189)

*Large, oval shell, up to 125 mm in length; shell gapes at anterior and posterior ends. Beaks slightly anterior. Chondrophores conspicuous; ∧-shaped ridge in front of chondrophore of left valve. Sculpturing of concentric lines. Outer surface white, pale orange, pale brown with thick brownish-green periostracum; inner surface white. Pallial line separate from edge of pallial sinus. Margin smooth. Not to be confused with* Mya arenaria *(p. 320).*

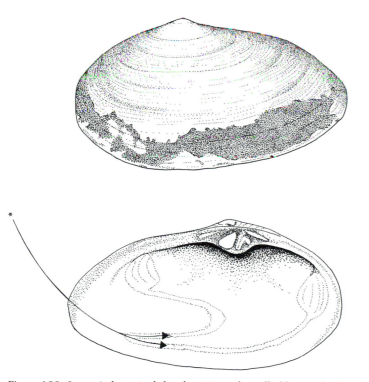

Figure 189  *Lutraria lutraria*, left valve.* Note the pallial line and pallial sinus are separate.

*L. lutraria* is widely distributed in north-west Europe and Britain from the lower shore to about 90 m. It lives buried in muddy-sand and gravel at depths of up to 300 mm, maintaining contact with the surface by the very long, united siphons which have a tough, outer covering. The foot is small and the animal more or less sedentary, penetrating deeper into the sediment with age. It is a suspension feeder.

Two other species of *Lutraria*, **Lutraria magna** (da Costa) and **Lutraria angustior** Philippi, occur in south and south-west Britain. They are sublittoral. Shells of *L. lutraria* can be distinguished from these two species by examination of the pallial sinus. In *L. lutraria* the pallial line and the lower margin of the pallial sinus are not joined (Fig. 189).

Family Solenidae – razor-shells
Shell long and narrow, fragile; vertical groove at anterior end. Beaks at anterior. Brownish-green periostracum. Shell gaping at anterior and posterior. Hinge line with cardinal teeth; lateral teeth absent. Anterior adductor muscle scar larger than posterior. Pallial sinus present.

**Solen marginatus** Pulteney    Grooved razor-shell (Fig. 190)

*Fragile, elongate shell, up to 125 mm in length; straight dorsally and ventrally. Beaks small, at anterior of shell. Valves gaping at both ends; narrow, vertical groove at anterior end. Shell smooth, yellowish with pale brownish-green periostracum; inner surface white. Foot dull red.*

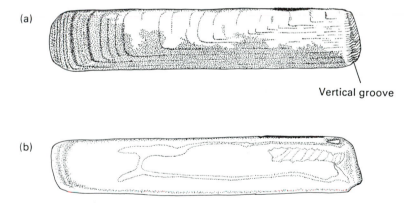

Figure 190  *Solen marginatus*, (a) right valve, (b) left valve.

*S. marginatus* is widely distributed in north-west Europe but in Britain appears to be confined to south and west coasts. It is found in sand or muddy-sand on the lower shore and into the sublittoral.

Family Cultellidae — razor-shells

Shell long, narrow and fragile. Beaks at anterior. Conspicuous olive-green periostracum. Shell gaping at anterior and posterior. Hinge line with cardinal and lateral teeth. Anterior adductor muscle scar larger than posterior. Pallial sinus present.

1 Dorsal and ventral margins of shell straight ..................................... *Ensis siliqua* (p. 306)
  Ventral margin of shell curved ........................................................................ 2
2 Dorsal margin of shell curved .............................................................. *Ensis ensis* (p. 305)
  Dorsal margin of shell straight ..................................................... *Ensis arcuatus* (p. 306)

*Ensis ensis* (Linnaeus)  Razor-shell (Fig. 191)

*Narrow, fragile shell, up to 125 mm in length; dorsal and ventral margins both curved. Beaks small, at anterior of shell. Anterior and posterior ends gaping. Outer surface smooth, whitish colour with vertical and horizontal reddish-brown or purplish-brown markings separated by diagonal line. Periostracum olive-green. Inner surface white with purple tinge. Foot pale red-brown.*

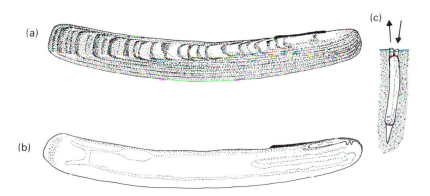

Figure 191 *Ensis ensis*, (a) right valve, (b) left valve, (c) position in sediment when feeding. Arrows indicate direction of water flow.

*E. ensis* is widely distributed in north-west Europe and is found all around Britain in sand, sometimes muddy-sand, from extreme low water into the shallow sublittoral. All species of *Ensis* have a large foot and are rapid, deep burrowers, the shell moving rapidly down into the vertical burrow when disturbed. Their presence in the sand is indicated by keyhole-shaped openings made by the short, united siphons which extend just above the sediment surface when the animal is suspension feeding. Breeding in *E. ensis* probably occurs during spring and the veliger larva has a pelagic life of about a month.

**Ensis arcuatus** (Jeffreys)    Razor-shell (Fig. 192)

*Fragile, elongate shell, up to 150 mm in length; dorsal margin straight, ventral margin curved. Beaks small, at anterior of shell. Valves gaping at both ends. Shell smooth; outer and inner surfaces and periostracum similar to* E. ensis *(above). Foot creamy-white with brown lines.*

Figure 192  *Ensis arcuatus*, right valve.

E. arcuatus is widely distributed in north-west Europe and Britain in coarse sand and fine gravel from the lower shore to about 35 m. It lives in a coarser sediment than either *E. ensis* (above) or *E. siliqua* (below). There is a free-swimming veliger larva.

**Ensis siliqua** (Linnaeus)    Pod razor-shell (Fig. 193)

*Fragile, elongate shell, up to 200 mm in length; dorsal and ventral margins straight. Beaks small, at anterior of shell. Valves gaping at both ends. Shell smooth; outer and inner surfaces and periostracum as* E. ensis *(above). Foot creamy-white with brown lines.*

Figure 193  *Ensis siliqua*, right valve.

E. siliqua is widely distributed in north-west Europe and around Britain in clean sand on the lower shore and in the shallow sublittoral. There is a free-swimming veliger larva. Longevity is up to 10 years.

Family Tellinidae – tellins
Thin, mostly flattened, relatively smooth shell. Beaks central or slightly posterior. Hinge line usually with cardinal and lateral teeth. Adductor muscle scars roughly the same size. Pallial sinus present.

1  Shell rounded, rather globular. Without lateral teeth  ...............  ***Macoma balthica*** (p. 308)
   Shell oval, flattened; rather fragile. Lateral teeth present  ...................................................  2

2  Shell sculpturing of concentric lines; faint diagonal lines on right valve only  .................
................................................................................................................ *Fabulina fabula* (p. 307)
   Shell sculpturing of concentric lines; no diagonal lines  ................ *Angulus tenuis* (p. 307)

**Angulus tenuis** (da Costa)    *(Tellina tenuis),* Thin tellin (Fig. 194a)

*Fragile, oval shell, up to 25 mm in length; flattened; right valve slightly larger than left.*
*Beaks a little to the posterior. Sculpturing of concentric lines. Outer surface shiny, usually*
*pink but often orange or white; inner surface similar. Margin smooth. Not to be confused*
*with Macoma (below).*

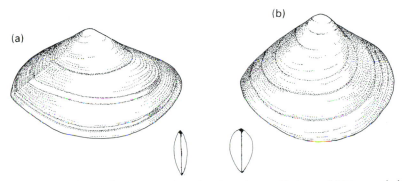

Figure 194  (a) *Angulus tenuis,* right valve. Anterior profile inset. (b) *Macoma balthica,* right
valve. Anterior profile inset.

A. *tenuis* is widely distributed in north-west Europe and Britain, where it occurs in fine,
clean sand from the middle shore into the shallow sublittoral. In some areas, it is abundant. *Angulus* usually lies on the left valve a few centimetres below the surface of the
sediment. The siphons are long and separate and, despite earlier reports that it was a
deposit feeder, it is now suggested that it is mainly a suspension feeder. Young flatfishes
often feed on the tips of the inhalent siphons when they are exposed above the surface
of the sediment, but the bivalve is not killed and regeneration of the siphons occurs. The
sexes are separate and breeding occurs during summer. There is a free-swimming veliger
larva and longevity is believed to be about five years.

**Fabulina fabula** (Gmelin)    *(Tellina fabula)* (Fig. 195)

*Fragile, oval shell, up to 20 mm in length; flattened; right valve slightly more convex than*
*left. Beaks a little to the posterior. Sculpturing of concentric lines on both valves; faint,*
*diagonal lines on right valve. Outer surface shiny, whitish; inner surface similar. Margin*
*smooth. Not to be confused with Macoma (below).*

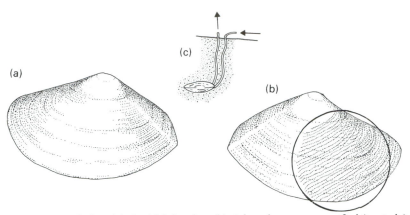

Figure 195  *Fabulina fabula*, (a) left valve, (b) right valve, part magnified (in circle) to show diagonal striations, (c) in sediment. Arrows indicate direction of water flow.

*F. fabula* is widely distributed in north-west Europe and Britain in fairly clean sand but, in contrast to *A. tenuis* (above), it is essentially a sublittoral species extending from the lower shore to about 55 m. It is found up to 100 mm deep in the sediment and lies in a more-or-less horizontal position on the left valve. It is both a suspension and a deposit feeder. The sexes are separate and in the eastern North Sea breeding occurs from March to September, with those animals which breed in March, breeding again in summer. There is a free-swimming veliger larva. Longevity is believed to be four or five years.

### *Macoma balthica* (Linnaeus)    Baltic tellin (Fig. 194b)

*Rounded, rather globular shell, up to 25 mm in length; valves almost equal. Beaks mid-way between anterior and posterior ends of shell. Sculpturing of fine, concentric lines. Hinge line without lateral teeth. Outer surface dull, usually pinkish-purple but sometimes yellowish or white and banded; inner surface similar. Margin smooth. Not to be confused with* Angulus tenuis *(above) and* Fabulina fabula *(above).*

*M. balthica* is widely distributed in north-west Europe and Britain, living a few centimetres below the surface of sand, mud and muddy-sand. It is found from the middle shore into the sublittoral, and occurs in estuaries, often with *Cerastoderma edule* (p. 300). It survives in salinities down to 5‰. The siphons are long and separate and although generally considered to be mainly a deposit feeder, it switches from deposit to suspension feeding when there is a high density of suspended particles in the water. During feeding, the extended siphons are frequently cropped by predators such as flatfishes. The siphons are quickly regenerated but in some situations they are so frequently cropped that the growth rate of the bivalve is inhibited. The sexes are separate and on the west coast of Wales the main breeding period is during February and March, with a second spawning in autumn. There is a free-swimming veliger larva which is pelagic for up to seven or eight weeks. Where growth is fast, longevity is about three years, but in slow-growing populations specimens live for six or seven years.

Family Donacidae – wedge shells
Shell wedge-shaped. Beaks towards posterior. Hinge line with cardinal and lateral teeth. Margin of shell finely to strongly crenulate. Adductor muscle scars roughly the same size. Pallial sinus present.

**Donax vittatus** (da Costa)    Banded wedge shell (Fig. 196)

*Shell up to 38 mm in length. Beaks quite close to posterior end of shell. Sculpturing of very fine, radiating lines, absent near anterior margin. Outer surface shiny, orange, purple, yellow or brown; colours often in bands; inner surface shiny, white with purple or yellow areas. Margin strongly crenulate.*

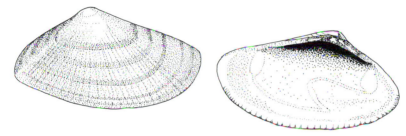

Figure 196  *Donax vittatus*, right valve.

*D. vittatus* is widely distributed in north-west Europe and Britain on exposed sandy beaches, from the lower shore to about 20 m. It lives just under the surface of the sediment and is often dislodged by pounding surf, but the presence of a large, powerful foot enables it to reburrow immediately and so reduce the dangers of predation and desiccation. It is a suspension feeder. The sexes are separate and on the west coast of Scotland breeding has been recorded during summer. Where growth is rapid, *Donax* lives for two to three years but where it is slow, length of life may be up to seven years.

Family Psammobiidae
SUBFAMILY PSAMMOBIINAE – SUNSET SHELLS
Oblong shell. Beaks slightly anterior or slightly posterior. Hinge line with cardinal teeth, but no laterals. Adductor muscle scars roughly the same size. Pallial sinus present.

**Gari fervensis** (Gmelin)    Faroe sunset shell (Fig. 197)

*Oblong shell, up to 50 mm in length; right valve slightly more convex and slightly larger than left. Beaks slightly anterior. Posterior margin truncate. Sculpturing of concentric lines and a conspicuous, diagonal posterior ridge. Outer surface pink, white or light brown with whitish rays; periostracum greenish-brown; inner surface white, purple or pink. Margin smooth.*

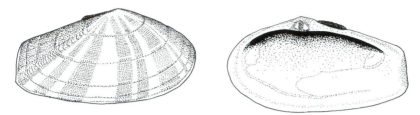

Figure 197  *Gari fervensis*, right valve.

*G. fervensis* is widely distributed in north-west Europe and Britain from the lower shore into the sublittoral. It is found in the top few centimetres of coarse sand and gravel. *Gari* is a suspension feeder.

**Gari depressa** (Pennant) is occasionally found on the very low shore, particularly in the south-west. The shell is more oval than that of *G. fervensis* and the posterior margin is not truncate.

SUBFAMILY SOLECURTINAE

Rectangular, somewhat elongate shell. Beaks slightly anterior. Shell gaping at anterior and posterior. Hinge line usually with cardinal and lateral teeth. Adductor muscle scars roughly the same size or anterior larger than posterior. Pallial sinus present.

**Pharus legumen** (Linnaeus)    (Fig. 198)

*Narrow, fragile shell, up to 120 mm in length. Beaks small and slightly anterior. Wide gape at anterior and posterior ends of shell. Sculpturing of concentric lines. Outer surface off-white to light brown with yellowish-green periostracum; inner surface white. Margin smooth. Very similar in appearance to razor-shells (p. 304) but beaks are near mid-line of shell, not at anterior end.*

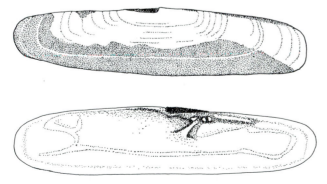

Figure 198  *Pharus legumen*, left valve.

*P. legumen* is a southern species extending into south-west Britain about as far as North Wales, and living in fairly clean sand on the lower shore and shallow sublittoral. It has a large, pink foot and is a rapid vertical burrower, being very difficult to catch. Unlike the true razor-shells, the siphons are separate throughout their length and *Pharus* is a deposit feeder. There is a free-swimming veliger larva and settlement has been recorded in summer. Length of life is up to six years. The siphons of *Pharus* are known to be cropped extensively by fishes.

Family Semelidae

Rounded, thin shell, flattened. Beaks slightly anterior. Chondrophores conspicuous. Hinge line with cardinal teeth; lateral teeth present or absent. Anterior adductor muscle scar slightly larger than posterior. Pallial sinus present.

***Scrobicularia plana*** (da Costa)    Peppery furrow shell (Fig. 199)

*Rounded, thin shell; much flattened; up to 63 mm in length. Beaks slightly anterior. Chondrophores conspicuous, semi-circular. Without lateral teeth. Sculpturing of fine, concentric lines. Outer surface white, pale grey or yellowish; inner surface white. Margin smooth.*

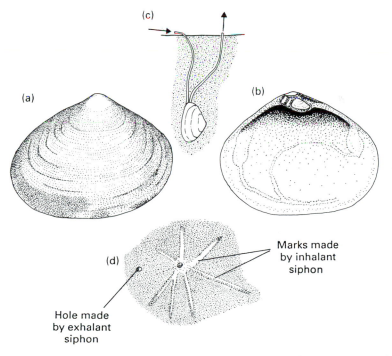

Figure 199 *Scrobicularia plana*, (a), (b) right valve, (c) in sediment (arrows indicate direction of water flow), (d) marks left on surface of sediment by siphons.

*S. plana* is a common intertidal bivalve in north-west Europe and occurs all around Britain, often in high densities. It is essentially estuarine and intertidal, surviving in salinities down to 10‰ in thick mud or muddy-sand, where it lives up to 200 mm below the surface of the sediment. It is mainly a deposit feeder and the star-shaped markings left on the surface of the sediment by the long inhalant siphon indicate the presence of *Scrobicularia*. When the siphons protrude above the sediment during feeding, they are often cropped by fishes, crabs and birds, but lost tissue is regenerated within four or five days. *Scrobicularia* is also known to suspension feed. The sexes are separate; on the coast of North Wales, breeding occurs during July and August but may be earlier in other areas. The free-swimming veliger larvae settle after a planktonic life of two or three weeks. Some specimens live for up to 18 years.

Family Arcticidae – cyprinas
Thick, heavy shell; beaks towards anterior. Periostracum thick and shiny. Hinge line with three cardinal teeth and a posterior lateral tooth. Anterior adductor muscle scar somewhat smaller than posterior. Pallial sinus absent.

*Arctica islandica* (Linnaeus) (*Cyprina islandica*)    (Fig. 200)

*Thick, heavy shell, oval to rounded; up to 120 mm in length; sculptured with concentric lines. Beaks well to the anterior. Periostracum thick, shiny and pale brown to black in colour, often peels away in dead specimens to reveal white to pale brown outer surface of shell beneath; inner surface white.*

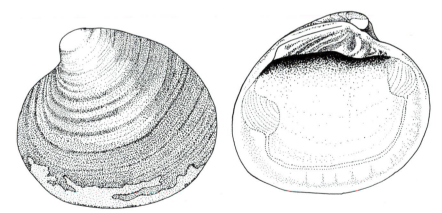

Figure 200  *Arctica islandica*, left valve.

*A. islandica* is widely distributed in north-west Europe and occurs all around Britain from the very low shore well into the sublittoral. It lies buried in sandy and muddy sediments and is a suspension feeder. The sexes are separate and breeding occurs in spring and summer. Studies on the growth of *Arctica* from the Atlantic have shown that it is slow growing, and it has been suggested that some specimens live for 105 years!

Family Veneridae — carpet shells, venus shells

Rounded to oval shell. Beaks towards anterior. Hinge line with cardinal teeth, sometimes lateral teeth. Heart-shaped region, the lunule, in front of beaks in all species described in this text. Adductor muscle scars roughly the same size. Pallial sinus present.

1   Inner margin of shell crenulate, at least in part ................................................. 2
    Inner margin of shell without crenulations ...................................................... 5
2   Shell sculpturing of 40–50 radiating ribs and many concentric grooves. Has some
    resemblance to a cockle ...................................................... *Timoclea ovata* (p. 315)
    Shell sculpturing predominantly of concentric ridges and lines ........................................... 3
3   Shell sculpturing of concentric ridges with conspicuous raised tubercles near anterior
    and especially towards posterior ................................................ *Venus verrucosa* (p. 313)
    Shell without raised tubercles ......................................................................... 4
4   Sculpturing of concentric lines between thick, prominent, concentric ridges .................
    .................................................................................... *Clausinella fasciata* (p. 315)
    Sculpturing of many narrow, concentric ridges. Outer surface usually with 3 broad
    reddish-brown rays ..................................................... *Chamelea gallina* (p. 314)
5   Shell rounded, whitish in colour, lunule conspicuous ....................................................... 6
    Shell oval or elongate, lunule usually indistinct ........................................................... 7
6   Region anterior to lunule convex but relatively low ................... *Dosinia lupinus* (p. 318)
    Region anterior to lunule markedly convex, raised ...................... *Dosinia exoleta* (p. 317)
7   Shell sculpturing without radiating lines; with many concentric ridges and lines.
    Lunule indistinct in larger shells .......................................... *Tapes rhomboides* (p. 316)
    Shell sculpturing of radiating lines, although sometimes faint, and concentric grooves
    and ridges ...................................................................................................... 8
8   Sculpturing of concentric ridges and faint, radiating lines   *Venerupis senegalensis* (p. 317)
    Sculpturing of concentric grooves and distinct radiating lines ... *Tapes decussatus* (p. 316)

*Venus verrucosa* Linnaeus    Warty venus (Fig. 201)

*Stout, rounded shell, up to 63 mm in length. Beaks slightly anterior; lunule brown, conspicuous. Sculpturing of thick, concentric ridges with tubercles near anterior and especially towards posterior. Outer surface off-white or brown; inner surface white, sometimes purple or brown near posterior adductor muscle scar. Inner margin crenulate except at posterior.*

*V. verrucosa* is a southern species which extends as far north as the west coast of Scotland. It is often found on the south and south-west coasts of Britain just below the surface in sand and fine gravel, from very low on the shore to about 100 m. It is a suspension feeder.

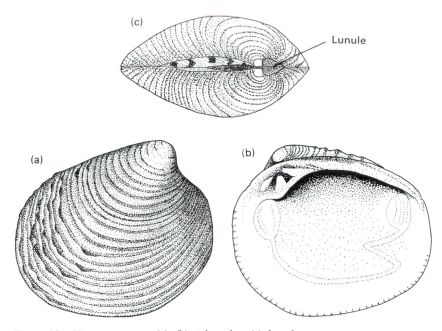

Figure 201  *Venus verrucosa*, (a), (b) right valve, (c) dorsal view.

**Chamelea gallina** (Linnaeus)    (*Chamelea striatula*, *Venus striatula*), Striped venus (Fig. 202)

*Shell triangular, up to 45 mm in length. Beaks slightly anterior; lunule conspicuous. Sculpturing of many narrow, concentric ridges. Outer surface off-white, usually with 3 broad, reddish-brown rays; inner surface whitish. Inner margin crenulate towards posterior.*

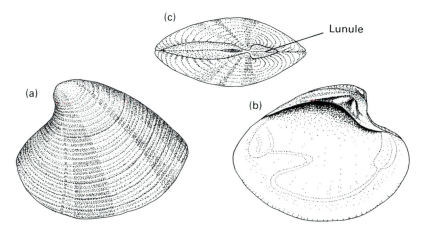

Figure 202  *Chamelea gallina*, (a), (b) left valve, (c) dorsal view.

C. *gallina* is widely distributed in north-west Europe and is commonly found in Britain buried some little way below the surface in sand and muddy-sand from the lower shore to depths of about 55 m. It is a suspension feeder. The sexes are separate, fertilization external and on the west coast of Scotland breeding has been recorded during spring and summer. Free-swimming veliger larvae are common in the plankton during autumn. Longevity is believed to be 10 or 11 years.

### *Clausinella fasciata* (da Costa)   (*Venus fasciata*), Banded venus (Fig. 203b)

*Thick, triangular shell, up to 25 mm in length. Beaks towards the anterior; lunule distinct. Sculpturing of concentric lines between thick, prominent, concentric ridges. Outer surface white, yellow, pinkish-brown with darker rays or markings; inner surface white to whitish-purple. Inner margin crenulate in anterior region only.*

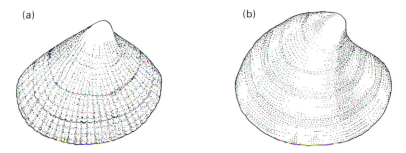

(a)                                        (b)

Figure 203   (a) *Timoclea ovata*, right valve. (b) *Clausinella fasciata*, right valve.

C. *fasciata* is widely distributed in north-west Europe and is common all around Britain from extreme low water down to about 110 m. It lives just below the surface in sand and gravel and is a suspension feeder. On the west coast of Scotland breeding has been recorded from February to July.

### *Timoclea ovata* (Pennant)   (*Venus ovata*), Oval venus (Fig. 203a)

*Shell thin; oval to triangular in shape; up to 20 mm in length. Beaks slightly anterior; lunule faint. Sculpturing of 40–50 radiating ribs and many concentric grooves. Outer surface off-white to light brown; inner surface whitish-orange with purple tinge. Inner margin crenulate for most of length. Bears some resemblance to a cockle.*

T. *ovata* is widely distributed in north-west Europe and is found all around Britain. It lives just below the surface in sand, muddy-sand and gravel, at depths ranging from about 3 to 180 m. It is a suspension feeder. On the west coast of Scotland breeding has been recorded in May.

***Tapes rhomboides*** (Pennant)    (*Paphia rhomboides*, *Venerupis rhomboides*), Banded carpet shell (Fig. 204)

*Elongate, oval shell, up to 60 mm in length. Beaks quite close to anterior end of shell; lunule narrow, indistinct in larger shells. Sculpturing of many concentric ridges and lines; without radiating lines, cf.* Venerupis senegalensis *(below) and* Tapes decussatus *(below). Outer surface creamy-yellow with purple, brown markings, occasionally no markings; inner surface glossy and white. Margin smooth.*

(a)                                                        (b)

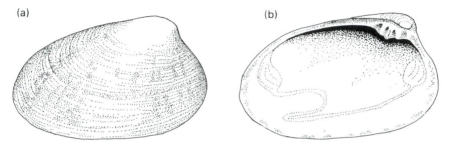

Figure 204  *Tapes rhomboides*, (a) right valve, (b) left valve.

*T. rhomboides* is widely distributed in north-west Europe and Britain from the very low shore to about 180 m in gravel and sand. It is a suspension feeder.

***Tapes decussatus*** (Linnaeus)    (*Venerupis decussata*), Carpet shell (Fig. 205)

*Oval shell, up to 70 mm in length. Beaks quite close to anterior end of shell; lunule inconspicuous. Sculpturing of concentric grooves and distinct, radiating lines. Outer surface white, yellowish, pale brown, often with dark brown marks; inner surface shiny, whitish-orange, sometimes with purple marks. Margin smooth. Distinguished from* T. rhomboides *(above) and* V. senegalensis *(below) by the distinct radiating lines.*

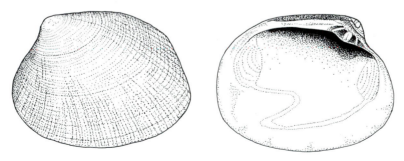

Figure 205  *Tapes decussatus*, left valve.

*T. decussatus* is a southern species reaching its northern limit in the south and west of Britain. It lives buried in sand or gravel on the lower shore and in the sublittoral. It is a suspension feeder. On the west coast of Scotland breeding has been recorded in July.

**Venerupis senegalensis** (Gmelin)    (*Venerupis pullastra, Venerupis saxatilis*), Pullet carpet shell (Fig. 206)

*Oval shell, up to 50 mm in length. Beaks quite close to anterior end of shell; lunule indistinct, especially in larger specimens. Sculpturing of many concentric ridges and faint, radiating lines. Outer surface off-white, cream or grey, sometimes with purple or brown markings; inner surface shiny white, occasionally with purple tinges. Margin smooth. Distinguished from* T. rhomboides *(above) by the sculpturing and the more angular appearance of the posterior part of the shell.*

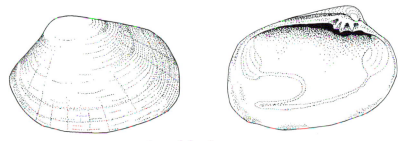

Figure 206  *Venerupis senegalensis*, left valve.

*V. senegalensis* is widely distributed in north-west Europe and Britain from the middle shore to depths of about 35 m. It lives in sand and gravel, often at the base of rocks, and is sometimes buried in the sediment to depths of 30–50 mm, attached to small stones or shells by the byssus. It is also found in rock crevices. *V. senegalensis* is a suspension feeder. The sexes are separate and on the west coast of Scotland breeding occurs from May to September, with free-swimming veliger larvae appearing in the plankton from June onwards. Larval life is about 30 days and individuals live for up to eight years.

**Dosinia exoleta** (Linnaeus)    Rayed artemis (Fig. 207e)

*Rounded shell, up to 60 mm in length. Similar to* D. lupinus *(below). Beaks slightly anterior; lunule conspicuous. Region anterior to lunule markedly convex, raised. Sculpturing of many concentric ridges. Outer surface white to light brown, with brown or pink rays; inner surface white. Margin smooth.*

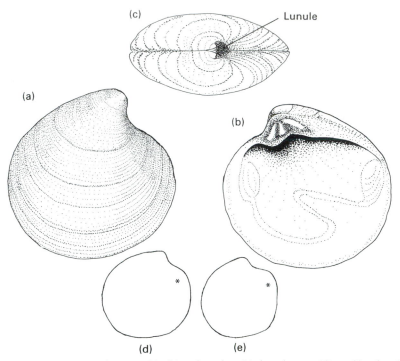

Figure 207  *Dosinia lupinus*, (a), (b) right valve, (c) dorsal view. (d) Profile of *D. lupinus*. (e) Profile of *Dosinia exoleta*. Note shape of shell at *.

*D. exoleta* is widely distributed in north-west Europe and around Britain, where it is found from the lower shore to about 70 m in gravel, burrowing deep into the sediment. It is a suspension feeder.

**Dosinia lupinus** (Linnaeus)    Smooth artemis (Fig. 207a, b, c, d)

*Rounded shell, up to 38 mm in length. Beaks slightly anterior; lunule conspicuous. Region anterior to lunule convex but relatively low. Sculpturing of very many narrow, concentric ridges giving smooth appearance. Outer surface of shell off-white, yellowish, pale brown; inner surface white. Margin smooth.*

*D. lupinus* is widely distributed in a variety of substrata from fine sand and muddy-sand to fine shell gravel. It is found from the lower shore to about 125 m and burrows deep into the sediment. It is a suspension feeder.

Family Turtoniidae
Small, oval shell, fragile. Beaks towards anterior. Lunule narrow. Hinge line with cardinal teeth; without lateral teeth. Adductor muscle scars, pallial line and pallial sinus difficult to see. Attached by byssus.

***Turtonia minuta*** (Fabricius)    (*Venus minuta*) (Fig. 208)

*Oval shell, up to about 3 mm in length. Rather plump. Beaks towards anterior. Lunule narrow and may be indistinct. Teeth of hinge line not well developed; 3 cardinal teeth but in each valve one of these is not distinct. Without lateral teeth. Pallial line may be difficult to see. Pale brown colour with pinkish tinge. Attached by byssus in crevices, among barnacles and seaweeds on rocky shores. Some shell characteristics difficult to detect. Shell size and habitat are important distinguishing features.*

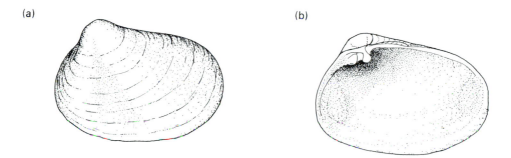

(a)                                                        (b)

Figure 208  *Turtonia minuta*, (a) left valve, (b) right valve.

*T. minuta* is widely distributed in north-west Europe. It has been recorded throughout the intertidal region and is sometimes abundant on the high shore in crevices, among barnacles and seaweeds, in situations similar to those occupied by *Lasaea adansoni* (p. 296), with which it might be confused. It is a suspension feeder. Reproduction occurs in summer. The sexes are separate and the eggs are fertilized in the mantle cavity of the female. The female produces gelatinous egg capsules which are attached to what have been described as modified byssus threads. Each capsule contains on average about four eggs, which develop directly into young bivalves, hatching with a shell length of about 0.3 mm.

Family Myidae – gapers

Shell valves dissimilar, right more convex than left. Beaks slightly anterior or slightly posterior. Brownish periostracum. Shell valves gape at posterior. Hinge line without teeth but in the species described in this text, a spoon-shaped chondrophore projects from the left valve; chondrophore of right valve not projecting. Anterior adductor muscle scar narrow, posterior scar rounded. Pallial sinus present.

Posterior margin of shell truncate ..................................................... *Mya truncata* (p. 320)
Posterior margin of shell slightly elongate, rounded ....................... *Mya arenaria* (p. 320)

***Mya truncata*** Linnaeus    Blunt gaper (Fig. 209)

*Shell up to 75 mm in length; valves dissimilar, right more convex than left. Beaks slightly posterior. Posterior margin truncate with a large gape. Hinge of left valve with conspicuous, spoon-shaped projection, the chondrophore, to which ligament is attached. Sculpturing of concentric lines. Outer surface off-white, with brownish periostracum; inner surface white.*

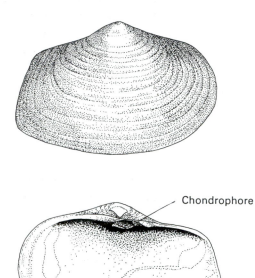

Figure 209  *Mya truncata*, right valve.

*M. truncata* is a northern species, widely distributed in north-west Europe and Britain. It lives buried at considerable depths in muddy-sand, from the middle shore to 70 m. *Mya* is a suspension feeder. There is a free-swimming veliger larva.

***Mya arenaria*** Linnaeus    Sand gaper, Soft-shelled clam (Fig. 210)

*Shell up to 150 mm in length; valves dissimilar; right more convex than left. Beaks slightly anterior. Posterior part of shell somewhat elongate, rounded, with wide gape. Hinge of left valve with large, spoon-shaped projection, the chondrophore, to which ligament is attached. Sculpturing of concentric lines. Outer surface off-white, greyish or brownish, with pale brown periostracum; inner surface whitish-brown. Not to be confused with* Lutraria lutraria *(p. 303).*

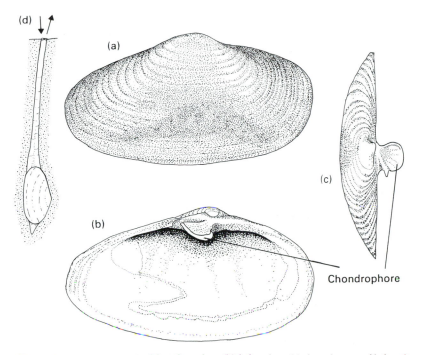

Figure 210  *Mya arenaria*, (a) right valve, (b) left valve, (c) dorsal view of left valve, (d) in
sediment. Arrows indicate direction of water flow.

*M. arenaria* is a northern species widely distributed in north-west Europe and Britain. It
lives in sandy-mud and mud from the lower shore to about 70 m, and is common on
mud-flats and in estuaries where it is known to feed in salinities down to 15‰, as well
as surviving for some considerable time at salinities as low as 4‰. Adult specimens live
up to 500 mm below the surface of the sediment and the siphons, joined along their
length, are capable of enormous extension, making contact with the surface for suspension feeding and respiration. The presence of *Mya* on mud-flats is betrayed by a keyhole-
shaped opening left on the surface of the sediment by the siphons. When contracted into
the shell, the size of the siphons causes the shell valves to gape posteriorly. The foot of
the adult is small and it is doubtful whether the animal can reburrow if dislodged. The
sexes are separate, and fertilization probably takes place in the mantle cavity, sperm
entering in the inhalant current of the female. The veliger larvae are planktonic for up to
three weeks. Settlement, which may be very dense in some areas, occurs on sand grains
and algae, the young animals attaching by a byssus. If the substratum is unsuitable they
can detach and move to another locality. Young *Mya* live in the top few centimetres of
sediment and have a large foot, enabling them to move easily. With increasing age they
move deeper into the sediment and byssal attachment is lost. Breeding occurs during
spring and summer and veligers are present in the plankton until early autumn. Length
of life is up to 17 years. It has recently been demonstrated that seasonal growth patterns

can be detected in the chondrophore on the hinge line of the shell. *Mya* is much prized as food, especially in America where it forms the basis of the famous clam chowder.

Family Hiatellidae – rock-borers
Shell more or less rectangular. Beaks to the anterior. Without lateral teeth. Right valve with single projecting cardinal tooth. Pallial line continuous or as separate scars. Pallial sinus present. Margin of shell smooth.

*Hiatella arctica* (Linnaeus)    Wrinkled rock-borer (Fig. 211)

*Shell up to 35 mm in length; more or less rectangular, but very irregular. Beaks close to anterior end of shell. Pallial line of separate scars. Posterior end gaping. Sculpturing of more or less concentric lines. Outer surface white with yellowish-brown periostracum; inner surface white.*

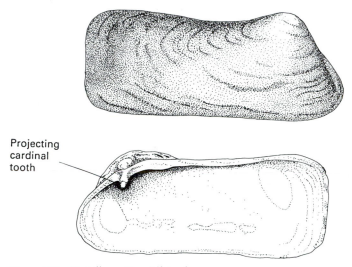

Projecting
cardinal
tooth

Figure 211  *Hiatella arctica*, right valve.

*H. arctica* is widely distributed in north-west Europe. It is common around Britain from the lower shore to about 50 m and occurs either boring into soft rock, or attached by a byssus and nestling in existing holes or crevices in rock or laminarian holdfasts. The siphons are long and reddish-coloured at the tips. It is a suspension feeder. The sexes are separate and on the west coast of Scotland breeding has been recorded from December to April. There is a free-swimming veliger larva and after metamorphosis the young bivalves are mobile for up to six months before selecting a more permanent habitat. The type of substratum selected determines whether the young become burrowers or nestlers; those settling on soft rock becoming burrowers, those on hard rock attach by the byssus and become nestlers. The boring action of the adults is achieved by mechanical abrasion using the valves of the shell, but the initial penetration of the rock by the young animal may involve chemical as well as mechanical means.

Family Pholadidae – piddocks
Shell elongate, oval; modified for boring into substrata such as wood, clay and rock. Beaks towards anterior of shell. Shell usually with pronounced gape. Up to four accessory plates in dorsal part of shell with part of the shell, known as the umbonal reflection, reflected over them. Hinge line lacks teeth but often has slender projection, the apophysis. Adductor muscle scars present, but position, size and number vary. Pallial sinus present.

1  Shell fragile, not gaping anteriorly; with 2 dorsal accessory plates   *Barnea candida* (p. 324)
   Shell gapes anteriorly and posteriorly; anterior gape reveals large, sucker-like foot  .......... 2
2  Shell elongate, of fragile appearance; sculpturing of concentric ridges crossed anteriorly
   by about 40 ribs, the anterior ones with spines  ........................... *Pholas dactylus* (p. 323)
   Shell oval, robust; sculpturing of uneven concentric ridges; crossed by vertical groove
   in mid-line  ................................................................. *Zirfaea crispata* (p. 325)

*Pholas dactylus* Linnaeus    Common piddock (Fig. 212)

*Elongate shell of fragile appearance, up to 150 mm in length. Beaks near anterior end of shell, with part of shell, the umbonal reflection, reflected over them. Four dorsal accessory plates. Shell gapes anteriorly and posteriorly, anterior gape revealing large sucker-like foot. Hinge line with slender projection, the apophysis (the point of attachment of foot muscles) just below beak of each valve. Sculpturing of concentric ridges crossed anteriorly by about 40 ribs; spines present at anterior end. Outer surface greyish-white with yellowish periostracum; inner surface shiny white. Anterior margin of shell crenulate, remainder smooth.*

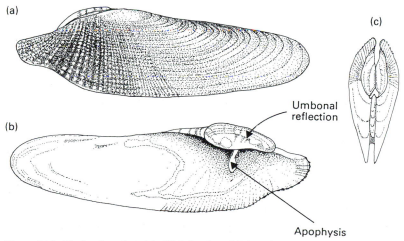

Figure 212  *Pholas dactylus*, (a), (b) left valve, (c) dorsal view.

*P. dactylus* is a southern species extending as far north as the south and south-west of Britain. It is found on the lower shore and in the shallow sublittoral boring into peat, soft rock and sometimes wood. *Pholas* is a suspension feeder. Boring is achieved solely by mechanical means; the shell valves are scraped against the rock surface and rotated to produce circular burrows which occasionally criss-cross one another. Interestingly, *Pholas* is phosphorescent, shining with a greenish-blue light while in its burrow. Breeding probably occurs during summer. There is a free-swimming veliger larva which attaches by a byssus at settlement, the byssus later being lost.

***Barnea candida*** (Linnaeus)    White piddock (Fig. 213)

*Fragile, oval shell, up to 63 mm in length. Beaks near anterior end of shell. Two dorsal accessory plates. Shell gapes posteriorly. Hinge line with slender projection, the apophysis (the point of attachment of the foot muscles) just below beak of each valve. Sculpture of concentric ridges crossed by ribs with spines, the spines well developed at anterior end of shell. Outer surface white; inner surface shiny white. Anterior margin of shell crenulate, remainder smooth.*

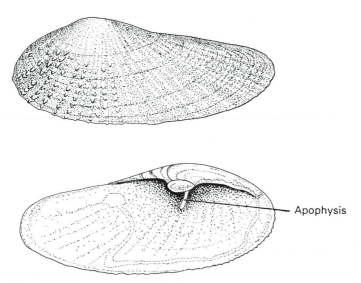

Figure 213  *Barnea candida*, left valve.

*B. candida* is widely distributed in north-west Europe and has been recorded all around Britain. Like *Pholas* (above), it bores by mechanical means into soft substrata including peat, wood and shale, from the middle shore into the shallow sublittoral, extending into estuaries in salinities down to about 20‰. It is a suspension feeder. The sexes are separate, fertilization external and on the south coast of England breeding occurs in early autumn. There is a free-swimming veliger larva.

**_Zirfaea crispata_** (Linnaeus)     Oval piddock (Fig. 214)

_Robust, oval shell, up to 90 mm in length. Beaks near anterior end of shell, part of shell reflected over them. Single dorsal accessory plate. Hinge line with large, flat projection, the apophysis (the point of attachment of the foot muscles), just below beak of each valve. Shell gapes anteriorly and posteriorly. Sculpturing of concentric ridges, very uneven in anterior region; crossed by vertical groove in mid-line, most conspicuous in young specimens. Outer surface whitish with brown periostracum; inner surface white. Anterior margin crenulate, remainder smooth._

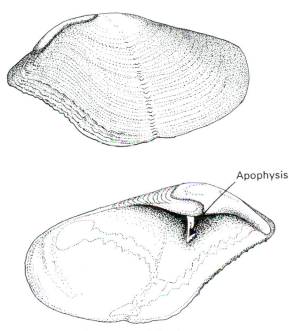

Apophysis

Figure 214  _Zirfaea crispata_, left valve.

_Z. crispata_ is widely distributed in north-west Europe and Britain. It is found from the lower shore to depths of a few metres boring into clay, shale and soft rocks. Like _Pholas_ (above) it burrows by mechanical means. It is a suspension feeder. The siphons are long, and close to the shell they are covered by the thick, dark-coloured periostracum. The sexes are separate and, on the north-east coast of England, breeding occurs from March to October with peaks from March to May and August to October. Free-swimming veliger larvae are present in the plankton for most of the year. Animals living intertidally survive for about five years, compared with seven years for sublittoral populations.

Family Teredinidae – shipworms
Shell very small, covers anterior end of long, worm-like body. Posterior end of body with pair of calcareous pallets. Bores into submerged timbers. Tunnels in timber lined by calcareous tube.

*Teredo navalis* Linnaeus    Shipworm (Fig. 215a–d)

*Shell small, up to 10 mm in length. Valves divided into number of regions, each with differ-ent sculpturing. Beaks near anterior end of shell. Outer surface white, with pale brown perios-tracum. Shell covers only the anterior region of the long, soft body which lies in calcareous tube, up to 150 mm in length. Posterior end of body with a pair of calcareous pallets; pallets off-white, paddle-shaped with forked blade, up to 5 mm in length.*

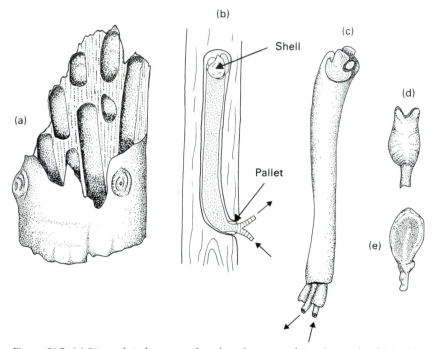

Figure 215   (a) Piece of timber opened to show burrows of *Teredo navalis*. (b) *Teredo navalis* in timber (after Yonge, 1949). (c) *Teredo navalis* removed from calcareous tube (after Eales, 1950). (d) Pallet of *Teredo navalis*. (e) Pallet of *Nototeredo norvegica*. ((d), (e). After Tebble, 1976.) Arrows indicate direction of flow of water.

*T. navalis* is widely distributed in north-west Europe, where it is found boring in sub-merged timber and is often seen in pieces of timber washed ashore. The ravages wrought by shipworms have been known since the time of the early Greeks, but it was not until 1733 that it was established by Sellius that they were bivalve molluscs. Shipworms burrow into piles, piers, wooden ships, bridges and dykes. The only external evidence of attack is the presence of fine holes (1–2 mm across) on the surface of the timber marking the site of the original penetration, but the centre of the timber is likely to be riddled with circular tunnels up to about 20 mm across. Boring is achieved by mechanical means, the shell valves cutting into the wood, with seawater forced into the tunnelling to lubricate and cool the valves. Shipworms usually bore along the grain of the wood, neither entering the tunnel of another specimen nor leaving the wood. As they bore into the wood, all but the anterior end of the tunnel is lined by a calcareous tube secreted by the mantle, and which protects the soft body of the animal. The siphons are attached just inside the external opening to the tunnel by the pallets, and can extend a considerable distance from the entrance. When they contract they are drawn into the tunnel and the opening is closed by the pallets, enabling the animal to survive for some time when the wood is out of water. Plankton drawn in with the inhalent current and fragments of wood from the tunnelling serve as food; digestion of the wood is brought about by cellulases.

It is believed that the gonad of *T. navalis* passes through male and female phases. Fertilization is internal; sperm is discharged into the water and enters the inhalant siphon of another animal. The larvae develop in the mantle cavity and are liberated through the exhalant siphon as veligers which have been recorded in the plankton of Danish waters from March to December. Larval life is of about three to five weeks duration and when contact is made with a suitable substratum, the larvae attach by means of a byssus and boring starts immediately. Growth rate is fast and length of life is two to three years. Many treatments have been used to protect wood from attack, the most effective being creosote.

A second species of shipworm, **Nototeredo norvegica** (Spengler), is common in north-west Europe. It is larger than *T. navalis* and is readily separated by the shape of the pallets. In *N. norvegica* they are paddle-shaped but the oval blade is not forked (Fig. 215e).

REFERENCES
Polyplacophora
Jones, A. M. & Baxter, J. M. (1987). *Molluscs: Caudofoveata, Solenogastres, Polyplacophora and Scaphopoda*. Keys and notes for the identification of the species. Synopses of the British fauna (New Series), no. 37. Leiden: E. J. Brill/Dr W. Backhuys.

Matthews, G. (1953). A key for use in the identification of British chitons. *Proceedings of the Malacological Society of London*, 29, 241–8.

Gastropoda
Prosobranchia
Fretter, V. & Graham, A. *The prosobranch molluscs of Britain and Denmark*.
  Part 1 Pleurotomariacea, Fissurellacea and Patellacea (1976).
  Part 2 Trochacea (1977).
  Part 3 Neritacea, Viviparacea, Valvatacea, terrestrial and freshwater Littorinacea and Rissoacea (1978).
  Part 4 Marine Rissoacea (1978).
  Part 5 Marine Littorinacea (1980).
  Part 6 Cerithiacea, Strombacea, Hipponicacea, Calyptraeacea, Lamellariacea, Cypraeacea, Naticacea, Tonnacea, Heteropoda (1981).

Part 7 'Heterogastropoda' (Cerithiopsacea, Triforacea, Epitoniacea, Eulimacea) (1982).
Part 8 Neogastropoda (undated).
Part 9 Pyramidellacea (1986) (with E. B. Andrews).
Supplements to the *Journal of Molluscan Studies.*

Graham, A. (1988). *Molluscs: prosobranch and pyramidellid gastropods.* Keys and notes for the identification of the species. Synopses of the British fauna (New Series), no. 2, 2nd edn. Leiden: E. J. Brill/Dr W. Backhuys.

Hannaford Ellis, C. (1979). Morphology of the oviparous rough winkle, *Littorina arcana* Hannaford Ellis, 1978, with notes on the taxonomy of the *L. saxatilis* species-complex (Prosobranchia: Littorinidae). *Journal of Conchology,* 30, 43–56.

Muus, B. J. (1963). Some Danish Hydrobiidae with the description of a new species, *Hydrobia neglecta. Proceedings of the Malacological Society of London,* 35, 131–8.


Opisthobranchia

Alder, J. & Hancock, A. (1845–55). *A monograph of the British nudibranchiate mollusca,* parts 1–7. London: Ray Society.

Thompson, T. E. (1976). *Biology of opisthobranch molluscs,* vol. I. London: Ray Society.

Thompson, T. E. (1988). *Molluscs: benthic opisthobranchs (Mollusca: Gastropoda).* Keys and notes for the identification of the species. Synopses of the British fauna (New Series), no. 8, 2nd edn. Leiden: E. J. Brill/Dr W. Backhuys.

Thompson, T. E. & Brown, G. H. (1984). *Biology of opisthobranch molluscs,* vol. II. London: Ray Society.


Pulmonata

Macan, T. T. (1977). A key to the British fresh- and brackish-water gastropods with notes on their ecology. *Freshwater Biological Association Scientific Publication,* no. 13, 4th edn. Ambleside.


Bivalvia

Christensen, J. M. & Dance, S. P. (1980). *Seashells. Bivalves of the British and northern European seas.* Harmondsworth: Penguin.

Eales, N. B. (1950). *The littoral fauna of Great Britain.* Cambridge: Cambridge University Press.

Forbes, E. & Hanley, S. (1848–53). *A history of British Mollusca, and their shells,* vols. I–IV. London: Van Voorst. (Also includes Polyplacophora and Gastropoda.)

Smith, S. M. & Heppell, D. (1991). *Checklist of British marine Mollusca.* National Museums of Scotland Information Series, no. 11. Edinburgh: National Museums of Scotland. (Also includes Polyplacophora and Gastropoda.)

Tebble, N. (1976). *British bivalve seashells. A handbook for identification,* 2nd edn. Edinburgh: HMSO.

Yonge, C. M. (1949). *The sea shore.* London: Collins.

# Arthropoda

The arthropods include such well-known representatives as the insects, spiders, shrimps and crabs; they are a large grouping of invertebrate animals which have bilateral symmetry and show metameric segmentation, although the latter is not always clear. They are characterized by paired, jointed appendages and it is from this character that the group gets its name, *arthropoda*, meaning jointed feet. The appendages show great variation and serve a wide range of functions, including sensory perception, feeding and locomotion. The body is further characterized by the possession of a semi-rigid exoskeleton which in places is thin enough to allow movement. Periodically, the exoskeleton is shed by a process of ecdysis or moulting. A new skeleton develops under the existing one and during moulting the animal frees itself from the old skin, which splits along lines of weakness. The new skin is soft and expands rapidly before hardening. Growth increments at moulting can be substantial: for example, the common shore crab increases in carapace breadth by an average of about 30%. The arthropods form the largest group in the animal kingdom and the possession of a hard exoskeleton is a major factor in their colonization of land, sea and air.

The classification of the group is open to question. Some authors retain the rank of phylum for the group as a whole while others regard the different groups of arthropods, for example the trilobites, the chelicerates, the crustaceans and the uniramians, as having different lines of ancestry and assign phylum status to each of these. This is the position adopted here.


## ARTHROPODA

Phylum TRILOBITOMORPHA   A fossil group of marine arthropods.

Phylum CHELICERATA   Arthropods in which the first pair of appendages (chelifores) are used in feeding.

Class Merostomata   Includes the horseshoe crabs of tropical waters.

Class Arachnida   Includes scorpions, spiders, mites and ticks.

Class Pycnogona   The sea-spiders.

**Phylum CRUSTACEA**    Arthropods with two pairs of antennae; appendages usually biramous; includes barnacles, shrimps, crabs and lobsters.

Class Cephalocarida    A small group of primitive, shrimp-like crustaceans.

Class Branchiopoda    A predominantly freshwater group including the water fleas and fairy shrimps.

Class Ostracoda    The mussel or seed shrimps, widely distributed in the sea and fresh water.

Class Copepoda    Small crustaceans abundant in marine and freshwater habitats. Planktonic or benthic. Some are parasitic.

Class Mystacocarida    A small group of tiny crustaceans adapted for living between sand grains.

Class Remipedia    A small group of primitive crustaceans which live in marine caves.

Class Tantulocarida    Ectoparasites of deep water crustaceans.

Class Branchiura    Blood-sucking ectoparasites of marine and freshwater fish.

**Class Cirripedia**    The barnacles.

**Class Malacostraca**    Includes shrimps, crabs and lobsters.

**Phylum UNIRAMIA**    Arthropods in which the appendages are uniramous.

**Subphylum Hexapoda**    The insects.

Subphylum Myriapoda

Class Chilopoda    The centipedes. Essentially terrestrial, some specimens found under stones on the upper shore.

Class Diplopoda    The millipedes. Terrestrial.

Class Symphyla, Class Pauropoda    Arthropods living in leaf mould and soil.


## Phylum Chelicerata

### Class Pycnogona

The pycnogonids, or sea-spiders, are marine arthropods superficially resembling the true spiders. They are found intertidally among sponges, hydroids and bryozoans, which along with seaweed holdfasts should be examined carefully in dishes of seawater. Some species are abundant sublittorally. The body is generally very slim and divided into a head and trunk, and has a number of segments, which are clearly visible in most species. The head has a proboscis and two pairs of feeding appendages, the chelifores (also known as chelicerae) and the palps (Fig. 216), although both pairs are sometimes absent.

The mouth is at the anterior end of the proboscis and is equipped with chitinous jaws. There are two pairs of eyes situated mid-dorsally and a pair of appendages known as ovigerous legs (ovigers). The trunk usually has four pairs of legs, which are often long and slender and give a spider-like appearance (Fig. 216). The posterior end of the trunk bears a projection known as the abdomen. Pycnogonids are generally regarded as carnivores, feeding on hydroids, sea-anemones and bryozoans. The sexes are separate; when the eggs are about to be laid, male and female cling together and the eggs are fertilized externally as they are released from the reproductive openings at the base of the legs of the female. The eggs are brooded by the male and are held in a mass on the ovigerous legs, presumably with the aid of sticky secretions, and here they develop and hatch as larvae. The larval pycnogonid, known as a protonymphon, is similar to the parent but with only three pairs of appendages. The protonymphon grows by a series of moults into the adult form.

### Key to species of pycnogonids

Body slender; legs long, slender. Chelifores and palps present; palps with 5 segments  .. ................................................................................................. *Nymphon* spp. (p. 331)
Body stout; legs about equal to body length. Chelifores and palps absent  ...................... ................................................................................. ***Pycnogonum littorale*** (p. 332)

*Nymphon* spp.   (Fig. 216a)

*Body slender; up to 10 mm length. Legs very long and slender, 3 or 4 times body length. Both chelifores and palps present; chelifores with chelae, palps with 5 segments. Ovigerous legs in males and females.*

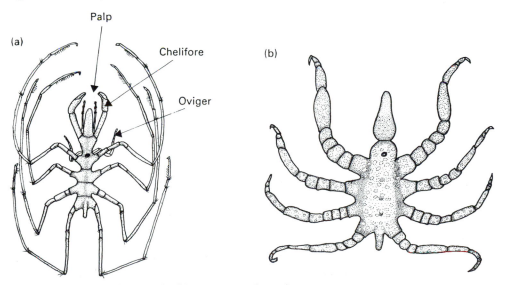

Figure 216  (a) *Nymphon gracile.* (b) *Pycnogonum littorale.*

There are three, possibly four, distinct species of *Nymphon* on British shores and the genus is well known from the shores of north-west Europe. **Nymphon gracile** Leach is perhaps the best known and is found under stones and seaweed on the middle and lower shore. In addition to crawling over the substratum, it is capable of swimming. It feeds on a wide range of animals, including sea-anemones, hydroids, bryozoans, small poly-chaetes, egg capsules of the dog whelk, *Nucella lapillus* (p. 255), and detritus. Most individuals migrate into the sublittoral during winter when breeding occurs. The animals return to the intertidal zone in March/April, the males often carrying eggs. Adult size is reached in about five months and longevity is believed to be one year.

*Pycnogonum littorale* (Ström)    (Fig. 216b)

*Body stout; up to 20 mm in length. Dorsal and lateral surfaces covered with tubercles. Chelifores and palps absent; ovigerous legs, with 9 segments, in males only. Legs short and stout, about equal to body length.*

*P. littorale* is widely distributed in north-west Europe and is the only species of the genus to be found in Britain, where it lives on the lower shore and in the sublittoral, under stones and seaweed, and on sea-anemones. It crawls slowly over the substratum but does not swim. It feeds predominantly on sea-anemones, and has a narrower range of food items than *Nymphon gracile* (above). The larvae feed on hydroids. Breeding occurs from spring to autumn and males and females may remain clinging together in the mating position for as long as five weeks, after which the developing eggs are carried by the male for about 10 weeks. Maturity is believed to be reached after one year.

## Phylum Crustacea

The crustaceans are a very large group of arthropods which are widely represented in the sea. They are found in the plankton, in the deep sea and on the shore and include bar-nacles, sand hoppers, prawns, shrimps, lobsters and crabs. As is to be expected in a group with such diversity of habits, there is considerable variation in external mor-phology, but the body can usually be divided into head, thorax and abdomen (see Fig. 222), although in some groups, for example the barnacles, the distinction is difficult to make. The body is covered by an exoskeleton which may be impregnated with calcium salts. Typically, the head has five pairs of appendages, the most anterior being the two pairs of antennae, known as antennae 1 (antennules) and antennae 2 (antennae), a fea-ture not seen in other arthropods. Posterior to the antennae are the mandibles, maxillae 1 (maxillules) and maxillae 2 (maxillae). These appendages are used in feeding and are adapted to suit a wide variety of feeding methods including carnivorous, and suspension and deposit feeding. The appendages of the thorax and abdomen vary from one group of crustaceans to the next, but the basic appendage is biramous with an inner (endopod) and an outer (exopod) ramus, the two rami often being very different morphologically and may serve different functions. In many crustaceans a dorsal shield or carapace is present, which to varying degrees affords protection to the body. Most crustaceans have separate sexes (but note the exception of the barnacle, below). After copulation the fertilized eggs may be brooded by the female, in some cases for the duration of embryonic life or, more commonly, there is at some stage in the life-cycle a pelagic larva which feeds in the plankton.

### Class Cirripedia

Barnacles are perhaps the best known of intertidal organisms and are so common and widespread in their distribution that they have been described as ubiquitous. They are exclusively marine. On the upper reaches of most rocky shores, encrusting acorn barnacles are so numerous that they form a distinct grey-white barnacle line which marks the upper limit of the eulittoral zone (Fig. 4). Although their crustacean affinities had been noted as early as 1819, it was not until 1830 that this view was firmly established through the work of an army surgeon, Mr J. V. Thompson. Prior to Thompson's research, the sessile habit of the adult barnacle and its shell of hard, calcareous plates (superficially resembling a mollusc) had caused taxonomic confusion. Thompson's great contribution was an account of the life-cycle of the barnacle in which he described a larval stage (a nauplius) showing unmistakable crustacean characters. Much of our knowledge of this group is based on the classic research of Charles Darwin.

In addition to the familiar acorn barnacles of the rocky shores, the class includes stalked barnacles and one group is highly specialized for a parasitic way of life. Stalked barnacles are occasionally washed ashore on floating debris; they are believed to be among the most primitive members of the class. Acorn barnacles do not have a stalk and the wall is made up of a number of plates which in some species has been reduced from the original number of eight by loss and/or fusion. The number and arrangement of these plates are important in identification. Parasitic barnacles (p. 342) are extensively modified but are recognized as cirripede crustaceans by their larval stages.

Typically, barnacles have six pairs of biramous thoracic appendages which are used, not in locomotion, but to filter suspended food particles from seawater. The common acorn barnacles of the shore are hermaphroditic, and in most cases copulation and cross fertilization take place between neighbouring individuals. After fertilization, the embryos are held within the body of the barnacle where they develop into a characteristic larval stage known as the nauplius. Nauplii are released from the adult barnacle and swim in the surface waters where they feed and grow by a series of moults, eventually developing into a second larval type, the cyprid. This is non-feeding, surviving on fat reserves laid down during the naupliar stages. The cyprid is a unique larval stage that has evolved for the all-important role of selecting a suitable site for attachment. Cyprids explore the nature of the substratum with the first pair of antennae and are attracted to rocks already colonized by adults. When a suitable site for settlement has been identified, a secretion pours from the ducts of the first antennae cementing the larva to the substratum. Metamorphosis into a juvenile barnacle then proceeds quickly, generally within 24 hours. The life-cycle of an acorn barnacle is shown in Fig. 217. During spring and early summer brown-coloured cyprids of *Semibalanus balanoides* (p. 341), just visible to the naked eye but clearly discernible with a hand lens, and newly metamorphosed barnacles are common on rocky shores around Britain.

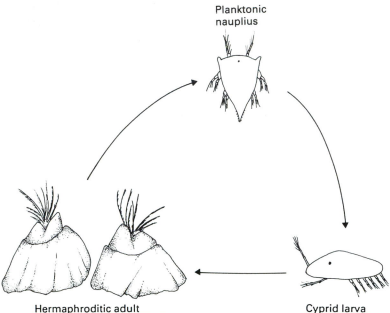

**Planktonic nauplius**

**Hermaphroditic adult barnacles. Cross fertilization**

**Cyprid larva seeks site for attachment and metamorphosis**

Figure 217  Life-cycle of an acorn barnacle (not to scale).

### Key to families of barnacles

1  Body divisible into soft, brown stalk and a head, the capitulum, with 5 calcareous plates. Generally found attached to floating objects ...................................................................................................................... **(Stalked barnacles) Lepadidae** (p. 335)
Body without stalk; with rigid wall of calcareous plates capped by movable opercular plates .................................................................................................... **(Acorn barnacles) 2**

2  Shell wall of 4 unequal ridged plates, opercular aperture off centre. Asymmetrical in appearance .............................................................................................. **Verrucidae** (p. 337)
Shell wall not of 4 unequal ridged plates, opercular aperture in central position. Symmetrical in appearance ...................................................................................... 3

3  Rostral plate large, broader than width of opercular aperture (Fig. 219a). Opercular aperture diamond-shaped ...................................................................... **Balanidae** (p. 340)
Rostral plate narrow, about equal to width of opercular aperture (Fig. 219b,c). Opercular aperture kite-shaped or rounded .................................... **Chthamalidae** (p. 338)

**Parasitic barnacles** are commonly found on decapod crustaceans. They are recognized as a smooth yellow-brown mass on the undersurface of the crab. The best-known example in British waters belongs to the family **Sacculinidae** (p. 342).

In addition to the barnacles belonging to the families listed above, there are others which bore into the shells of molluscs. They do this by chemical and mechanical means, using teeth-like structures on the mantle. The adult barnacles are tiny, being a few millimetres in length. Members of the family **Alcipidae** bore into the columella of the shell of the whelks *Buccinum* and *Neptunea* (p. 258) housing the hermit crab *Pagurus bernhardus* (p. 395). The presence of the barnacle is revealed by tiny slit-like holes in the shell.

### Stalked barnacles – goose barnacles

Stalked barnacles are primitive cirripedes characterized by the possession of a stalk and a 'head' (also known as the capitulum) bearing a number of calcareous plates. The best-known examples are pelagic, attached to drifting timbers, which from time to time are washed ashore. The common name of goose barnacles derives from the myth that barnacle geese hatched from barnacles growing on trees at the water's edge.

### Family Lepadidae

Stalk soft, flexible, often very long; without calcareous plates. Capitulum with five calcareous plates, the paired terga and scuta, and a single carina.

### *Lepas* spp.    Goose barnacles (Fig. 218)

*Body clearly divisible into soft brown stalk and capitulum bearing 5 white plates which have blue sheen. May reach an overall body length of 200 mm or more.*

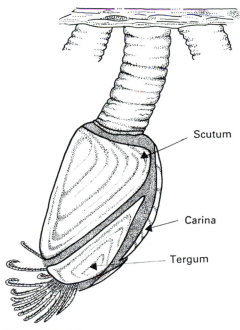

Figure 218  *Lepas* sp.

There are a number of species of *Lepas*, **Lepas anatifera** Linnaeus being probably the best known from European waters. It is occasionally washed ashore on driftwood. Although essentially a tropical and subtropical species, both adults and larvae have been found as far north as the Norwegian coast. It is a cross-fertilizing hermaphrodite apparently producing a succession of broods at optimum temperatures of 19–25 °C and reaching sexual maturity at these temperatures after about four to five weeks. **Lepas fascicularis** Ellis & Solander is also found washed onto the shore, particularly in the north and west of Britain. It is readily identified by a spongy white float secreted by the barnacle, and to which it is attached.

### Acorn barnacles

Adult acorn barnacles are sessile and characterized by a rigid wall of calcareous plates arranged in the shape of a flat-topped cone. The apex of the cone is known as the opercular aperture, and it is covered by movable plates, the terga and scuta (Fig. 219). When the latter are closed the animal is effectively housed within a calcareous box, which

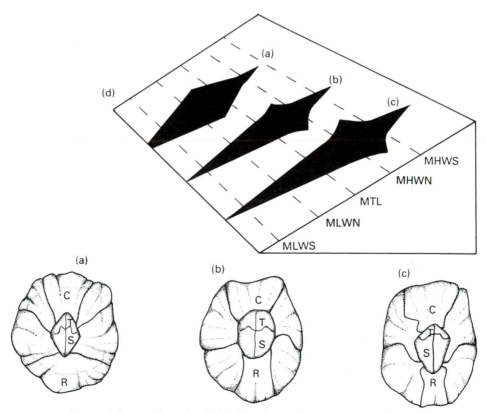

Figure 219  (a) *Semibalanus balanoides*. (b) *Chthamalus stellatus*. (c) *Chthamalus montagui*.
(d) Typical intertidal distribution of the species, drawn from data from the west coast of Ireland. C – carinal plate; R – rostral plate; S – scutum; T – tergum.

affords some protection against desiccation during low water. The large basal area of acorn barnacles gives firm adhesion to the substratum, enabling them to survive in wave-beaten situations. On the rising tide, the terga and scuta open and the thoracic appendages are extended and repeatedly drawn through the water, collecting suspended food particles from the plankton. Acorn barnacles are preyed on extensively by the gastropod mollusc *Nucella lapillus* (p. 255) and shore fishes such as the shanny, *Lipophrys pholis* (p. 503). Empty barnacle cases are the habitat of a range of fauna including the small periwinkle, *Littorina neglecta* (p. 241), the juveniles of other periwinkles, small bivalves such as *Lasaea adansoni* (p. 296) and the isopod *Campecopea hirsuta* (p. 359). Examination of barnacle plates with a hand lens often reveals the presence of the crustose lichen *Pyrenocollema halodytes* (p. 81).

Acorn barnacles can be identified in the field with the aid of a good-quality hand lens by the number, shape and arrangement of the plates and by the shape of the opercular aperture. The base of the shell is either soft and membranous or hard and calcareous and in the latter case a distinct white scar is left on the rock when the barnacle is removed. This is a useful character in identification. Old, worn specimens are often distorted and the individual plates obscured, making identification difficult.

### Family Verrucidae

Opercular aperture covered by a single, movable plate formed from one scutum and one tergum. Asymmetrical arrangement of plates in outer shell which comprises one scutum and one tergum together with the rostral plate and carina, cf. the Balanidae and the Chthamalidae. Shell base usually membranous.

### *Verruca stroemia* (O. F. Müller)  (Fig. 220e)

*Shell wall of 4 unequal, ridged plates. Opercular aperture contains single tergum and scutum forming 1 plate. Arrangement of plates not the same on either side of mid-line; shell is therefore asymmetrical compared with* Semibalanus, Balanus, Chthamalus *and* Elminius, *all of which are symmetrical. Shell dirty white colour, may reach diameter of 10 mm, but usually much less. Shell base membranous.*

*V. stroemia* is widely distributed in north-west Europe on the underside of stones and shells on the lower shore. It has also been recorded from laminarian holdfasts and extends into the sublittoral to a depth of about 70 m. Although self-fertilization has been recorded in isolated individuals, it is believed that cross-fertilization usually takes place. On the west coast of Scotland, fertilization takes place in winter and the main release of nauplii coincides with the spring diatom increase in the plankton. A series of smaller broods is released during summer. Cyprids settle from April to September. Sexual maturity is reached at a diameter of about 2 mm.

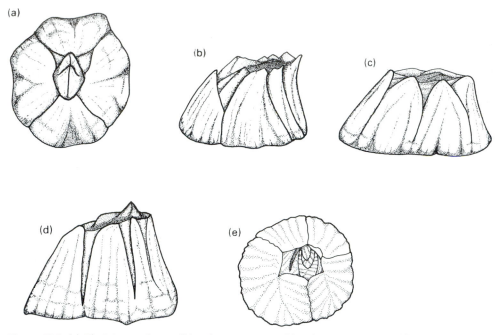

Figure 220  (a) *Elminius modestus*. (b) *Balanus crenatus*. (c) *Balanus improvisus*. (d) *Balanus perforatus*. (e) *Verruca stroemia*.

### Family Chthamalidae

Acorn barnacles with six plates in shell wall. Rostral plate relatively narrow, about equal to width of opercular aperture. The opercular aperture is kite-shaped or rounded and closed by the paired terga and paired scuta. Shell base membranous.

### *Chthamalus* spp.

Prior to 1976 British chthamalid barnacles were all regarded as members of a single species with wide morphological variation. More recent work has established that there are, in fact, two species, *Chthamalus stellatus* and a 'new' species, *Chthamalus montagui*, first described by Southward in 1976. Both species are widely distributed in north-west Europe and can be separated in the field with the aid of a hand lens. Together with *Semibalanus* (below) they are the most common intertidal barnacles in Britain.

Opercular aperture oval or more or less circular. Junction between terga and scuta convex towards rostral plate ............................................. *Chthamalus stellatus* (p. 339)
Opercular aperture kite-shaped, angular. Junction between terga and scuta concave towards rostral plate ........................................................ *Chthamalus montagui* (p. 339)

### *Chthamalus stellatus* (Poli)    (Fig. 219b)

*Shell wall of 6, light grey plates, often difficult to distinguish because of corrosion and over-growth of algae and lichens. Opercular aperture oval or more-or-less circular. Junction between terga and scuta forms curve convex towards rostral plate. Rostral plate narrow (cf. Semibalanus, below). Tissue inside opercular aperture generally bright blue with black and orange marks. Diameter up to 10 mm or more. Shell base membranous.*

*C. stellatus* is a southern species extending as far north as Shetland. It is found on western coasts of Britain extending around the north of Scotland, south to about as far as Aberdeen. Within this range it is most abundant on wave-beaten, open-coast situations exposed to the south-west while apparently being absent from many Irish Sea coasts. The species appears to favour clear, oceanic waters (cf. *C. montagui*, below). In the English Channel it extends eastwards as far as the Isle of Wight but is absent from the eastern part of the Channel and the North Sea coast. The vertical distribution of *C. stellatus* on the shore overlaps with that of *C. montagui* and *S. balanoides*; it is generally most abundant between these two (Fig. 219d).

### *Chthamalus montagui* Southward    (Fig. 219c)

*Shell wall of 6, light grey or brownish plates often difficult to distinguish because of corrosion and overgrowth of algae and lichens. Opercular aperture kite-shaped and angular; junction between terga and scuta forms curve concave towards rostral plate. Rostral plate narrow (cf. Semibalanus, below). Tissue inside opercular aperture bluish with brown and black marks, usually paler than in* C. stellatus *(above). Similar in size to* C. stellatus. *Shell base membranous.*

The geographical distribution of *C. montagui* is similar to that of *C. stellatus* (above) but, unlike *C. stellatus*, it has not been recorded from Shetland. It is more widely distributed on Irish Sea coasts and in contrast to *C. stellatus* apparently favours more turbid waters. It overlaps with *C. stellatus* and *Semibalanus* (below) in its vertical distribution, although it is more abundant higher on the shore (Fig. 219d).

*Chthamalus* spp. breed throughout the summer and at Plymouth breeding commences when sea temperatures reach 10 °C. They are multiple brooders capable of self-fertilization, with the number of broods released per season being dependent on shore level. In the Plymouth area, one or two broods are produced on the high shore, increasing to four on the lower shore. The first brood is produced in the first year after settlement. Development of the embryos takes about three weeks.

Family Balanidae

Acorn barnacles with four or six plates in shell wall. Rostral plate broader than width of opercular aperture. The opercular aperture is diamond-shaped and closed by the paired terga and paired scuta. Shell base calcareous or membranous.

1 Shell wall of 4 plates ................................................ *Elminius modestus* (p. 340)
  Shell wall of 6 plates ........................................................................................ 2
2 Shell base membranous .................................................. *Semibalanus balanoides* (p. 341)
  Shell base calcareous, left as a shelly plate when barnacle is removed ............................. 3
3 Shell plates of wall purplish colour, often ridged vertically. Apex of terga pointed; opercular aperture relatively small. Large barnacle, up to 30 mm in height and diameter. Tissue inside opercular aperture brightly coloured, purple, pink, blue .................
  ........................................................................ *Balanus perforatus* (p. 342)
  Shell plates of wall grey-white colour .............................................................. 4
4 Shell inclined to one end when viewed in profile. Opercular aperture a broad diamond shape; upper edge of plates toothed. Tissue inside opercular aperture striped with purple and yellow or purple and white ...................................... *Balanus crenatus* (p. 341)
  Shell not inclined. Opercular aperture an elongated diamond shape. Tissue inside opercular aperture with purple and white markings .............. *Balanus improvisus* (p. 342)

Another species, *Balanus amphitrite* Darwin, may be found in areas warmed by cooling water discharges from power stations. It is an introduced, tropical species.

*Elminius modestus* Darwin    (Fig. 220a)

*Shell wall of 4, grey-white plates. Junction of terga and scuta noticeably pointed. Diameter up to 10 mm. Shell base membranous.*

*E. modestus* is an Australasian barnacle which is believed to have become established in Britain from ship-borne populations in the early 1940s. The rapid spread of the species since that time has been followed with interest by shore ecologists; it is now known to be established all around Britain and on southern North Sea coasts of north-west Europe. It is a cross-fertilizing hermaphrodite breeding almost continuously throughout the year. When conditions are favourable it has been known for broods to be released every 10 days. Cyprids are generally found on the shore from about May to October and the newly metamorphosed barnacles grow rapidly, reaching maturity in about eight weeks. *Elminius* apparently thrives in water of a high suspended silt content and is more tolerant of reduced salinity than *Semibalanus* (below) and *Chthamalus* spp. (above), being common in estuaries and harbours. It is also found on open coasts where it occurs on the middle shore, but it appears to be less tolerant of wave exposure than *Semibalanus* and *Chthamalus*.

***Semibalanus balanoides*** (Linnaeus)    (*Balanus balanoides*) (Fig. 219a)

*Shell wall of 6, grey-white plates, generally easy to distinguish. Individual specimens up to 15 mm diameter but, when crowded together, tall columnar individuals are common in which separate plates may be difficult to distinguish. Opercular aperture diamond-shaped. Rostral plate broad (cf. Chthamalus spp., above). Tissue inside opercular aperture usually white, may be pinkish-white. Shell base membranous.*

S. *balanoides* is widespread in north-west Europe and occurs extensively on British coasts, but may be absent or rare in parts of Cornwall and the Scilly Isles. It is found on shores of all degrees of exposure and tolerates reduction in salinity to about 20‰, penetrating the lower reaches of estuaries. Together with *Chthamalus* spp. (above), *Semibalanus* forms the characteristic barnacle zone seen on most rocky shores. When the two occur on the same shore, they overlap in distribution, but generally *Semibalanus* is found lower than *Chthamalus* spp. (Fig. 219d). *Semibalanus* is less tolerant of desiccation than *Chthamalus* spp. and this has been shown to be related to the greater permeability of its shell plates. In areas exposed to strong wave action, the upper limit of distribution of *Chthamalus* and *Semibalanus* on the shore is raised (see Fig. 4). On shores dominated by fucoid seaweeds, the density of barnacles is reduced, the seaweed preventing establishment of the barnacle larvae. In contrast to *Chthamalus* spp., *Semibalanus* has a single breeding season with copulation and cross-fertilization taking place in the autumn, but the nauplii are held in the barnacle to be released into the sea from about February to May of the following year when diatom production in the plankton is high. The pelagic phase lasts some weeks, during which time the larvae feed on the rich stocks of diatoms. There are six naupliar stages, each separated by a moult. Cyprids settle on the shore during spring and early summer, depending on locality. The number of eggs produced varies according to position on the shore and age of the barnacle, and estimates as high as 8000 per individual per year have been made. Longevity depends not only on locality but also on beach level; those on the upper shore live for up to eight years compared with about two to three years for individuals on the lower shore. Maturity is generally reached after one to two years.

***Balanus crenatus*** Bruguière    (Fig. 220b)

*Shell wall of 6, grey-white plates; up to 25 mm diameter. Shell inclined to one end when viewed in profile. Opercular aperture a broad diamond shape; upper edge of shell plates toothed. Tissue inside opercular aperture with yellow and purple stripes. Shell base calcareous, often left as scar on rock when barnacle removed.*

B. *crenatus* is a northern species extending as far south as the Atlantic coast of France. It is essentially a sublittoral species that can be found under stones on the lower shore, less frequently on open rock faces. It is often attached to shells which have been washed ashore. Nauplii are released from February through to about September, the cyprids settling from April onwards with a peak in April and May. Length of life is about 18 months.

### *Balanus improvisus* Darwin    (Fig. 220c)

*Shell wall of 6 smooth, whitish plates; up to 15 mm diameter. Opercular aperture narrow, elongated diamond shape. Tissue inside opercular aperture with white and purple markings. Shell base calcareous.*

B. *improvisus* is widespread in north-west Europe. It is a lower shore species extending into the sublittoral. It tolerates low salinities and is found in estuaries. Around Britain it is most common in the larger estuaries at salinities of about 15‰. It is capable of self-fertilization. Nauplii are found in the plankton during summer and take about two weeks to develop to the cyprid.

### *Balanus perforatus* Bruguière    (Fig. 220d)

*Shell wall of 6, purplish plates, often vertically ridged. Apex of terga pointed. Tissue inside opercular aperture marked with bright colours of purple, pink, blue. Shell sometimes up to 30 mm diameter and 30 mm tall. Shell base calcareous.*

B. *perforatus* is a southern species extending to south-west England and South Wales. It is found on the lower shore, sometimes being common in exposed situations. Breeding occurs during summer, the cyprids settling on the shore during August and September. The isopod crustacean *Dynamene* (p. 358) is often found in the empty cases of B. *perforatus*.

### Parasitic barnacles
Parasitic barnacles are extensively modified but are recognized as cirripede crustaceans by their larval stages.

Family Sacculinidae
### *Sacculina carcini* Thompson    (Fig. 221)

*Found on undersurface of abdomen of shore crab,* Carcinus maenas *(p. 401), and other crabs of the family Portunidae, as conspicuous, yellow-brown mass which bears superficial resemblance to the crab's eggs.* Sacculina *has a smooth, apparently amorphous nature compared with granular appearance of egg mass.*

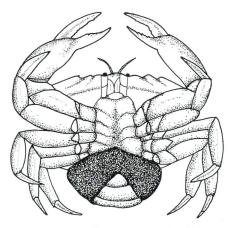

Figure 221 *Sacculina carcini* on the abdomen of the shore crab.

*S. carcini* is widely distributed in north-west Europe along with its host *Carcinus maenas*. Nauplii of the parasite are liberated into the plankton and, unlike those of other cirripedes, they do not feed but survive on stored yolk. After four or five moults, the nauplii metamorphose into cyprids of two sizes, the smaller being female and producing kentrogon larvae, while the larger are male and produce trichogon larvae. The kentrogon larvae are the infective stages and after attaching to small setae on the surface of a young *Carcinus*, cells pass from the kentrogon into the crab where they develop and spread through the tissues of the host. This is known as the soma or body of the parasite and this stage has been estimated to be anything from five to 34 months' duration. Eventually, the parasite makes its appearance on the external surface of the crab and is known as the externa. This is effectively a brood sac which is seen under the abdomen of the crab. It contains mantle tissue and ovaries, and the eggs are fertilized by a trichogon larva developed from a large male cyprid. Some six weeks or so after fertilization, nauplii are produced. At Plymouth, breeding occurs throughout the year and the ratio of male to female cyprids varies seasonally. It appears that during summer the cyprids produced are mainly female, the kentrogon larvae, while the trichogon larvae predominate in the plankton during autumn and winter.

A similar life-cycle is shown by species of the genus *Peltogaster*, parasites of hermit crabs.

### Class Malacostraca

The malacostracans are the largest group of crustaceans and include the prawns, shrimps, crabs and lobsters. The generalized body form is shown in Fig. 222 and comprises a head, a thorax of eight segments (of which at least one is fused with the head) and an abdomen of six segments. In some groups, e.g. the decapods, the head and thorax are protected by a dorsal shield, the carapace. The appendages of the thorax are generally

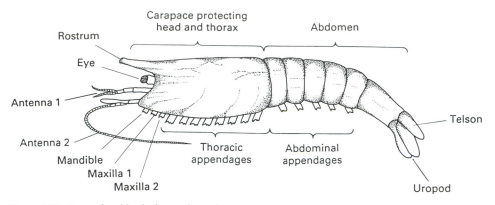

Figure 222  Generalized body form of a malacostracan.

called pereopods and, depending on species, the first one, two or three pairs are associated with feeding; the remaining pairs are for walking and grasping. In the typical malacostracan, the first five pairs of abdominal appendages, called pleopods, are usually similar to each other. They are involved in a variety of functions, including swimming, producing currents, as gills or for carrying eggs. The last abdominal segment bears a pair of appendages known as uropods; together with the terminal telson they often form a tail fan.

The malacostracans of the shore are grouped into a number of orders which show modifications of the generalized body plan. Those orders featured in the text are recognized using the characters given below.

(a) Small (up to about 20 mm in length), shrimp-like; often transparent. Thoracic appendages biramous, lack pincers. (Mysids or opossum shrimps.) .............................
..................................................................................... Order Mysidacea (p. 344)
*or*
(b) Usually dorso-ventrally flattened. Thoracic appendages similar in appearance. (Sea-lice and sea-slaters.) ......................................................... **Order Isopoda** (p. 351)
*or*
(c) Usually laterally flattened. First 3 pairs of abdominal appendages with pleopods; last 3 pairs with uropods. (Sand-hoppers.) ........................................ **Order Amphipoda** (p. 362)
*or*
(d) Thorax with 5 pairs of 'walking legs'. Well-developed carapace. (Prawns, shrimps, lobsters and crabs.) ................................................................. **Order Decapoda** (p. 382)

### Order Mysidacea
Widely known as the opossum shrimps, the mysids are a group of small, often transparent, shrimp-like animals. A carapace is present but in most species it does not cover the last one or two thoracic segments. Antennae 1 are conspicuous, biramous appendages. The outer ramus of antenna 2 takes the form of a flattened plate known as the

antennal scale (Fig. 223b). It is bordered along part or all of its margin with setae and is an important diagnostic character. The eyes are large and carried on stalks; the thoracic appendages are biramous and lack chelae, a feature which distinguishes them from prawns and shrimps. The first one or two pairs are generally used for feeding, while the remainder are used for a variety of functions such as catching prey, and locomotion. The abdomen has six segments, the first five of which usually have small appendages, the pleopods, while the last has a pair of uropods. The inner ramus of each uropod of most mysids has a conspicuous, rounded statocyst which is believed to be concerned with balance. During mating, sperm are deposited in a ventral brood chamber formed from brood plates (oostegites) arising from the bases of some thoracic appendages. The eggs are fertilized as they are laid, and development takes place in a brood chamber. Young mysids emerge after about two weeks. When feeding, the beating of the thoracic append-ages creates a current of water from which suspended matter is filtered. Mysids feed extensively on detritus and diatoms, but will also take small crustaceans.

Mysids are usually found in shallow coastal waters but some species occur at greater depths. They are not generally common intertidally but some are abundant in inshore waters and can be collected from rock pools and among seaweed. Some extend into estuaries where they are often seen in dense swarms, the individuals characteristically 'hovering' in the water. They are important in estuarine food webs.

Mysids are best collected by sweeping a fine-meshed net through the fringes of sea-weeds around the margins of rock pools and gullies, and particularly through the creeks and channels in estuaries at low water, where some species are found in surprisingly high densities.

The mysids included in this text belong to a single family, the Mysidae.

Family Mysidae

### Key to species of mysids

1 Telson divided at tip ..................................................................................................... 2
  Telson not divided at tip ............................................................................................. 5
2 Outer ramus of uropod with spines on outer margin; 5th segment of abdomen narrow in dorsal view and with longitudinal ridge; posterior margin of segment with spine ......
  ................................................................................ *Gastrosaccus spinifer* (p. 347)
  Outer ramus of uropod with setae on outer margin ......................................... 3
3 Antennal scale long, fairly narrow, with single spine on outer margin about $\frac{2}{3}$ way along length. Eyes on long, narrow stalks, extending well beyond margin of carapace ...
  ................................................................................ *Schistomysis spiritus* (p. 350)
  Antennal scale with single spine on outer margin close to tip of scale ............................. 4
4 Antennal scale long and narrow, about 8 times longer than broad. Lateral margins of telson each with 22–28 spines ................................................. *Praunus flexuosus* (p. 349)
  Antennal scale about 4 times longer than broad. Lateral margins of telson each with 15–17 spines ................................................................ *Praunus inermis* (p. 350)

5   Telson long, triangular in outline with narrow, truncated tip; short spines around margin ........................................................................ *Neomysis integer* (p. 348)

    Telson not as above ........................................................................................ 6

6   Telson short, bluntly rounded; with short spines around margin and single longer lateral spine on each side. Eye stalks very long ................ *Mesopodopsis slabberi* (p. 348)

    Telson not as above ........................................................................................ 7

7   Carapace with long rostrum, extending in front of eyes. Outer ramus of uropod with proximal spines and distal setae on outer margin. Telson elongate; larger spines around margin separated by smaller ones; 2 large spines on posterior margin separated by 4 small spines and 2 setae ............................................................... *Siriella armata* (p. 346)

    Carapace with short rostrum, not extending in front of eyes. Outer ramus of uropod with setae all around margin. Spines on margin of telson increase in size towards tip ....
............................................................................. *Leptomysis lingvura* (p. 347)

**Siriella armata** (Milne-Edwards)    (Fig. 223a)

*Body up to 22 mm in length, slender. Carapace with long, pointed rostrum extending in front of eyes. Eyes on long stalks. Antenna 2 with scale lacking setae on outer margin. Outer ramus of uropod with proximal spines and distal setae on outer margin. Telson not divided, elongate, rounded at tip; larger spines around margin separated by smaller ones, 2 large spines at tip separated by 4 small spines and 2 setae.*

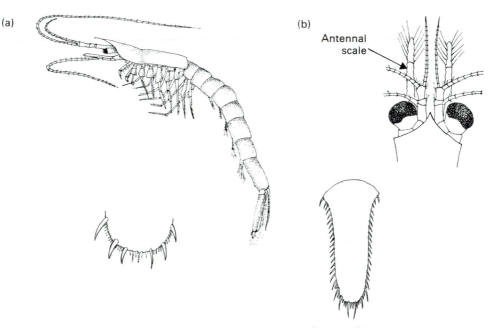

(a)              (b)    Antennal scale

Figure 223   (a) *Siriella armata*, with detail of posterior region of telson. (b) *Leptomysis lingvura*, detail of anterior region and telson. For clarity setae are shown only on the apical segment of the antennal scale (after Tattersall & Tattersall, 1951).

*S. armata* is found all around Britain from the lower shore to depths of about 20 m. It is often found in shoals in rock pools on the lower shore among seaweeds. Breeding has been recorded from March to October at Plymouth.

**Gastrosaccus spinifer** (Goës)   (Fig. 224)

*Body up to 21 mm in length; 5th segment of abdomen narrow in dorsal view, and with longitudinal ridge and spine on posterior margin. Eyes small. Antenna 2 with long scale lacking setae on outer margins. Posterior margin of carapace with fringe of spines. Outer ramus of uropod with spines on outer margin. Telson deeply divided at tip.*

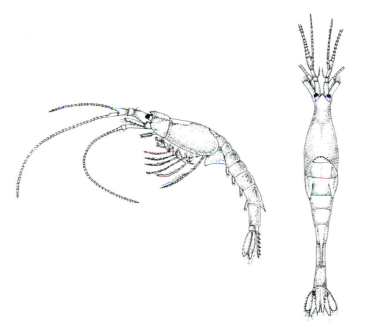

Figure 224  *Gastrosaccus spinifer*, lateral and dorsal views (after Tattersall & Tattersall, 1951).

G. *spinifer* is widely distributed in north-west Europe and occurs all around Britain. It is often found in dense swarms in brackish water and among *Zostera* (p. 85) at salinities down to about 20‰. It has been recorded at depths of over 200 m.

**Leptomysis lingvura** (G. O. Sars)   (Fig. 223b)

*Body transluscent or reddish in colour; up to 17 mm in length. Carapace with short rostrum, not extending in front of eyes. Antenna 2 with narrow scale with setae on all margins; distal part of scale with 5 or 6 setae on each side. Outer ramus of uropod with setae all around margin. Telson more than twice as long as broad, rounded; spines on margin increase in size towards tip.*

*L. lingvura* is widely distributed in north-west Europe and common all around Britain. It is found on the lower shore and in shallow waters, often in shoals. Breeding occurs in summer and autumn.

**Mesopodopsis slabberi** (van Beneden)    (Fig. 225a)

*Body slender and colourless; up to 15 mm in length. Eyes on very long stalks. Antenna 2 with long, narrow scale, with setae on all margins. Telson short, bluntly rounded with single lateral spine on each side.*

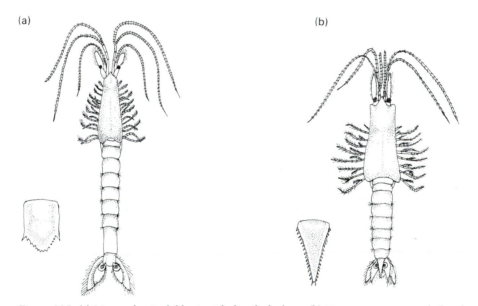

(a)                                                                                                    (b)

Figure 225   (a) *Mesopodopsis slabberi*, with detail of telson. (b) *Neomysis integer*, with detail of telson. (Both after Tattersall & Tattersall, 1951.)

*M. slabberi* is widely distributed in north-west Europe and around Britain, but is believed to be rare on the west coast of Scotland. It tolerates very low salinities (0.5‰) and is found in estuaries where it is sometimes abundant. *Mesopodopsis* is omnivorous. In British waters breeding has been recorded in spring and summer.

**Neomysis integer** (Leach)    (Fig. 225b)

*Body almost transparent; up to 18 mm in length. Eyes on stalks. Antenna 2 with long, pointed scale, with setae on all margins. Telson long, triangular in outline with narrow truncated tip; short spines around margin.*

*N. integer* is widely distributed in north-west Europe and common on the coasts of Britain, where it is the only species of the genus. It withstands salinities down to about 1‰

and is mainly found in estuaries, where it frequently occurs in swarms. It also tolerates high salinity and is found in saltmarsh pools. Breeding has been recorded throughout the year on the west coast of Scotland, with spring, summer and autumn generations. Breeding females carry up to about 50 young in the brood chamber. Longevity is about nine months for the overwintering generation and about four months for animals spawned in spring. *Neomysis* is an omnivore, feeding on a wide variety of material including detritus, diatoms, filamentous algae and small crustaceans.

### *Praunus flexuosus* (Müller)    Chameleon shrimp (Fig. 226a–c)

*Body up to 25 mm in length; colour varied, depending on light intensity and background, from dark grey-black to colourless. Abdomen distinctly bent when seen in side view. Eyes large, on long stalks. Antenna 2 with long, narrow scale about 8 times longer than broad, lacking setae on outer margin but with small spine close to tip. Outer ramus of uropod with setae on outer margin. Telson long, divided to about $\frac{1}{6}$ length; lateral margins with 22–28 spines.*

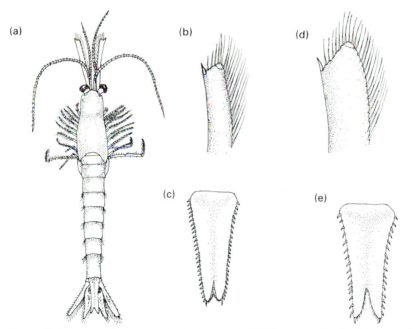

Figure 226 *Praunus flexuosus*, (a) whole animal, (b) left antennal scale, (c) telson. *Praunus inermis*, (d) left antennal scale, (e) telson. (All after Tattersall & Tattersall, 1951.)

*P. flexuosus* has been recorded from the coasts of Scandinavia to northern France and is common on the shores of Britain. It is found in large numbers in rock pools and among *Laminaria* (p. 43) and *Zostera* (p. 85) and can tolerate salinities down to about 5‰. It has the characteristic habit of 'hovering' in the water. *Praunus* breeds in spring and summer, some of the young maturing quickly to breed in the autumn of the same year.

Females mature at about 13–14 mm body length and produce up to three broods, the eggs being carried for about three weeks. Females carry up to 60 eggs in the brood chamber. Longevity is about one year. *Praunus* is omnivorous, feeding on organic debris and small crustaceans. It is probably the most abundant species of mysid in Britain, but can be easily confused with other species of the genus, particularly *Praunus neglectus* (G. O. Sars). The latter species tends to have a more greenish coloration.

**Praunus inermis** (Rathke)    (Fig. 226d, e)

*Body up to 15 mm in length. Antenna 2 with scale about 4 times longer than broad; lacking setae on outer margin but with small spine on outer margin close to tip. Outer ramus of uropod with setae on outer margin. Telson divided at tip; lateral margins with 15–17 spines.*

P. inermis is widely distributed and common in north-west Europe among seaweed in shallow water, extending into the mouths of estuaries. It occurs all around Britain, but is most common in the north, extending to Scandinavia. Breeding has been recorded from spring to autumn in British waters.

**Schistomysis spiritus** (Norman)    Ghost shrimp (Fig. 227)

*Body slender, transparent, glassy appearance; up to 18 mm in length. Eyes on long, narrow stalks projecting well beyond anterior margin of carapace. Antenna 2 with long, narrow scale lacking setae on outer margin but with spine about ⅔ way along length. Outer ramus of uropod with setae on outer margin. Telson divided at tip, noticeably constricted at base.*

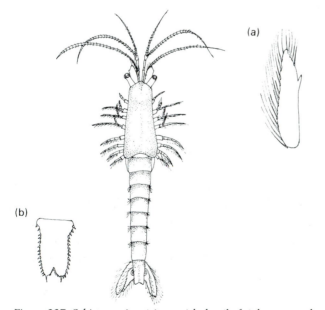

Figure 227 *Schistomysis spiritus*, with detail of right antennal scale (a), and telson (b) (after Tattersall & Tattersall, 1951).

*S. spiritus* is widely distributed in north-west Europe, often being recorded in dense swarms. It is found in shallow water among seaweed and extends into estuaries, feeding on animal and plant detritus. *Schistomysis* breeds during summer.

Order Isopoda

Marine isopods occur from the intertidal to the deep sea. The body is usually dorso-ventrally flattened and the head and first thoracic segment are fused, forming a cephalo-thorax leaving seven, externally visible, thoracic segments. There is no carapace. Antennae 1 are small and antennae 2 larger and conspicuous; the eyes are not stalked. There are usually seven pairs of pereopods of similar appearance. The pleopods are flattened plates which are respiratory and also beat to create a current of water for swimming. One or more abdominal segments are fused with the telson. Isopods show a diversity of feeding habits and include herbivores, carnivores and scavengers.

The sexes are separate and sperm are transferred to the female in capsules known as spermatophores. The eggs are held on the undersurface of the female in a brood chamber, formed from overlapping oostegites. Here, the embryos develop and emerge as juveniles, closely resembling the adult but in some species having only six pairs of pereopods. In a few species, embryonic development takes place in internal pouches leading from the brood chamber. There is no pelagic larva. Many isopods are parasitic, some living ecto-parasitically on fish; others, which are not dealt with in this text, are highly modified in structure and parasitize crustaceans.

Key to families of isopods

1 With 5 pairs of pereopods of similar appearance ................................ Gnathiidae (p. 352)
  With 7 pairs of pereopods of similar appearance ............................................... 2
2 Uropods lie under the telson (ventral) or along the sides of the telson (lateral) ................ 3
  Uropods project in mid-line from the posterior margin of the telson (terminal) ............... 6
3 Uropods ventral, large, flattened, covering the pleopods ...................... Idoteidae (p. 360)
  Uropods lateral. Animal typically short, compact ........................................... 4
4 Rami of uropods stick-like, not flattened; outer ramus very short and bears apical claw
  (Fig. 230a). Animal wood boring ...................................................... Limnoriidae (p. 354)
  Rami of uropods flattened, outer ramus not bearing an apical claw ................................. 5
5 Abdomen with 5 clearly demarcated segments. Does not curl up tightly when
  disturbed ............................................................................ Cirolanidae (p. 355)
  Abdomen not so. Curls up tightly when disturbed ..................... Sphaeromatidae (p. 356)
6 Abdominal segments not demarcated. Segments of pereopods not flattened. Thoracic
  segments much the same size as each other. Antennae long ................... Janiridae (p. 361)
  Abdominal segments clearly demarcated; animals terrestrial and in the littoral fringe ...
  ......................................................................................... Ligiidae (p. 361)

Family Gnathiidae

Isopods in which the first pair of pereopods (the pylopods) are modified and closely associated with the mouth parts, and the last pair are absent, leaving only five typical pairs. There is marked sexual dimorphism; the male has a straight, slender body and a pair of large mandibles which project between the eyes, while the female has a much-expanded thoracic region. There are two juvenile stages, the zuphea and the praniza.

First pair of pereopods (pylopods) of male 5-segmented. Found in mud banks in estuaries ........................................................................ *Paragnathia formica* (p. 352)
First pair of pereopods (pylopods) of male, 2- or 3-segmented; anterior margin of head between mandibles gently concave but with small median lobe. Found in rock crevices, empty shells of barnacles and the holdfasts of *Laminaria* ....... **Gnathia maxillaris** (p. 354)

**Paragnathia formica** (Hesse)   (Figs. 228, 229a)

*Body up to 5 mm in length; male, female and juvenile differ markedly in appearance. First pereopod in the male is 5-segmented, flattened and expanded and known as a pylopod; in the juvenile it is 7-segmented, not expanded and terminates in a hooked claw.*

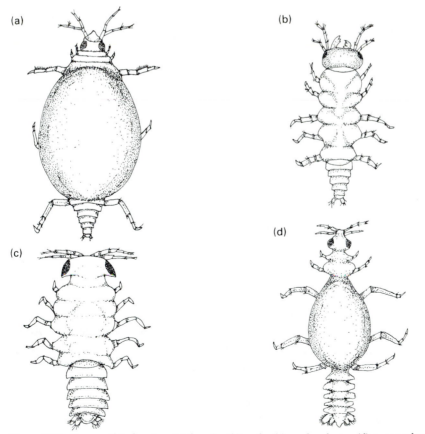

Figure 228 *Paragnathia formica*, (a) female, (b) male, (c) zuphea larva, (d) praniza larva.

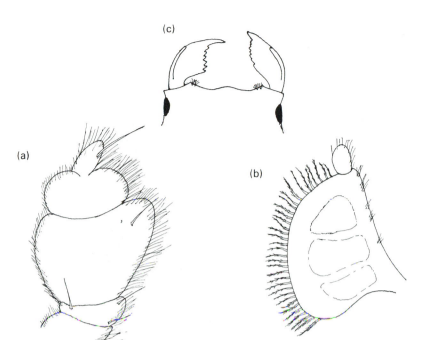

Figure 229   (a) Pylopod of male *Paragnathia formica*. (b) Pylopod of male *Gnathia maxillaris*. (c) Anterior of head of male *Gnathia maxillaris*. ((b), (c) After Naylor, 1972.)

*P. formica* reaches its northern limit of distribution in Britain and extends as far south as Morocco. It occurs in the muddy banks of intertidal saltmarsh creeks in sheltered estuaries tolerating salinities down to about 18‰. Males and females live in burrows in the sediment, a single male with as many as 25 females. The males are usually in the narrow entrance to the burrow while the females are deeper inside. Here, up to 140 embryos develop within the body of each female during the summer months and are released in the autumn. Release occurs on high tides, when the burrows are submersed. After release of the young the female dies, and the dead remains are removed from the burrow by the male. The newly released larva measures about 1 mm in body length and is known as a zuphea. It is an active swimmer and once it locates a fish host, typically an estuarine species such as *Pomatoschistus microps* (p. 510), it attaches itself using the hooks on the first pair of pereopods, and sucks blood. The thorax becomes much expanded and the larva, now called a praniza with a body length of about 1.5 mm, leaves the host and returns to the sediment where it remains until the blood meal has been digested. After moulting it again becomes a zuphea and swims off to find another fish host. After three zuphea stages and three praniza stages the larva moults to become a young adult. Young females are captured by mature males using the large mandibles and taken into the burrow where breeding takes place. Males and females do not feed but survive on food reserves. Females live for about one year, breeding only once; males live for about two years.

*Gnathia maxillaris* (Montagu)    (Fig. 229b, c)

*Body up to 5 mm in length; male, female and juvenile differ markedly in appearance. First pereopod (pylopod) of the male is 2- or 3-segmented, flattened and expanded. In the juvenile it is 7-segmented, not expanded and terminates in a hooked claw. In the male the anterior margin of the head between the mandibles is gently concave with a small median lobe; anterior corners of head are square; without longitudinal ridge over eyes.*

*G. maxillaris* is common and widespread in southern Britain and Ireland, where it occurs intertidally in crevices, under stones, in empty barnacle shells and in laminarian hold-fasts. Up to 100 or so embryos develop within the body of the female and are released as larvae known as zupheae. They attach themselves to the body surface of fish such as *Lipophrys pholis* (p. 503), *Taurulus bubalis* (p. 493) and *Symphodus melops* (p. 496). After a blood meal they leave the fish host; the thoracic region is much expanded and the larvae are known as pranizae. It has been suggested that there are three praniza stages after which the larva moults to the adult. Adult *G. maxillaris* are believed not to feed, but little is known of their life.

Family Limnoriidae
Isopods which bore into timber. The body is narrow and straight-sided and the abdomen has five distinct segments. The uropods are small and lie along the sides of the telson; the outer ramus is short and bears an apical claw.

*Limnoria lignorum* (Rathke)    Gribble (Fig. 230a)

*Body up to 3.5 mm in length, rather narrow. Antennae 1 and 2 short, about equal length. Abdomen with 5 distinct segments. Uropods small, the shorter, outer ramus bearing an apical claw. Dorsal surface of telson without tubercles, marked with inverted Y-shape.*

(a)                                                                                      (b)

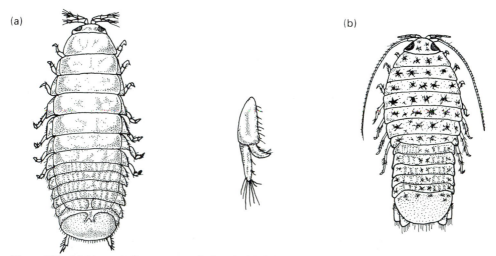

Figure 230  (a) *Limnoria lignorum*, with detail of right uropod (after Sars, 1899). (b) *Eurydice pulchra.*

L. *lignorum* is a northern species extending as far as the south coast of Britain, and is well known for its habit of boring into timbers and causing extensive damage. The adult colonizes new timber, excavating a burrow using the mandibles, and feeds on fragments of wood digested by cellulolytic enzymes. Burrows are generally less than 20 mm deep and are of the same width throughout their length, communicating with the exterior via a number of pores through which a respiratory current is drawn. There is generally a male and a female in each burrow, the female producing up to 30 eggs which are incubated in a brood chamber. Breeding occurs predominantly in winter. After emergence the young establish their own burrows leading from the walls of the parent burrow. Gribbles have also been recorded swimming and crawling; at this time they presumably colonize new timbers.

There are two other British species of *Limnoria*, **Limnoria quadripunctata** Holthuis, which is usually dominant on south and west coasts of Britain, and **Limnoria tripunctata** (Menzies), found only on the south coasts of England and Wales. The species are separated from *L. lignorum* by the presence of tubercles on the mid-dorsal surface of the telson. *L. quadripunctata* has four tubercles and *L. tripunctata* three. In addition, *L. tripunctata* has a row of small tubercles around the posterior margin of the telson. When the three species occur on the same timbers, they show separation of habitat – *L. tripunctata* occupies the upper, *L. quadripunctata* the middle and *L. lignorum* the lower part of the infected timber. Treatment of timber with creosote does not prevent attack and research on *L. tripunctata* suggests that bacteria in the gut of the isopod may contribute to creosote resistance by detoxification of the creosote.

### Family Cirolanidae

Isopods in which all the thoracic segments except the first have conspicuous lateral expansions known as coxal plates. The abdomen has five clearly demarcated segments. The uropods are flattened and somewhat triangular and lie along the sides of the telson.

### *Eurydice pulchra* Leach    (Fig. 230b)

*Thorax broader than abdomen giving oval outline; up to 8 mm in length. Large lateral eyes. Antennae 1 short, antennae 2 about ⅔ body length, with 4 segments in basal region. Abdomen with 5 distinct segments. Uropods lie along the sides of telson. Posterior margin of telson broad, convex. Pale grey to brown colour, conspicuously marked with black chromatophores on all surfaces of body.*

*E. pulchra* is widely distributed in north-west Europe and Britain on open coast and estuarine sandy beaches. Its intertidal distribution is complicated by a migration up the shore during spring tides, falling to lower levels during neap tides. This migration is controlled by a complex swimming rhythm, with the animal leaving the sand to swim in the water at high tide during certain phases of the tidal cycle. At such times it can be caught in large numbers in the intertidal plankton, particularly at night. *Eurydice* is a voracious carnivore feeding on the wide range of invertebrates associated with sandy shores and the intertidal migration is an adaptation which extends its feeding range. The sexes are separate and mating apparently takes place during the swimming phase in spring and summer, after which embryos develop in the brood chamber for some seven to eight weeks. Maturity is reached after 12 to 20 months and the animals have a life-span of about two years.

A second intertidal species of *Eurydice*, *Eurydice affinis* Hansen, is also found on British shores and in this species the black chromatophores occur only on the dorsal surface of the body, giving overall a paler appearance. In Britain it has a more restricted distribution than *E. pulchra* and apparently extends from south-west England to North Wales. It occurs on the coast of The Netherlands and on the Atlantic coast of France.

Family Sphaeromatidae
Isopods which curl up tightly when disturbed. Abdominal segments indistinct; the uropods are flattened and lie along the sides of the telson. Some genera show marked sexual dimorphism.

1  Uropod with large, paddle-shaped outer ramus; without inner ramus. May (male) or may not (female) have large backwardly directed mid-dorsal spine on 6th thoracic segment ............................................................ ***Campecopea hirsuta*** (p. 359)
   Uropod with 2 rami ................................................................................................ 2
2  Posterior margin of telson with a single, median indentation. May (male) or may not (female) have 2 large backwardly pointing projections on 6th thoracic segment ............
   ...................................................................................... ***Dynamene bidentata*** (p. 358)
   Posterior margin of telson smooth, without any indentation ........................................ 3
3  Outer edge of outer ramus of uropod serrated; dorsal surface of telson smooth .............. 4
   Outer edge of outer ramus of uropod not serrated or only barely so; dorsal surface of telson with tubercles either in 2 longitudinal rows or scattered ................................. 5
4  Outer edge of outer ramus of uropod with 4–7 conspicuous serrations; long setae on pereopod 1 with many fine side branches ........................... ***Sphaeroma serratum*** (p. 356)
   Outer edge of outer ramus of uropod with 4–7 poorly defined serrations; long setae on pereopod 1 smooth ................................................................. ***Lekanesphaera levii*** (p. 358)
5  Dorsal surface of telson with 2 longitudinal rows of tubercles ...........................................
   .................................................................................. ***Lekanesphaera hookeri*** (p. 357)
   Dorsal surface of telson with many small tubercles ...... ***Lekanesphaera rugicauda*** (p. 358)

***Sphaeroma serratum*** (Fabricius)    (Fig. 231a)

*Oval-shaped body up to 11 mm in length. Antennae 1 about ½ length of antennae 2; antennae 2 about ¼ body length. Abdominal segments fused and indistinct. Laterally placed uropods have 4–7 conspicuous serrations on outer edge of outer ramus. Long setae on pereopod 1 with many fine side branches. Dorsal surface of telson smooth; posterior margin without indentation. Curls up tightly when disturbed.*

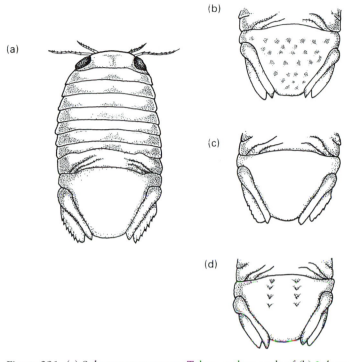

Figure 231  (a) *Sphaeroma serratum*. Telson and uropods of (b) *Lekanesphaera rugicauda*, (c) *Lekanesphaera levii* and (d) *Lekanesphaera hookeri*. ((b), (c), (d) After Naylor, 1972.)

*S. serratum* is a southern species extending north to Anglesey, North Wales. It is common under stones from about the middle shore and below on south-west shores of Britain, just penetrating into estuaries. The breeding season is restricted to the summer and embryos are incubated in internal pouches opening from the brood chamber. On the coast of South Wales, the majority of young are released in August and have a life-span of two and a half to three years, reaching maturity after about two years. In south-west Britain the commensal *Jaera hopeana* (p. 361) may be found on the undersurface of *S. serratum*.

**Lekanesphaera hookeri** (Leach)    (*Sphaeroma hookeri*) (Fig. 231d)

*Similar body shape to* S. serratum *(above), but with smooth outer margin to outer ramus of uropod. Mid-dorsal surface of telson with 2 longitudinal rows of tubercles.*

*L. hookeri* is locally distributed around Britain and although recorded from East Anglia, appears to be absent from much of the east coast of England and Scotland. It extends northwards to southern Sweden. It is generally found in salinities ranging from 1 to 10‰ in the upper reaches of estuaries, but is also found in higher salinities.

*Lekanesphaera levii* (Argano & Ponticelli)    (*Sphaeroma monodi*) (Fig. 231c)

*Similar body shape to S. serratum (above), but outer margin of outer ramus of uropod with 4–7 poorly defined serrations. Long, smooth setae on pereopod 1. Dorsal surface of telson smooth.*

*L. levii* is a southern species extending as far north as southern Britain. It penetrates well into estuaries, usually in salinities down to 14‰, but is also found under stones on the open coast, particularly in areas of freshwater seepage.

*Lekanesphaera rugicauda* (Leach)    (*Sphaeroma rugicauda*) (Fig. 231b)

*Similar body shape to S. serratum (above) but with smooth outer margin to outer ramus of uropod. Dorsal surface of telson covered with many small tubercles.*

*L. rugicauda* is widely distributed in north-west Europe. It withstands salinities down to about 4‰ and is widely distributed on the upper shore in estuaries and saltmarshes. The species is characterized by a number of different colour varieties, for example, 'yellow', 'red', 'grey', the frequency of which varies seasonally, possibly in response to environmental conditions. Breeding occurs in spring and summer. Maturity is reached in one year and the animals live for about one and a half years.

*Dynamene bidentata* (Adams)    (*Naesa bidentata*) (Fig. 232)

*Oval-shaped body up to 7 mm in length. Abdominal segments fused and indistinct. Shows pronounced sexual dimorphism; 6th thoracic segment of male with 2 large, backwardly pointing projections, absent from female. Posterior margin of telson with a single median indentation, a feature which readily distinguishes females from* Sphaeroma *(above) and* Lekanesphaera spp. *(above).*

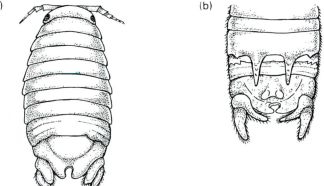

Figure 232  *Dynamene bidentata*, (a) female, (b) posterior region of male.

*D. bidentata* has been recorded from the north and west coasts of France and from south and west Britain, extending northwards to south-west Scotland. In the English Channel it is found as far east as the Isle of Wight. Adult *Dynamene* are found in crevices and the empty cases of the barnacle *Balanus perforatus* (p. 342) with the juveniles occurring among seaweeds such as *Fucus serratus* (p. 49) and *Ascophyllum nodosum* (p. 47). Embryos develop in the brood chamber of the female and on the south coast of Wales, juveniles are liberated from May to July. During summer, the juveniles live and feed on seaweed; by the autumn they have reached maturity and move to rock crevices and empty barnacle shells. Here, a single male may be found with a number of females. After fertilization, the embryos are incubated for some three to four months over winter and spring. The female produces a single brood and dies after release of the juveniles. Males, however, remain in the crevices and empty barnacle shells to survive for a second breeding season. Mature specimens of *Dynamene* do not feed but survive on lipid food reserves, the females for about nine months and the males for up to 21 months.

### *Campecopea hirsuta* (Montagu)    (Fig. 233)

*Oval body up to 4 mm in length. Abdominal segments fused and indistinct. Shows pronounced sexual dimorphism; large, backwardly directed, mid-dorsal spine on 6th thoracic segment of male, absent from female. In both sexes, uropod with large paddle-shaped outer ramus, but without inner ramus. Posterior margin of telson without indentation. Curls up tightly when disturbed.*

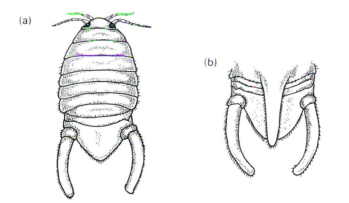

(a)

(b)

Figure 233  *Campecopea hirsuta*, (a) female, (b) posterior region of male.

*C. hirsuta* is the only species of the genus. It is found in south-west Britain, about as far north as Anglesey, North Wales, and on the west coast of France. It lives in the empty shells of *Semibalanus balanoides* (p. 341), *Chthamalus* spp. (p. 338) and among *Lichina* (p. 80), and on exposed shores is often found in large numbers. It is readily collected by taking barnacle scrapings for subsequent examination in the laboratory. The eggs are incubated in internal pouches opening from the brood chamber, and on the coast of

South Wales one brood of eggs per year is produced. Females carrying embryos are recorded in summer with release of young in late summer or autumn. The life-span is about one and a half years.

Family Idoteidae
Isopods with a more or less elongate body. The abdominal segments are indistinct. The uropods are ventral and usually have only one ramus which is large and flattened and covers the pleopods.

### *Idotea granulosa* Rathke    (Fig. 234a)

*Elongate, somewhat oval body, narrow at posterior. Males up to 20 mm in length; females 13 mm. Antennae 2 about ⅖ body length. Abdomen with 2 distinct segments; 3rd segment only partially demarcated. Uropods ventral and cover the pleopods. Sides of telson slightly concave; posterior margin with rounded median extension; without a mid-dorsal keel. Colour varied, includes shades of brown, green, red; some specimens have longitudinal white markings.*

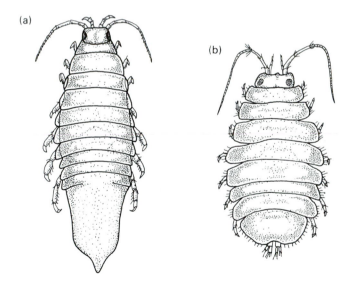

(a)

(b)

Figure 234  (a) *Idotea granulosa* (after Sars, 1899). (b) *Jaera albifrons*, female (after Sars, 1899).

*I. granulosa* is widespread in north-west Europe, where it is found on a variety of inter-tidal seaweeds. In Britain it is probably the commonest intertidal species of *Idotea*. Adults occur mainly on the large brown seaweeds such as *Ascophyllum* (p. 47) and *Fucus* (p. 50) (but avoid *Laminaria*), while young specimens are found on *Polysiphonia* (p. 71) and *Cladophora* (p. 37). *I. granulosa* is omnivorous and the colour of the animal is dependent on the seaweed on which it feeds, the gut contents being visible through the semi-transparent body wall. Breeding females have been found throughout the year with a peak in early summer on the Isle of Man; breeding in winter has been recorded off the south-east coast of Ireland. Longevity varies from a few months to about one year.

Several other species of *Idotea* occur in British waters. **Idotea pelagica** Leach often outnumbers *I. granulosa* on exposed shores where it lives among barnacles and fucoid seaweeds. **Idotea chelipes** (Pallas) is found in estuaries, and is able to withstand salinities down to about 8‰. Other species, such as **Idotea baltica** (Pallas), normally found in the sublittoral, are occasionally washed ashore in large numbers on decaying seaweed.

### Family Janiridae

Isopods generally with an oval body. The abdominal segments are not demarcated. The uropods are narrow and stick-like and project from the posterior margin of the telson. The anterior pleopods are modified and cover the other pleopods in the female, or function as a copulatory organ in the male.

### *Jaera* spp.   (Fig. 234b)

*Oval-shaped body fringed with setae; margins of thorax deeply notched. Females up to 6 mm in length, usually larger than males. Antennae 1 short, antennae 2 about ⅓ length of body. Abdominal segments not visible dorsally. Telson large, with smooth, non-serrated margin. Uropods very small; project from indentation in posterior margin of telson.*

The genus *Jaera* is widely distributed in north-west Europe and Britain, and includes a number of species. Generally speaking, they are common in moist situations on the middle regions of the shore. Some species withstand reduced salinity and are found under stones where freshwater streams run across the beach, and in estuaries. The species overlap in distribution but each has its own ecological niche and there appears to be little evidence of hybridization. One species, **Jaera hopeana** Costa, is no more than 2.5 mm in length and lives ectocommensally on the undersurface of *Sphaeroma serratum* (p. 356). It is a southern species found in south-west Britain, very occasionally free-living but always where *S. serratum* is found.

### Family Ligiidae

Terrestrial and littoral fringe isopods. Antenna 1 inconspicuous, antenna 2 large with at least 10 small distal segments. Abdominal segments clearly demarcated. Uropods large, each with two long rami, project from the posterior margin of the telson.

### *Ligia oceanica* (Linnaeus)   Sea-slater (Fig. 235)

*Oval-shaped body more than twice as long as broad; up to 30 mm in length. Antennae 1 inconspicuous, antennae 2 about ⅔ body length. Abdominal segments clearly demarcated. Uropods large, each with 2 long processes which project from posterior margin of telson. Various shades and intensities of grey, olive.*

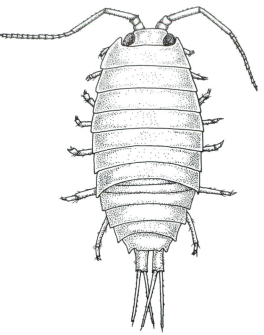

Figure 235 *Ligia oceanica.*

*L. oceanica* is widely distributed in north-west Europe. It is the only species of the genus in Britain and is common in crevices and under stones on the upper shore. It is omnivorous, emerging at night to feed on detritus, decaying seaweed and encrusting diatoms. Breeding occurs in spring and summer. On the east coast of England maturity has been recorded in the first year, although most animals do not breed until they are two years of age. The majority of individuals breed only once, and length of life is two and a half to three years.

### Order Amphipoda

Amphipods have many characters in common with the isopods (above), such as the lack of a carapace. As a general rule, however, the body of the amphipod is laterally flattened and one important distinguishing feature is the fact that the first three segments of the abdomen typically each bear a pair of flattened pleopods, while the last three have narrow, backwardly directed uropods. There are seven externally visible thoracic segments and the appendages of the first two are known as gnathopods and are used in feeding and grasping. The remaining five segments of the thorax bear pereopods. In many amphipods, the terminal claw of the gnathopods closes against the preceding segment, a condition known as subchelate (Fig. 236c). This is an important character in identification. Other characteristics useful in identification include the form of the coxal plates (lateral expansions along the sides of the thorax), the presence or absence of an accessory flagellum on antenna 1 and the shape of the telson (Fig. 239). The range of feeding habits shown by amphipods includes suspension feeding, scavenging and detritus feeding. The sexes are generally separate and some species show precopula behaviour, during which the male holds the female using the gnathopods, and carries her for some days before mating. Sperm are deposited in the brood chamber of the female, where external fertilization takes place, and where the eggs develop to the juvenile stage. The brood chamber is formed of brood plates (oostegites) arising from the bases of some thoracic appendages (Fig. 241a).

Amphipods are generally benthic in habit and many burrow into sand and mud. Others are found among shingle, under stones and seaweed, and in the holdfasts of laminarians. Characteristically, they move away quickly on their sides when disturbed. Amphipods swim using the pleopods and some are caught in high densities in the intertidal plankton, especially at night. Over 100 species have been recorded from British shores; only some of the most common are included here. For more detailed information the reader is referred to Lincoln (1979).

The Order Amphipoda is subdivided into suborders, each of which shows a modification of the basic amphipod design. The suborders can be recognized using the following characters:

(a) Body usually, but not always, laterally compressed; with pleopods and uropods. Found buried in sandy and muddy sediments, or among shingle, and under stones and seaweed. The typical amphipods of the shore .................. **Suborder Gammaridea** (p. 363)
*or*
(b) Body rounded; with pleopods and uropods. Eyes very large, bulbous. Pelagic or found on the undersurface of jellyfishes ........................................ **Suborder Hyperiidea** (p. 380)
*or*
(c) Body very elongate and narrow; abdomen tiny. Some pereopods absent or reduced in size; pleopods and uropods absent. Found clinging to seaweeds, hydroids, sponges; move by characteristic looping movements ........................ **Suborder Caprellidea** (p. 381)

### Suborder Gammaridea

There is considerable discussion in the literature regarding the classification of gammaridean amphipod families. Family groupings used in the present text follow Costello, Holmes, McGrath & Myers (1989).

### Key to families of gammaridean amphipods

Identification of amphipod families is difficult and often relies on the combination of a number of characteristics. These are shown in Fig. 239. A selection of the most common families is included here. In a small number of cases it has proved most convenient to key to the level of genus.

1 Antenna 2 stout and stumpy in appearance; uropods dissimilar, uropod 3 very large ....
.................................................................................................... **Cheluridae** (p. 379)
Not so ........................................................................................................ 2

2 Body stout, broad, solid; of very distinctive appearance (Fig. 242). Pereopods 5, 6, 7 very much expanded with many setae and spines; without terminal claw .......................... ...................................................................................................... **Haustoriidae** (p. 372)
   Not so ....................................................................................................................... 3

3 Antenna 1 with very large, rectangular basal segment; remaining segments arising at right angles (Fig. 241). Basal segments of pereopods 5, 6, 7 flattened and expanded ..... ................................................ **Pontoporeiidae (Genus *Bathyporeia*)** (p. 370)
   Not so ....................................................................................................................... 4

4 Head with small rostrum; antenna 1 with accessory flagellum; pereopod 5 with horizonal rows of stout spines, quite different in appearance from other pereopods ............. .................................................................................................... **Urothoidae**(p. 372)
   Not so ....................................................................................................................... 5

5 Head with large, downwardly pointing rostrum; eyes large, extend towards dorsal surface of head. Antennae shortish; antenna 1 without accessory flagellum. Pereopod 7 elongate and narrow, much longer than preceding pereopods ........ **Oedicerotidae** (p. 373)
   Not so ....................................................................................................................... 6

6 Body dorso-ventrally flattened; coxal plates small, not touching one another. Eyes small. Antenna 2 large and stout. Pereopod 7 narrow, longer than preceding pereopods ......................... **Corophiidae (Genus *Corophium*)** (p. 377)
   Not so ....................................................................................................................... 7

7 Antenna 1 much shorter than antenna 2, without accessory flagellum. Uropod 3 uniramous, small. Amphipods which are capable of jumping considerable distances ....... 8
   Not so ....................................................................................................................... 9

8 Antenna 1 shorter than large basal segments of antenna 2. Common on the strandline . ................................................................................................. **Talitridae** (p. 365)
   Antenna 1 longer than large basal segments of antenna 2 but less than total length of antenna 2 ...................................................................................... **Hyalidae** (p. 367)

9 Last 3 segments of abdomen with conspicuous clusters of spines and setae on dorsal surface ........................................................................................ **Gammaridae** (p. 368)
   Last 3 segments of abdomen without conspicuous clusters of spines and setae, or at most with only a few spines. Dorsal surface of some segments of body may be toothed . 10

10 Antenna 1 considerably longer than antenna 2; antenna 1 with accessory flagellum. Eyes small. Dorsal surface of abdomen usually toothed. Gnathopod 2 larger than gnathopod 1. Telson bilobed .............................................................. **Melitidae** (p. 369)
   Not so ..................................................................................................................... 11

11 Antenna 1 shorter than antenna 2, without accessory flagellum. Abdomen with only 5 visible segments. Abdominal segment 4 with 2 dorsal teeth, the posterior one the larger. Dorsal surface of other abdominal segments may have tooth-like posterior projections. Telson bilobed .................................................................. **Atylidae** (p. 375)
   Not so ..................................................................................................................... 12

12 Uropod 3 large, biramous; rami elongate, flattened, fringed with setae and spines. Gnathopods 1 and 2 small, of similar size. Telson not divided, large ............................. .................................................................... **Calliopiidae (Genus *Apherusa*)** (p. 374)

Uropod 3 usually biramous; rami short, shorter than basal region; outer ramus with terminal hooks and/or small teeth. Gnathopods 1 and 2 well developed, gnathopod 2 larger than 1 ............................................................................................ 13

13 Basal segment 3 of antenna 1 much smaller than basal segment 2 .... **Ampithoidae** (p. 376)
Basal segment 3 of antenna 1 nearly as long as basal segment 2. Antennae usually fringed with long setae .................................................................. **Ischyroceridae** (p. 379)

Family Talitridae

Stout-bodied amphipods; laterally compressed. Head without a rostrum; antenna 1 shorter than the large basal segments of antenna 2; antenna 1 without an accessory flagellum. Antennae 2 longish; basal joints stout in male. Gnathopods subchelate or not; gnathopod 2 sometimes large in males. Uropod 3 small, uniramous. Telson usually divided only at tip.

   Generally among debris and decaying seaweed on the strandline; known as sand hoppers because of their ability to jump considerable distances.

Small terminal segments of antenna 2 with teeth on inner margin, giving a rough, spiky appearance. Gnathopod 2 of male small, not subchelate. Uropod 3 terminating in long spine ............................................................ *Talitrus saltator* (p. 366)
Small terminal segments of antenna 2 without teeth on inner margin. Gnathopod 2 of male large and subchelate .............................................................. *Orchestia* spp. (p. 365)

*Orchestia* spp.   Sand-hopper (Fig. 236)

*Greenish-brown body up to 22 mm in length. Eyes black. Antennae 1 small, shorter than large basal segments of antennae 2; antennae 2 about ⅓ body length. Small terminal segments of antenna 2 without teeth on inner margin. Gnathopod 1 small and subchelate; gnathopod 2 small in female, large and subchelate in male.*

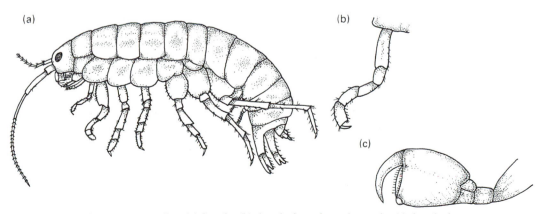

Figure 236  *Orchestia gammarellus*, (a) female, (b) detail of gnathopod 1, male, (c) detail of
          gnathopod 2, male (after Sars, 1895).

The genus *Orchestia* is widely distributed in north-west Europe, with **Orchestia gammar-ellus** (Pallas) being the commonest species. It is found under stones and rotting seaweed, particularly on shingle shores on the upper-shore–terrestrial fringe, and is also found in estuaries. It is an omnivorous scavenger and is particularly active at night. Breeding has been recorded from April to about September at a number of localities and there is some evidence to suggest that young released in early summer will mature by the autumn of the same year. Longevity is estimated to be about 18 months. When disturbed, *Orchestia* jumps considerable distances.

**Talitrus saltator** (Montagu)    Sand-hopper (Fig. 237)

*Greyish-green body up to 25 mm in length. Eyes black. Antennae 1 small, much shorter than large basal segments of antennae 2. Small terminal segments of antenna 2 with teeth on inner margin giving a rough, spiky appearance, especially in male. Gnathopods 1 and 2 small, not subchelate. Uropod 3 with long, slender spine on posterior margin.*

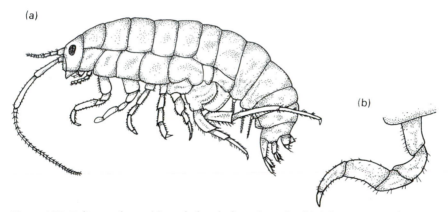

Figure 237  *Talitrus saltator* (a), with detail of gnathopod 1 (b) (after Sars, 1895).

*T. saltator* is the only species of the genus to be found in north-west Europe. It lives on the upper-shore–terrestrial fringe and is widely distributed on sandy beaches around Britain where it is associated with decaying seaweed. It is an omnivorous scavenger, usually spending the day-time buried in the sediment, emerging at night to feed on detritus on the strandline. Although essentially nocturnal, *Talitrus* has been shown to be capable of navigating by means of a sun compass. If for some reason the animals are transported from the upper shore, they can return by orientating to the angle of the sun and other features. The common name sand-hopper (which includes *Orchestia*, above) is associated with the animal's ability to jump up to 300 mm by rapid backwards extension of the abdomen and telson. On Irish Sea coasts *T. saltator* breeds during the summer months, with peak activity from May to August. Females are believed to produce more than one brood and the incubation period varies from eight to 20 days, depending on temperature. Longevity is between 18 months and two years.

*Talorchestia deshayesii* (Audouin) is another sand-hopper occurring amongst shingle and strandline debris around Britain. It is similar to *Talitrus* in that the small terminal segments of antennae 2 are toothed, but differs in that gnathopod 2 of the male is very large and pincer-like, and the posterior margin of uropod 3 does not bear a long slender spine.

Family Hyalidae
Stout-bodied amphipods, laterally compressed. Head without a rostrum; antennae 1 and 2 short; antenna 1 longer than large basal segments of antenna 2, but less than total length of antenna 2. Antenna 1 without an accessory flagellum. Gnathopods 1 and 2 subchelate, more or less the same size in female; in male, gnathopod 2 much larger than gnathopod 1. Uropod 3 small and uniramous. Telson bilobed.

   Live intertidally among seaweed and in crevices; able to jump considerable distances.

*Hyale* spp.    (Fig. 238)

*Brownish-green, usually up to about 8 mm in length. Eyes black or reddish colour. Antennae 1 shorter than antennae 2, but longer than the large basal segments of antennae 2 (cf.* Orchestia *(above) and* Talitrus *(above)). Gnathopods 1 and 2 subchelate. Gnathopod 2 of male much larger than gnathopod 1. Gnathopods 1 and 2 about same size in female.*

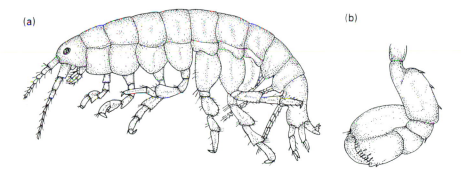

(a)                                                                          (b)

Figure 238  *Hyale prevostii*, (a) female, (b) detail of gnathopod 2, male.

The genus *Hyale* is widely distributed in north-west Europe. Some four species have been recorded from Britain, of which **Hyale prevostii** (Milne-Edwards) (*Hyale nilssoni*) is widely distributed and probably the best known. Found among seaweeds such as *Fucus* and *Ascophyllum* on rocky shores, it occurs from the upper shore into the sublittoral and also penetrates into estuaries. It is able to jump considerable distances. Specimens on the high shore have been shown to exhibit a preference for feeding on the fertile tips of the brown seaweed *Pelvetia canaliculata* (p. 51). Breeding has been recorded on the west coast of Scotland during spring and summer; the eggs are green coloured.

Family Gammaridae

Mostly stout-bodied amphipods; laterally compressed. Dorsal surface smooth, not toothed; last three segments of abdomen with clusters of spines and setae on the dorsal surface. Head with small rostrum; eyes mostly large and kidney-shaped. Antennae 1 and 2 longish, and of similar length; antenna 1 with conspicuous accessory flagellum. Gnathopods 1 and 2 subchelate; well developed, much the same size. Uropod 3 biramous, large; rami with many marginal spines and setae. Telson bilobed.

Amphipods *par excellence*. A large family, common in marine and freshwater habitats. Representatives are found in all salinity concentrations, typically under stones and sea-weed and in shingle; seen wriggling on their sides when disturbed.

*Gammarus* spp.    (Fig. 239)

*Body usually brownish-green; up to 33 mm in length. Eyes rounded or kidney-shaped. Antenna 1 with accessory flagellum; antenna 2 of similar length to antenna 1. Gnathopods 1 and 2 subchelate. Last 3 segments of abdomen have clusters of spines or spines and setae on posterior, dorsal margins giving spiny appearance. Inner ramus of 3rd pair of uropods more than ⅔ length of outer ramus.*

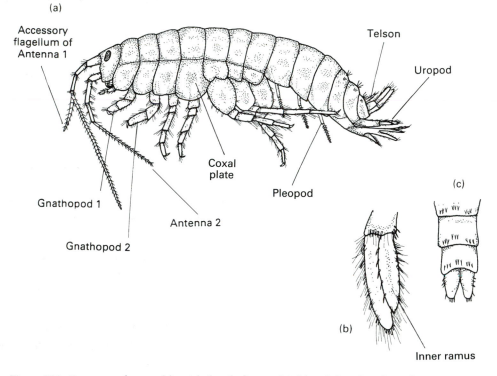

Figure 239  *Gammarus locusta* (a), with detail of uropod 3 (b) and dorsal surface of
               posterior end of abdomen (uropods omitted) (c).

Many species of *Gammarus* have been described from marine and freshwater habitats. In Britain there are some 12 species, mostly intertidal, extending from fully marine through brackish water to freshwater environments. Individual species are often used as indicators of salinity.

*Gammarus locusta* (Linnaeus), one of the species which is widespread and often abundant in Britain and north-west Europe, is found in damp situations under stones and seaweed on the middle and lower shore, surviving in salinities down to about 4‰. It feeds mainly on detritus and seaweed. The male carries the female for a number of days before fertilization, and females with embryos in the brood chamber are found throughout most of spring and summer.

The genus *Gammarus* can be separated from the three other genera in the family, *Eulimnogammarus* (formerly *Marinogammarus*), *Echinogammarus* (formerly *Chaetogammarus*, *Marinogammarus*) and *Pectenogammarus*, using the relative lengths of the inner and outer rami of the third uropod. In *Gammarus* (Fig. 239) the inner ramus is more than $\frac{1}{3}$ of the length of the outer ramus, while in the other genera it is less. All three genera occur intertidally. The genus *Pectenogammarus* has only one species, **Pectenogammarus planicrurus** Reid. This is found on open coast gravel and shingle beaches and is very restricted in its distribution, having been recorded from only a few localities in north-west Europe. The only species of *Eulimnogammarus* widely occurring in Britain is **Eulimnogammarus obtusatus** (Dahl), a common and widespread species among seaweed and in gravel and shingle. Of the British species of *Echinogammarus*, **Echinogammarus marinus** (Leach) is probably the most common and widespread, occurring under seaweed and stones on the upper and middle shore. Reference to Lincoln (1979) is recommended for further study.

Family Melitidae

Mostly slender-bodied amphipods; laterally compressed. Dorsal surface of thorax smooth; abdomen usually toothed dorsally; last three segments sometimes with a few spines on dorsal surface, but without distinct clusters of spines and setae (cf. Gammaridae). Head without a rostrum; eyes small; antennae 1 and 2 longish, antenna 1 usually considerably longer than antenna 2; antenna 1 with accessory flagellum. Gnathopods 1 and 2 subchelate, gnathopod 1 small, gnathopod 2 large, much larger in male. Uropod 3 biramous. Telson bilobed.

Commonly found among stones on sandy and muddy sediments, and among seaweed.

*Melita* spp.   (Fig. 240)

*Body rather slender, up to about 16 mm in length. Colour varied, reddish-brown, grey, whitish. Eyes rather small, rounded or oval. Antennae 1 with accessory flagellum; antennae 1 longer than antennae 2. Gnathopods 1 and 2 subchelate; gnathopod 2 much larger than gnathopod 1 in male with segment next to claw triangular in shape. Dorsal surface of abdomen usually, but not always, toothed. Last 3 segments of abdomen may have a few spines but without clusters of spines and setae. Outer ramus of 3rd pair of uropods elongate, longer than basal region and with marginal spines; inner ramus very small.*

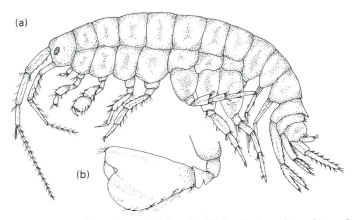

Figure 240  *Melita palmata*, (a) female, (b) detail of gnathopod 2, male.

The genus *Melita* is widely distributed in north-west Europe and of the intertidal species, **Melita palmata** (Montagu) is the most widespread. It is found among stones on sandy and muddy sediments.

Family Pontoporeiidae

GENUS *BATHYPOREIA*

Slender-bodied amphipods, laterally compressed. Head without rostrum. Antennae 1 shorter than antennae 2. Antenna 1 with very large basal segment; with accessory flagellum. Gnathopods 1 and 2 not strongly developed. Pereopods 3–7 with small terminal claw. Basal segments of pereopods 5, 6, 7 flattened and expanded. Uropod 3 biramous. Telson bilobed.

The only genus of the family found in Britain. Found in clean, sandy sediments; common and widespread and extends into estuaries.

**Bathyporeia** spp.    (Fig. 241)

*Body up to 8 mm in length; semi-transparent to white, with varying degrees of red pigment associated with abdomen. Dorsal surface of abdominal segment 4 notched, with seta or seta and spines. Eyes red. Antennae 1 with accessory flagellum and shorter than antennae 2. Basal segment of antenna 1 very large, more or less rectangular; other segments smaller and arise at right angles to it, a feature known as geniculate and characteristic of the genus. Antennae 2 about twice the length of antennae 1 in females but equal to, and often exceeding, body length in mature males.*

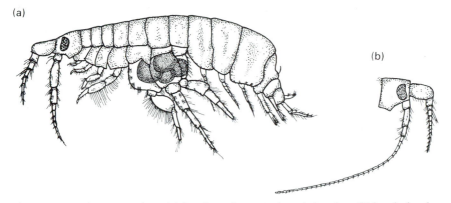

Figure 241 *Bathyporeia pilosa*, (a) female with eggs in brood chamber, (b) head of male.

The genus *Bathyporeia* is well represented on European shores and eight species have been described from British waters. Morphologically they are very similar. **Bathyporeia pilosa** Lindström is one of the commonest species, often reaching densities of several thousand per square metre. It is widely distributed in north-west Europe on the upper shore of sheltered sandy beaches and is also common in estuaries, tolerating salinities down to about 5‰. It is an epistrate feeder, individual sand grains being rotated by the mouth parts and organic matter removed. Although *Bathyporeia* lives buried in sandy sediments, it leaves the protection of the sand to swim in the surface waters during spring tides; large numbers can be taken in intertidal plankton hauls, particularly during darkness. On the west coast of Wales, breeding has been recorded throughout the year, with peaks of reproductive activity in spring and autumn, while in some localities a single peak of breeding has been recorded in summer. Males and females pair while swimming, but there is no prolonged precopula behaviour. During the breeding season, gravid females are readily identified by the presence of blue eggs in the brood chamber, each female producing a succession of broods. In summer, development takes about 14 days and is synchronized with spring tides. Longevity is about one to one and a half years.

Other species of *Bathyporeia* may be common on sandy shores, at some localities showing a zonation of species down the shore. **Bathyporeia pelagica** (Bate) is widely distributed around Britain and has been recorded from The Netherlands and the Channel coast of France. It is found from about mean tide level into the sublittoral on open coasts in clean, sandy deposits and, although unable to tolerate such low salinities as *B. pilosa*, it is often found at the mouths of estuaries. It has a similar life-cycle to *B. pilosa*, with a peak of breeding activity during the summer. **Bathyporeia sarsi** Watkin is common on the coast of Wales, south-west England and the North Sea at about the level of mean high water of neap tides and below. Where the species co-exist, it tends to be distributed between *B. pilosa* and *B. pelagica*. **Bathyporeia elegans** Watkin and **Bathyporeia guilliamsoniana** (Bate) are widely distributed on the lower shore and at some localities may be common. Both are found from about the level of mean low water neap tides into the sublittoral. Separation of these species is difficult and requires microscopic examination. Reference to Lincoln (1979) is recommended.

Family Haustoriidae
Solid, stout-bodied amphipods of very distinctive appearance. Pereopods 5, 6, 7 with segments, except the most distal one, flattened and very much expanded, with many setae and spines. Head with small rostrum. Antennae 1 and 2 short; antenna 1 with accessory flagellum. Gnathopods 1 and 2 not strongly developed. Pereopods 3–7 without terminal claw. Uropod 3 biramous. Telson bilobed.

Only one British species. Found buried in clean, sandy sediments; able to burrow into the sediment with great rapidity.

*Haustorius arenarius* (Slabber)    (Fig. 242)

*Body broad, stout, up to 13 mm in length; whitish colour. Eyes small, difficult to see. Antennae 1 slightly shorter than antennae 2, both relatively short. Pereopods 3–7 without a terminal claw; pairs 5, 6, 7 very much flattened. Sexes difficult to distinguish, but brood plates present in mature female. Animal of unmistakable and distinctive appearance.*

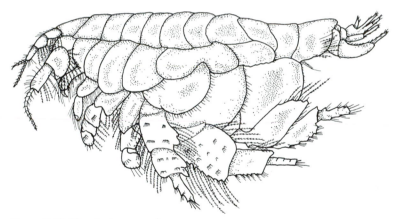

Figure 242 *Haustorius arenarius.*

*H. arenarius* is widespread in north-west Europe and is the only species of the genus found in Britain. It is commonly recorded in clean, sandy deposits on open coasts and at the mouths of estuaries, being most abundant on the lower shore. It is primarily a suspension feeder. When *Haustorius* leaves the sediment, it swims with the ventral surface uppermost and mating is believed to take place while the animals are swimming. Little is known of the breeding patterns of populations in Britain, but females with eggs are found in the summer months and longevity is believed to be two, possibly three, years.

Family Urothoidae
Solid, compact-bodied amphipods; laterally compressed. Head with small rostrum. Antennae 1 short; antennae 2 usually short in female, elongate in male. Antenna 1 with

accessory flagellum. Gnathopods 1 and 2 not strongly developed. Pereopods 3–7 with terminal claw. Pereopod 5 with horizontal rows of stout spines. Basal segments of pereopods 6 and 7 flattened and expanded. Uropod 3 biramous. Telson usually bilobed.

There is a single genus in Britain, found in clean, sandy sediments. It is common and widespread.

*Urothoe* spp.    (Fig. 243)

*Body mostly yellowish-white, with superficial resemblance to* Bathyporeia *(above) but without geniculate antenna 1; up to 8 mm in length. Head with very small rostrum. Eyes large in male and, in some species, they almost merge on dorsal surface of head. Antenna 1 with accessory flagellum. Antennae 1 and 2 shortish in female but antennae 2 nearly as long as body in male. Pereopod 5 spiny, noticeably different in appearance from other pereopods.*

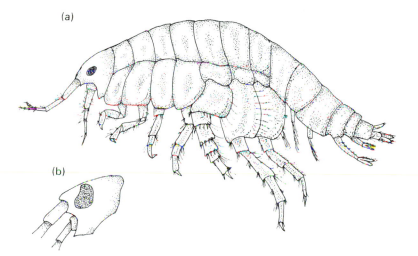

Figure 243  *Urothoe brevicornis*, (a) female, (b) head of male.

The genus *Urothoe* is widely distributed in clean, sandy sediments, some species extending into deep water. Of the intertidal species *Urothoe brevicornis* Bate is probably the best known. It is found from the middle shore to depths of 40 m and has also been recorded from The Netherlands and the north and west coasts of France. It reproduces in summer and autumn and, although little is known of its life-cycle, it probably lives for one to two years.

Family Oedicerotidae
Somewhat slender-bodied amphipods, slightly laterally compressed. Head usually with a large downwardly pointing rostrum; eyes large, extending onto dorsal surface of head. Antennae 1 and 2 shortish; antenna 1 without an accessory flagellum. Gnathopods usually subchelate, sometimes gnathopod 2 with thin, elongate terminal segments. Pereo-

pod 7 elongate and narrow, much longer than preceding pereopods. Uropod 3 biramous. Telson not divided.

Live buried in clean, sandy deposits, widespread and common.

**Pontocrates** spp.    (Fig. 244)

*Whitish body up to 7 mm in length. Head with conspicuous downwardly pointing rostrum. Eyes large, rounded. Antennae 1 slightly shorter than antennae 2 in female. Antennae 2 of male long, at least equal to body length in mature specimens. Gnathopod 1 well developed, subchelate; gnathopod 2 thin, elongate, with narrow process extending beyond the lower edge of the terminal claw. Pereopod 7 very long and narrow.*

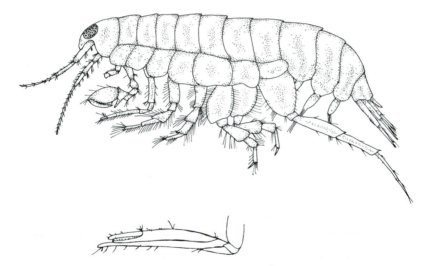

Figure 244  *Pontocrates altamarinus* female, with detail of gnathopod 2.

The genus *Pontocrates* is widely distributed in north-west Europe. Two species, **Pontocrates arenarius** (Bate) and **Pontocrates altamarinus** (Bate & Westwood), are commonly recorded from Britain. They are found from the middle shore into the sublittoral on clean, sandy beaches, and often occur in high densities. In populations of *P. arenarius* from Irish Sea coasts, breeding has been recorded throughout the year.

Family Calliopiidae

GENUS *APHERUSA*
Stout-bodied amphipods, laterally compressed; dorsal surface sometimes with tooth-like projections. Head with rostrum; eyes small or large. Antennae 1 shorter than antennae 2; antenna 1 without accessory flagellum. Gnathopods 1 and 2 subchelate, small; of similar size. Uropod 3 biramous; rami elongate, flattened, fringed with setae and spines. Telson large, not divided.

Living among seaweed; widespread and common.

*Apherusa jurinei* (Milne-Edwards)   (Fig. 245)

*Colour varied, pink, reddish, yellowish-brown; may be banded; sometimes with white patches. Up to 8 mm in length. Dorsal surface of body smooth. Eyes large, kidney-shaped. Antennae 1 slightly shorter than antennae 2; without an accessory flagellum. Gnathopods 1 and 2 subchelate, small.*

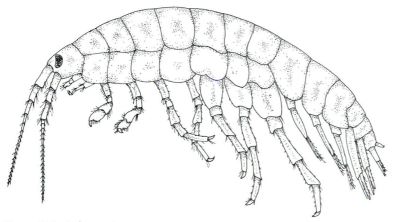

Figure 245   *Apherusa jurinei.*

A. *jurinei* is widely distributed in north-west Europe and Britain and is found among seaweed on the lower shore and in the shallow sublittoral.

Other species of *Apherusa* have been recorded from among seaweed on the lower shore and may be locally common. These can generally be distinguished from A. *jurinei* by the presence of tooth-like, dorsal projections on the posterior margins of the first two abdominal segments.

Family Atylidae

Stout-bodied amphipods, laterally compressed; dorsal surface of at least some segments of abdomen with tooth-like projections. Abdomen with only five visible segments. Head with downwardly pointing rostrum; eyes large, kidney-shaped. Antenna 1 shorter than antenna 2; antenna 1 without accessory flagellum. Gnathopods 1 and 2 small, subchelate, more or less equal in size or gnathopod 1 bigger than gnathopod 2. Uropod 2 markedly shorter than uropod 1. Uropod 3 biramous, large. Telson bilobed.

Live on lower shore in sandy deposits or among seaweed.

*Atylus swammerdami* (Milne-Edwards)   (*Nototropis swammerdami*) (Fig. 246)

*Body whitish in colour with brown patches; up to 10 mm in length. Head with small rostrum. Eyes large, kidney-shaped. Antennae 1 shorter than antennae 2. Two dorsal teeth on abdominal segment 4, the posterior one the larger; other thoracic and abdominal segments smooth dorsally. Second pair of uropods shorter than 1st.*

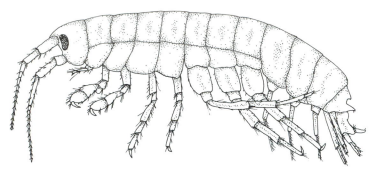

Figure 246  *Atylus swammerdami.*

*A. swammerdami* is widely distributed in north-west Europe and Britain. It is found mainly in sandy deposits on the lower shore and sublittorally.

Family Ampithoidae
Slender-bodied amphipods, slightly compressed laterally. Head without rostrum; eyes small. Antennae moderately long; antennae 1 and 2 same length or antennae 2 shorter; basal segment 3 of antenna 1 much smaller than basal segment 2. Antenna 1 without accessory flagellum or flagellum very small. Gnathopods 1 and 2 subchelate; gnathopod 2 the larger, more so in male. Uropod 3 biramous, rami shorter than basal region; inner ramus with terminal setae, outer ramus with one to three large hooked spines. Telson not divided; sometimes with hooks.

  Live in tubes of algae and detritus among seaweed and stones.

**Ampithoe rubricata** (Montagu)    (Fig. 247)

*Body reddish, brown, green, sometimes mottled with small, white spots; up to 20 mm in length. Eyes small, red. Antennae 1 about equal to or slightly longer than antennae 2. Gnathopods 1 and 2 large, subchelate; gnathopod 2 larger than 1, more so in males. Outer ramus of uropod 3 with 2 hooked spines and a group of setae. Telson without large hooked spines; posterior margin convex.*

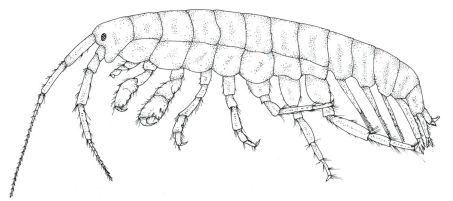

Figure 247  *Ampithoe rubricata.*

*A. rubricata* is widely distributed, often commonly occurring on the shores of north-west Europe and Britain. It lives in tubes which it builds from algae and detritus, and attaches to stones and seaweed on the middle and lower shore. Most tubes house only one animal but a few contain a male and a female and it is believed that fertilization occurs within the tube. Breeding occurs during the summer. *Ampithoe* feeds mainly on algae.

Family Corophiidae

GENUS *COROPHIUM*

Narrow, elongate body, not laterally compressed but dorso-ventrally flattened; coxal plates small, not touching one another. Last three abdominal segments often indistinct. Head with or without a rostrum; eyes small. Antennae well developed; antenna 1 narrow, shorter than or the same length as antenna 2; antenna 2 large and stout, especially so in male. Antenna 1 without accessory flagellum. Gnathopods 1 and 2 small, densely fringed with setae; gnathopod 2 with middle two segments joined along their length. Pereopod 7 much longer than preceding pereopods. Uropod 3 uniramous. Telson small, not divided.

The genus *Corophium* includes many species which live in burrows in soft sediments, or construct tubes among hydroids, sponges and laminarian holdfasts. They are found in marine, brackish and fresh water and may be locally abundant. Two species characteristic of mud and sand are described here.

Outer margin of basal region of uropod 1 with 10–12 stout spines; inner margin with 3–4 slender spines (Fig. 248b) .......................................... *Corophium volutator* (p. 377)
Outer margin of basal region of uropod 1 with 11–15 spines and 1–5 long, proximal setae; inner margin with a single spine (Fig. 248c) ........... *Corophium arenarium* (p. 378)

*Corophium volutator* (Pallas)    (Fig. 248a, b, d)

*Body not laterally compressed; up to 10 mm in length. Head with very small pointed rostrum. Antennae 1 without accessory flagellum, shorter than antennae 2 which are large and stout, particularly in males, and characteristically extended in front of animal. Basal segment 3 of antenna 1 much shorter than basal segment 1. Gnathopods 1 and 2 small, with many setae. Coxal plates small and separate. Last pair of pereopods much longer than preceding pairs. Last 3 abdominal segments distinct. Outer margin of basal segments of uropod 1 with 10–12 stout spines; inner margin with 3–4 slender spines.*

*C. volutator* is widely distributed in north-west Europe and Britain. It lives in U-shaped burrows in muddy, estuarine sediments, tolerating salinities down to about 2‰, and when present in high densities the apertures of the burrows are clearly visible on the surface of the sediment. *Corophium* is a selective deposit feeder but also feeds by filtering particles from the respiratory current drawn through the burrow by the action of the pleopods. On the west coast of Wales, breeding takes place from April to October and although the animals are known to swim on spring tides, mating takes place in the

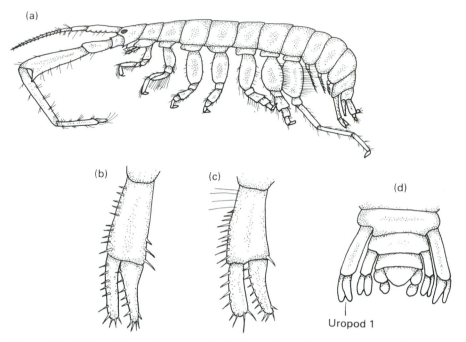

Figure 248  (a) *Corophium volutator*, male. (b) *Corophium volutator*, uropod 1. (c)
*Corophium arenarium*, uropod 1. (d) Posterior region of abdomen of
*Corophium arenarium/volutator*, setae omitted.

burrow. Adult males crawl over the surface of the moist sediment as the tide recedes in
search of burrows occupied by mature females. Juveniles are released from the brood
chamber after about 14 days, and development is synchronized with spring tides. Each
female produces a number of broods.

### *Corophium arenarium* Crawford    (Fig. 248c, d)

*Body not laterally compressed; up to 7 mm in length. Closely resembles* C. volutator *(above)
but outer margin of basal region of uropod 1 has 11–15 spines and 1–5 long setae; inner
margin with single spine.*

*C. arenarium* is similar in its distribution, ecological requirements and life-cycle to *C.
volutator*, but lives in muddy-sand, compared with the more glutinous muds preferred
by the latter species. Records of its distribution round Britain are patchy, largely as a
result of misidentification. Nevertheless, it is believed to be widely distributed, having a
slightly narrower salinity tolerance than *C. volutator*. The vertical distribution of the two
species appears to be determined by the presence or absence of a suitable sediment rather
than height on the shore. Both *C. volutator* and *C. arenarium* are frequently recorded in
very high densities on mud-flats, where they are preyed on extensively by wading birds.

Family Cheluridae

Amphipods of very distinctive appearance; flattened dorso-ventrally. Antenna 1 shorter than antenna 2; antenna 2 large, stout, of stumpy appearance. Antenna 1 with very small accessory flagellum. Gnathopods small; gnathopods 1 and 2 similar size. Uropods modified, and very different from one another, uropod 3 very large. Telson not divided. Found boring in timber.

**Chelura terebrans** Philippi    (Fig. 249)

*Body not laterally compressed. Light brown; up to 6 mm in length. Antennae 1 shorter than antennae 2; antennae 2 stout and densely covered with setae. Posterior margins of body segments with many setae giving overall 'hairy' appearance. Third abdominal segment extended backwards into curved mid-dorsal projection, larger in males than females. Uropods 1, 2 and 3 dissimilar, 2nd and 3rd pairs much modified and show pronounced sexual dimorphism.*

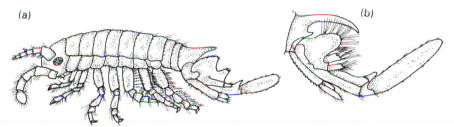

Figure 249   *Chelura terebrans*, (a) female, (b) posterior region of male (after Sars, 1895).

The genus *Chelura* is represented in north-west Europe by a widely distributed single species, *C. terebrans*. It is found in timbers which have been attacked by *Limnoria* (p. 354), and populations of *Chelura* will not survive any length of time if *Limnoria* is not present. It feeds predominantly on its own faecal pellets but also ingests wood particles and the faeces of *Limnoria*.

Family Ischyroceridae

Amphipods somewhat flattened dorso-ventrally. Head usually without a rostrum. Antennae well developed, usually densely fringed with setae; basal segment 3 of antenna 1 nearly as long as basal segment 2; antenna 1 with or without small accessory flagellum. Gnathopods well developed; gnathopod 2 larger than gnathopod 1 and in male usually very well developed with a distinctive claw. Uropod 3 biramous or uniramous, rami much shorter than basal region; without terminal setae; with hooked terminal spine or small teeth or both. Telson usually not divided; without terminal hooks.

    Live in tubes among hydroids, seaweeds, mooring ropes and buoys.

*Jassa* spp.   (Fig. 250)

*Body slender; up to 13 mm in length. Colour varied, often greyish with brownish and red-dish patches. Antennae 1 about ⅔ length of antennae 2 and with small accessory flagellum. Antennae 2 large, particularly in male. Gnathopod 2 larger than gnathopod 1, terminating in characteristic claw which differs in size and form between sexes. Uropod 3 biramous; outer ramus with hooked spine and 1–3 small teeth.*

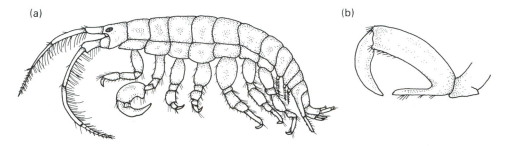

Figure 250  *Jassa falcata*, (a) female, (b) gnathopod 2, male (after Sars, 1895).

The genus *Jassa* has a worldwide distribution and in some localities occurs in high densities. It is widely distributed in north-west Europe. ***Jassa falcata*** (Montagu) is one of the most familiar species but its polymorphism has caused taxonomic problems. Although recorded from rock pools, typically it is found attached to mooring ropes, buoys, hydroids and seaweeds. Males and females live in separate tubes constructed of pieces of debris cemented together, the tubes often forming dense mats, particularly in warm water discharge pipes from power stations. Reproduction occurs throughout the year, and males are believed to seek out mature females. Females produce a succession of broods.

Suborder Hyperiidea

***Hyperia galba*** (Montagu)   (Fig. 251)

*Dorsal margin of body markedly curved, less so in male; up to 14 mm in length. Eyes very large, bulbous, greenish in colour. Antennae 1 and 2 very small in female; long and whip-like in male and reaching to at least ½ body length.*

*H. galba* is widely distributed in north-west Europe and extends south to about the English Channel. It is widely distributed in Britain and can often be found in the cavities on the undersurface of jellyfishes, particularly *Rhizostoma* (p. 124), and occasionally on the ctenophore *Beroë* (p. 143), where it clings by means of the last three pairs of thoracic appendages. Its diet includes the tissues of the host, including the gonads of mature jellyfishes, as well as prey captured by the host. In Danish waters breeding has been recorded from July to December. Mature females carry large numbers of embryos; when

(a)

(b)

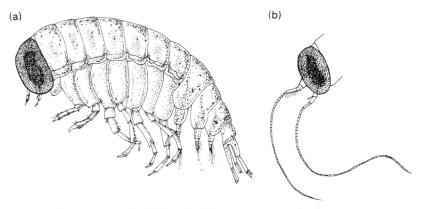

Figure 251  *Hyperia galba*, (a) female, (b) anterior region of male.

newly released, the juveniles are unable to swim and remain within the protection of the jellyfish. After a few moults, the pleopods and uropods are fully developed and the juveniles are capable of swimming away to seek a new host. While in association with the jellyfish, the body of the amphipod is almost transparent, but when free-swimming it is a reddish-brown colour.

Suborder Caprellidea

**Caprella** spp.   Ghost shrimp (Fig. 252)

*Body very elongate and narrow; up to 20 mm in length; reddish-brown. Antennae 1 longer than antennae 2. Gnathopod 2 often larger in male than female and differing in shape. Female less elongate than male and carries brood chamber on thoracic segments 3 and 4. In both sexes these segments lack pereopods. Abdomen minute.*

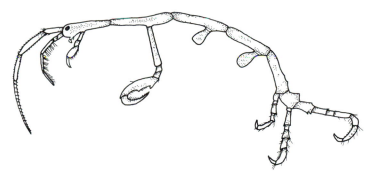

Figure 252  *Caprella linearis*, male (after Sars, 1895).

**Caprella linearis** (Linnaeus) is probably the best-known caprellid in British waters. It is widely distributed on the lower shore among seaweeds, hydroids and sponges, where it is found clinging by means of the well-developed posterior, thoracic appendages. It crawls in a characteristic looping movement achieved by alternately grasping and releasing the substratum with anterior and posterior appendages. Breeding probably occurs during autumn and winter; longevity is believed to be about one year.

### Order Decapoda

The decapods are perhaps the best-known crustaceans and include the commercially important prawns, shrimps, lobsters, crayfish and crabs. The anterior three pairs of thoracic appendages, the maxillipeds, are used in feeding, while the other five pairs are the pereopods, or 'walking legs'. The first pair of pereopods are known as chelipeds and often terminate in large pincers or chelae. The head and thorax are protected by a well-developed carapace which extends laterally to protect the gills. As a rule the first five abdominal segments each have a pair of pleopods, which are sometimes known as swimmerets, and, as the name implies, are in some groups used in locomotion. In females the eggs are held on the pleopods until hatching. In prawns, shrimps and lobsters, the uropods form a tail fan with the telson, while in the true crabs the abdomen is reduced in size and folded under the thorax.

Most decapods have separate sexes and elaborate courtship behaviour often takes place. In many cases the female undergoes a moult, the precopulatory moult just before mating. In prawns, shrimps and lobsters fertilization is external, the eggs being fertilized as they leave the female, but in the true crabs it is internal, sperm being transferred to the spermathecae of the female. The fertilized eggs are attached by sticky secretions to fine hairs on the pleopods of the female where they may be held for several months. During this time the female is referred to as 'berried'. The eggs hatch as pelagic larvae.

### Key to the different groups of decapods

The decapods are a large, diverse order of crustaceans. A useful first step in identification is separation on the basis of habit. The prawns and shrimps, which have a swimming habit, are known as the Natantia, while the lobsters and crabs, of characteristic walking habit, are known as the Reptantia. Classification of the order is complex but the species described here can be subdivided into a number of readily identifiable groupings using the characters given below.

1 Exoskeleton relatively thin and light; not heavily calcified. Abdomen extended. Swimming habit – the Natantia ............................................. **Prawns and shrimps** (p. 383)
Exoskeleton generally hard; usually heavily calcified. Abdomen extended or folded under thorax. Walking habit – the Reptantia. Some occupy empty gastropod shells ........ 2

2 Abdomen extended, terminates in a well-developed tail fan; or abdomen extended, asymmetrical, soft and twisted ................................................................................. 3
Abdomen reduced and folded, either tightly or loosely, under thorax; symmetrical .......... 4

3 Abdomen extended, terminates in a well-developed tail fan. First 3 pairs of pereopods with chelae, those of the 1st pair very large ............................................. **Lobsters** (p. 389)
Abdomen extended; asymmetrical, soft and twisted. Animal lives in empty gastropod shell ................................................................................. **Hermit crabs** (p. 393)

4 Fifth pair of pereopods much the same size as preceding 3 pairs. Abdomen small, tightly folded under thorax ................................................................... **True crabs** (p. 397)

Fifth pair of pereopods very much smaller than preceding pairs. Abdomen tightly or loosely folded under thorax ................................................................................... 5

5 Abdomen much reduced; tightly folded under thorax. Carapace rounded ...................... ................................................................................................................ **Porcelain crabs** (p. 392)

Abdomen slightly reduced; folded loosely under thorax. Carapace longer than broad .... ................................................................................................................ **Squat lobsters** (p. 391)

In addition to the groups described above, the order also includes stone crabs of deep water and the burrowing shrimps found in burrows at low water and into the sublittoral. Neither of these groups is included here.

### Prawns and shrimps

The words 'prawn' and 'shrimp' lack precise definition but popular usage has led to the larger species being referred to as 'prawns' and the smaller as 'shrimps'. They are distinguished from other decapods by their swimming habit. The exoskeleton is relatively thin and light and the abdomen is extended. The second antennae each bear a plate-like extension which gives stability during swimming. The carapace is often extended between the eyes to form a spine or rostrum, particularly in those animals referred to as prawns. The pleopods are the organs of propulsion. Most prawns and shrimps have separate sexes, and after mating the eggs are carried by the female for varying periods of time before hatching into planktonic larvae known as zoeae. Prawns and shrimps eat a wide variety of food, both plant and animal, and are preyed on extensively by fishes and birds.

### Key to families of prawns and shrimps

1 First pair of pereopods subchelate (Fig. 257); rostrum reduced to a spine or is very short ................................................................................................. **Crangonidae** (p. 388)

First pair of pereopods with or without chelae ................................................................ 2

2 First pair of pereopods with chelae. Pereopod 2 without 2 or more small segments next to chela ................................................................................................. **Palaemonidae** (p. 383)

Pereopod 2 with 2 or more small segments next to chela ................................................. 3

3 Pereopods 1 and 2 clearly chelate; chela of pereopod 1 not greatly enlarged .................. ................................................................................................................ **Hippolytidae** (p. 385)

Pereopod 1 with tiny chela (seen under magnification) or chela absent; pereopod 2 with tiny chela ................................................................................................. **Pandalidae** (p. 387)

### Family Palaemonidae

Rostrum conspicuous (in specimens included here) with teeth on dorsal and ventral margins. Pereopods 1 and 2 terminate in chelae, chela of pereopod 2 being the larger.

1  Rostrum curves upwards; teeth on dorsal margin do not extend to tip of rostrum but
   are restricted to basal ⅔  ........................................................ *Palaemon serratus* (p. 385)
   Rostrum straight, or only slightly curved upwards; teeth all along dorsal margin ............. 2
2  Rostrum with 4–6 dorsal teeth, one of which is posterior to the orbit; 2 ventral teeth  ...
   .......................................................................... *Palaemonetes varians* (p. 385)
   Rostrum with 7–9 dorsal teeth, 3 (sometimes 2) of which are posterior to the orbit; 3
   (occasionally 2 or 4) ventral teeth  .............................................. *Palaemon elegans* (p. 384)

**Palaemon elegans** Rathke    (*Leander squilla*), Prawn (Fig. 253a)

*Semi-transparent, often with brownish markings. Up to 60 mm in length. Rostrum more or
less straight with 7–9 teeth along dorsal edge, 3 (sometimes 2) of which are posterior to the
orbit; (3 occasionally 2 or 4) teeth along ventral edge of rostrum. Tip of rostrum usually
divided into two. Antennae 1 triramous. First and 2nd pereopods terminate in chelae, those
of the 2nd pair being the larger. Lateral margin of telson with 2 pairs of spines.*

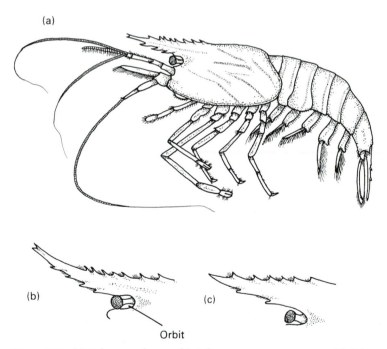

(a)

(b)

(c)

Orbit

Figure 253  (a) *Palaemon elegans*. (b) *Palaemon serratus*, rostrum. (c) *Palaemonetes varians*,
        rostrum.

*P. elegans* is widely distributed round Britain and is found in rock pools from the middle
shore into the shallow sublittoral, but is known to tolerate salinities down to 6‰ and
has been recorded from lagoons. In northern parts of Britain it migrates into deeper water
in winter. Breeding has been recorded from April to September and the eggs are carried

for about six weeks. Some females produce two broods a year and longevity is two to two and a half years. *Palaemon* feeds on a wide range of organisms, including algae and crustaceans.

**Palaemon serratus** (Pennant)    (*Leander serratus*), Common prawn (Fig. 253b)

*Similar to P. elegans (above). Pinkish-brown, marked with reddish lines and spots. Up to 100 mm in length, generally about 60 mm. Rostrum curves upwards, usually with tip divided into two; 6 or 7 teeth along dorsal edge of rostrum restricted to basal ⅔ of length; 2 of dorsal teeth posterior to orbit. Four or 5 teeth along ventral edge of rostrum. Antennae 1 triramous. First and 2nd pairs of pereopods terminate in chelae, those of 2nd pair the larger. Lateral margin of telson with 2 pairs of spines.*

*P. serratus* extends as far north as Denmark, commonly occurring in Britain on south and west coasts but less so on the east coast. It is found among seaweed in rock pools on the lower shore, where it can often be caught in large numbers on the rising tide using a hand net. It occurs offshore to depths of about 40 m and is fished commercially by trawl on the south and west coasts of Britain. Breeding occurs from about November to June; a single female can carry as many as 4000 eggs which are brooded for up to four months. Young settle from the plankton in July/August and reproduce the following February. Longevity is about three to four years. Most of the population move offshore in winter and return in spring. It is omnivorous. Some individuals of this species and of *P. elegans* (above) are infected by a parasitic isopod, **Bopyrus squillarum** Latreille, which gives rise to pronounced swellings on the sides of the carapace, a condition known as 'face ache'.

**Palaemonetes varians** (Leach)    (Fig. 253c)

*Body almost transparent; faint brown pigment lines. Up to 50 mm in length, generally 20–30 mm. Rostrum straight, tip usually undivided. Number of teeth on dorsal and ventral margins of rostrum varies, but majority have 4–6 dorsal teeth, one of which is posterior to the orbit, and 2 ventral teeth. Antennae 1 triramous. First 2 pairs of pereopods with chelae, those of 2nd pair the larger. Lateral margin of telson with 2 pairs of spines.*

*P. varians* is widely distributed in north-west Europe but rare in parts of northern Scotland. It is found only in areas of reduced salinity, down to 1‰, and is characteristic of brackish water and estuaries. It is omnivorous, feeding on detritus, diatoms and invertebrates. Breeding occurs from April to July.

Family Hippolytidae
Rostrum conspicuous, with dorsal and ventral teeth. Pereopods 1 and 2 with chelae. Chela of pereopod 1 not greatly enlarged. Pereopod 2 with two or more small segments next to chela.

Three small segments next to chela of pereopod 2. Rostrum about equal to carapace length; with 2 widely separated teeth on dorsal margin and 2 teeth on ventral margin .. ................................................................................ *Hippolyte varians* (p. 386)
Seven small segments next to chela of pereopod 2. Rostrum short; with 2–5 teeth on dorsal margin ................................................................ *Eualus pusiolus* (p. 386)

### *Eualus pusiolus* (Krøyer)    (Fig. 254)

*Semi-transparent, flecked with various colours. Up to 28 mm in length. Rostrum short; straight or slightly curved downwards, tip almost always undivided; 2–5 teeth along dorsal margin. Antennae 1 short, biramous. Carapace with 1 spine beneath eye. First and 2nd pair of pereopods terminate in chelae; 1st pair shorter than others; 7 small segments next to chela of pereopod 2. Lateral margins of telson with 4 or 5 pairs of spines.*

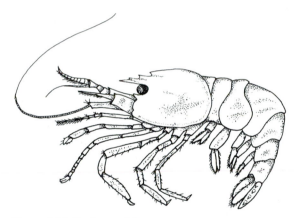

Figure 254  *Eualus pusiolus.*

*E. pusiolus* is widely distributed in north-west Europe and Britain, where it is commoner in the north. It occurs under stones and among seaweed (especially *Laminaria*) on the lower shore and into the subittoral to depths of about 500 m. Breeding occurs throughout the year.

### *Hippolyte varians* Leach    Chameleon prawn (Fig. 255)

*Green, red, brown in colour, but sometimes transparent with red, yellow markings; up to 32 mm in length, usually less. Rostrum nearly equal to carapace length; more or less straight with 2 widely separated teeth on dorsal margin; 2 teeth on ventral margin; tip undivided. Antennae 1 short, biramous. Carapace with single spine above eye, 2 spines below eye. First pair of pereopods shorter than others; 1st and 2nd pairs with chelae; 3 small segments next to chela of pereopod 2. Lateral margins of telson with 2 pairs of spines.*

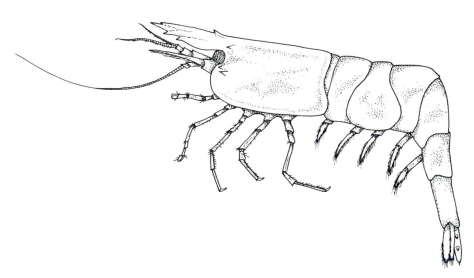

Figure 255  *Hippolyte varians.*

*H. varians* is widely distributed in north-west Europe but is commoner in the south than the north of Britain. It occurs on the lower shore in rock pools and under seaweed, extending to depths of 50 m. Breeding occurs throughout the year.

Family Pandalidae
Rostrum conspicuous, in some cases longer than carapace and with marked upward curve; with dorsal and ventral teeth. Pereopod 1 with tiny chela or chela absent; pereopod 2 with tiny chela and with two or more, sometimes many more, small segments next to chela.

***Pandalus montagui*** Leach    Aesop prawn, Pink shrimp (Fig. 256)

*Semi-transparent, pink colour; generally 40–50 mm in length; may be much longer. Rostrum long, curves markedly upwards, tip divided into two; 10–12 teeth on basal half of dorsal edge of rostrum, 4 of which usually posterior to the orbit; 5 or 6 teeth on ventral edge of rostrum. Carapace with spine just beneath eye. Antennae 1 biramous; antennae 2 longer than body and marked with light and dark bands. First pair of pereopods with tiny chelae; 2nd pair of pereopods of unequal size, left larger than right. Right pereopod 2 with 20–22 small segments next to chela. Lateral margin of telson with 5–7 pairs of spines.*

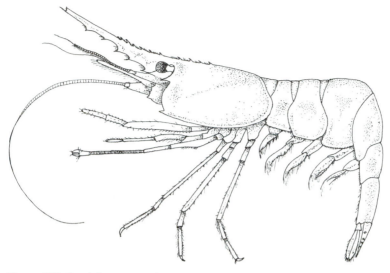

Figure 256  *Pandalus montagui.*

*P. montagui* is a northern species extending as far south as the English Channel. It occurs all around the coast of Britain extending to depths of 100 m or more, but is occasionally found in rock pools on the lower shore. Although some individuals are found in relatively shallow water throughout the greater part of the year, offshore migration takes place in most areas in October/November, followed by a migration into shallow water, including the mouths of estuaries, in the spring. In the sand and mud in shallow waters off the Wash, the south-east coast, Morecambe Bay and the Solway Firth, densities are high enough to support commercial exploitation. About 50% of a population are protandrous hermaphrodites becoming female at an age of about 13–16 months; the remainder are female and stay as such throughout life. Individuals mature and breed in the first year of life and in the North Sea eggs are laid from November to February and hatch in April/ May; young settle in July/August. Each female lays a single batch of eggs in a breeding season and length of life is usually three to four years. *Pandalus* feeds mainly on hydroids, small crustaceans and polychaetes.

Family Crangonidae – shrimps
Rostrum reduced to a small spine (as in species described here) or very short. Pereopod 1 more heavily built than others; subchelate.

**Crangon crangon** (Linnaeus)    (*Crangon vulgaris*), Common shrimp (Fig. 257)

*Brownish in colour, flecked; generally 30–50 mm in length, but up to 90 mm. Rostrum reduced to small spine. Antennae 1 biramous; antennae 2 about ¾ body length. First pair of pereopods more heavily built than others, subchelate; 2nd pair reach to ¾ the distance along large penultimate segment of pereopod 1 (Fig. 257), with tiny chelae. Telson with 2 pairs of small, lateral spines; dorsal surface of 6th abdominal segment smooth, not grooved.*

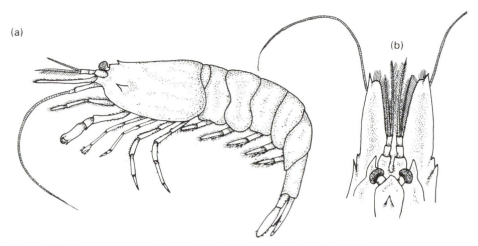

Figure 257 *Crangon crangon* (a), with dorsal view of anterior of head (b).

*C. crangon* is very common in European waters, buried in sand from the middle shore to depths of 150 m. It extends into estuaries and can be caught in large numbers using a push net on the rising tide, and in the gullies and creeks left after the tide has receded. It is fished commercially in some areas, for example, Morecambe Bay. The sexes are separate, although there are suggestions that the species is a protandrous hermaphrodite. Maturity is reached after one to two years and the breeding period differs according to locality. In the Bristol Channel breeding has been recorded in spring and summer, while in other localities three breeding periods have been recorded. Eggs are carried for three to four weeks during summer and 10 weeks in spring, with larval life lasting approximately five weeks. Average longevity is three years and it has been estimated that a female spawning three times a year will produce over 30 000 eggs. *Crangon* feeds on a wide variety of invertebrates, such as polychaete worms, crustaceans and molluscs, and it is itself preyed on by a wide range of fishes.

A second species of *Crangon*, **Crangon allmanni** Kinahan, is found offshore over sandy and muddy sediments. It is very similar to *C. crangon* but can be distinguished by the presence of longitudinal grooves on the dorsal surface of the 6th abdominal segment.

Lobsters
Family Nephropidae
Exoskeleton heavily calcified. Abdomen extended and ends in a well-developed tail fan. The first three pairs of pereopods have chelae, the first pair large.

**Homarus gammarus** (Linnaeus)    Common lobster (Fig. 258)

*Body up to 400 mm or more in length; specimens found intertidally generally much smaller. Dark blue in colour. First pair of pereopods very large, held forward, terminate in very large chelae; unequal in size, one used for cutting, the other for crushing prey; 2nd and 3rd pereopods with small chelae. Uropods and telson form prominent tail fan.*

*G. squamifera* is the most common of the intertidal squat lobsters and is widely distrib-uted in north-west Europe. It is a suspension feeder living under stones on the lower shore and extending to depths of about 70 m. In the English Channel breeding occurs in winter and spring.

Other species of *Galathea* are sublittoral, occasionally being found under stones on the lower shore. **Galathea strigosa** (Linnaeus) is a large, red-coloured species (up to 90 mm body length) with striking blue lines across the carapace. The first pair of pereopods have very many more spines than *G. squamifera*.

### Porcelain crabs
### Family Porcellanidae
Abdomen much reduced, tightly folded under thorax. Carapace rounded. Porcelain crabs move sideways and have general appearance of true crabs (p. 397) but can be dis-tinguished by the much reduced fifth pair of pereopods which are folded under the cara-pace, and the long antennae 2.

First pair of pereopods long and broad, with well-developed chelae. Pereopods, edge of carapace and outer edge of chelae with many setae ............ **Porcellana platycheles** (p. 392)
First pair of pereopods long and slender. Pereopods, edge of carapace and outer edge of chelae without setae ............................................................. **Pisidia longicornis** (p. 393)

**Porcellana platycheles** (Pennant)    Broad-clawed porcelain crab (Fig. 260a)

*Rounded carapace about 10 mm across; brownish colour. Antennae 2 long. First pair of pereopods long, broad, flattened; have well-developed chelae; 5th pair of pereopods much reduced. Abdomen folded under thorax. Pereopods, edge of carapace, and outer edge of chelae with many setae.*

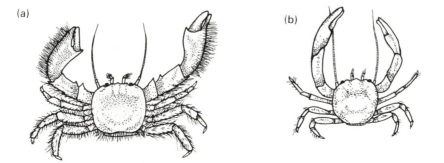

Figure 260  (a) *Porcellana platycheles*. (b) *Pisidia longicornis*.

*P. platycheles* is widely distributed, reaching its northern limit in Shetland. It is common under stones on the middle and lower shore, particularly those among mud and gravel; although essentially an intertidal species, it is found in the shallow sublittoral. It is a suspension feeder. In the English Channel and South Wales, egg-bearing females have been recorded from March/April to August. Maturity is reached at a carapace width of about 4 mm and females carry eggs during their second year on the shore.

**Pisidia longicornis** (Linnaeus)  (*Porcellana longicornis*), Long-clawed porcelain crab (Fig. 260b)

*Similar in appearance to* P. platycheles *(above), but smaller; body shiny, not heavily clothed with setae. First pair of pereopods long and slender.*

*P. longicornis* is widely distributed in north-west Europe and is common under stones at low water and in laminarian holdfasts. It is found in the sublittoral to depths of about 90 m. Egg-bearing females have been recorded in spring and summer. Maturity is reached at a carapace width of about 3.5 mm and females carry eggs during their second year on the shore, possibly surviving to breed two or three times. It is a suspension feeder.

### Hermit crabs

Hermit crabs live in the empty shells of a variety of prosobranch molluscs. The abdomen is soft and twisted, allowing it to fit into the coiled shell. The crab is held securely by the specially adapted left uropod which is hook-like and grips the inside of the shell, while the fourth and fifth pereopods are held against the wall of the shell to give further purchase. The pleopods have been lost in males but are retained on the left side of the body in females and are used to carry the eggs. The first pair of pereopods are generally large and equipped with chelae. The fourth and fifth pairs are usually small. Only the anterior part of the carapace is calcified and this is often referred to as the hard carapace (Fig. 262). When disturbed, hermit crabs retreat completely into the gastropod shell. When the crab becomes too big for the shell, it changes to another. A new shell is not occupied until it has been thoroughly tested for weight and size, and in choosing a new shell the crab goes through a complex ritual of behaviour which can be observed in aquarium tanks if the crabs are kept with a supply of empty shells. Empty shells on the shore are clearly a valuable resource.

There are 2 families of hermit crabs in British waters. These can generally be identified according to the size of the chelae.

Chelae about equal in size, or left chela larger than right .................... **Diogenidae** (p. 394)

Right chela larger than left ................................................................ **Paguridae** (p. 394)

Family Diogenidae
***Diogenes pugilator*** (Roux)    (Fig. 261)

*Bluish in colour. Carapace up to 10 mm in length; surface with a few groups of setae. Antennae 2 with many setae. First pair of pereopods with chelae, the right with many setae. Left chela larger than right and each with white tip.*

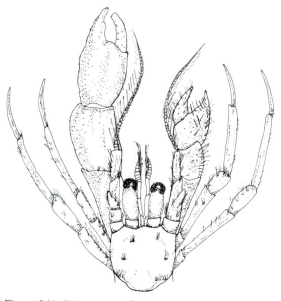

Figure 261  *Diogenes pugilator*, anterior of body.

*D. pugilator* reaches its northern limit at Anglesey, North Wales. It is found at extreme low water and below, in clean, sandy sediments. In south-west Britain it is locally abundant during some years and living specimens are occasionally washed ashore in large numbers at the edge of the incoming tide; the crabs quickly bury into the sediment. Egg-bearing females are found from June to September, each carrying two broods of eggs per year.

Family Paguridae

1  Hard carapace with almost straight lateral margins; roughly triangular in shape; smooth surface ............................................................ ***Anapagurus hyndmanni*** (p. 397)
   Hard carapace with convex lateral margins ..................................................................... 2
2  Anterior (rostral region) of carapace pointed. Right chela with 2 longitudinal rows of large tubercles plus many smaller tubercles. Often with the sea-anemone *Calliactis parasitica* (p. 137) fixed to the shell ....................................... ***Pagurus bernhardus*** (p. 395)
   Anterior (rostral region) of carapace rounded. Right chela with fine tubercles. Almost always found with the sea-anemone *Adamsia carciniopados* (p. 137) wrapped around the shell ..................................................................... ***Pagurus prideaux*** (p. 396)

***Pagurus bernhardus*** (Linnaeus)    (*Eupagurus bernhardus*), Common hermit crab (Fig. 262)

*Carapace red, yellow or brownish colour, lateral margins of hard carapace convex; rostral region pointed; up to 35 mm in length, intertidal specimens usually about 12 mm. Well-developed chelae on 1st pair of pereopods, right larger than left and with many tubercles. Red line down centre of right chela, to each side of which is row of large tubercles. Terminal segment of 2nd and 3rd pairs of pereopods sharply pointed.*

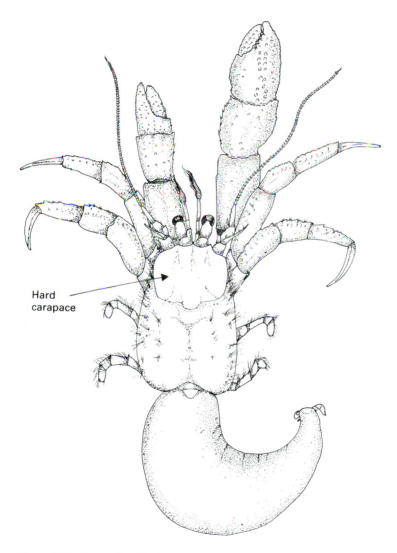

Figure 262  *Pagurus bernhardus.*

*P. bernhardus* is widely distributed in north-west Europe extending to depths of 140 m. Small specimens are abundant on most British shores during spring and summer, their frequent occurrence in the shells of *Littorina littorea* (p. 238), *Littorina obtusata* (p. 239) and *Nucella lapillus* (p. 255) no doubt reflecting the availability of these shells. In deeper waters offshore, shells of other species are occupied. Egg-bearing females have been recorded throughout the year in sublittoral populations, with a peak in January and February. Small females probably carry a single brood per year, but larger specimens are believed to produce multiple broods. The eggs are carried for up to two months and the pelagic larval stage lasts several weeks. Most individuals reach maturity in the first year at about 4 mm carapace length and although longevity is difficult to ascertain, the largest specimens are believed to be at least three years old. The hermit crab is a scavenging omnivore but also filters small particles from the water. Shells occupied by *P. bernhardus* are often colonized by a variety of animals such as barnacles, polychaete worms and sponges. Familiar associations are those with *Hydractinia echinata* (p. 104) and *Nereis fucata* (p. 179), but by far the best known is that between the crab and the anemone *Calliactis parasitica* (p. 137). A parasitic barnacle of the genus *Peltogaster* (p. 343) is sometimes found under the abdomen of the crab, and appears as a smooth, yellowish mass. It should not be confused with the crab's eggs.

**Pagurus prideaux** Leach    (*Pagurus prideauxi*) (Fig. 263b)

*Reddish-brown carapace up to 20 mm in length, lateral margins of hard carapace convex; rostral region rounded. Right chela with fine tubercles; larger than left. Second and 3rd pairs of pereopods grooved longitudinally, terminal segments sharply pointed.*

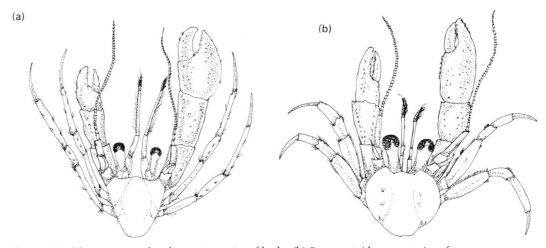

Figure 263  (a) *Anapagurus hyndmanni*, anterior of body. (b) *Pagurus prideaux*, anterior of body (modified after Ingle, 1992).

*P. prideaux* reaches its northern limit in southern Norway but is apparently absent from Denmark and the southern North Sea. It is very occasionally found on the shore but is more common offshore to depths of about 30 m. This hermit crab is almost always found in association with the sea-anemone *Adamsia carciniopados* (p. 137). Apparently, the sea-anemone cannot reach sexual maturity without the crab, although the crab can live without the anemone, at least under aquarium conditions. Breeding females have been found throughout the year and most produce two broods a year.

### *Anapagurus hyndmanni* (Bell)    (Fig. 263a)

*Carapace up to about 7 mm in length; hard carapace roughly triangular in shape, with straight lateral margins, surface smooth. Well-developed chelae on 1st pair of pereopods; right larger than left and without, or with very few, setae. Right chela usually whitish in colour; much wider than other segments of that appendage.*

*A. hyndmanni* extends as far north as Shetland and is widely distributed in Britain, but is apparently absent from southern North Sea coasts. It occurs on the lower shore and extends into the sublittoral. On the west coast of Ireland, breeding has been recorded from March to September and most females carry two broods of eggs a year.

**True crabs**

In true crabs the abdomen is tightly folded under the carapace. The first pair of pereopods is armed with chelae; the fifth pair is similar in size to the preceding three pairs. The sexes are separate and the abdomen of the female is usually broader than that of the male. Only the first two pairs of pleopods are present in males and these are modified for the transference of sperm during copulation. Sperm are stored in the spermathecae of the female and fertilization is internal. The eggs are carried on the pleopods of the female for a varying length of time before hatching as planktonic larvae known as zoeae. These undergo several moults, the exact number depending on species, before developing into a second larval type, the megalopa. The megalopa is planktonic and moults into a benthic, juvenile crab.

Soft crabs, known as peelers, are occasionally found on the shore, the soft, pliable exoskeleton indicating that the crab has recently moulted. The skeleton requires a few days to harden; while soft the crab seeks shelter under rocks and stones to avoid predators.

### Key to families of true crabs

The measurements of carapace length and width are shown in Fig. 264.

1   Carapace longer than wide; with or without bifid rostrum between eyes .......................... 2
    Carapace wider than long; without bifid rostrum  ............................................... 3

2 Carapace with pronounced, bifid rostrum; roughly triangular in shape; broader towards posterior. Antennae 2 not densely fringed with setae. May be spider-like in appearance ...................................................................................... **Majidae** (p. 405)

Carapace with 3, sometimes 4, sharp teeth on lateral margins; 2 teeth between eyes. Antennae 2 longer than width of carapace, densely fringed with setae     **Corystidae** (p. 398)

3 Terminal segment of pereopod 5 flattened and usually fringed with setae; often segment is very much expanded and used in swimming. From 5 to 9 sharp teeth on antero-lateral margins of carapace ......................................... **Portunidae** (p. 399)

Terminal segment of pereopod 5 not flattened ................................................. 4

4 Region of carapace between eyes with 3 lobes. Carapace much wider than long, with up to 10 rounded lobes on anterior and lateral margins, giving pie-crust effect ...................
................................................................................................ **Cancridae** (p. 402)

Carapace not as above ........................................................................... 5

5 Carapace rounded, smooth; region between eyes not lobed or toothed. Small crab living as a commensal in mantle cavity of bivalve molluscs, particularly the common mussel (*Mytilus edulis*) (p. 283) ...................................... **Pinnotheridae** (p. 405)

Region of carapace between eyes broad, with 2 broad lobes, sometimes finely serrated. Crab not living as a commensal ......................................... **Xanthidae** (p. 403)

Family Corystidae – masked crabs
Carapace longer than broad, with two teeth between the eyes and three or four teeth on lateral margins. Antenna 2 longer than carapace width, densely fringed with setae.


*Corystes cassivelaunus* (Pennant)    Sand crab, Masked crab (Fig. 264)

*Reddish-brown to yellow carapace; up to 40 mm in length, 25 mm across; 2 teeth between eyes and 3, sometimes 4, teeth on lateral margins of carapace. Antennae 2 longer than carapace width, heavily clothed with setae. First pair of pereopods with chelae and about twice length of carapace in male, equal to carapace length in female.*

*C. cassivelaunus* is the only member of the genus. It is widely distributed on sandy beaches from low water of spring tides to depths of about 55 m. Large numbers of dead and dying specimens are often found on the strandline after storms. It extends as far north as the west coast of Sweden although absent from much of the north of Scotland. If exposed and placed on the surface of the sand the crab will burrow quickly by digging with legs 2 to 5, going into the sediment backwards. Once in the sediment the long, second pair of antennae make contact with the surface and are held parallel to one another so that the setae interlock and form a tube, down which an inhalant current of water is drawn for respiration. Egg-bearing females have been recorded in June and July and the eggs are carried for about 10 months. Females breed successively for a number

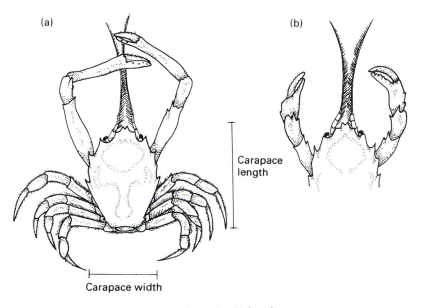

Figure 264  *Corystes cassivelaunus*, (a) male, (b) female.

of years. *Corystes* feeds on a wide range of burrowing invertebrates, particularly bivalve molluscs, polychaete worms and amphipods.

Family Portunidae – swimming crabs
Carapace wider than long. Terminal segment of pereopod 5 flattened and usually fringed with setae; often segment is very much expanded, and used in swimming. Usually five to nine sharp teeth on antero-lateral margins of carapace.

A large family of crabs. In addition to the common species described here, others are found in the shallow sublittoral and occasionally on the lower shore. Reference to Ingle (1983) is recommended for further study.

1 Pereopod 5 with flat, but narrow, pointed terminal segment (Fig. 267), slightly wider than terminal segments of pereopods 2–4. Carapace with 5 sharp teeth on antero-lateral margins; 3 small, rounded teeth between eyes. Widely distributed, common . .................................................................................. *Carcinus maenas* (p. 401)
Pereopod 5 with broad, flat terminal segment  .............................................................. 2
2 Seven to 10 unequal-sized teeth between eyes, central pair often the largest. Eyes red; joints between segments of pereopods bluish with red markings ..... *Necora puber* (p. 400)
Three similar-sized, rounded teeth between eyes. Dorsal surface of carapace with short transverse rows of small tubercles with setae ................... *Liocarcinus depurator* (p. 400)

***Necora puber*** (Linnaeus)   (*Liocarcinus puber*, *Portunus puber*, *Macropipus puber*)
Velvet swimming crab (Fig. 265)

*Reddish-brown carapace; broader than long; up to 90 mm across; dorsal surface covered with fine setae. Joints between segments of pereopods bluish, with red markings. Eyes red; anterior margin of carapace with 7–10 small teeth, central pair largest; 5 sharp teeth on antero-lateral margins of carapace. First pair of pereopods with well-developed chelae; 5th pair with broad, flattened, terminal segment.*

Figure 265  *Necora puber.*

*N. puber* is widely distributed in north-west Europe. Small specimens are common at low water on rocky shores, particularly in the south and west of Britain, while larger specimens are found to depths of 80 m. It is a powerful swimmer and an aggressive crab. Maturity is reached at a carapace width of about 40 mm and on British shores egg-bearing females are found throughout most of the year. *Necora* feeds largely on brown algae such as *Laminaria* and *Fucus* and although crustaceans and molluscs are important food items, they are less so in intertidal populations. The diet of juveniles is mostly crustaceans such as barnacles and small crabs. The species is fished commercially using baited pots. It is a relatively small fishery in British waters but the catch has increased rapidly since the early 1980s.

***Liocarcinus depurator*** (Linnaeus)   Swimming crab (Fig. 266)

*Carapace broader than long, up to 40 mm across; dorsal surface with short transverse rows of small tubercles with setae. Reddish-brown colour; terminal segment of last pair of pereopods violet in larger crabs. Three similar-sized rounded lobes between eyes. Antero-lateral margins of carapace with 5 pointed teeth. First pair of pereopods with well-developed chelae; 5th pair with broad, flat, terminal segment.*

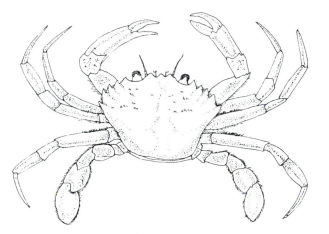

Figure 266  *Liocarcinus depurator.*

*L. depurator* is widely distributed in north-west Europe and in some localities may be common. It is found from the lower shore to depths of about 400 m on sandy deposits.

**Carcinus maenas** (Linnaeus)    Shore crab, Green crab (Fig. 267)

*Carapace broader than long; up to 80 mm across. Dark green, drab colour to orange-red; young forms often conspicuously marked with white patches. Three small, rounded lobes or teeth between eyes, sometimes difficult to detect, particularly in young. Antero-lateral margins of carapace with 5 pointed teeth. First pair of pereopods with well-developed chelae. Terminal segment of 5th pair of pereopods flat, but pointed and narrow.*

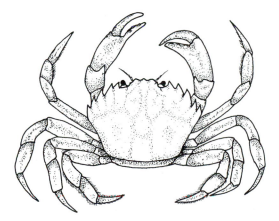

Figure 267  *Carcinus maenas.*

*C. maenas* is one of the commonest crabs on the shores of north-west Europe. Found on all types of shore from high water into the subittoral to depths of about 60 m, it extends into estuaries, the adults surviving salinities down to about 4‰. *Carcinus* shows a wide range of carapace colours from dark green to orange and red. It would appear that the green crabs are actively growing and moulting, while the orange-red specimens are in an extended intermoult. Red crabs are generally most common in the sublittoral, the green dominating on the shore and on saltmarshes, and this has led to the suggestion that colour reflects differing abilities to withstand low salinity. Maturity is generally reached after about one year at a carapace width of 25–30 mm (males) and 15–31 mm (females). The male seeks out a female crab about to moult and carries her under his body for a few days. When the female moults, copulation takes place. The female then burrows in sand and forms a large cavity beneath her. While in this position the eggs are laid and attached to the pleopods. Females carrying eggs are found on the shore throughout the year and in south-west Britain there is a peak of breeding activity in summer. Although adult crabs survive low salinities, the eggs will not develop at salinities less than 20‰. The eggs are held by the female for several months and it has been estimated that as many as 185 000 eggs can be carried at any one time. After a larval life of two to three months in the plankton young crabs settle on the bottom and reach maximum size after about four years. Very small crabs are commonly found on mud-flats and among *Spartina* (p. 85). The shore crab feeds on a wide variety of invertebrates including polychaete worms, molluscs and crustaceans; young crabs feed on barnacles and small molluscs. Predators include fishes and gulls.

The abdomen of female specimens of *Carcinus* is broader than that of the male and has seven segments compared with five visible segments in the male. Male crabs parasitized by *Sacculina carcini* (p. 342) show a broadening of the abdomen, and may be mistaken for females.

***Pirimela denticulata*** (Montagu) may be confused with *Carcinus maenas*. It is found on the lower shore, reaching a carapace width of about 20 mm. It is not as abundant as *Carcinus* and can be recognized by the presence of an anterior projection between the eyes, which has three sharp points. The last segment of pereopod 5 is not flattened.

Family Cancridae
Carapace much broader than long; margins with many rounded lobes giving characteristic appearance. Region of carapace between eyes with three blunt lobes.

***Cancer pagurus*** Linnaeus    Edible crab (Fig. 268)

*Carapace much broader than long; may be more than 200 mm across; reddish-brown in colour. Carapace has 3 small lobes between eyes and about 10 larger, rounded lobes on antero-lateral margins giving what is often referred to as a 'pie-crust' effect. First pair of pereopods with very large chelae, with black tips; 2nd to 5th pairs with many setae.*

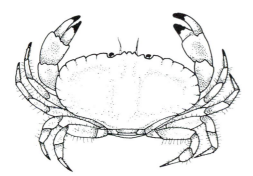

Figure 268  *Cancer pagurus.*

*C. pagurus* is widely distributed in north-west Europe in pools and gullies from the lower shore to depths of about 90 m. The larger specimens are usually found offshore where they are fished commercially, using pots baited with pieces of fish. Most crabs reach maturity at about 110 mm (male) and 127 mm (female) carapace width when they are about three to five years old. Prior to copulation, the male holds the female and has been observed to help her cast the exoskeleton during the precopulatory moult. Immediately after moulting, copulation occurs; the eggs are laid some months later. The female seeks a sandy substratum where she excavates a small depression which she occupies during egg laying. Breeding takes place in winter and the eggs are carried for seven to eight months, hatching in spring and summer. The edible crab lives for up to 20 years and it has been estimated that large females carry more than two million eggs at any one time. The larvae are planktonic for 23–30 days. Adults take a wide variety of food and like *C. maenas* are important predators of molluscs.

Family Xanthidae
Carapace broader than long; region between eyes broad, with two broad lobes, sometimes finely serrated. First pair of pereopods with large broad chelae. Fifth pair of pereopods shorter than first pair.

Antero-lateral margins of carapace with 5 well-defined teeth. Carapace and pereopods with long, club-shaped setae .................................................. *Pilumnus hirtellus* (p. 403)
Antero-lateral margins of carapace with 5 blunt lobes. Carapace and pereopods without club-shaped setae, but margins of pereopods 2–5 may have a few simple setae
................................................................................................. *Xantho incisus* (p. 404)

*Pilumnus hirtellus* (Linnaeus)    Hairy crab (Fig. 269a)

*Carapace broader than long; up to 30 mm across; reddish-brown colour; very finely serrated between eyes; antero-lateral margins with 5 teeth, anterior 2 small. First pair of pereopods with chelae, large and dissimilar in size, the smaller with rough tuberculate surface.*

*Pereopod 1 larger in male than female. Pereopods 2–5 banded purple and white. Carapace and appendages covered with long, club-shaped setae.*

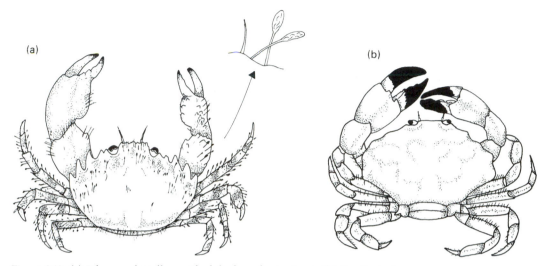

Figure 269  (a) *Pilumnus hirtellus*, with club-shaped setae inset. (b) *Xantho incisus*.

*P. hirtellus* is widely distributed in north-west Europe and in Britain is most frequent in the south and west. It is found under stones and among laminarian holdfasts from the lower shore to depths of about 75 m. Silt accumulates among the setae on the body making the crab difficult to detect. Egg-bearing females are found from April to August. *Pilumnus* eats a wide variety of plant and animal material, including small mussels.

**Xantho incisus** Leach    Montagu's crab (Fig. 269b)

*Carapace much broader than long; up to 70 mm across. Yellowish or reddish-brown; chelae black. Antero-lateral margins of carapace with 5 blunt lobes. First pair of pereopods with large chelae. Margins of pereopods 2–5 with few setae.*

*X. incisus* is common under stones on the lower shore in the south and west of Britain, where it is sometimes abundant. It extends into the sublittoral to depths of about 35 m and reaches the northern limit of its distribution on the west coast of Scotland. Egg-bearing females are found from March to July and the larvae are present in the plankton for most of the summer. *Xantho* is a herbivore feeding on a variety of algae.

A second species of *Xantho*, **Xantho pilipes** A. Milne Edwards, is found on the lower shore of sandy and stony beaches. It is most readily separated from *X. incisus* by dense fringes of setae (not club-shaped) on the lateral and posterior margins of the carapace and the second to fifth pairs of pereopods.

Family Pinnotheridae
Carapace rounded, smooth; often soft. Region between eyes not lobed or toothed. Eyes small, widely separated. Antennae very small. Small crabs living as commensals in the mantle cavity of bivalve molluscs, particularly the edible mussel, *Mytilus edulis*.

*Pinnotheres pisum* (Linnaeus)    Pea crab (Fig. 270)

*Carapace rounded and smooth; up to 6 mm across in male, 13 mm in female. Carapace of male, but not female, projects in front of orbits. Eyes small, particularly in female. Male yellowish in colour with dark patches; female semi-translucent, with pale colour of internal organs showing through. In some specimens carapace hard, in others soft (see below).*

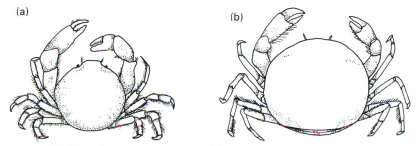

Figure 270  *Pinnotheres pisum*, (a) male, (b) female.

*P. pisum* is widely distributed in north-west Europe, where it can be found living in the mantle cavity of a range of bivalve molluscs, including the edible mussel, *Mytilus* (p. 283). The life-cycle usually involves two hosts. The larvae are planktonic. The newly metamorphosed crab has a hard cuticle and swims with the aid of setae on some of the pereopods. It invades the bivalve *Spisula solida* (p. 302) where it undergoes a series of moults, during which time the cuticle is soft and the pereopods lack setae. Eventually, a hard-shelled stage is formed, developing long setae on the third and fourth pairs of pereopods. The crab then leaves *Spisula*, mates and finds the second host, which is most often *Mytilus edulis* (p. 283), but can sometimes be *Modiolus modiolus* (p. 285) or *Tapes decussatus* (p. 316). In the second host the crabs have a soft shell and it is here that the female breeds. In some populations of *Pinnotheres* newly metamorphosed crabs are found in *Mytilus* and it has been suggested that a change of host does not always take place. In south-west England, the main breeding period lasts from May to August and females are capable of producing several broods. The male crabs apparently pass through a number of hard and soft stages. The crab feeds on strings of food collected by the bivalve and may cause erosion of the gill surfaces of its bivalve host.

Family Majidae — spider crabs
Carapace longer than broad, roughly triangular in shape, broader towards posterior; usually with bifid rostrum between eyes. First pair of pereopods usually shorter than other pereopods. Pereopods 2–5 sometimes long and slender, and spider-like.

1 Rostrum slender, bifid, the 2 parts touching along length, not curving upwards or downwards. Eyes not retractile. Pereopods 2–5 long, slender, spidery .................................
................................................................................... *Macropodia rostrata* (p. 408)
Not as above ................................................................................................ 2
2 Rostrum bifid, the 2 parts touching at tips .................................... *Hyas araneus* (p. 407)
Rostrum not as above .................................................................................... 3
3 Rostrum short, bifid, the 2 parts well separated. Four tubercles in a row across anterior part of carapace. Small, delicate, up to 30 mm carapace width ..............................................
.................................................................................... *Inachus dorsettensis* (p. 408)
Rostrum strong, bifid, the 2 parts diverging or parallel. Carapace with many sharp spines, particularly on the antero-lateral margins. Large crab, up to 150 mm across carapace ....................................................................... *Maja squinado* (p. 406)

*Maja squinado* (Herbst)    Common spider crab (Fig. 271)

*Carapace longer than broad; narrows anteriorly; up to 150 mm across; the largest of British spider crabs. Reddish-brown colour; tip of terminal segment of pereopods 2–5 black. Rostrum strong, bifid, the 2 parts usually diverging, or nearly parallel in larger specimens. Eyes retract into orbit. Many sharp spines over carapace, those on antero-lateral margins well developed. First pair of pereopods with small chelae; 5th pair of pereopods relatively short. Often with seaweed attached to carapace.*

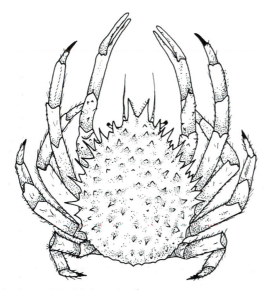

Figure 271  *Maja squinado.*

*M. squinado* reaches its northern limit in the south of Britain. There are records of its presence in the southern North Sea and on the coasts of The Netherlands, and it is sometimes abundant in the south and west of Britain. Found under rocks and seaweed on the lower shore, it has also been recorded from sandy beaches and extends into the sublittoral to depths of about 75 m. Off the south coast of England, copulation takes place in July and August and the eggs are laid about six months later. Egg-bearing females have been recorded throughout spring and summer at a range of localities. The eggs are carried for up to nine months, each female producing a single brood. The common spider crab forms heaps or mounds of individuals in shallow water during late summer and autumn. These can be over 600 mm high and 900 mm in diameter and contain as many as 100 crabs. The crabs remain together for some months and presumably the heaps offer protection to the soft-shelled, recently moulted crabs which are found in the centre. This behaviour pattern may also be important as a means of bringing together the sexes for reproduction. *Maja* is omnivorous, feeding on a wide range of both plants and animals such as *Enteromorpha* (p. 35), *Corallina* (p. 61), bryozoans and hydroids.

### *Hyas araneus* (Linnaeus)   Spider crab (Fig. 272)

*Reddish-brown carapace, narrow anteriorly, broad and rounded posteriorly; generally 30–50 mm across; tubercles on dorsal surface. Rostrum bifid, the 2 parts meeting at tips. Eyes not completely concealed when withdrawn. First pair of pereopods thick, with chelae, others long and thin. Carapace often covered with encrusting invertebrates.*

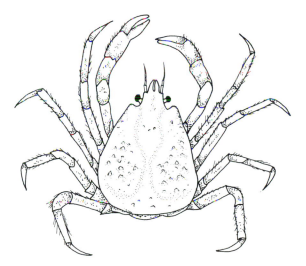

Figure 272  *Hyas araneus.*

*H. araneus* is widely distributed in north-west Europe, reaching its southern limit on the south and west coasts of Britain. It is found on the lower shore among rocks, seaweed and on sand, extending to depths of about 80 m. Egg-bearing females have been recorded throughout the year at a number of localities in Britain. It is omnivorous and includes a large proportion of seaweed in its diet.

A second species of *Hyas*, **Hyas coarctatus** Leach, may also be found on the very low shore. It is similar in appearance to *H. araneus* but the carapace is described as lyre-shaped, compared with the more rounded, triangular carapace of *H. araneus*.

*Inachus dorsettensis* (Pennant)    Scorpion spider crab (Fig. 273)

*Reddish-brown carapace, triangular in outline; up to 30 mm across; 4 tubercles lie in a row across anterior part of carapace, other tubercles posterior to them. Short, bifid rostrum, the 2 parts well separated. Eyes retractile. First pair of pereopods with chelae, larger in male than female; 2nd pair very long, longer and stouter than remaining pairs. Carapace often covered with encrusting organisms.*

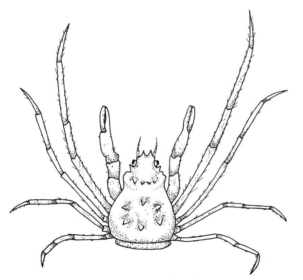

Figure 273  *Inachus dorsettensis*, female.

*I. dorsettensis* is widely distributed in north-west Europe extending to depths of about 180 m. It is occasionally found among rocks and seaweed on the lower shore and also on muddy-sand. Egg-bearing females have been found throughout the year on the west coast of Britain and aquarium observations suggest that the eggs take about 15 weeks to develop to pelagic larvae. It has been estimated that each female carries three batches of eggs. *Inachus* is omnivorous.

*Macropodia rostrata* (Linnaeus)    Long-legged spider crab (Fig. 274)

*Yellow-brown to red carapace; triangular in shape, up to about 15 mm across; with 8 or so tubercles on surface. Bifid rostrum, the 2 parts touching along their length; less than $\frac{1}{2}$ length of large basal segments of antenna 2; more or less straight, not curving markedly upwards or downwards. Long eye stalks, clearly visible, without prominent tubercle on anterior margin; eyes not retractile. First pair of pereopods with chelae, larger in male than female; 2nd to 5th pairs slender and spider-like, with 4th and 5th pairs having curved terminal segments.*

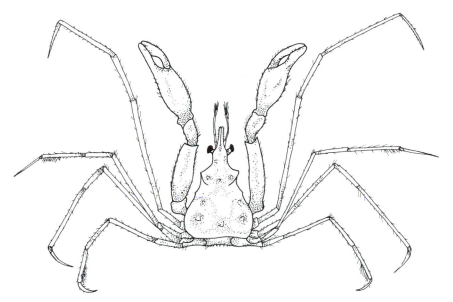

Figure 274  *Macropodia rostrata*, male.

*M. rostrata* is widely distributed in north-west Europe and is the commonest spider crab in Britain. It is found under seaweed on the lower shore and on muddy-sand and sand, extending to depths of about 120 m. Egg-bearing females have been found throughout the year. Aquarium observations suggest that the eggs are carried for about 14 weeks and it has been estimated that each female carries up to three broods of eggs per year. *Macropodia* is omnivorous, feeding mainly on algae and crustaceans.


## Phylum UNIRAMIA

### Subphylum Hexapoda
The body of the insect is divided into head, thorax and abdomen, although these are not always clear. The head has a single pair of antennae and there are three pairs of legs (and usually two pairs of wings) on the thorax. Insects are essentially terrestrial but some do extend onto the shore. Two genera are of particular interest to the shore ecologist.


*Anurida maritima* (Guérin)    (*Lipura maritima*) (Fig. 275a)

*Body roughly cigar-shaped; dull, dark blue; about 3 mm in length. Antennae short. Thorax has 3 pairs of short legs; no wings.*

A. *maritima* is widely distributed in north-west Europe, where it is found on the upper shore. Characteristically, it is seen floating on the surface film of sheltered rock pools, often in dense aggregations. A film of air saturated with water is trapped around the cuticle, reducing desiccation and allowing it to extend its distribution onto the high shore. It has also been recorded from saltmarshes. When the tide comes in, *A. maritima* takes shelter in crevices in the rock and among seaweeds. Breeding occurs in summer. The male leaves a spermatophore on the rock surface and this is picked up by the female. Groups of pale yellow-coloured eggs are laid in August in cracks and crevices in the rock where they overwinter, becoming a dark orange colour and hatching the following spring. The adults die before winter and longevity is believed to be less than one year. *Anurida* feeds on a variety of invertebrates and has been described as one of the most important scavengers on the shore.

*Petrobius* spp.    Bristle tail (Fig. 275b)

*Grey coloured, body about 10 mm in length. Head bears pair of very long antennae, and pair of palps. Thorax has 3 pairs of legs, no wings. Abdomen with 3 conspicuous bristles at posterior end; central bristle very long.*

(a)                                    (b)

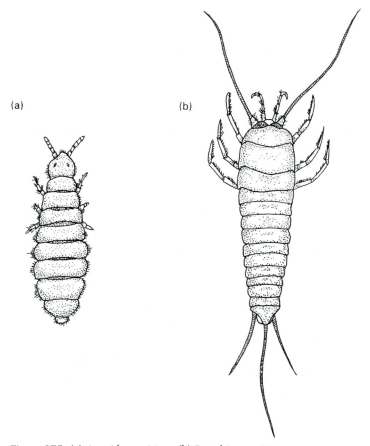

Figure 275   (a) *Anurida maritima.* (b) *Petrobius maritimus.*

Two species of the genus are widespread in north-west Europe. These are *Petrobius maritimus* (Leach) and *Petrobius brevistylis* Carpenter, while a third species, *Petrobius lohmanderi* Agrell has been recorded from the east coast of Sweden. Where *P. maritimus* and *P. brevistylis* co-exist there is some separation of the species. *Petrobius* spp. occur higher on the shore than *Anurida* (above) and are widely distributed in the littoral fringe, extending into the terrestrial environment. Shiny black eggs are laid in crevices in groups of up to 100 in October/November and hatch in May/June of the following year. *Petrobius* is active at night, feeding on microorganisms and lichens; longevity is believed to be between one and two years.

## REFERENCES
### Chelicerata
King, P. E. (1974). *British sea spiders. Arthropoda: Pycnogonida.* Keys and notes for the identification of the species. Synopses of the British fauna (New Series), no. 5. London: Academic Press.

King, P. E. (1986). Sea spiders. A revised key to the adults of littoral Pycnogonida in the British Isles. *Field Studies*, 6, 493–516.

### Crustacea
### Cirripedia
Rainbow, P. S. (1984). An introduction to the biology of British littoral barnacles. *Field Studies*, 6, 1–51.

Southward, A. J. (1976). On the taxonomic status and distribution of *Chthamalus stellatus* (Cirripedia) in the north-east Atlantic region: with a key to the common intertidal barnacles of Britain. *Journal of the Marine Biological Association of the United Kingdom*, 56, 1007–28.

### Malacostraca
### Mysidacea
Makings, P. (1977). A guide to the British coastal Mysidacea. *Field Studies*, 4, 575–95.

Mauchline, J. (1980). The biology of mysids. *Advances in Marine Biology*, 18, part 1, 1–369. (Includes a key to identification and a taxonomic list of Mysidacea.)

Tattersall, W. M. & Tattersall, O. S. (1951). *The British Mysidacea.* London: Ray Society.

### Isopoda
Jacobs, B. J. M. (1987). A taxonomic revision of the European, Mediterranean and NW African species generally placed in *Sphaeroma* Bosc, 1802 (Isopoda: Flabellifera: Sphaeromatidae). *Zoologische Verhandelingen*, no. 238, 3–71.

Naylor, E. (1972). *British marine isopods.* Keys and notes for the identification of the species. Synopses of the British fauna (New Series), no. 3. London: Academic Press.

Sars, G. O. (1896–9). *An account of the Crustacea of Norway*, vol. 2. *Isopoda.* Christiania: Cammermeyers.

Amphipoda

Barnard, J. L. & Barnard, C. M. (1983). *Freshwater Amphipoda of the world*, vols. 1 & 2. Mt Verona, Virginia: Hayfield Associates.

Barnard, J. L. & Karaman, G. S. (1991). *The families and genera of marine gammaridean Amphipoda (except marine gammaroids)*. Records of the Australian Museum. Supplement 13 (parts 1 & 2).

Bousefield, E. L. (1973). *Shallow-water gammaridean Amphipoda of New England*. Ithaca: Cornell University Press.

Bousefield, E. L. (1983). An updated phyletic classification and palaeohistory of the Amphipoda. In *Crustacean Issues*, 1. *Crustacean phylogeny*, ed. F. R. Schram, pp. 257–77. Rotterdam: A. A. Balkema.

Costello, M. J., Holmes, J. M. C., McGrath, D. & Myers, A. A. (1989). *A review and catalogue of the Amphipoda (Crustacea) in Ireland*. Irish Fisheries Investigations Series B (Marine), no. 33.

Lincoln, R. J. (1979). *British marine Amphipoda: Gammaridea*. London: British Museum (Natural History).

Sars, G. O. (1895). *An account of the Crustacea of Norway*, vol. 1. *Amphipoda*. Christiania; Cammermeyers.


Decapoda

Allen, J. A. (1967). *The fauna of the Clyde Sea area. Crustacea: Euphausiacea and Decapoda*. Millport: Scottish Marine Biological Association.

Crothers, J. & Crothers, M. (1983). A key to the crabs and crab-like animals of British inshore waters. *Field Studies*, **5**, 753–806.

Ingle, R. W. (1980). *British crabs*. London: British Museum (Natural History).

Ingle, R. W. (1983). *Shallow-water crabs*. Keys and notes for the identification of the species. Synopses of the British fauna (New Series), no. 25. Cambridge: Cambridge University Press.

Ingle, R. W. (1985). Northeastern Atlantic and Mediterranean hermit crabs (Crustacea: Anomura: Paguroidea: Paguridae). 1. The genus *Pagurus* Fabricius, 1775. *Journal of Natural History*, **19**, 745–69.

Ingle, R. W. (1992). *Hermit crabs of the north-eastern Atlantic Ocean and Mediterranean Sea. An illustrated key*. London: Chapman & Hall.

Smaldon, G. (1993). *Coastal shrimps and prawns*. Keys and notes for identification of the species. Synopses of the British fauna (New Series), no. 15, 2nd edn. Revised and enlarged by L. B. Holthuis & C. H. J. M. Fransen. Shrewsbury: Field Studies Council.


Unirama

Cheng, L. (ed.) (1976). *Marine insects*. Amsterdam: North-Holland Publishing.

# Sipuncula

Sipunculans are worm-like, marine animals occurring both intertidally and sublittorally. They are found in burrows in sandy and muddy sediments, in the empty tubes of polychaetes, the empty shells of molluscs and in rock crevices. Sipunculans have an unsegmented body in which the anterior part, or introvert, is narrow and the posterior part, the trunk, is wider and more or less cylindrical. The introvert can be withdrawn; in relaxed specimens the mouth is seen at the anterior end of the introvert, surrounded completely, or in the dorsal aspect, by lobes or tentacles. The anus is at the anterior end of the trunk in a mid-dorsal position. Sipunculans are mainly deposit feeders, feeding on mud, sand and detritus. Almost all species have separate sexes with external fertilization. There is a free-swimming larva and the length of pelagic life varies from a few days to a month or so.

Identification to species can be difficult and often requires examination of internal anatomy. The species described here can be recognized on external morphology and habitat.

*Golfingia elongata* (Keferstein)    (Fig. 276a)

*Slender body up to 150 mm in length. Anterior end with up to 36 tentacles (usually 24–28) encircling mouth. Anterior region of introvert with rings of hooks. Yellow-brown in colour.*

G. elongata extends as far north as Denmark and is widely distributed in north-west Europe. It lives in vertical burrows up to 500 mm below the surface in muddy-sand on the lower shore and sublittorally. On the Channel coast of France breeding has been recorded in summer and the larva has a pelagic life of about one week.

A second species of *Golfingia*, *Golfingia vulgaris* (de Blainville) is widespread in north-west Europe in mud, sand and gravel from the lower shore to depths of about 2000 m. It is similar in length to *G. elongata* (150–200 mm), but is distinguished by the irregularly arranged hooks on the introvert (Fig. 276b).

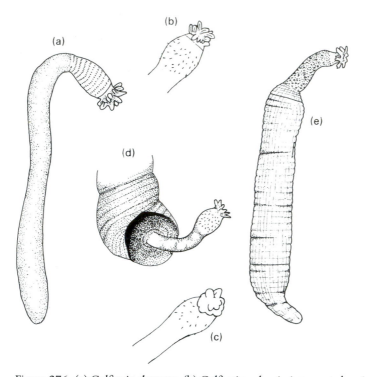

Figure 276  (a) *Golfingia elongata*. (b) *Golfingia vulgaris*, introvert showing irregularly
         arranged hooks (after Gibbs, 1977). (c) *Nephasoma minuta*, introvert (after
         Gibbs, 1977). (d) *Phascolion strombi* in empty shell of *Turritella* (after Cuénot,
         in Grassé, 1959). (e) *Sipunculus nudus* (after Gibbs, 1977).

### *Nephasoma minuta* (Keferstein)    (*Golfingia minuta*) (Fig. 276c)

*Small, up to 15 mm in length and about 1.5 mm wide. Anterior end with short tentacles
(lobes) encircling mouth. Anterior region of introvert with irregularly arranged hooks, particularly in small specimens, but may be few or absent in larger specimens.*

N. minuta is widespread in north-west Europe in sandy and muddy deposits from the
middle shore, where it is also found in rock crevices and among the reefs of *Sabellaria*
(p. 197), into the sublittoral to depths of about 50 m. It is hermaphroditic and on the
south coast of England breeding has been recorded from November to January.

### *Phascolion strombi* (Montagu)    (Fig. 276d)

*Elongate body up to 50 mm in length. Anterior end with up to 50 tentacles encircling
mouth. Anterior region of introvert with many hooks, irregularly arranged. Body covered
with papillae. Usually found in empty prosobranch shells; also lives freely in sediment.
Greyish coloured.*

*P. strombi* usually lives in mud and sand accumulated in empty shells, especially those of *Aporrhais* (p. 248) and *Turritella* (p. 235). It is widely distributed and sometimes common in north-west Europe from the shallow sublittoral to depths of over 3000 m and shells containing *Phascolion* are often washed ashore. When *Phascolion* is present, a small round hole can be seen in the plug of sediment in the mouth of the shell, and from this the introvert of the sipunculid projects. Breeding probably occurs during most of the year; the larva has a pelagic life of about one week.

### *Sipunculus nudus* Linnaeus    (Fig. 276e)

*Cylindrical body up to 350 mm in length. Anterior end with convoluted lobes surrounding mouth. Introvert with many papillae over entire length. Trunk marked with rectangular pattern. Whitish, greyish in colour.*

*S. nudus* is a southern species occurring as far north as the southern North Sea. It is occasionally found in sandy sediments on the lower shore and sublittorally to depths of 700 m. Breeding has been recorded during summer and the larva has a pelagic life of about one month.

REFERENCES

Gibbs, P. E. (1977). *British sipunculans.* Keys and notes for the identification of the species. Synopses of the British fauna (New Series), no. 12. London: Academic Press.

Gibbs, P. E. & Cutler, E. B. (1987). A classification of the phylum Sipuncula. *Bulletin of the British Museum Natural History (Zoology),* **52**, 43–58.

Grassé, P.-P. (ed.) (1959).*Traité de zoologie. Anatomie, systématique, biologie,* 5. Paris: Masson.

Stephen, A. C. & Edmonds, S. J. (1972). *The phyla Sipuncula and Echiura.* London: British Museum (Natural History).

# Echiura

Members of the phylum Echiura are marine, worm-like animals found burrowing in mud and sand, under rocks and in crevices, both on the shore and sublittorally. The body is divided into two parts; an anterior proboscis with a longitudinal ventral groove, and a posterior, cylindrical or sac-like trunk, which has two large, ventral setae near its anterior end. The mouth is at the posterior end of the proboscis. The species vary in length, from a few millimetres to a metre or more in those species in which the proboscis is long and highly extensible. Echiurans feed on detritus collected on a mucous sheet secreted by the proboscis. The sexes are separate and fertilization is usually external; there is a free-swimming trocophore larva.

*Echiurus echiurus* (Pallas)    (Fig. 277a)

*Trunk cylindrical, up to 200 mm in length. Anterior end with large proboscis with conspicuous ventral groove. Trunk covered with small papillae; 2 anterior, ventral setae; 2 rings of setae at posterior. Proboscis orange with brown stripes, trunk yellowish-grey.*

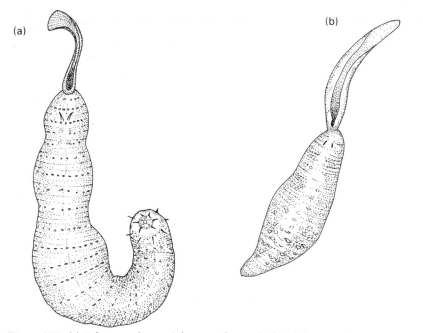

(a)

(b)

Figure 277  (a) *Echiurus echiurus* (after Greef, 1879). (b) *Thalassema thalassemum* (after Rietsch, 1886).

*E. echiurus* is widely distributed in north-west Europe, where it is found in burrows in sand and mud on the lower shore and sublittoral. Breeding occurs during winter.

**Thalassema thalassemum** (Pallas)    (*Thalassema neptuni*) (Fig. 277b)

*Trunk sac-like in shape, pointed posteriorly; up to 70 mm in length. Anterior end with long, highly extensible proboscis with conspicuous ventral groove; may be 200 mm in length when extended. Trunk with many papillae on surface, especially on posterior region; 2 anterior, ventral setae. Colour varied, blue, grey, yellow, pink, with white line on mid-ventral surface.*

*T. thalassemum* is a southern species found as far north as south-west Britain, where it is sometimes common on the lower shore but has recently been recorded off south-west Scotland. It lives in cracks and crevices in rocks and among stones but has also been found sublittorally burrowing in soft sediments. It is a deposit feeder using the extensible proboscis to collect particles from the surface of the sediment. Breeding occurs during summer.

REFERENCES

Greef, R. (1879). Die Echiuren (*Gephyrea armata*). *Nova Acta Academiae caesareae Leopoldino-Carolinae*, **41**, 1–172.

Rietsch, M. (1886). Étude sur les Gephyriens armés ou Échiurens. *Recueil Zoologique Suisse*, **3**, 313–515.

Stephen, A. C. & Edmonds, S. J. (1972). *The phyla Sipuncula and Echiura*. London: British Museum (Natural History).

# Bryozoa

Bryozoans are small, colonial animals found mainly in marine environments, although a few species are present in fresh water. They are almost all sessile and occur in a variety of shapes and sizes on substrata such as seaweeds, rocks, stones and shells, both inter-tidally and sublittorally, sometimes in high densities. Each individual in the colony is known as a zooid and is usually no more than a millimetre in length. The zooids are surrounded by a cuticle which is often strengthened by calcareous material. One of the most characteristic features of bryozoans is the lophophore, a group of ciliated tentacles which encircles the mouth and collects microorganisms such as bacteria and diatoms on which the animals feed. Ciliated tentacles distinguish bryozoans from hydroids, with which they may be confused. In hydroids the tentacles lack cilia. The lophophore pro-jects through an aperture in the cuticle and in some species this is closed by an opercu-lum, or by contraction of muscles. The alimentary canal is U-shaped with the anus out-side the lophophore.

Most bryozoans exhibit polymorphism and, in addition to the feeding zooids, there are zooids adapted for anchoring the colony to the substratum and others for brooding developing embryos. The two most modified types of zooid are the avicularium and the vibraculum. In the avicularia, which are somewhat like a bird's head in appearance (see Fig. 284a), the operculum is in the form of a beak. Avicularia pick up and remove larvae and other invertebrates which settle on, and tend to smother, the bryozoan. In the vibracu-ula, the operculum is in the form of a seta which is moved over the surface of the colony keeping it free of debris. The role of avicularia and vibracula has been likened to that of the pedicellariae of echinoderms (p. 432).

Bryozoan colonies are hermaphroditic but cross-fertilization is believed to occur. In some species, testes and ovaries develop in the same zooid, in others there are separate male and female zooids within a colony. A few species have a free-swimming planktonic larva with a triangular, bivalve shell. This is known as a cyphonautes larva and has a pelagic life of several weeks. In most species, however, the eggs are brooded in special chambers in the colony and the larvae have only a short free-swimming life. The larvae of bryozoans are very selective in their choice of substratum and some species will settle only on a specific area of an algal frond. The first zooid formed after settlement is known as an ancestrula; growth of the colony occurs by budding. Length of life varies from a few months to 10 or more years. The main predators of bryozoans are sea-slugs, and to a lesser extent, sea-spiders, sea-urchins and fishes.

Classification follows Hayward & Ryland (1985).

## Phylum BRYOZOA

Class Phylactolaemata   Freshwater bryozoans.

Class Stenolaemata   Marine bryozoans, cylindrical zooids; cuticle calcified. Includes
 many fossil species.

Order Cyclostomata   Aperture of zooid rounded, terminal and always open.
 Operculum absent. Colony usually upright or cushion-like.

Class Gymnolaemata   Nearly all are marine; zooids vase-shaped and upright, or
 flattened and box-like; cuticle sometimes calcified. Aperture closed by operculum
 or muscular contraction.

Order Ctenostomata   Cuticle not calcified. Zooids vase-shaped or flat. Aperture closed
 by muscular contraction. Colony upright or as a soft encrustation.

Order Cheilostomata   Cuticle calcified to varying degrees. Zooids box-like. Aperture
 closed by operculum. Colony upright or as a firm encrustation.

### Key to species of bryozoans

Bryozoans are widely distributed on the shore attached to a variety of solid substrata,
and may be common in some situations. The form of the colonies is highly varied, from
encrusting to upright, to creeping; some are calcified and hard to the touch, or partly
calcified; others are gelatinous. These features are important in identification and are
used in conjunction with other details as revealed by examination under a
stereo-microscope. For further detail reference should be made to the four volumes
covering bryozoans in the Synopses of the British Fauna (see p. 428).

1 Colony upright, sometimes creeping; attached to seaweed, stones, shells ........................ 2
  Colony encrusting; as a gelatinous mass, a glassy, shiny patch or a lacework; on
  seaweed, stones, shells ............................................................................................... 7
2 Colony upright, a brownish-lobed gelatinous mass; may be much branched. Zooids
  appear as tiny blunt areas on the surface. Very low shore and sublittoral; often washed
  ashore ...................................................... *Alcyonidium diaphanum* (p. 421)
  Colony upright and tufted, sometimes creeping, *or* an upright, much-divided flattened
  frond ............................................................................................................................. 3
3 Colony upright and tufted, sometimes creeping ........................................................... 4
  Colony an upright, much-divided, flattened frond. Common on the strandline;
  frequently misidentified as a 'dried seaweed' ............................................................... 6
4 Colony upright, tufted with branches curving inwards, jointed; 5–7 zooids between
  each joint. Zooids calcified and cylindrical with rounded, open aperture, without
  spines ............................................................. *Crisia eburnea* (p. 420)
  Colony not so ................................................................................................................ 5

5 Colony upright, tufted, sometimes creeping; zooids large, semi-transparent and vase-shaped, not calcified, arranged in groups along the branches but not spirally arranged. Free end of zooids squarish. Without avicularia   *Bowerbankia imbricata* (p. 424)
Colony upright, tufted, branches spirally arranged; zooids predominantly in 2 rows; rectangular and box-like with 1 spine at each corner of distal end. Avicularia large, rounded with short, sharply bent beak ...................................... *Bugula turbinata* (p. 427)

6 Frond bluntly rounded at ends. Zooids rectangular, with short spines. Pale greyish-brown in colour. Common on the strandline ................... *Flustra foliacea* (p. 426)
Frond narrow, truncate at ends. Zooids rectangular, without spines. Pale yellow in colour. On the strandline ................................................ *Securiflustra securifrons* (p. 427)

7 Colony a purplish-brown, coarse-textured mass, usually on *Fucus serratus*, *Mastocarpus stellatus* and *Chondrus crispus*. Zooids rectangular, with spines .......................................
........................................................................... *Flustrellidra hispida* (p. 423)
Colony not so ...................................................................................... 8

8 Colony a gelatinous mass, *or* a brittle, shiny, rounded, glassy patch; zooids without spines ...................................................................................... 9
Colony a whitish/silvery lacework; zooids with spines .................................... 11

9 Colony a shiny, glassy patch, especially in depressions on fronds of *Laminaria saccharina* ............................................................. *Celleporella hyalina* (p. 427)
Colony a gelatinous mass, on seaweeds, mostly *Fucus serratus*; also on stones and shells ...................................................................................... 10

10 (a) Colony a thin, transparent, gelatinous mass, nearly always on *Fucus serratus*. Surface of colony smooth, without papillae ................. *Alcyonidium gelatinosum* (p. 422)
*or*
(b) Very similar but on stones and shells; not on seaweed ...... *Alcyonidium mytili* (p. 423)
*or*
(c) Colony a thick, brownish gelatinous mass on *Fucus serratus*, *Chondrus crispus* and other red algae. Surface of colony with many small papillae .................................................
........................................................................ *Alcyonidium hirsutum* (p. 422)

11 Colony whitish, mostly even-edged; usually on fronds of *Laminaria*, often covering large areas. Zooids rectangular, with blunt spine at each corner. Front of zooids without pores ...................................................... *Membranipora membranacea* (p. 424)
Colony silver-grey, usually uneven in outline; on seaweeds, stones and shells. Zooids oval, surrounded by 4–12 spines, 1 of which is often long. Front of zooids with numerous pores (Fig. 282b) ............................................................. *Electra pilosa* (p. 425)

**Class Stenolaemata**
**Order Cyclostomata**

***Crisia eburnea*** (Linnaeus)   (Fig. 278)

*Colony upright, tufted, jointed. Branches curving inwards. Zooids calcified, alternate; cylindrical, without spines; usually 5–7 between each joint. Eight ciliated tentacles. White in colour, joints yellow or pale brown. Up to 20 mm in height.*

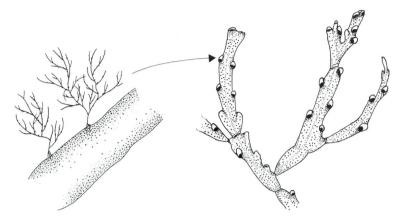

Figure 278  *Crisia eburnea.*

*C. eburnea* is widely distributed in north-west Europe and Britain on the lower shore and into the sublittoral, attached to red seaweeds, hydroids, the bryozoan *Flustra foliacea* (p. 426), stones and shells. It is often found washed up on the shore. On Irish Sea coasts, embryos can be found in the colonies from April to August. After fertilization each egg divides by a process known as polyembryony into a number of embryos which develop in large, swollen zooids known as gonozooids. On release, the larvae swim rapidly and settlement and metamorphosis occur within minutes.

**Class Gymnolaemata**
**Order Ctenostomata**
The genus *Alcyonidium*
Species of *Alcyonidium* are very difficult to separate and recent studies on the genetics of the genus suggest that the different 'species' may in fact be aggregates of species. *Alcyonidium diaphanum* is believed to be two species, *Alcyonidium hirsutum* five species, *Alcyonidium gelatinosum* two species and *Alcyonidium mytili* three species. The nomenclature applied to species of the genus has undergone revision and for clarification the reader is referred to Hayward (1985).

***Alcyonidium diaphanum*** (Hudson)    (*Alcyonidium gelatinosum, Alcyonium gelatinosum*)
(Fig. 279a)

*Colony form very varied, generally a lobed gelatinous mass, may be extensively branched. Surface smooth. Always upright in form, not encrusting; attached to rocks, stones and shells. Zooids appear as tiny, blunt areas on the surface; 14–16 ciliated tentacles. Colour varied, often yellowish-brown, but also brown, grey, dark red. Up to 200 mm or more in height.*

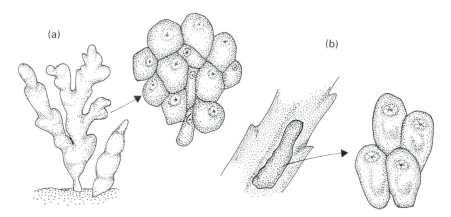

Figure 279  (a) *Alcyonidium diaphanum* (detail after Hayward, 1985). (b) *Alcyonidium gelatinosum* on *Fucus serratus*.

A. *diaphanum* is probably widely distributed in north-west Europe. It is common in Britain, where it is essentially sublittoral but is also found on the lower shore. It is often washed ashore. Little is known of its reproduction. Interestingly, it has been reported that contact with A. *diaphanum* can cause an allergic dermatitis, commonly referred to by fishermen as 'Dogger Bank itch'.

### *Alcyonidium gelatinosum* (Linnaeus)   (*Alcyonidium polyoum*) (Fig. 279b)

*Colony in the form of thin, encrusting, gelatinous mass, nearly always on* Fucus serratus *(p. 49), but occasionally on other seaweeds. Surface smooth, lacking papillae; 17–21 ciliated tentacles. Transparent.*

A. *gelatinosum* is probably widely distributed in north-west Europe and is common on the shores of Britain. It occurs only intertidally. Embryos develop within the zooids, each brooding zooid containing four to six embryos showing as white rings on the surface of the colony. Brooding of embryos has been recorded from a number of localities during spring or autumn. The larvae have only a brief pelagic life.

### *Alcyonidium hirsutum* (Fleming)   (Fig. 280)

*Colony in the form of thick, encrusting, gelatinous mass, usually on* Fucus serratus *(p. 49), but also on red seaweeds such as* Chondrus crispus *(p. 60)* and Mastocarpus stellatus *(p. 60). Surface covered with many conspicuous, small papillae; 14–19 ciliated tentacles. Colour brown or pale brown.*

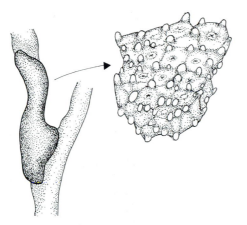

Figure 280 *Alcyonidium hirsutum* on stipe of *Fucus* sp.

A. *hirsutum* is widely distributed in north-west Europe and around Britain. In Irish Sea populations, embryos develop within the zooids during autumn and each brooding zooid contains four to eight embryos showing as white rings on the surface of the colony. It is believed that sperm are released through pores at the tip of the tentacles. Larvae are released in late winter and colonies are believed to live for one or two years.

*Alcyonidium mytili* Dalyell   (not illustrated)

*Colony in the form of thick, encrusting, gelatinous mass on stones and shells, also on shells housing hermit crabs; 17–21 ciliated tentacles. Colour brown or greenish-brown. Very similar to, and often mistaken for,* A. gelatinosum *(above), but not occurring on seaweed.*

A. *mytili* has been recorded from a number of localities in north-west Europe and Britain, both intertidal and sublittoral. On the north-east coast of England embryos are present in October and November. They are pink in colour and this is a useful character by which A. *mytili* can be separated from A. *gelatinosum* (above), with its white embryos.

*Flustrellidra hispida* (Fabricius)   (Fig. 281a)

*Colony in the form of thick mass, usually on* Fucus serratus *(p. 49), but also on other seaweeds including* Mastocarpus stellatus *(p. 60),* Chondrus crispus *(p. 60) and the holdfasts of* Laminaria *(p. 43), but not usually on rocks and stones. Spines present round the individual zooids (a condition known as hispid), giving coarse texture; 28–40 ciliated tentacles. Colour purplish-brown, reddish-brown.*

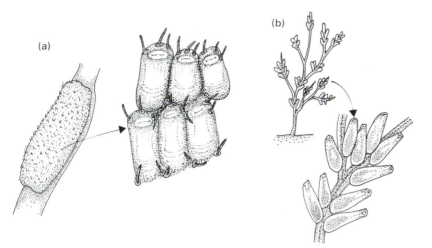

Figure 281  (a) *Flustrellidra hispida* on stipe of *Fucus* sp. (b) *Bowerbankia imbricata*.

*F. hispida* is widely distributed in north-west Europe and Britain, where it is predominantly intertidal. Embryos develop within the zooids, each brooding zooid containing as many as eight, yellowish-coloured embryos. Breeding occurs during winter. Larvae are released in spring and early summer and are similar to cyphonautes larvae in having a shell; they are, however, pelagic for only a short time.

**Bowerbankia imbricata** (Adams)    (Fig. 281b)

*Colony usually upright, growing in tufts; sometimes creeping; branches usually thick. Zooids large, 1 mm in height; vase-shaped, arising in groups along branches, but not spirally arranged. Free end of zooids squarish; 10 ciliated tentacles. Brownish-yellow in colour. Up to 70 mm in height.*

*B. imbricata* is widely distributed in north-west Europe and Britain, where it occurs from the middle shore into the shallow sublittoral, usually on *Ascophyllum nodosum* (p. 47), *Fucus vesiculosus* (p. 50), *Fucus serratus* (p. 49), other byrozoans such as *Flustra foliacea* (p. 426) and stones. The zooids are hermaphoditic; sperm are released through a tiny pore at the tip of the tentacles and enter a neighbouring zooid in the feeding current. Only one embryo is brooded within a zooid, the embryos giving a yellowish tinge to the colony. After release, the larvae have only a brief pelagic phase. Breeding occurs during summer.

**Order Cheilostomata**

**Membranipora membranacea** (Linnaeus)    Sea-mat (Fig. 282a)

*Colony in the form of an encrusting, mostly even-edged, lacework, usually on fronds of* Laminaria (p. 43), *but occasionally on* Fucus serratus (p. 49) *and other seaweeds. May cover very large areas. Zooids rectangular in shape with short, blunt spine at each corner. In some colonies there are tall columns, known as towers, projecting from the front of the zooids. About 17 ciliated tentacles. Colour whitish.*

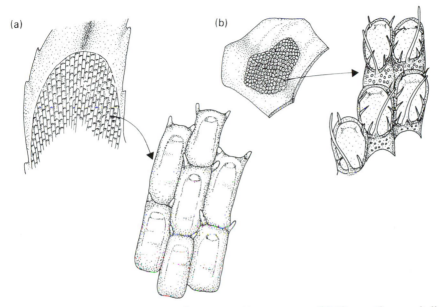

Figure 282  (a) *Membranipora membranacea* on *Fucus serratus*. (b) *Electra pilosa* on shell.

*M. membranacea* is one of the best-known bryozoans. It is widely distributed in north-west Europe and Britain on the lower shore and shallow sublittoral, breeding during spring and early summer. Sperm are released into the sea through pores at the tip of the tentacles and drawn into a neighbouring zooid, where cross-fertilization is believed to take place. The cyphonautes larvae have a planktonic life of several weeks and in the Irish Sea they are found in the plankton from February to November. *Membranipora* lives for several years. The colonies are flexible and able to withstand flexure of the algal frond. They grow towards the base of the frond, so ensuring that not all the colony is lost when the laminarian is damaged and eroded at the tip.

*Electra pilosa* (Linnaeus)   Hairy sea-mat (Fig. 282b)

*Colony in the form of an encrusting mat, often uneven in outline; occasionally upright. Found on variety of seaweeds, including* Fucus serratus *(p. 49), but also on* Laminaria *(p. 43),* Mastocarpus stellatus *(p. 60) and* Palmaria palmata *(p. 64); also on stones, shells, hydroids and often on the shells of the whelk* Buccinum *(p. 258) inhabited by hermit crabs. Zooids more or less oval, surrounded by 4–12 spines, one of which (the median proximal) is often long and conspicuous; 11–15 ciliated tentacles. Front of zooids with numerous pores. Colour silvery-grey.*

*E. pilosa* is widely distributed in north-west Europe and Britain, where it is one of the most common, if not the commonest, species of bryozoan. It is found from the middle shore into the sublittoral, breeding during August and September. Sperm are released into the sea through pores at the tip of the tentacles and it is believed they are drawn into a neighbouring zooid where fertilization occurs. Cyphonautes larvae have been recorded in the plankton during all months of the year, but they settle mainly in the autumn. Drifting, spherical colonies of *Electra* up to 70 mm across are occasionally washed ashore on the coasts of the North Sea. They are made up of branches of *Electra* attached to an object such as a stone or shell fragment.

Two other species of *Electra*, **Electra crustulenta** (Pallas) and **Electra monostachys** (Busk) occur around Britain but are found only in estuarine areas. In these species, the proximal medial spine is short and the front of the zooid is without pores.

**Flustra foliacea** (Linnaeus)    Hornwrack (Fig. 283a)

*Colony upright; much divided, flattened fronds, bluntly rounded at ends. Zooids more or less rectangular with 4–5 short spines at one end; 13–14 ciliated tentacles. Greyish-brown in colour, dead specimens paler. Up to 200 mm in height, generally smaller.*

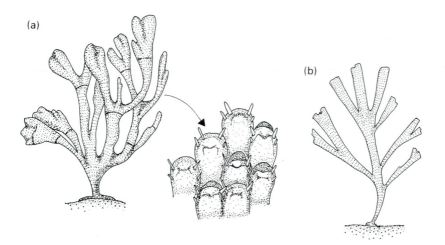

(a)

(b)

Figure 283   (a) *Flustra foliacea.* (b) *Securiflustra securifrons.*

*F. foliacea* is widely distributed in north-west Europe and Britain, growing sublittorally attached to rocks, shells and gravel. It smells strongly of lemons, although the smell has been likened to that of violets, roses or fish. Large quantities are often washed ashore after storms and may be misidentified as 'dried seaweed'. Breeding occurs during autumn and winter; the colonies contain separate male and female zooids. Embryos, which are pale orange in colour, develop in an outgrowth of the zooid known as an ovicell and are present from October to February in colonies on Irish Sea coasts. The larvae are released in early spring and have a short pelagic life. During the first year after settlement, the colony is flat and encrusting and only assumes the upright form in its second year. Colonies of *Flustra* only grow in spring and summer, growth ceasing for the rest of the year. As a result, annual growth lines can be detected on the surface of the frond, enabling an estimate of age to be made. Longevity is up to 12 years. Other byrozoans such as *Crisia eburnea* (p. 420), hydroids and sedentary polychaetes are often found living on *Flustra*.

### *Securiflustra securifrons* (Pallas)   (Fig. 283b)

*Colony upright; similar to* Flustra foliacea *(above) but fronds narrower and truncate at ends. Zooids rectangular, lacking spines. Pale yellow in colour. Up to 150 mm in height.*

*S. securifrons* is a northern species widespread in north-west Europe and occurring all around Britain, but is most abundant in the north. It is sublittoral, attached to rocks, stones and shells, and is often washed ashore in large quantities. Off the north-east coast of England, breeding has been recorded in November, March and April. The developing embryos are pinkish-orange in colour.

### *Bugula turbinata* Alder   (Fig. 284a)

*Colony upright, tufted, firm to the touch. Branches arranged spirally on central axis. Zooids in 2 rows at base of branches, 3 or more distally. Zooids rectangular with 1 short spine at each corner of distal end. Avicularia (p. 418) squat and rounded, with short, sharply bent beaks; conspicuous among zooids; 13 ciliated tentacles. Orange-brown in colour. Up to 60 mm in height.*

*B. turbinata* is a southern species extending northwards to south-west Scotland. It is found on the lower shore and in the shallow sublittoral under rocky overhangs and on stones and shells. Breeding has been recorded during August and September. The embryos are yellowish in colour and develop in outgrowths of the zooids known as ovicells. The larvae have only a brief pelagic life. Several species of *Bugula* are found around Britain; *B. turbinata* is probably the most common intertidally.

### *Celleporella hyalina* (Linnaeus)   (Fig. 284b)

*Colonies in rounded patches; rather brittle; on seaweed, especially in the depressions along the fronds of* Laminaria saccharina *(p. 45). Zooids more or less oval with rounded aperture; without spines; 12 ciliated tentacles. Shiny and glassy in appearance, sometimes whitish.*

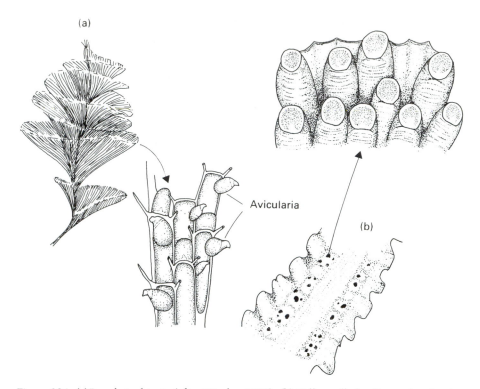

Figure 284  (a)*Bugula turbinata* (after Hincks, 1880). (b) *Celleporella hyalina* on *Laminaria saccharina*.

*C. hyalina* is probably widely distributed in north-west Europe and Britain, where it occurs on the lower shore and into the sublittoral. There are separate male and female zooids within a colony, fertilization is internal and yellowish-coloured embryos develop in outgrowths of the female zooids known as ovicells. Development takes three to four weeks and the larvae are released almost always during daylight. They have a brief pelagic life of up to some four hours before settling and metamorphosing. In Irish Sea populations, breeding occurs during most of the year. Length of life is believed to be only three to four months.

REFERENCES

Hayward, P. J. (1985). *Ctenostome bryozoans*. Keys and notes for the identification of the species. Synopses of the British fauna (New Series), no. 33. Leiden: E. J. Brill/Dr W. Backhuys.

Hayward, P. J. & Ryland, J. S. (1979). *British ascophoran bryozoans*. Keys and notes for the identification of the species. Synopses of the British fauna (New Series), no. 14. London: Academic Press. (Includes *Celleporella hyalina*.)

Hayward, P. J. & Ryland, J. S. (1985). *Cyclostome bryozoans*. Keys and notes for the identification of the species. Synopses of the British fauna (New Series), no. 34. Leiden: E. J. Brill/Dr W. Backhuys.

Hincks, T. (1880). *A history of the British marine Polyzoa* (2 vols.). London: Van Voorst.

Ryland, J. S. (1962). Biology and identification of intertidal Polyzoa. *Field Studies*, 1, 33–51.

Ryland, J. S. (1974). A revised key for the identification of intertidal Bryozoa (Polyzoa). *Field Studies*, 4, 77–86.

Ryland, J. S. & Hayward, P. J. (1977). *British anascan bryozoans Cheilostomata: Anasca.* Keys and notes for the identification of the species. Synopses of the British fauna (New Series), no. 10. London: Academic Press. (Includes all cheilostome bryozoans described in this text, with the exception of *Celleporella hyalina*.)

# Phorona

Phoronids are small, worm-like marine animals, usually no more than 200 mm in length, occurring both on the shore and sublittorally. They inhabit chitinous tubes to which sediment particles are often attached, and are found in sandy and muddy sediments, or encrusting or burrowing into rocks and shells. The body is narrow, elongate and unsegmented, the anterior end being characterized by a lophophore, a group of ciliated tentacles surrounding the mouth. The tentacles are clothed with cilia and increase in number as the animal grows. They are used in suspension feeding, the cilia drawing in plankton and detritus. As in the Bryozoa, the alimentary canal is U-shaped and the anus opens outside the lophophore. The sexes are either separate or the animals are hermaphroditic, but in both cases cross-fertilization occurs. Fertilization is internal. The eggs are either brooded in the parental tube or on the lophophore and have only a short pelagic life, or are liberated into the sea. The latter develop as larvae known as actinotrochs, with a pelagic life of several weeks. Asexual reproduction by budding also occurs. The animals have considerable powers of regeneration and can regenerate the anterior end in as little as two or three days.

A single genus, containing several species, has been recorded from north-west Europe. Separation of the species is difficult, sometimes requiring preparation of material for sectioning and examination under a microscope. Two species are likely to be encountered on the shore.

### Phoronis hippocrepia Wright    (Fig. 285)

*Body long and narrow; up to 100 mm in length and 1.5 mm wide; horseshoe-shaped lophophore with up to 150 tentacles of 2–3 mm in length. Green-grey, or yellowish in colour.*

P. hippocrepia has been recorded from a few widely separated localities in north-west Europe and Britain on the lower shore and sublittorally to depths of 10 m or more. It sometimes occurs in high densities and is found either encrusting rocks and stones or burrowing into calcareous rock, shells and encrusting red algae. The species is hermaphroditic and breeds throughout the year. The fertilized eggs are brooded on the lophophore and develop into free-swimming larvae.

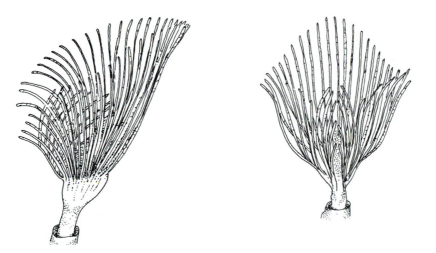

Figure 285  *Phoronis hippocrepia* (after Dyster, 1859).

A second species, **Phoronis ovalis** Wright, is also found intertidally. It is much smaller than *P. hippocrepia*, usually being about 6 mm in length, occasionally up to 15 mm, and is no more than 0.3 mm in width. The lophophore is oval-shaped and has up to 28 small tentacles, each about 1 mm in length. It is found from the lower shore into the sublittoral, often in extremely high densities, burrowing in empty mollusc shells such as *Arctica* (p. 312) and *Ostrea* (p. 287), and into barnacle shells and limestone rocks. Larval development takes place in the tube of the worm and the larva has only a short pelagic life. *P. ovalis* also reproduces asexually by budding and transverse fission.

REFERENCES

Dyster, F. D. (1859). Notes on *Phoronis hippocrepia. Transactions of the Linnean Society of London*, 22, 251–6.

Emig, C. C. (1979). *British and other phoronids.* Keys and notes for the identification of the species. Synopses of the British fauna (New Series), no. 13. London: Academic Press.

# Echinodermata

The echinoderms include starfishes, brittle-stars, sea-urchins, sea-cucumbers, feather-stars and sea-lilies and are one of the most distinct and well-known groups of shore animals. All are marine. The surface of the body is often brightly coloured and is generally spiny or warty. Echinoderms are characterized by a unique five-rayed or pentamerous radial symmetry very clearly displayed in the starfishes and brittle-stars and usually readily detected in the globular sea-urchins and the elongate sea-cucumbers. Calcareous plates are present in the body wall. In starfishes these are held together by connective tissue rendering the body wall flexible; in sea-urchins the plates are fused to give a rigid test, and in sea-cucumbers they are isolated within the tissue. In all echinoderms, a delicate epidermis overlies the skeletal plates and the skeleton is therefore an endoskeleton. Spines are characteristic of the phylum, and in asteroids and echinoids specialized structures known as pedicellariae (see Figs. 292, 297) are also found on the body surface. These are pincer-like in appearance, sometimes on long stalks, and equipped with adductor muscles enabling the jaws to be snapped shut. Some are equipped with poisonous glands. They are used both in defence and as a protection against smothering by particles of detritus and small organisms such as settling larvae. They are important in identification. Echinoderms are characterized by a unique water vascular system made up of fluid-filled canals projecting to the exterior as rows of blind ending tube-feet (podia), usually found in grooves or ambulacra running across the surface of the body. These thin-walled projections are alternately extended and contracted by changes in hydrostatic pressure and are used in a variety of functions such as respiration, feeding and locomotion. The canals of the water vascular system make contact with the exterior at the madreporite (Fig. 292), a calcareous plate perforated by very fine canals and believed to be the site at which a balance is achieved between the hydrostatic pressure in the water vascular system and the exterior. In starfishes and sea-urchins it is often seen as a conspicuous button or plate-like structure on the upper surface. In the brittle-stars it is on the lower or oral surface, while in the majority of holothurians, the madreporite opens not on the surface, but into the body cavity. The crinoids lack a madreporite. Small, thin-walled protuberances on the upper surfaces of some echinoderms are sites of gaseous exchange.

Echinoderms show a wide range of feeding habits including scavenging, grazing, suspension and deposit feeding. Some are carnivorous. Most species have separate sexes and external fertilization, and although many have free-swimming pelagic larvae, some brood the young.

## Phylum ECHINODERMATA

**Class Crinoidea**    The sea-lilies and feather-stars. Central disc or calyx with 10 feathery arms. Mouth and anus on upper surface of disc. May be attached to substratum by a stalk, or unattached.

**Class Asteroidea**    The starfishes. Five or more arms radiate from a central disc. Arms thick, narrowing gradually from central disc.

**Class Ophiuroidea**    The brittle-stars. Slender, long arms. Distinct, rounded central disc, sharply demarcated from arms.

**Class Concentricycloidea**    A class of echinoderms described in 1986. Specimens from waterlogged wood have been recorded in deep water off New Zealand and off the Bahamas.

**Class Echinoidea**    The sea-urchins and heart-urchins. Body globular, oval or flattened; covered with spines. Rigid, calcareous test. No arms.

**Class Holothuroidea**    The sea-cucumbers. Elongate body, cucumber or sausage-like in shape; lacking arms. Branched tentacles (modified tube-feet) around mouth.

### Class Crinoidea

The crinoids are primitive echinoderms attached to the substratum permanently by a stalk (the sea-lilies) or attached only during early development (the feather-stars). The body is made up of a cup-shaped central disc or calyx with both mouth and anus on the upper surface, the anus often on a raised cone-like structure. The disc has five arms but these divide into two close to the disc to give the appearance of 10 arms. In some tropical species, further subdivision gives the appearance of many more arms. In the feather-stars, a number of cirri arise from the undersurface of the disc. These are prehensile and allow temporary attachment to the substratum. The arms are clothed with side branches or pinnules giving them a feathery appearance. These pinnules have tube-feet arranged in groups of three. Crinoids suspension feed by extending the arms and pinnules. Suspended particles of food are trapped on the mucus-covered tube-feet and transported along ciliated grooves to the central mouth. Feather-stars are found clinging to rocks, hydroids and seaweed by the prehensile cirri on the undersurface of the disc. They crawl over the surface on their arms, and swim by moving the arms up and down through the water. Crinoids are dioecious; the gametes develop in the pinnules at the base of the arms and are shed by rupture of the walls of the pinnules. In some species the young are brooded, but in the majority there is a free-swimming larva known as the doliolaria or vitellaria.

Around Britain, the stalked sea-lilies are found in deep water and are not included here. One family of feather-stars, the Antedonidae, is found in shallow waters and representatives may occur on the very low shore.

*Antedon bifida* (Pennant)    Rosy feather-star (Fig. 286)

*Without stalk. Central disc; 5 pairs of feathery arms; up to 150 mm diameter. Up to 25, rarely 30, short cirri on undersurface of disc. Colour varied, red, pink, orange, yellow; sometimes banded.*

Figure 286  *Antedon bifida.*

A. *bifida* is found from the shallow sublittoral to depths of 450 m but is occasionally seen at the level of the lowest spring tides. It is free-moving but is most often found clinging by means of the cirri to the walls of gullies and on seaweeds, particularly the stipes of *Laminaria*, in situations where there is fast current flow. It is widely distributed in north-west Europe, and is found around most of Britain except the southern part of the east coast, mainly in shallow water and often in densities of over 1000 per square metre. *Antedon* feeds on fine particles in suspension and takes a wide range of detritus, plankton and inorganic particles. The food particles are trapped in mucus on the tube-feet which are of three sizes and arranged on the pinnules in groups of three. The tube-feet are wiped against rows of cilia and the food particles transferred to ciliated grooves leading to the mouth. The sexes are separate. Mature individuals can be recognized by swollen genital papillae at the base of the arms. Eggs escape through splits which appear in the pinnule walls, and adhere in groups to the external wall of the pinnule where fertilization takes place. The embryos are held on the pinnules in a mucous net during which time the female holds its arms close together in what has been described as brooding behaviour, allowing spawned females to be readily identified. The embryos hatch as free-swimming larvae which, after a short pelagic phase, attach to the substratum and develop a short stalk. At this stage they are known as pentacrinoid larvae. The pentacrinoids eventually detach and by this time the small, prehensile cirri have developed on the undersurface of the disc. Sexual maturity is reached in the second year, and in populations in the English Channel eggs and early larvae are brooded during May to July. The corkwing wrasse, *Symphodus melops* (p. 496), has been observed to feed extensively on the genital pinnules.

A second species of *Antedon*, **Antedon petasus** (Düben & Koren), has a more northerly distribution around the British Isles than *A. bifida* and is sometimes confused with the latter species. It is larger and has 50 or so short cirri on the undersurface of the disc.

## Class Asteroidea

The asteroids or starfishes are readily identified by the presence of arms which grade into the central disc. In some cases the arms are very short, and in some the animal has a flattened, star-shaped appearance. The undersurface has a central mouth, and rows of tube-feet running along the length of each arm. The anus is usually on the upper surface but is often very difficult to see. Spines are developed to varying degrees, and in some asteroids they are modified as pedicellariae (p. 432). These can often be seen under a hand lens, but in some cases it is necessary to take scrapings from the body surface and examine under a microscope. The tips of some spines end in a brush- or comb-like arrangement and these are known as paxillae (Fig. 290a). The presence or absence of pedicellariae and paxillae can be important in identification. The tube-feet are used in locomotion and, in some species, for feeding, and are often equipped with terminal suckers. In burrowing forms the tube-feet may be pointed. Starfishes are widely distributed in the sea, with some species moving sluggishly over the bottom, while others are adapted to burrowing in soft sediments. Many are carnivorous and feed on a wide range of invertebrates, sometimes by everting thin-walled lobes of the stomach through the mouth to engulf the prey. Cannibalism has been recorded for a number of asteroids. Some species are suspension feeders and extend the arms to trap suspended material in mucus.

A few asteroids are hermaphroditic, but most have separate sexes and external fertilization, with the development of a planktonic larva known as a bipinnaria. This has a free-swimming life of up to several weeks; eventually it develops short arms, tipped with adhesive cells and is known as a brachiolaria larva. The brachiolaria attaches temporarily to the substratum and metamorphoses into a tiny starfish. In some asteroids either the bipinnaria or brachiolaria stage is omitted; others brood their eggs and lack a free-swimming larval stage altogether. Asteroids are capable of regenerating lost arms.

### Key to species of asteroids

1 Marginal plates (see Fig. 287) along arms, large and conspicuous on upper and/or lower surfaces. Tube-feet without suckers ......................................................................... 2
 Marginal plates of arms inconspicuous ......................................................................... 3
2 Both upper and lower marginal plates of arms conspicuous; 5 arms .............................
 ................................................................................. *Astropecten irregularis* (p. 436)

Lower marginal plates of arms conspicuous but upper plates replaced by brush-like spines (paxillae); 7 arms ................................................................ *Luidia ciliaris* (p. 437)

3    More than 5 arms (usually 9–12) ...................................................................... 4

     With 5 arms (rarely 4 or 6) ............................................................................... 5

4    Eight to 14 (usually 10–12) arms. Brush-like spines (paxillae) along margins of arms in single conspicuous row ...................................... *Crossaster papposus* (p. 439)

     Seven to 13 (usually 9–10) arms. Brush-like spines (paxillae) along margins of arms, inconspicuous, in double row ................................... *Solaster endeca* (p. 440)

5    Very short, blunt arms; flattened pentagonal shape, but upper surface of disc rather rounded; usually olive-green to brown in colour; up to 50 mm across ............................. 6

     Arms not short; characteristic starfish appearance .......................................... 7

6    Upper surface of central disc with dark, star-shaped area. Up to 15 mm across ............. ........................................................................ *Asterina phylactica* (p. 439)

     Upper surface of central disc not marked with star-shaped area. Up to 50 mm across ... ........................................................................ *Asterina gibbosa* (p. 438)

7    Upper surface not conspicuously spiny. Arms rather stiff, cylindrical in section. No pedicellariae ........................................................................ *Henricia* spp. (p. 441)

     Upper surface of arms with longitudinal rows of spines. Pedicellariae present ................. 8

8    Upper surface with short spines; row of whitish spines along mid-line of arms. Arms broad at base, tapering .............................................. *Asterias rubens* (p. 442)

     Upper surface very spiny, with conspicuous rows of spines on upper and lateral margins of arms. Pedicellariae grouped around base of spines. Arms long, narrow, tapering gradually .............................................. *Marthasterias glacialis* (p. 443)

*Astropecten irregularis* (Pennant)    (Fig. 287)

*Flattened body, stiff; 5 arms with conspicuous marginal plates on upper and lower surfaces with spines arising from lower marginal plates; up to 120 mm across, rarely up to 200 mm. Pointed tube-feet without suckers. Upper surface varied in colour, pink, yellow, orange, brown, sometimes with purple marks; undersurface pale.*

Figure 287   *Astropecten irregularis.*

A. *irregularis* is a sublittoral species, widely distributed in north-west Europe and found all around Britain in sand and gravel down to 1000 m. It lies buried in the sediment and contact with the surface is maintained by the tips of the arms. Specimens are often washed ashore, particularly after storms. It is carnivorous and uses the tube-feet to push food into the mouth. It feeds mainly on young bivalves, often digging down into the sediment to locate prey, and is believed to be an important predator of benthic animals. Brittle-stars, polychaete worms and crustaceans are also eaten. The sexes are separate and breeding apparently takes place during the summer months. There is a bipinnaria larva but no brachiolaria in the life-cycle.

### *Luidia ciliaris* (Philippi) (Fig. 288)

*Body with 7 arms; usually up to 400 mm across but occasionally 600 mm. Marginal plates on undersurface of arms conspicuous and bear spines; upper marginal plates replaced by brush-like spines known as paxillae. Tube-feet long, without suckers. Upper surface red, lower surface paler.*

Figure 288 *Luidia ciliaris.*

*L. ciliaris* is widely distributed in north-west Europe from the shallow sublittoral to depths of 400 m and is common around Britain. It is occasionally found on the lower shore, particularly on south and west coasts but is not found off southern North Sea coasts. *Luidia* burrows into sandy and gravelly sediments and feeds mainly on other echinoderms. It is a fragile species and specimens are often found with one or more regenerating arms. Off the south coast of England it breeds in summer.

A second species of *Luidia*, **Luidia sarsi** Düben & Koren, is similar to *L. ciliaris*, but is found in deeper water in muddy deposits. It has five arms.

### *Asterina gibbosa* (Pennant)    Cushion-star (Fig. 289a)

*Body with 5 (rarely 4 or 6) stubby arms; up to 50 mm across. Ventral surface flat, dorsal surface rather rounded. Colour uniform but varied, olive-green, brown, orange.*

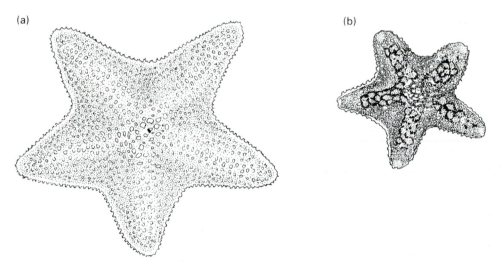

(a)

(b)

Figure 289  (a) *Asterina gibbosa*. (b) *Asterina phylactica*.

*A. gibbosa* is a southern species reaching the northern limit of its distribution on west and south-west coasts of Britain. There are only scattered records of the species north of the Isle of Man and it has not been recorded from North Sea coasts. Found under boulders and stones on the lower shore on sheltered and semi-exposed shores, and also in rock pools, it extends into the sublittoral to depths of about 100 m. Specimens from the lower shore and sublittoral are generally pale orange or grey in colour, compared with olive-green for specimens from rock pools higher on the shore. *Asterina* is an omnivorous scavenger feeding on microorganisms, decaying seaweed and dead invertebrates by everting thin-walled lobes of its stomach through the mouth. It is a protandrous hermaphrodite and in south-west Britain matures at a diameter of about 10 mm when two years old, first as a male, then changes to female when about four years old, at a diameter of about 20 mm. In May, females lay as many as 1000 orange-coloured eggs on the undersurface of rocks, and after 2–3 weeks these hatch as tiny starfishes. They reach a diameter of about 5 mm in the first year and live for about seven years, producing between three and seven broods.

### *Asterina phylactica* Emson & Crump    (Fig. 289b)

*Body with 5 (rarely 4 or 6) stubby arms; up to 15 mm across. Ventral surface flat, dorsal surface rather rounded. Colour usually dark olive green; very similar to* A. gibbosa *(above) but with dark star-shaped area in centre of disc.*

*A. phylactica* was first described from British waters in 1979 and has now been recorded from a number of sites on the south and west coasts of Britain, where it appears to favour rock pools with a rich growth of seaweed. It has also been recorded from the sublittoral. Small specimens of up to about 4.5 mm diameter (one year old) cannot reliably be separated from *A. gibbosa*. Above this size the colour differences are established. *A. phylactica* is an omnivorous scavenger. The species is a simultaneous hermaphrodite. Sexual maturity is reached at a diameter of 5–6 mm, at an age of two years, and in May aggregations of up to 10 animals can be found in crevices on the shore where the eggs are laid. The adults stay together and, unlike *A. gibbosa*, protect the developing embryos for about three weeks until they hatch as tiny starfishes. Aggregation during egg laying is believed to ensure cross-fertilization, but the animals are capable of self-fertilization. Longevity is up to four years and up to three broods are produced.

### *Crossaster papposus* (Linnaeus)    (*Solaster papposus*), Common sun-star (Fig. 290a)

*Body with 8–14 (usually 10–12), short, thick arms; up to about 300 mm across but usually 100–200 mm. Conspicuous brush-like spines (paxillae) in single row along margins of arms. Dorsal surface covered with spines and thinly distributed paxillae. Reddish-purple in colour; arms may be banded, giving overall pattern of white rings on body. Undersurface whitish.*

*C. papposus* is a northern species, widely distributed in north-west Europe and common around the coasts of Britain, reaching its southern limit in the English Channel. It is occasionally found on rocks and stones on the lower shore and has been recorded in the sublittoral to depths of about 70 m. Sun-stars feed on a wide range of food items, including other echinoderms, particularly the common starfish *Asterias rubens* (p. 442). There is no larval stage in the life-cycle.

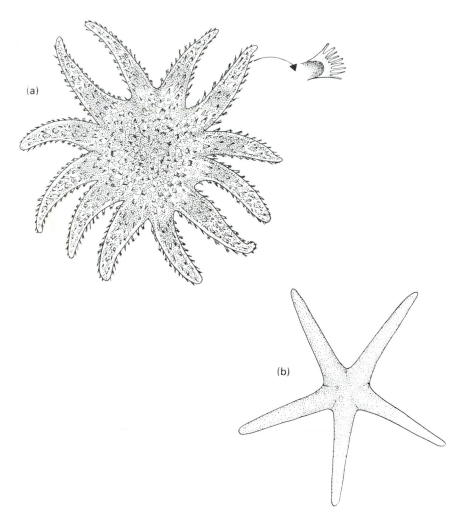

Figure 290  (a) *Crossaster papposus*, with detail of paxilla. (b) *Henricia* sp.

***Solaster endeca*** (Linnaeus)    Purple sun-star (Fig. 291)

*Body with 7–13 (usually 9–10) tapering arms; up to 400 mm across. Brush-like spines (paxillae), inconspicuous, in double row along margin of arms. Upper surface very hard and rough with somewhat inconspicuous spines and small paxillae. Cream coloured or shades of violet, purple, orange in colour.*

*S. endeca* is a northern species, widely distributed in north-west Europe and absent from English Channel coasts and the southern North Sea. It is found from the lower shore to depths of about 450 m. It feeds on other echinoderms. There is no larval stage in the life-cycle.

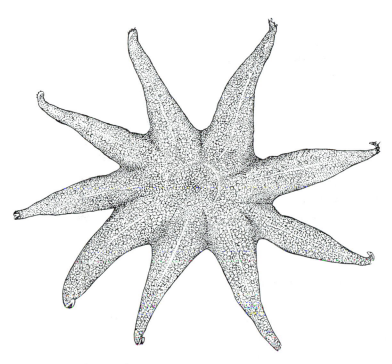

Figure 291  *Solaster endeca.*

*Henricia* spp.   (Fig. 290b)

*Body with 5 rounded, rather stiff arms, cylindrical in section; usually up to 120 mm across, occasionally much larger. No pedicellariae. Dorsal surface usually dark purple, red, occasionally yellowish; ventral surface pale.*

Separation of the different species of *Henricia* is difficult and requires microscopic examination of the calcareous plates of the endoskeleton.

*Henricia oculata* (Pennant) has been recorded from a wide range of localities around Britain, particularly the English Channel and west coast. It is found on hard substrata such as stones, shell and gravel, occasionally on the lower shore. *H. oculata* is a suspension feeder, extending the arms in the water to trap suspended particles in mucus. It also feeds on sponges, hydroids and detritus by stomach eversion, and scavenging on dead fish has been recorded. The sexes are separate and breeding occurs in spring, the gametes being liberated directly into the sea. The larva is believed to be pelagic.

A related species, *Henricia sanguinolenta* (O. F. Müller), is also recorded from Britain, particularly the north-east coast. In this species the female broods the eggs under the central disc until they hatch. There is no larval stage.

*Asterias rubens* Linnaeus    Common starfish (Fig. 292a)

*Body with 5 arms (occasionally 4 or 6), broad at base, tapering; usually up to 150 mm across but much larger specimens have been recorded. Arms round in section. Upper surface rough with row of whitish spines along mid-line of arms; pedicellariae present, especially on lateral walls of arms. Colour varied, upper surface from yellowish-brown to red, sometimes violet. Lower surface pale with conspicuous rows of tube-feet.*

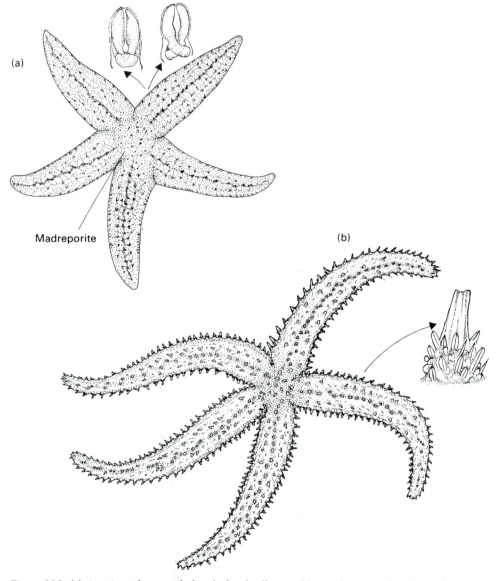

(a)

Madreporite

(b)

Figure 292  (a) *Asterias rubens*, with detail of pedicellariae. (b) *Marthasterias glacialis*, with detail of a single spine to show clusters of pedicellariae at base.

A. *rubens* is widely distributed in north-west Europe and abundant in many areas. It is commonly found on British shores, especially among mussels and barnacles, extending into the sublittoral to depths of about 400 m. In many areas it shows an onshore migration during spring. Although infrequently recorded from estuaries, laboratory studies suggest that North Sea populations tolerate salinities down to about 23‰. The tube-feet of *Asterias* have terminal suckers enabling it to attach to the substratum, so that when the animal is moving across a flat rock surface it is essentially stepping on its tube-feet. *Asterias* preys on a wide range of food items including molluscs, polychaete worms and other echinoderms. Small crustaceans are occasionally caught on the suction discs of the tube-feet, but the common starfish preys particularly on bivalve molluscs. The valves of the shell are forced open by the tube-feet, the tips of which attach to the bivalve shell by suction and pull the valves apart. As soon as a tiny gap, believed to be as small as 0.1 mm, is established between the valves of the shell, the starfish everts its stomach lobes which pass into the bivalve and commences digestion. Dense aggregations of starfishes are often seen feeding on the shore: one such aggregation on the west coast of England, measuring 1.5 km long and 15 m wide, was estimated to have cleared about 50 hectares (about 3500 to 4500 tonnes) of young of the edible mussel in four months. Starfishes are also important predators of oysters.

The sexes of *Asterias* are separate and fertilization external, and it has been estimated that a starfish of 140 mm diameter spawns as many as 2.5 million eggs. Breeding takes place in spring and summer and bipinnaria larvae are often abundant in summer plankton samples. The bipinnaria is followed by a brachiolaria larva and larval life may be as long as three months. On the south coast of Britain, sexual maturity has been recorded in the second year, when the starfish has a diameter of about 50 mm.

### *Marthasterias glacialis* (Linnaeus)    Spiny starfish (Fig. 292b)

*Body with 5 arms, long, narrow; tapering gradually; up to 600 mm across but usually 250–300 mm. Upper surface of arms very rough, with conspicuous spines; spines also present on lateral margins. Many pedicellariae round base of spines. Colour varied, yellowish, reddish, bluish-green.*

M. *glacialis* is widely distributed in north-west Europe; around Britain it is restricted to west and south-west coasts, where it is essentially a sublittoral species extending to depths of about 180 m, but occasional specimens are found at extreme low water on rocky shores. It feeds on a wide range of organisms, including algae, crustaceans, molluscs and other echinoderms. The sexes are separate, fertilization external and breeding aggregations have been recorded in summer. It is suggested that spawning takes place in shallow water. Both bipinnaria and brachiolaria larval stages are present in the life-cycle.

Class Ophiuroidea

The ophiuroids, or brittle-stars, are well represented on the lower shore where they are found in crevices, under stones and among algae. They extend into the sublittoral to the deep sea and are often found on gravelly bottoms or burrowing into soft sediments. In both situations they may be present in very high densities. Most are relatively small and have five long, slender arms and a distinct, rounded central disc, a character which is important in the separation of ophiuroids from asteroids. The arms of ophiuroids are made up of articulating calcareous ossicles which generally allow considerable twisting in the lateral plane. These ossicles, also known as vertebrae or segments, are each encased in a regular arrangement of four plates, known as the arm plates. There is a dorsal plate, a ventral plate and two lateral plates, the latter carrying spines known as the arm spines. The dorsal surface of the disc is also covered with plates and these are of varying size. At the base of each arm there is a pair of plates known as the radial shields (Fig. 293a). These are sometimes large and conspicuous. Plates known as arm combs may also be present at the base of each arm (Fig. 295a). The undersurface of the disc has a complex arrangement of plates and papillae and of these the plates surrounding the mouth, known as the oral shields, may be conspicuous (Fig. 295b). The mouth lies centrally on the undersurface of the disc; there is no anus. The tube-feet are without suckers; they are used in feeding but not locomotion. Pedicellariae are lacking. Ophiuroids are very active and when disturbed move quickly in characteristic fashion, the arms pushing against the substratum. They have a wide range of feeding habits including scavenging, deposit and suspension feeding. In most species the sexes are separate and fertilization external, leading to the development of a pelagic larva, the ophiopluteus. In a few, however, the embryos develop in a thin-walled sac within the body of the female until they emerge as juveniles.

### Key to species of ophiuroids

Ophiuroids can be difficult to identify to species. In most cases detailed examination of skeletal plates on the disc and the arms using a microscope or hand lens is required.

1  Arms very long, up to 15 times the diameter of the disc ....... *Amphiura brachiata* (p. 447)
    Arms up to about 8 times the diameter of the disc .......................................................... 2
2  Spines along margins of arms lie almost flat against arms. Disc covered by distinct
    scales. Conspicuous oral shields on undersurface (Fig. 295b) ........................................... 3
    Spines along margins of arms stand out, set at right angles to arm ................................... 4
3  Arm combs at base of each arm, each with about 30 papillae. Ventral plates at base of
    arm separated by grooves ......................................................... *Ophiura ophiura* (p. 448)
    Arm combs at base of each arm, each with less than 20 papillae. Ventral plates at base
    of arm not separated by grooves .................................................. *Ophiura albida* (p. 449)
4  Upper surface of disc with pair of conspicuous radial shields at base of each arm (Fig.
    293a) ......................................................................................................................... 5
    Radial shields obscured ............................................................................................... 6

5 Disc spiny. Radial shields large, up to ⅔ radius of disc; triangular in shape and
separate along length .............................................................. *Ophiothrix fragilis* (p. 445)
Disc covered with small scales. Radial shields conspicuous, joined for almost entire
length .............................................................. *Amphipholis squamata* (p. 448)

6 Disc covered with scales and small granules. Dorsal plates of arms oval, ringed by
small plates .............................................................. *Ophiopholis aculeata* (p. 446)
Disc completely covered with fine granules, smooth. Dorsal plates of arm not ringed by
small plates .............................................................. *Ophiocomina nigra* (p. 446)

***Ophiothrix fragilis*** (Abildgaard)    Common brittle-star (Fig. 293a)

*Central disc up to 20 mm diameter, spiny; specimens found on shore generally much
smaller. Arms about 5 times diameter of disc, slender, with many serrated spines. Arms very
fragile. Upper surface of central disc with 5-rayed pattern of spines; radial shields large,
triangular in shape; up to ⅔ radius of disc; separate along length and without spines. Very
varied in colour; red, white, brown, sometimes banded.*

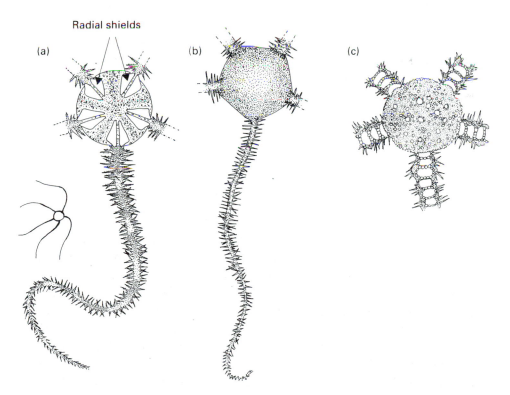

Figure 293  (a) *Ophiothrix fragilis*, whole animal inset. (b) *Ophiocomina nigra*.
(c) *Ophiopholis aculeata.*

O. *fragilis* is widely distributed in north-west Europe and often abundant around British coasts, extending to depths of about 150 m. Sublittorally, it forms dense aggregations, in which densities as high as 10 000 individuals per square metre have been recorded. It is from studies on such populations that most of our knowledge of the biology of this species has been gained. Intertidally it is found in crevices and under stones and seaweed. *Ophiothrix* is a suspension feeder: the arms are raised above the substratum and the tube-feet extended to filter suspended material from the water. Food passes along the arms to the mouth and consists of detritus and diatoms. Scavenging on decaying material has also been recorded. The sexes are separate and mature males and females have been recorded at disc diameters of over 3 and 5 mm, respectively. In the Irish Sea, breeding occurs in summer. Electron microscope studies on the ossicles of the arms suggest that the banding patterns which can be seen result from annual deposition of calcareous material, allowing age to be determined. It is suggested that O. *fragilis* is a long-lived species, living for 10 years or more. There is a planktonic larval stage, the ophiopluteus, and in dense, offshore aggregations larvae settle on the adults. It has been suggested that they feed semi-parasitically for a short while before migrating to other areas. *Ophiothrix* is often preyed on extensively by starfishes.

### *Ophiocomina nigra* (Abildgaard)   (Fig. 293b)

*Central disc up to 30 mm diameter. Arms about 5 times diameter of disc, with many long, smooth spines. Upper surface of central disc covered with many fine granules obscuring radial shields. Colour usually uniform black, grey or brown; may be pink or whitish.*

O. *nigra* is widely distributed in north-west Europe and around Britain but appears to have a patchy distribution on North Sea coasts. It lives on the lower shore under stones and seaweed, extending into the sublittoral to depths of about 400 m. Dense offshore aggregations have been recorded but, despite the high density of animals, individuals are spaced out, barely touching one another. *Ophiocomina* shows a range of feeding habits including deposit feeding, herbivorous browsing, carrion feeding and suspension feeding in which fine material is trapped in a mucous net and carried to the mouth by the tube-feet. The sexes are separate and fertilization external, with the development of a pelagic larva.

### *Ophiopholis aculeata* (Linnaeus)   Crevice brittle-star (Fig. 293c)

*Disc up to 15 mm diameter; covered with scales and small granules obscuring radial shields. Arms about 4 times diameter of disc. Dorsal arm plates oval-shaped, each surrounded by a circle of much smaller plates; 6–7 spines on either side of each arm segment. Colour red, purple; arms characteristically banded with darker colour.*

*O. aculeata* reaches the southern limit of its distribution in Britain, where it is common in the north but much less so in the south and the English Channel. It is found on the lower shore under stones and particularly in crevices with the arms often protruding. It has been recorded sublittorally to depths of about 200 m.

**Amphiura brachiata** (Montagu)    (*Acrocnida brachiata*) (Fig. 294a)

*Disc up to 12 mm diameter. Arms very long, slender, up to 15 times diameter of disc; with short spines. Upper surface of central disc covered with small scales; radial shields separate, conspicuous. Arm plates with many short spines, the number decreasing from base to tip of arm. Brownish-grey in colour.*

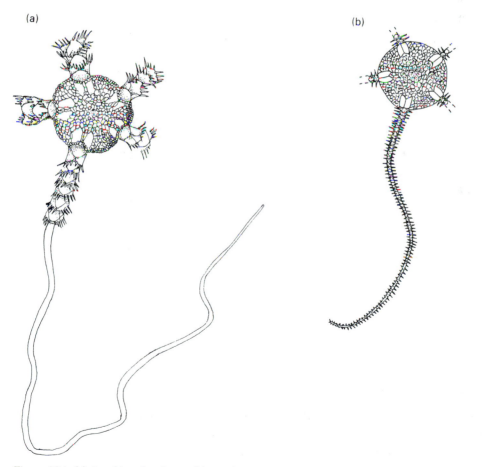

(a)

(b)

Figure 294  (a) *Amphiura brachiata*. (b) *Amphipholis squamata*.

*A. brachiata* is widely distributed on the shores of north-west Europe, extending sublittorally to depths of about 40 m. In Britain it appears to be most frequently found in the south and west, buried in sandy sediments, often in dense aggregations, with the tips of the arms extending above the surface. A bivalve mollusc, **Mysella bidentata** (Montagu), and the young stages of a polychaete worm of the genus *Harmothoë* (p. 168) are sometimes commensal with this brittle-star. *Amphiura* is a suspension feeder. It spawns during summer. Development is not known; it has been suggested that it has a brief pelagic phase.

### *Amphipholis squamata* (delle Chiaje)    (Fig. 294b)

*Small, central disc up to about 5 mm diameter. Arms about 4 times diameter of disc, with short spines. Upper surface of central disc with small scales; radial shields conspicuous, joined for almost entire length. Grey, white, blue-grey in colour.*

*A. squamata* is widely distributed in north-west Europe and around Britain, extending into the sublittoral to depths of about 250 m. It is found under stones, in sand, gravel and rock pools, often associated with *Corallina* (p. 61) and bryozoans. *A. squamata* is a deposit feeder using the tube-feet around the mouth to take a wide range of plant and animal material; it also suspension feeds, raising the arms above the sediment to trap fine material on mucus-coated spines. Food is transported to the mouth by the tube-feet. It is a phosphorescent species. *Amphipholis* is hermaphroditic; there is no pelagic larval stage, young are brooded by the parent. Maturity is believed to be reached at a disc diameter of about 1.5 mm, when the brittle-stars are in the first year of growth. Juveniles are released from the brood pouch during the summer at disc diameters of 0.8–1.2 mm. Studies on a rock pool population in Devon suggest a longevity of 12–18 months, but in the Firth of Forth some adults survive a second winter to reproduce in the following spring.

### *Ophiura ophiura* (Linnaeus)    (*Ophiura texturata*) (Fig. 295a, b)

*Disc up to about 35 mm diameter. Arms about 4 times diameter of disc. Upper surface of disc with coarse scales. Arm spines lie almost flat against arms. Radial shields large, and lie close to pair of smaller plates (the arm combs) which have comb-like edge with about 30 papillae along the margin. Oral shields, on undersurface of disc, large. Grooves between ventral plates at the base of the arms. Dorsal surface of disc red/brown, mottled, ventral side paler; radial shields red/brown.*

*O. ophiura* is widely distributed in north-west Europe and common all around Britain in sand and muddy-sand. It extends from the lower shore to depths of about 200 m and in the shallow sublittoral has been recorded in densities as high as 700 per square metre. It is a suspension feeder taking a wide range of food. The sexes are separate, fertilization external and there is a pelagic larva in the life-cycle. Newly metamorphosed juveniles

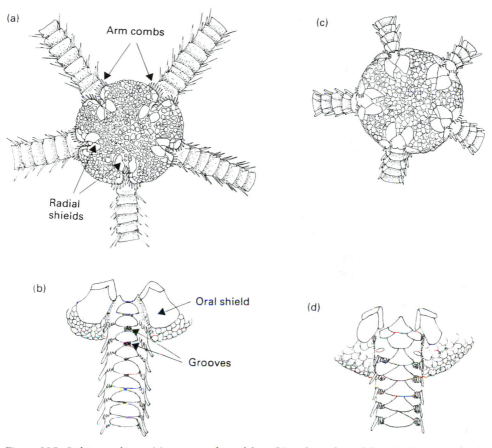

Figure 295  *Ophiura ophiura*, (a) upper surface of disc, (b) undersurface of disc. *Ophiura albida*, (c) upper surface of disc, (d) undersurface of disc. (Based on Mortensen, 1927.)

have a disc diameter of 0.2–0.3 mm, and studies on the microscopic growth bands on the arm ossicles suggest a longevity of five or more years.

*Ophiura albida* Forbes    (Fig. 295c, d)

*Disc up to 15 mm diameter. Arms about 4 times diameter of disc. Upper surface of disc with coarse scales. Arm spines lie almost flat against arms. Radial shields lie close to pair of smaller plates (the arm combs) which have comb-like edge with less than 20 papillae along the margin. Oral shields on undersurface of disc conspicuous. Ventral plates at base of arms not separated by grooves. Dorsal surface of disc red/brown; radial shields often whitish.*

*O. albida* is widely distributed in north-west Europe. It is common all around Britain in soft sediments from the lower shore into the sublittoral. It feeds on microorganisms, organic debris and carrion. The sexes are separate, fertilization external and there is a pelagic larva in the life-cycle. It has been suggested that longevity is up to three years.

## Class Echinoidea

The echinoids or urchins are characterized by a hard, calcareous test, which is globular, heart-shaped or flattened. There are no arms and the test is covered in movable spines. Urchins with a globular test are known as regular echinoids. In these forms, for example the common sea-urchin *Echinus* (p. 452), the spines are generally long and well developed. Those with an oval or heart-shaped, or much-flattened test are known as irregular echinoids and include the heart-urchins and sand dollars. These generally have small spines. In the regular echinoids the mouth lies on the undersurface of the test and the anus is on the upper surface. In irregular urchins the mouth is on the undersurface and the anus either on the posterior edge of the test or on the undersurface. In some echinoids the mouth is equipped with a complex feeding apparatus, the Aristotle's lantern, which includes five hard, calcified teeth capable of tearing algae and removing encrusting organisms from the rock surface. These teeth can generally be seen protruding from the mouth. The body of an echinoid has five double rows of tube-feet and in living specimens many of these can be seen to be long and highly extensible. On the dry, denuded test, the position of the tube-feet is indicated by pairs of tiny pores in the ambulacral plates (Fig. 297e) and through which the tube-feet emerge in life. Among the spines and tube-feet are large numbers of pedicellariae consisting of a long stalk bearing a head region. Different types of pedicellariae are shown in Fig. 297. The stalk is supported by a delicate calcareous rod which in some cases stops short of the head leaving a flexible region, the neck: in others, the skeletal rod runs up the base of the head. The head is made up of usually three delicate calcareous jaws, and in the globiferous pedicellariae the dentition of these is important in identification. Globiferous pedicellariae secrete a toxin.

Regular echinoids move slowly using the tube-feet and the spines, the former pulling and the latter pushing the animal along. Some species are adapted for boring into soft rocks such as limestone. Irregular echinoids are generally found burrowing into soft sediments and some of the spines have flattened ends and are used in burrowing. Other small, club-shaped spines are ciliated and create a flow of water through the burrow. They lie in definite tracks which in some species form conspicuous rings (fascioles).

Many echinoids graze on plant and animal material, and a number, including the common sea-urchin *Echinus* (p. 452), remove encrusting organisms by means of the Aristotle's lantern. Some of the irregular urchins or heart-urchins lack an Aristotle's lantern and use modified tube-feet to feed on particles of detritus. As echinoids grow, individual skeletal plates increase in size by deposition of calcite round their margins. In some species, for example *Echinus esculentus* (p. 452) and *Paracentrotus lividus* (p. 454), growth rings can be detected as contrasting light and dark bands in individual plates which have been carefully ground down with fine sand paper and examined under the microscope. Echinoids have separate sexes. Fertilization is external and the majority have a free-swimming pelagic larva known as an echinopluteus. In some, the eggs are brooded among the spines or in depressions on the surface of the test until they hatch as juveniles.

Key to species of echinoids

Although many echinoids can be identified by the shape and coloration of the test, the nature and colour of the spines and the habit or habitat, in some cases further criteria are needed to confirm identification. The shape of the jaws of the *globiferous* pedicellariae and the arrangement of the tube-feet are two important characteristics. Pedicellariae have stalks measuring up to about 4 mm in length and are conspicuous when the test is seen under a stereo-microscope. They can be removed with fine forceps without serious damage to the specimen and should be carefully examined to ensure that it is the globiferous pedicellariae that have been removed (see Fig. 297c); they should then be compared with the drawings given for the particular species. In order to determine the arrangement of the tube-feet, the spines have to be removed and the test cleaned with a bleach solution. This reveals the arrangement of the tiny pores in the plates (ambulacral plates) of the ambulacral region of the test (Fig. 297e) through which the tube-feet project. In the following key, the shape and coloration of the test and colour of the spines are given first and these are followed, where appropriate, by the nature of the globiferous pedicellariae and arrangement of tube-feet to be used for confirmation as required.

1  Test rounded, radially symmetrical – 'regular' echinoids ................................................ 2
   Test oval, heart-shaped – 'irregular' echinoids ......................................................... 5
2  Test somewhat flattened, greenish-brown in colour. Spines green, red or violet, often with white tips. Restricted to Shetland and the east coast of Britain. Jaws of globiferous pedicellariae without teeth below terminal tooth ..........................................
   ................................................ *Strongylocentrotus droebachiensis* (p. 455)
   Coloration of test and spines not as above. Jaws of globiferous pedicellariae with 1 or more teeth below terminal tooth ............................................................................. 3
3  Test flattened, greenish in colour; with long violet- or olive-coloured spines. Bores into rock. Found on west coast of Ireland, rarely south-west England. In denuded test, each ambulacral plate has 5 or 6 pairs of pores ........................... *Paracentrotus lividus* (p. 454)
   Test coloration and habitat not as above. In denuded test, each ambulacral plate has 3 pairs of pores ............................................................................................. 4
4  Test slightly flattened; overall greenish colour. Tips of spines with violet tinge. Jaws of globiferous pedicellariae with row of teeth below terminal tooth .....................................
   ................................................ *Psammechinus miliaris* (p. 452)
   Test may be slightly flattened at poles, but overall globular; reddish colour. Jaws of globiferous pedicellariae with 1 tooth below terminal tooth ... *Echinus esculentus* (p. 452)
5  Feeding apparatus (Aristotle's lantern) present. Mouth and anus on lower surface, mouth central. Very small, no more than 15 mm in length, distinctive ..........................
   ................................................ *Echinocyamus pusillus* (p. 455)
   Aristotle's lantern not present. Mouth on undersurface, towards anterior; anus on posterior edge of test. Large, up to 120 mm in length ........................................... 6

6  Violet colour. Spines mostly directed backwards; long curved spines on upper surface
   paler than other spines, may be white ............................... *Spatangus purpureus* (p. 456)
   Yellow-brown colour; close-set spines directed backwards ................................................. 7
7  Conspicuous anterior furrow on test. In profile, highest point of test towards posterior
   ...................................................................... *Echinocardium cordatum* (p. 456)
   No anterior furrow on test. In profile, highest point of test towards centre ....................
   ...................................................................... *Echinocardium pennatifidum* (p. 457)

**Psammechinus miliaris** (Gmelin)    (Fig. 296)

*Test up to 50 mm diameter; somewhat flattened; greenish in colour. Spines green with*
*purple tips. Aristotle's lantern present. Jaws of globiferous pedicellariae with row of teeth*
*below terminal tooth. In denuded test, each ambulacral plate has 3 pairs of pores.*

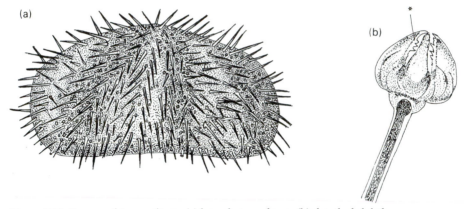

Figure 296  *Psammechinus miliaris*, (a) lateral view of test, (b) detail of globiferous
           pedicellaria.

*P. miliaris* is widely distributed in north-west Europe and occurs all around Britain under
stones and in crevices on the lower shore. Often found with seaweeds, it may be camouflaged by pieces of weed trapped among the spines. *Psammechinus* is omnivorous and
consumes a wide variety of plant and animal material including barnacles, bivalve molluscs and ascidians. The sexes are separate, fertilization external and breeding occurs
during spring and summer. The larva is pelagic for about four weeks before metamorphosing and settling on the shore at 0.5–1 mm diameter. Sexual maturity is reached
after one or sometimes two years when the urchins are generally between 6 and 10 mm
diameter. Longevity is believed to be up to 10 years.

**Echinus esculentus** Linnaeus    Common sea-urchin (Fig. 297)

*Test almost globular; usually 100–120 mm diameter, may be larger. Aristotle's lantern*
*present. Pinkish-red colour, also shades of green, yellow or purple. Spines rather short; often*
*coloured red and white with lilac tips. Jaws of globiferous pedicellariae with 1 tooth below*
*terminal tooth. Denuded test with white tubercles; each ambulacral plate with 3 pairs of*
*pores.*

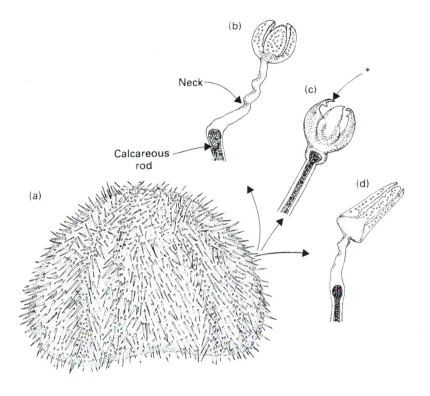

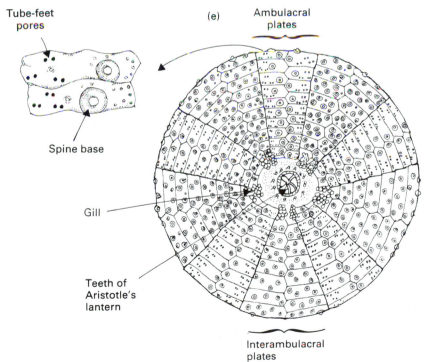

Figure 297 *Echinus esculentus*, (a) lateral view of test, with detail of (b) ophiocephalous,
(c) globiferous (note teeth on jaws) and (d) tridentate pedicellariae. (e) Oral
view of denuded test with detail of ambulacral plates showing spine bases and
tube-feet pores.

*E. esculentus* is widely distributed in north-west Europe. Although sometimes found on the lower shore, highest densities are recorded offshore and estimates of as many as 2000 individuals per hectare have been made for populations in the English Channel. It has been found in depths of 100 m and more. The common sea-urchin shows considerable variation in the shape of the test. In some areas specimens from shallow water have a relatively flat test, while those from deeper water are taller. Coloration is also varied: in some specimens the spines are slender, in others they are more robust. *Echinus* browses on seaweed and encrusting invertebrates using the sharp teeth of the Aristotle's lantern and is capable of clearing large areas of rock. The sexes are separate and breeding occurs in spring when male and female gametes are shed into the sea where fertilization takes place. It has been estimated that large specimens contain as many as 20 million eggs. The fertilized eggs develop into echinopluteus larvae which are planktonic for seven to eight weeks before metamorphosing into tiny urchins about 1 mm in diameter. Rate of growth depends on locality and it has been calculated that urchins can reach a diameter of 40 mm in the first year. Studies on the west coast of Scotland suggest that urchins are slow growing and may survive for more than 16 years.

### *Paracentrotus lividus* (Lamarck)    (Fig. 298a–d)

*Test somewhat flattened; up to 70 mm diameter. Aristotle's lantern present. Brownish-green in colour; closely covered with long spines. Spines long, sharply pointed; purple, brown or green in colour. Jaws of globiferous pedicellariae with 1 tooth below terminal tooth. In denuded test, each ambulacral plate has 5 or 6 pairs of pores.*

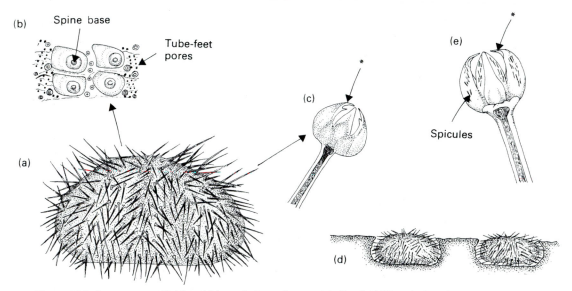

Figure 298 *Paracentrotus lividus*, (a) lateral view of test, with detail of (b) ambulacral
plates, (c) globiferous pedicellaria. (d) Specimens in characteristic depressions
in rock. (e) Globiferous pedicellaria of *Strongylocentrotus droebachiensis*.

*P. lividus* is found as far north as the Channel Islands and the west coast of Ireland. There are a few records of its presence in south-west England and it is on the west coast of Ireland that it is widespread, being abundant in some areas, such as the shores of County Clare. It has some commercial importance as the roe is considered a delicacy. It is found on the lower shore in rock pools and typically bores into rock, often limestone, by means of its spines, and the teeth of the Aristotle's lantern. Each urchin fits tightly into its cavity and thus derives some protection from wave action and from desiccation during low tide. Some are believed never to leave the cavity. They are found in the sublittoral to depths of 30 m. *Paracentrotus* grazes on seaweed, detritus and algal sporelings, keeping in check, and in some situations eliminating, the growth of seaweeds. An important source of food, particularly in very dense aggregations, is believed to be the fragments of seaweed brought in by the tide and trapped on the spines of the urchin. The sexes are separate and fertilization external with the development of a pelagic larva. On the west coast of Ireland, breeding has been recorded in January–March and August–September, individuals reaching a diameter of 35–50 mm in about four years and living for six to nine years.

### *Strongylocentrotus droebachiensis* (O. F. Müller)    (Fig. 298e)

*Test somewhat flattened; up to 80 mm diameter. Aristotle's lantern present. Greenish-brown. Spines green, red or violet, often with white tips. Jaws of globiferous pedicellariae without teeth below terminal tooth; with spicules on sides of jaws. In denuded test, each ambulacral plate has 5 or 6 pairs of pores.*

*S. droebachiensis* is widespread and often abundant in north-west Europe. While very similar to *P. lividus* (above), it is, however, only found on the east coast of Britain. It is found on the lower shore where it often bores into rock and extends well into the sub-littoral. *Stronglyocentrotus* grazes on the rock surface and feeds on seaweeds and encrusting invertebrates. The sexes are separate, fertilization external and breeding occurs in spring. There is a pelagic larva.

### *Echinocyamus pusillus* (O. F. Müller)    Green urchin, Pea-urchin (not illustrated)

*A distinctive, very small echinoid. Oval, much-flattened test with short, closely set spines; up to 15 mm in length; mouth and anus on lower surface, mouth central; Aristotle's lantern present. Greenish-grey in colour, becoming brighter green when damaged. In denuded test the ambulacral plates form a petal-like arrangement.*

*E. pusillus* is widely distributed in north-west Europe and found all around Britain in coarse sand or gravel from extreme low water into the subittoral. It feeds on a wide variety of small organisms and detritus. The sexes are separate, fertilization external and there is a pelagic larva. Breeding occurs in summer.

*Spatangus purpureus* O. F. Müller    Purple heart-urchin (Fig. 299)

*Heart-shaped test broad and low; up to 120 mm in length; covered with large number of shortish spines, mostly directed backwards. Mouth on undersurface towards anterior. Conspicuous anterior furrow on test. Violet in colour; long, curved spines on upper surface paler or white in colour. In denuded test the ambulacral plates form a petal-like arrangement.*

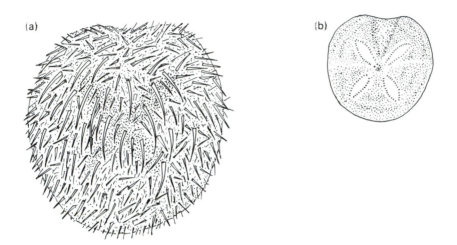

Figure 299  *Spatangus purpureus*, (a) dorsal view of test, (b) denuded test showing
petal-like arrangement of ambulacral plates. ((b) After Forbes, 1841.)

S. *purpureus* has a wide but patchy distribution in north-west Europe. It is found all around Britain buried in coarse sand and gravel on the lower shore, extending into the sublittoral to depths of about 900 m. It is sometimes locally common. The bivalve **Montacuta substriata** (Montagu) is often commensal with *Spatangus*. The sexes are separate, fertilization external and breeding occurs in summer. There is a pelagic larva.

*Echinocardium cordatum* (Pennant)    Sea-potato, Heart-urchin (Fig. 300)

*Heart-shaped test, usually 40–50 mm in length, but up to 90 mm; covered with large number of closely set spines, mostly directed backwards. Mouth on undersurface towards anterior. Conspicuous anterior furrow on test. In profile, highest point of test towards posterior. Yellow-brown colour. In denuded test the ambulacral plates form a petal-like arrangement.*

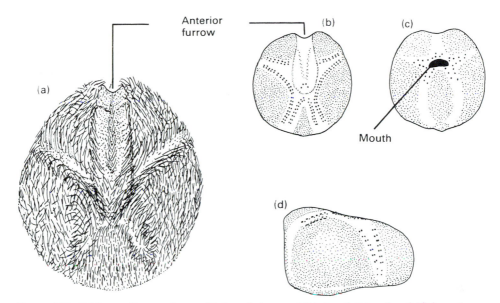

Figure 300 *Echinocardium cordatum*, (a) dorsal view, and (b) dorsal, (c) oral and (d) lateral views of denuded test.

*E. cordatum* is widely distributed in north-west Europe and occurs all around the coasts of Britain, extending to depths of over 200 m. The denuded tests of dead specimens are frequently washed ashore, particularly after storms. Living specimens are found on the lower shore buried to depths of about 150–300 mm in sandy sediments. Contact with the surface is maintained by a respiratory channel that leaves a funnel-shaped opening on the surface of the sediment and reveals the position of the urchin. *Echinocardium* is a deposit feeder. Tube-feet pick up particles of sand and detritus from in front of the mouth and, in addition, long, highly extensible tube-feet extend up the respiratory funnel and sometimes pick up particles of sediment and detritus. The sexes are separate and fertilization external, with the development of a pelagic larva. Breeding occurs in summer with individuals from a littoral population on the north-east coast of England breeding for the first time when three years old. On the west coast of Scotland, breeding has been recorded at the end of the second year. Longevity is believed to be about 10 years. The bivalve *Tellimya ferruginosa* (p. 297) is often commensal with *Echinocardium*.

**Echinocardium pennatifidum** Norman    (not illustrated)

*Heart-shaped test, up to 70 mm in length; covered with large number of closely set spines, mostly directed backwards. Mouth on undersurface towards anterior. No anterior furrow on test; in profile, highest point of test is central, cf.* E. cordatum *(above). Yellow-brown colour. In denuded test the ambulacral plates form a petal-like arrangement.*

*E. pennatifidum* is widely distributed in north-west Europe and around Britain appears
to be absent from the south-east coast. Although generally not common, it is found on
the lower shore in large numbers in some areas, such as the Channel Islands. It extends
to depths of 250 m.

Class Holothuroidea

The holothurians or sea-cucumbers are elongate, often worm-like echinoderms. They are
without arms. The mouth is anterior, the anus posterior and the animal lies on one side,
the ventral or trivium, which is often marked by three, longitudinal rows of tube-feet.
The dorsal surface, or bivium, is often marked by two rows of tube-feet. In some holothurians, the tube-feet are scattered over the surface of the body, while in others they
are absent. When relaxed, the mouth is seen to be surrounded by a ring of retractile,
branched tentacles which are modified tube-feet. The skin is generally thick and tough.
When dried and smoked it is known as 'trepang' or 'bêche-de-mer', which is regarded
as a delicacy by the Chinese. The calcareous endoskeleton is much reduced, generally
consisting of microscopic, calcareous plates which in some species, such as *Leptosynapta*
(p. 461) and *Labidoplax* (p. 462), are in the form of wheels and anchors. These plates
are important in species identification. Holothurians are found buried in mud and sand,
sometimes in U-shaped burrows, or attached to stones and rock. Although not generally
common on the shore, they are sometimes abundant offshore, extending into the deep
sea. They are either suspension feeders, using mucus-coated tentacles to trap suspended
particles, or deposit feeders in which the tentacles push bottom sediments into the
mouth. The sexes are generally separate, although hermaphroditic species are known.
Many have a pelagic larva known as an auricularia, but in some the eggs are brooded,
either in pouches on the outside of the body or internally.

Key to species of holothurians

Holothurians are generally recognized by the shape of the body, the number and shape
of the tentacles, and the nature and distibution of the tube-feet. Specimens in which
the tentacles are retracted can be relaxed by immersing in seawater containing 7%
magnesium chloride. Where tube-feet are absent, the presence of calcareous spicules in
the body wall confirms that the specimen is a holothurian. Where necessary, these can
be isolated for examination under the microscope by digesting a small piece of skin in
bleach.

1  Tube-feet well developed ................................................................................ 2
   Without tube-feet. Worm-like body, may be tacky to the touch because of the presence
   of calcareous spicules projecting through skin ................................................ 5
2  Twenty short, stumpy, branched tentacles. Ventral tube-feet well developed, in 3,
   sometimes 4, longitudinal rows; reduced to conical papillae on dorsal surface. Up to
   250 mm in length ................................................ *Holothuria forskali* (p. 459)

Tentacles long, tree-like; usually 10 in number, 8 long, 2 short. Tube-feet in 5 longitudinal rows, single or double; some tube-feet may be scattered over surface ........... 3

3 Tube-feet in irregular, single rows. Skin thick, smooth, white in colour. No more than 40 mm in length ........................................................................ *Ocnus lacteus* (p. 461)

Tube-feet in 5 rows, rows may be single or double ......................................... 4

4 Skin tough, wrinkled. Tube-feet in 5 double, occasionally multiple, rows ..................... ...................................................................................... *Aslia lefevrei* (p. 460)

Skin relatively thin and smooth. Tube-feet in 5 rows, 3 ventral rows double, with cylindrical tube-feet; 2 dorsal rows irregular with conical, papilla-like tube-feet ............ ...................................................................................... *Pawsonia saxicola* (p. 459)

5 Twelve much-branched tentacles, the branches increasing in length to tip ................... ...................................................................................... *Leptosynapta inhaerens* (p. 461)

Twelve tentacles, each with 2 pairs of branches at tip ........... *Labidoplax digitata* (p. 462)

**_Holothuria forskali_ delle Chiaje    Cotton spinner (Fig. 301)**

*Cylindrical body up to 250 mm in length, flattened on ventral side. Skin soft and thick. Ventral tube-feet well developed, in 3 longitudinal rows; conical papillae on dorsal surface; 20 short stumpy, branched tentacles. Dorsal surface usually dark brown, black; ventral surface paler; tentacles yellowish.*

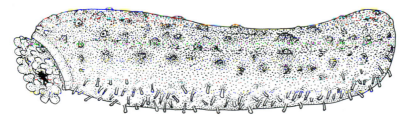

Figure 301  *Holothuria forskali.*

*H. forskali* is found on the Atlantic coast of France and south and west coasts of Britain. It is occasionally found at extreme low water among rocks and stones, but is more common in soft sublittoral sediments and has been recorded at 300 m. When roughly handled, it discharges a mass of sticky threads, the Cuvieran organs or cotton glands, from which it gets its common name. On discharge, the threads disorientate and entangle would-be predators. An annual reproductive cycle has been described, and in a sublittoral population off Brittany, spawning occurred in April. The sexes are separate, fertilization external and there is a pelagic larva. Breeding occurs in summer.

**_Pawsonia saxicola_ (Brady & Robertson)    (_Cucumaria saxicola_), Sea-gherkin (Fig. 302)**

*Cylindrical body up to about 150 mm in length. Skin relatively thin and smooth. Tube-feet in 5 rows; 3 ventral rows most distinct and double, with cylindrical tube-feet; 2 dorsal rows*

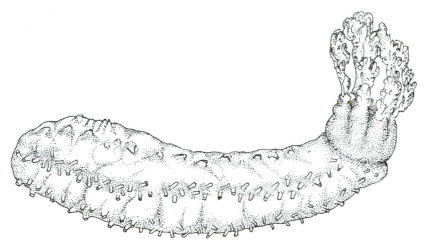

Figure 302  *Pawsonia saxicola.*

*irregular, single, with conical, papilla-like tube-feet; 10 tentacles, tree-like; 8 long, 2 short. Body white, grey, sometimes blackish; tentacles blackish. Darkens on exposure to light.*

*P. saxicola* occurs on the south coast of England, the south and west coast of Ireland and the Atlantic coast of France. It lives in crevices, under stones and on mussel beds, from the lower shore to 50 m. It is often found with *Aslia* (below) with which it may be confused, but *Pawsonia* has smoother skin and a different arrangement of tube-feet. Breeding is believed to occur during spring. There is a pelagic larva.

**Aslia lefevrei** (Barrois)    *(Cucumaria normani)* (Fig. 303)

*Cylindrical body up to about 150 mm in length. Skin tough, wrinkled. Tube-feet in 5 double, occasionally multiple, rows; 10 tentacles, tree-like; 8 long, 2 short. Body brownish colour, tentacles blackish; darkens on exposure to light.*

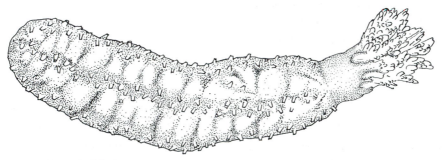

Figure 303  *Aslia lefevrei.*

A. *lefevrei* occurs on the Atlantic coast of France, the south coast of England, the west of Ireland and the west coast of Scotland to Orkney. It lives in crevices and under stones, and has been recorded in dense aggregations of more than 70 per square metre. It extends from the lower shore into the sublittoral to depths of about 20 m. *Aslia* is a suspension feeder. Suspended material is trapped on extended tentacles, equipped with adhesive papillae, and the tentacles are then pushed into the mouth where the food is removed. On the west coast of Ireland, breeding takes place in early spring. The sexes are separate and fertilization external. The larva is pelagic and, following metamorphosis, a size of 30–35 mm body length is achieved in the first year, with growth of up to 80 mm after two and a half years. A body length of 180 mm has been recorded for specimens living in the sublittoral, and it has been suggested that longevity may be more than 10 years.

***Ocnus lacteus*** (Forbes & Goodsir)  (*Cucumaria lactea*) (Fig. 304)

*Cylindrical body up to 40 mm in length, usually smaller. Skin thick, smooth. Tube-feet in irregular, single rows; 10 tentacles, tree-like; 8 long, 2 short. Body white in colour, tentacles yellowish.*

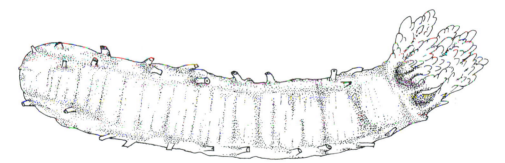

Figure 304 *Ocnus lacteus.*

*O. lacteus* is widely distributed in north-west Europe from Iceland to the Atlantic coast of France. In Britain it is apparently absent from southern North Sea coasts. It is found on the lower shore on stones and shells, often among hydroids and *Corallina* (p. 61), extending sublittorally to about 100 m.

***Leptosynapta inhaerens*** (O. F. Müller)  (Fig. 305a)

*Elongate, worm-like body, generally up to 150 mm in length, may reach 300 mm. Skin soft, semi-transparent; adhesive to touch due to presence of tiny, anchor-shaped calcareous plates which project through skin. Without tube-feet; 12 much-branched tentacles, the branches increasing in length to the tip. Body pale-pink to greyish-white in colour.*

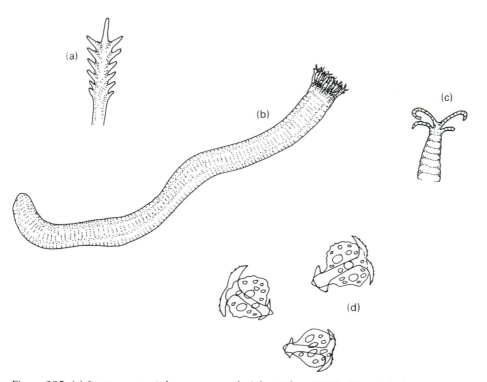

Figure 305  (a) *Leptosynapta inhaerens*, tentacle (after Eales, 1950). (b) *Labidoplax
digitata*. (c) *Labidoplax digitata*, tentacle. (d) *Labidoplax digitata*, calcareous
plates from body wall.

*L. inhaerens* is widely distributed in north-west Europe. Around Britain it is found on
the west coast as far north as the Shetlands, and on the north-east coast of Scotland.
It lives buried in sand and mud from the lower shore to depths of about 50 m. It is
hermaphroditic; development is direct.

### *Labidoplax digitata* (Montagu)   (Fig. 305b–d)

*Body elongate, worm-like, usually 150 mm in length, may be up to 300 mm. Skin soft,
adhesive to touch due to presence of tiny, anchor-shaped calcareous plates which project
through skin. Without tube-feet; 12 tentacles, each with 2 pairs of branches at tip. Pink or
brownish in colour.*

*L. digitata* is found on the Atlantic coast of France and south-west and western coasts
of Britain in muddy-sand from the lower shore sublittorally to depths of about 70 m.
The sexes are separate and fertilization external; the larvae are pelagic.

REFERENCES

Brun, E. (1976). Ecology and taxonomic position of *Henricia oculata* Pennant. *Thalassia Jugoslavica*, **12**, 51–64.

Clark, A. M. & Downey, M. E. (1992). *Starfishes of the Atlantic*. London: Chapman & Hall.

Crump, R. G. & Emson, R. H. (1983). The natural history, life history and ecology of the two British species of *Asterina*. *Field Studies*, **5**, 867–82.

Eales, N. B. (1950). *The littoral fauna of Great Britain*. Cambridge: Cambridge University Press.

Forbes, E. (1841). *A history of British starfishes, and other animals of the class Echinodermata*. London: Van Voorst.

McKenzie, J. D. (1991). The taxonomy and natural history of north European dendrochirote holothurians (Echinodermata). *Journal of Natural History*, **25**, 123–71.

Mortensen, Th. (1927). *Handbook of the echinoderms of the British Isles*. Oxford: Oxford University Press. (Reprinted 1977. Leiden: E. J. Brill.)

Picton, B. E. (1993). *A field guide to the shallow-water echinoderms of the British Isles*. London: Immel Publishing.

# Hemichordata

The hemichordates are marine animals with the body divided into three regions: proboscis, collar and trunk. Some are sessile and colonial, and live in deep water, but others are worm-like and found intertidally burrowing in soft sediments.

## Phylum HEMICHORDATA

**Class Enteropneusta**   The acorn worms. Solitary, unsegmented, worm-like hemichordates, often burrowing in mud and sand. Also found under rocks and stones. The body is clearly divided into a proboscis, a short collar and a long trunk.

Class Pterobranchia   Sessile hemichordates, no more than 12 mm in length, usually found in deep water. Colonial. Tube dwelling.

### Class Enteropneusta

The enteropneusts have a soft, unsegmented, worm-like body divided into three regions: a conical, ciliated proboscis, a short, cylindrical collar and a long trunk. The mouth lies on the ventral side between the proboscis and collar and the anus is terminal. The anterior region of the trunk has pairs of gill openings on the dorsal surface. Some species live in U-shaped burrows in sandy and muddy deposits and ingest large quantities of sediment. Others are found under rocks and stones, and suspension feed using the cilia on the proboscis to create feeding currents. The sexes are separate, fertilization external and the characteristic pelagic larval stage, known as the tornaria, resembles the larvae of some echinoderms. Some species show direct development and asexual reproduction by fragmentation has been recorded.

A small number of species is found in north-west Europe, and in Britain they have a patchy distribution, tending to be most common on west and south coasts. They are readily separated from other worm-like animals by the division of the body into three distinct regions. Enteropneusts are fragile and must be handled with care if fragmentation is to be avoided. They are not common on the shore; a single species is described here.

**Saccoglossus ruber** (Tattersall)    (*Saccoglossus cambrensis*) (Fig. 306)

*Proboscis long, up to 30 mm; pink to pale orange in colour; collar longer than broad, reddish in colour; trunk long and coiled; fragile, with 50–95 pairs of gill openings; orange, yellow colour, with small red spots. Whole animal up to 200 mm in length, generally smaller.*

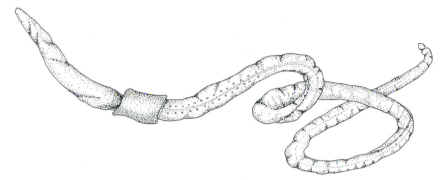

Figure 306  *Saccoglossus ruber* (after Brambell & Cole, 1939).

*S. ruber* has been recorded from a number of localities on the west coasts of Ireland, Scotland and Wales. It lives in U-shaped burrows on the lower shore in sand and muddy-sand. Coiled faecal casts are left on the surface of the sediment and the burrows extend to depths of about 200 mm. Breeding probably takes place in spring and summer.

REFERENCES

Brambell, F. W. R. & Cole, H. A. (1939). *Saccoglossus cambrensis*, sp. n., an enteropneust occurring in Wales. *Proceedings of the Zoological Society of London, Series B*, **109**, 211–36.

Burdon-Jones, C. & Patil, A. M. (1960). A revision of the genus *Saccoglossus* (Enteropneusta) in British waters. *Proceedings of the Zoological Society of London*, **134**, 635–45.

# Chordata

The chordates include such diverse forms as the sea-squirts found attached to rocks and stones on the lower shore, and the fishes, amphibians, reptiles, birds and mammals. Although there is great diversity of structure within the phylum, chordates are united by four features present in at least some stage in the life-cycle. These are a notochord, a hollow dorsal nerve cord, pharyngeal clefts and a post-anal tail containing extensions of the nerve cord and notochord. The notochord is a dorsal, flexible rod which has a skeletal function and may persist throughout life. It is often completely or partly replaced by cartilaginous or bony material to form a backbone. The nerve cord is dorsal to the noto-chord and often expanded anteriorly to form a brain, but in some groups, such as the sea-squirts, it is, like the notochord, lost in the adult. The pharyngeal clefts (sometimes known as gill slits), lead from the cavity of the pharynx to the exterior and have a variety of functions: in aquatic chordates they may be used in both feeding and respiration, but are generally reduced or lost in the adult terrestrial chordate. The post-anal tail persists in many aquatic vertebrates such as the fishes as a muscular, flexible structure which is important in locomotion.

## Phylum CHORDATA

**Subphylum UROCHORDATA**   The tunicates. All are marine. Larvae, but not adults, possess chordate characteristics.

**Class Ascidiacea**   Sessile tunicates, the sea-squirts.

Class Thaliacea   Translucent, planktonic tunicates. Includes the salps.

Class Larvacea   Very small, planktonic tunicates in which the adults show some larval characteristics.

Subphylum CEPHALOCHORDATA   The lancelets (amphioxus). Free-swimming, fish-like chordates.

**Subphylum VERTEBRATA**   The vertebrates. Characterized by a vertebral column made up of a large number of separate units, the vertebrae. Show wide diversity of form and habit.

Class Agnatha    Lampreys and hagfishes. Fish-like, primitive vertebrates lacking jaws
     and paired fins.

Class Chondrichthyes    Skates, rays and sharks. Fishes with a cartilaginous skeleton.

**Class Osteichthyes**    The bony fishes.

Class Amphibia    The frogs and toads and newts.

Class Reptilia    The snakes, lizards, turtles and crocodiles.

Class Aves    The birds.

Class Mammalia    The mammals.

## Subphylum UROCHORDATA

### Class Ascidiacea

Ascidians are widely distributed on the lower shore and are characteristically found in
crevices and attached to rocks, stones and seaweeds. They also occur on soft sediments
and some species are adapted to living in the tiny spaces between sand grains. The body
is surrounded by a jelly-like coating, or test, which varies considerably in thickness. It
may be hard or soft and is sometimes encrusted with sand and fragments of shell.
Ascidians show a wide range of colours as the test is sometimes pigmented; however, it
is often translucent, the colours of the internal organs being visible. Some species are
solitary, others colonial, and the generalized body structure is shown in Figure 307.

In the solitary species, the free end of the body has an inhalant or oral siphon through
which water enters, and an exhalant or atrial siphon through which water leaves the
body. The water current is maintained by the beating of cilia. Water entering the body
passes through the pharynx, the walls of which are perforated by rows of small slits or
stigmata, then into a space surrounding the pharynx, the atrium, before leaving through
the exhalent siphon. Ascidians are suspension feeders and as the water passes through
the pharynx, suspended detritus and plankton are trapped in mucus which is then carried by cilia to the oesophagus. When disturbed, rapid contractions of longitudinal
muscles in the body wall lead to the expulsion of a jet of water and it is from this that
the group derives its common name.

The basic organization of colonial ascidians is essentially as described above. In some,
the individuals or zooids are united at the base by a stolon. In others, the colony forms
flat encrusting growths in which the zooids are embedded in a common gelatinous test.
Here, the exhalant siphons of several zooids lead into a common chamber, the cloaca,
opening to the exterior via a common exhalant opening (Fig. 307). These forms are
known as compound ascidians.

Ascidians are generally hermaphroditic with self-fertilization, although some rely on
cross-fertilization. In most solitary forms eggs and sperm are liberated into the sea where

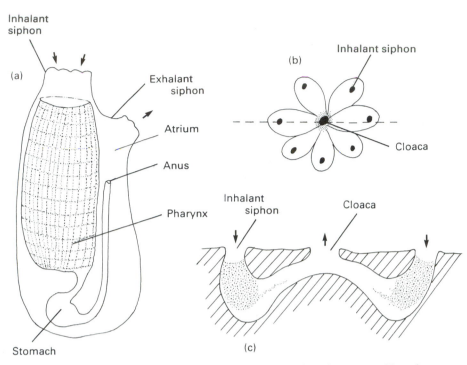

Figure 307   (a) Generalized body form of a solitary ascidian based on *Ciona*. (b) Surface
view of compound ascidian such as *Botryllus* showing arrangement of zooids
round a common cloaca. (c) Diagrammatic representation of a compound
ascidian seen in section along dotted line in (b). Arrows indicate flow of water.

fertilization and development to the larval stage take place, while in compound forms
the eggs are generally held in the atrium until the larvae are fully developed. The typical
larva is known as a tadpole larva and, as the name implies, it has a flexible tail as the
main organ of propulsion. It is the structure of the tail, in which a notochord and dorsal
nerve cord are present, which betrays the relationship between ascidians and other chordates. The larvae are non-feeding and have a free-swimming life varying from a few
minutes to a few days. The selection of a suitable site for settlement is their primary role
and they often settle among adults of the same species, leading to dense aggregations.
Asexual reproduction by budding is common and reproduction by fission occurs in some
colonial species.

  Ascidians are distributed from the intertidal region to the deep sea. Some species are
abundant on sheltered shores, while others thrive in wave-swept gullies, on rock faces
and on laminarian holdfasts. Encrusting colonial forms are often strikingly patterned
with bright colours. Some ascidians are tolerant of reduced salinity and are found in
estuaries, while others are well known as fouling organisms, often being present in large
numbers on ships' hulls, floating buoys and quays. Although it is not easy to determine
the age of ascidians, particularly the colonial forms, length of life is probably one or two

years. Ascidians have a number of predators, including gastropod molluscs, echinoderms and fishes, but the body wall of some species is highly acidic, rendering them unpalatable to predators.

### Key to species of ascidians

Although some ascidians can be identified on the basis of external morphology, accurate identification of others is difficult and requires dissection, followed by examination of the detail of the pharynx and the position of the gut and gonads. The following key to the common intertidal species is based on growth form and distinctive morphological features. Dissection is not required. It should be borne in mind that some individuals are highly contractile and this might render the characters difficult to see; specimens should be observed under water. In some species growth form and colour are variable.

1  Individuals or zooids occur singly or aggregated in clusters; sometimes attached to one another at base or along sides; may be joined at base by stolon  ........................................ 2
   Zooids embedded in common test; growth form encrusting or as upright lobes often encrusted with sand .......................................................................................... 10

2  Zooids transparent, very delicate; joined at base by stolon; zooids with white ring round pharynx; no more than 20 mm in height  ................ *Clavelina lepadiformis* (p. 470)
   Not as above  ...................................................................................................... 3

3  Individuals may occur singly or aggregated in dense clusters or sheet-like aggregations, attached at base or along sides. Firm to the touch; usually no more than 10–20 mm in height; reddish-brown colour, without adhering sand  ...................................................... 4
   Individuals occur singly; attached to substratum at base or along one side. Cylindrical or rounded in shape; may have adhering sand grains; up to 120 mm in height  ................ 5

4  Zooids in sheet-like aggregations; joined at base and along sides; siphons relatively inconspicuous. Up to 10 mm in height  ................................ *Distomus variolosus* (p. 478)
   Individuals may occur singly or aggregated in dense, rather bumpy clusters; body shape varies from flattened and rounded to elongate and cylindrical; siphons conspicuous. Up to 20 mm in height  ................................ *Dendrodoa grossularia* (p. 477)

5  Test very soft to the touch, gelatinous, translucent; highly contractile. Attached at base; siphons conspicuous, with red spots  .............................. *Ciona intestinalis* (p. 474)
   Test to varying degrees, rough; may have projections, papillae or fibrils; sometimes with adhering sand grains  ............................................................................................ 6

6  Body rounded; test thin, with fibrils and adhering sand; siphons at free end; attached to substratum at base, or unattached  ................................................. *Molgula* spp. (p. 479)
   Body more or less cylindrical; attached to substratum at base or along one side  .............. 7

7  Inhalant siphon at free end; exhalant siphon about $\frac{1}{2}$ to $\frac{2}{3}$ of way down body  .............. 8
   Inhalant siphon at free end; exhalant siphon no more than $\frac{1}{3}$ of way down body  ............. 9

8  Test thick, uneven, mostly attached by part of one side; translucent. Siphons small;

inhalant at free end, exhalant $\frac{1}{2}$ to $\frac{2}{3}$ of way down body    ............ *Ascidia mentula* (p. 476)
Test attached by most of one side; thin and rough, often with attached shell. Inhalant
siphon terminal, exhalant about $\frac{2}{3}$ of way down body    ............ *Ascidia conchilega* (p. 477)

9  Test stiff and rough to touch, with small papillae. Attached at base. Inhalant siphon at
free end; exhalant about $\frac{1}{3}$ of way down body    .......................... *Ascidiella aspersa* (p. 475)
Test stiff, reddish papillae at base of siphons, otherwise smooth. Inhalant siphon
terminal, exhalent nearly so. May be attached by most of one side .................................
................................................................................................................ *Ascidiella scabra* (p. 475)

10  Colonies rounded, club-shaped or flat-topped, up to 50 mm in height; may be
encrusted with sand ....................................................................................................... 11
Colonies encrusting on rocks and seaweed; may be brightly coloured and gelatinous or
hard, bluish, purplish, marked with dark lines ................................................................ 12

11  (a) Club-shaped lobes with distinct narrow stalk; stalk red in colour and encrusted
with sand. Rounded head without sand. Each zooid with 4 orange-red spots. Inhalant
siphons of zooids with 8 lobes    ................................................ *Morchellium argus* (p. 472)
*or*
(b) Colonies flat-topped lobes or club-shaped with narrow stalks; joined at base by
creeping stolon; test semi-transparent with conspicuous exhalant openings; inhalant
siphons of zooids with 8 lobes. Not encrusted with sand ..... *Sidnyum turbinatum* (p. 473)
*or*
(c) Colonies of rounded lobes, sometimes flat-topped, forming irregular masses usually
encrusted with sand; each lobe with one or more conspicuous exhalant openings; inhalant
siphons of zooids with 6 lobes    ........................................... *Polyclinum aurantium* (p. 471)
*or*
(d) Colony stalked, club-shaped or broad-based, flat-topped. Exhalant openings large;
inhalant openings with 6 lobes with white pigment ............ *Aplidium proliferum* (p. 473)

12  Colony thin, encrusting, sheet-like or thrown into small lobes; hard. Greyish, bluish,
purplish in colour; marked with dark lines ....................... *Didemnum maculosum* (p. 473)
Colony generally thinnish, gelatinous; individual zooids arranged around exhalant
openings in star-shaped or linear patterns; often brightly coloured ................................. 13

13  Zooids arranged in star-shaped patterns with common exhalant opening in centre .......
................................................................................................ *Botryllus schlosseri* (p. 478)
Zooids arranged around elongate exhalant opening ................. *Botrylloides leachi* (p. 479)

*Clavelina lepadiformis* (Müller)    (Fig. 308)

*Colonial; groups of very delicate zooids joined at base by stolon; up to 20 mm in height.*
*Pharyngeal region short. Siphons close together. Zooids transparent with pale yellow, pink*
*and brown lines; white ring around pharynx.*

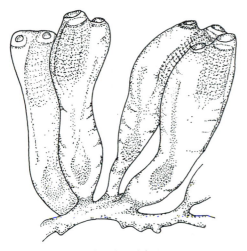

Figure 308 *Clavelina lepadiformis.*

*C. lepadiformis* is widely distributed in north-west Europe and occurs all around Britain. In some areas it is common on the lower shore attached to seaweed, rocks and stones, extending into the sublittoral to depths of about 50 m. It is tolerant of salinities as low as 14‰. On the west coast of Scotland breeding has been recorded during summer. The eggs are fertilized in the atrium, where development to the tadpole larval stage takes place. After release the larvae are free-swimming for only about three hours before settling to establish new colonies. Asexual reproduction by budding also takes place. Towards the end of summer, many of the adult zooids regress and the colony survives as 'winter buds' from which new zooids develop in spring.

**Polyclinum aurantium** Milne-Edwards    (Fig. 309a)

*Colonial; zooids embedded in common test and forming rounded lobes, each narrowing towards the base and often with a flat top. Up to 30 mm across and 15 mm in height. Zooids irregularly arranged round conspicuous exhalant openings. Inhalant siphons with 6 lobes. Yellow-brown colour, usually encrusted with sand.*

*P. aurantium* is widely distributed in north-west Europe. It is found on the lower shore in gullies and attached to rocks and seaweed, particularly in exposed areas, extending to depths of about 100 m. On the west coast of Scotland breeding occurs in spring and summer. The eggs are fertilized in the atrium where development to the tadpole larva takes place. After hatching, the larvae have a short free-swimming life of about two or three hours before settling. Asexual reproduction by budding also takes place, usually during winter.

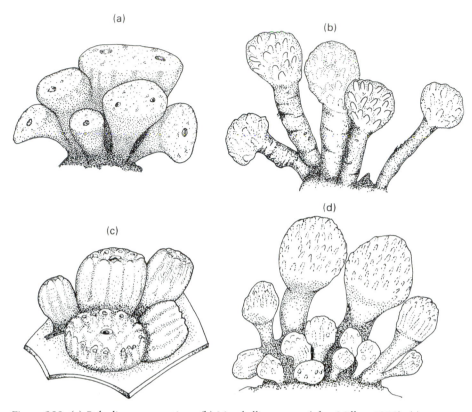

Figure 309  (a) *Polyclinum aurantium.* (b) *Morchellium argus* (after Millar, 1970). (c)
*Sidnyum turbinatum.* (d) *Aplidium proliferum.* ((a), (c), (d) After Alder &
Hancock, 1912.)

### *Morchellium argus* (Milne-Edwards)    (Fig. 309b)

*Colonial; zooids embedded in common test forming club-shaped lobes with distinct head*
*and stalk. Inhalant siphons with 8 lobes. Stalk up to 40 mm in height, encrusted with sand;*
*red in colour. Head up to 30 mm across, not encrusted with sand; yellowish colour, each*
*zooid with 4 orange-red spots around base of inhalant siphon.*

M. *argus* has been recorded from the west coast of France and south and west Britain,
extending as far north as the west coast of Scotland. It is found on the lower shore
attached to rocks and stones, often on the underside of rocky overhangs in wave-swept
situations where it is sometimes abundant. Breeding occurs in summer. Fertilized
embryos are held in the atrium where they develop to tadpole larvae and have a brief,
free-swimming life of two or three hours. Asexual reproduction by budding generally
takes place during late summer and autumn.

*Sidnyum turbinatum* Savigny    (Fig. 309c)

*Colonial; zooids embedded in common test to form rounded, sometimes club-shaped lobes, generally flat-topped, narrowing at base; 10–20 mm in height. Lobes joined at base by creeping stolon; exhalant openings conspicuous, surrounded by 6–12 zooids. Inhalant siphons with 8 lobes. Test semi-transparent; zooids with white markings, occasional red spots. Not encrusted with sand.*

*S. turbinatum* is widely distributed in north-west Europe and Britain, where it is most common in the south and west. It is found on rocks, stones and seaweed from the lower shore to depths of about 200 m. On the west coast of Scotland breeding has been recorded during spring and summer. The fertilized eggs are held in the atrium and released as tadpole larvae.

*Aplidium proliferum* (Milne-Edwards)    (*Aplidium nordmanni*) (Fig. 309d)

*Colonial; zooids embedded in common test. Two growth forms recognized: club-shaped lobes up to 50 mm in height, and broad-based, flat-topped colonies 5–10 mm thick and up to 60 mm across. Zooids irregularly arranged round large exhalant openings. Inhalant siphons of zooids with 6 lobes, distinctly white pigmented. Some colonies yellowish-white in colour, others pink.*

*A. proliferum* is widely distributed in north-west Europe. It is sometimes common on the west coast of Britain but is absent from North Sea coasts. It is found on rocks, stones and seaweed, extending from the lower shore into the shallow sublittoral.

Some authors regard *A. proliferum* and *Aplidium nordmanni* (Milne-Edwards) as separate species, the club-shaped colonies being *A. proliferum* and the flat-topped colonies *A. nordmanni*.

*Didemnum maculosum* (Milne-Edwards)    (*Didemnum candidum*) (Fig. 310a)

*Colonial; thin, encrusting, sheet-like or thrown into small lobes; surface hard, smooth, often with dark lines; individual zooids small, difficult to distinguish, irregularly arranged around a common exhalant opening. Siphonal openings inconspicuous in contracted specimens. Colour varied, white, grey, bluish, purple.*

*D. maculosum* is widely distributed around Britain, particularly on south and west coasts, and in some situations may be very common on the lower shore encrusting stones, rock, laminarian stipes and a wide range of seaweeds. It extends into the sublittoral.

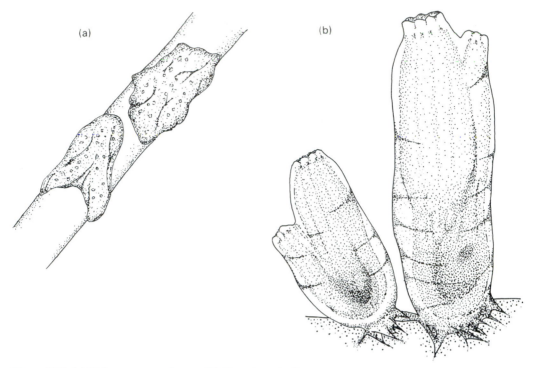

Figure 310  (a) *Didemnum maculosum*. (b) *Ciona intestinalis*.

### *Ciona intestinalis* (Linnaeus)    (Fig. 310b)

*Solitary, attached at base; body elongate with soft gelatinous, translucent test. Longitudinal bands of muscle in body wall. Up to 120 mm in height. Siphons large, conspicuous with delicate yellow margins; inhalant at free end, with 8 lobes with reddish spots between them; exhalant siphon just posterior to inhalant, 6-lobed, with reddish spots. Pharyngeal region extensive. Yellow-greenish colour, sometimes orange. Ciona is highly contractile and in contracted specimens some of the characters are difficult to see.*

C. *intestinalis* is widely distributed in north-west Europe and although generally not common, in some situations it may be abundant, reaching densities of up to 5000 per square metre. It is found on the lower shore in dense groups attached to the underside of stones, rocks and seaweed, and also commonly occurs in harbours attached to piers, and on ships' hulls and buoys. It extends into the sublittoral to depths of over 500 m. *Ciona* is tolerant of reduced salinity and breeding populations have been recorded at salinities of 11‰. It is a protandrous hermaphrodite. Spawning normally occurs just before dawn, the gametes being shed into the sea, where cross-fertilization takes place. In Britain breeding generally takes place in spring and summer, but breeding throughout the year has been recorded at some localities. The larvae are free-swimming for up to 36 hours. Two generations per year have been recorded in Scandinavia and Britain, with this reduced to a single generation at some localities. On the west coast of Scotland, *Ciona* reaches sexual maturity at a body height of about 25–30 mm and longevity is one and a half years. It does not undergo asexual reproduction by budding.

**Ascidiella aspersa** (Müller)   (Fig. 311a)

*Solitary, attached at base; body more-or-less oval-shaped; test with small papillae, stiff and rough to touch; with attached detritus. Up to 100 mm in height. Siphons obvious; inhalant at free end, sometimes frilled; exhalant about ⅓ of way down body. Greyish-black or brown in colour.*

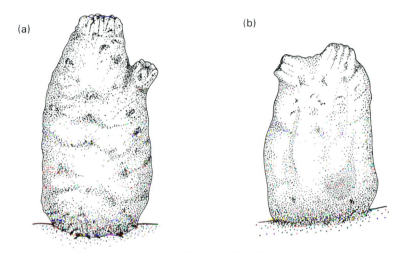

Figure 311  (a) *Ascidiella aspersa.* (b) *Ascidiella scabra.*

A. *aspersa* is widely distributed in north-west Europe and Britain, where it appears to be most common on south and west coasts. It is found on the lower shore and sublittorally to 80 m attached to stones, seaweed and pilings. It is tolerant of salinities down to 18‰ and is often common in estuaries. A. *aspersa* is a protandrous hermaphrodite. Gametes are discharged into the sea, where fertilization and development of the tadpole larvae take place. Larvae settle during summer. On the west coast of Scotland, sexual maturity has been recorded at body heights ranging from 25 to 50 mm and breeding takes place in the summer following settlement. On British coasts, longevity is one to one and a half years compared with two to three years in Norwegian waters. The small bivalve *Modiolarca tumida* (p. 287) is occasionally found living as a commensal in the test.

**Ascidiella scabra** (Müller)   (Fig. 311b)

*Solitary, may be attached by most of one side; body oval-shaped; test stiff, with reddish papillae at base of siphons. Up to 50 mm in height. Siphons obvious, inhalant at free end, exhalant nearly so. Semi-transparent with some reddish markings.*

*A. scabra* is widely distributed in north-west Europe and Britain. It lives on the lower shore attached to seaweeds, rocks and shells, extending to depths of about 300 m. *A. scabra* is a protandrous hermaphrodite. The gametes are discharged into the sea where fertilization and development of the tadpole larvae take place. Breeding occurs during summer; on the west coast of Sweden, one generation a year has been recorded. Longevity is about one year.

### *Ascidia mentula* Müller    (Fig. 312a)

*Solitary, mostly attached by part of one side; body oblong; test thick, uneven. Up to 100 mm in height, may be more. Siphons small; inhalant at free end, with 8 whitish lobes; exhalant $\frac{1}{2}$ to $\frac{2}{3}$ of way down body. Test translucent, pinkish in colour, sometimes greyish.*

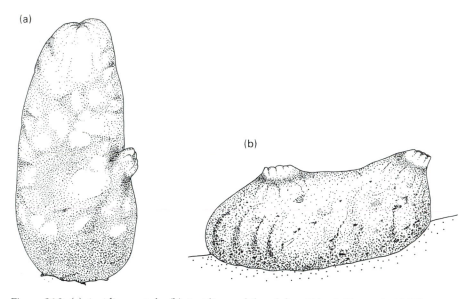

(a)

(b)

Figure 312  (a) *Ascidia mentula*. (b) *Ascidia conchilega* (after Alder & Hancock, 1905).

*A. mentula* is widely distributed in north-west Europe and around Britain, where it is most commonly found on south and west coats, extending sublittorally to depths of about 200 m. Intertidally it occurs on the lower shore, particularly in shaded gullies and crevices on rocky shores, sometimes attached to laminarian holdfasts. It is also found attached to shells and stones on sand and mud. It survives in salinities down to 20‰. At a number of localities breeding has been recorded throughout the year. Fertilization is external and the tadpole larva has a short pelagic life. A longevity of seven years has been recorded for a population on the west coast of Sweden. The small bivalve *Modiolarca tumida* (p. 287) is sometimes found living as a commensal in the test.

### *Ascidia conchilega* Müller    (Fig. 312b)

*Solitary, body flattened, oblong; attached by most of one side; test thin and rough; may be encrusted with sand and shell fragments. Up to 60 mm in length, generally smaller. Inhalant siphon at free end, with 8 red spots; exhalant siphon about ⅔ of way along body, with 6 red spots. Test translucent, greenish colour.*

*A. conchilega* is widely distributed in north-west Europe and Britain on stones, shells and, occasionally, seaweeds. It is found on the lower shore and extends into deep water. Fertilization is external. There is a tadpole larva in the life-cycle.

### *Dendrodoa grossularia* (van Beneden)    (Fig. 313a)

*Solitary or aggregated in dense, bumpy clusters. Firm to the touch; shape varied, flattened and rounded in solitary individuals; usually upright, elongate and cylindrical in aggregated individuals. Siphons conspicuous, noticeably extended upwards from free end, apertures rounded. Up to 20 mm in height. Reddish-brown colour. Easily confused with* Distomus variolosus *(below).*

Figure 313  (a) *Dendrodoa grossularia.* (b) *Distomus variolosus.*

*D. grossularia* is widely distributed in north-west Europe and Britain, occurring on the lower shore and extending sublittorally to depths of about 600 m. It is found on a variety of substrata, such as rocks, shells and seaweeds. Solitary individuals have a wide but patchy distribution in Britain and are found on most coasts, while aggregates are especially common in the south and west. Clusters of *Dendrodoa* have a superficial resemblance to *Distomus variolosus* (below). Dense groups of individuals result from the gregarious settlement of larvae, young often settling on older individuals. Asexual reproduction by budding has not been recorded in *Dendrodoa* (cf. *Distomus variolosus*, below). On the west coast of Scotland breeding occurs throughout summer and autumn, while spring and autumn peaks in breeding have been recorded in a sublittoral population off the south-east coast of Britain. Fertilization takes place in the atrium where the eggs are held and released as tadpole larvae. Newly settled young grow to a height of about 3—3.5 mm before their first winter, reaching maximum size the following year. Longevity is one and half to two years.

*Distomus variolosus* Gaertner    (Fig. 313b)

*Colonial, encrusting in sheet-like aggregations. Zooids firm to the touch, joined at base and along sides. Oval to cylindrical in shape; up to 10 mm in height. Reddish-brown colour, often sheen-like appearance. The siphons are less conspicuous than in* Dendrodoa grossularia *(above) but the 2 species are easily confused.*

*D. variolosus* is found on the lower shore and shallow sublittoral attached to rocks, stones and hydroids, and on laminarian holdfasts and stipes. It has been recorded on south coasts of Britain and the west of Ireland, but at more northerly localities it may have been confused with *Dendrodoa grossularia* (above) and confirmation of records is needed. Dense aggregations of *Distomus* result from asexual budding, but the colony does not form the thick clumps and masses so characteristic of *Dendrodoa*.

*Botryllus schlosseri* (Pallas)    Star ascidian (Fig. 314a)

*Colonial; zooids embedded in common test to form colonies which are very varied in form, usually thin and encrusting sheets, but thick, or lobed when living in deep or calm water. Test gelatinous. Individual zooids arranged in groups of 3—12 round exhalant opening to give star-shaped patterns. Colonies vary in size from a few millimetres to 150 mm and more across. Colour varied, includes shades of blue, yellow and brown; zooids often contrasting in colour with test.*

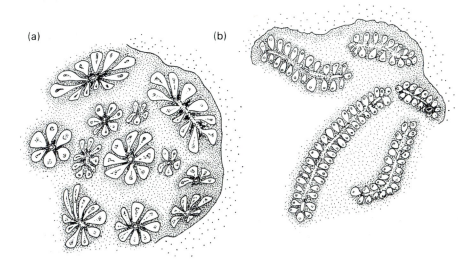

Figure 314  (a) *Botryllus schlosseri*. (b) *Botrylloides leachi*, both on rock.

*B. schlosseri* is widely distributed and sometimes abundant in north-west Europe. It is found all around the coast of Britain on stones and seaweeds on the lower shore and into the shallow sublittoral. Surviving in salinities down to about 25‰, it is found in harbours and estuaries. On the west coast of Scotland, breeding occurs during spring and summer. Eggs are fertilized in the atrium, where they develop to tadpole larvae which escape through the common exhalant openings. The free-swimming larvae survive for up to about 16 hours before settling to form new colonies. In Britain there are one or two generations each year. Asexual reproduction by budding also occurs and appears to take place at any time of year. Longevity is one to one and a half years. *Botryllus* and *Botrylloides* (below) are preyed on by cowries (*Trivia* spp., p. 251), which in addition to feeding on the ascidians also lay egg capsules in holes bitten out of the test.

### *Botrylloides leachi* (Savigny)    (Fig. 314b)

*Colonial; zooids embedded in common test to form thin, encrusting colonies. Test gelatinous. Individual zooids arranged round elongate, exhalant opening, and not in star-shaped patterns (cf. Botryllus, above). Colour varied but generally yellow, orange, grey.*

*B. leachi* has a similar distribution to that of *Botryllus schlosseri* (above) and is found on rocks and seaweed on the lower shore and in the shallow sublittoral. Colony form and colour are less varied than in *Botryllus*. On the west coast of Norway, colonies have been recorded in salinities as low as 16‰. Breeding occurs during summer and the tadpole larvae escape through the common exhalant openings. Asexual reproduction by budding also takes place. Longevity is about two years.

### *Molgula* spp.

Several species of *Molgula* are found on the lower shore and in the shallow sublittoral. They occur individually or in groups, attached to stones, shells and seaweed, while some species are unattached. The test bears adhering sand and shell to varying degrees and in some species only the siphons are free of adhering material. The species are difficult to separate.

### *Molgula manhattensis* (De Kay)    (Fig. 315)

*Solitary, body rounded, attached at base; siphons long, both at free end; inhalant 6-lobed, exhalant 4-lobed. Test relatively thin, with fine fibrils and often with encrusting sand grains. Up to 30 mm across. Greenish-grey in colour.*

*M. manhattensis* is widely distributed in north-west Europe and in Britain appears to be most common in the south and west. It is found from the lower shore to depths of about 90 m attached to stones, shells and seaweeds. It is also found on sand and survives salinities down to about 23‰. Fertilization is external, leading to the development of a

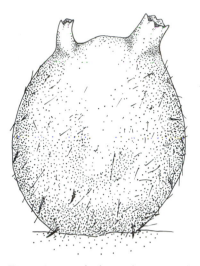

Figure 315 *Molgula manhattensis* (after Millar, 1970).

tadpole larva. On the south coast of England breeding has been recorded throughout the year.

## Subphylum VERTEBRATA

### Class Osteichthyes

Bony fishes are found throughout the world's oceans from the intertidal region to the greatest depths and belong to the group known as the teleosts. As the name implies, they usually have a bony skeleton and typically are covered by bony scales, although the latter are often absent or reduced in size in shore fishes because of the danger of them being rubbed off or damaged. When present, the number of scales counted in a row along the lateral line is sometimes important in identification. The mouth is terminal and the body has a number of fins, supported by fin rays. Dorsal, anal and tail (caudal) fins are single, the pectorals and pelvics paired. Pectoral fins are usually found on the sides of the body (Fig. 316) just behind the head, while the position of the pelvic fins varies. They have a ventro-lateral position and in the more primitive bony fishes are situated close to the anus, while in more advanced forms they lie in a much more anterior position, either slightly anterior or posterior to the pectoral fins. The fin rays of the primitive telosts are soft, while in more advanced forms stiff spines are often present on dorsal, anal and pelvic fins. The arrangement of the fins and the number of rays and spines are important in identification. A well-developed lateral line system, sensitive to vibrations and changes in pressure, is often present, but in shore fishes it is generally reduced, presumably as a result of them living in wave-beaten situations where the effectiveness of such a system would be limited. In some groups, for example the gobies, the lateral line system is absent and sensory pores are found on the head. Most teleosts have a swim bladder, an outgrowth of the anterior region of the alimentary canal which functions as a hydrostatic organ. In some shore fishes it is reduced or absent, resulting in negative buoyancy, a further adaptation to living in turbulent, inshore waters. The gill openings are covered by a bony plate, the operculum, which is free along its posterior margin to allow water to leave after entering through the mouth and passing over the gills. Body shape varies widely, but a typical teleost is shown in Fig. 316.

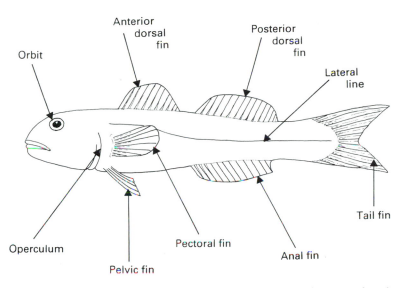

Figure 316 Generalized body form of a bony fish to illustrate features referred to in the text.

The majority of fishes have separate sexes. Fertilization is normally external and the embryos develop into pelagic larvae which feed in the plankton. In some species there is a complex behaviour pattern associated with breeding whereby the eggs are laid in nests where the young develop. These nests are often guarded by one or both parents.

The term 'shore fishes', as used here, includes species typical of rock pools and damp crevices, such as the blennies and gobies. They are truly intertidal and highly adapted to the sharply fluctuating conditions found there. Also included are fishes which, although more typical of the shallow sublittoral, migrate in with the tide and frequently take refuge in crevices, under stones and among seaweed during low water. Other fishes, not included in the text, feed over the intertidal region during high water and retreat with the falling tide. For example, flat fish such as plaice and flounders move into estuaries with the tide and feed on the invertebrates living in the sand and mud deposits. Along with the shore fishes, they are important predators of intertidal plant and animal communities.

The Class Osteichthyes is divided into a large number of families; phylogenetic sequence and nomenclature is based on Nelson (1984).

Key to families of fish

1  Body very slender, elongate; divided into rings along length. Head extended into tube-like snout. Dorsal fin conspicuous, other fins reduced or absent ................................. .......................................................................................... **Syngnathidae** (p. 488)
   Body not divided into rings. Head not extended into snout ....................................... 2

2  Body long, slender, slimy. Dorsal fin continuous with anal fin. Scales tiny or absent; no pelvic fins. The eels ......................................................................................... 3
   Not as above .................................................................................................. 4

3  Anterior margin of dorsal fin some way behind pectoral fins. Lower jaw longer than upper .................................................................................. **Anguillidae** (p. 483)
   Anterior margin of dorsal fin just behind pectoral fins. Upper jaw longer than lower ..... .......................................................................................... **Congridae** (p. 484)

4  Body elongate with forked tail fin. Dorsal fin long. No pelvic fins. Scales small .............. .......................................................................................... **Ammodytidae** (p. 504)
   Pelvic fins present. Sometimes reduced to spines or fused to form crescent-shaped disc (Fig. 337) or wholly or partly modified to form conspicuous sucker (Fig. 328) ................. 5

5  Pelvic fins modified to form sucker or joined to form crescent-shaped disc ...................... 6
   Pelvic fins not as above ................................................................................... 9

6  Pelvic fins joined to form crescent-shaped disc, enabling fish to attach loosely to substratum; scales present; eyes large, protruding, dorso-lateral in position. Two dorsal fins .......................................................................... **Gobiidae** (p. 507)
   Pelvic fins modified wholly or partly to form a conspicuous sucker ................................ 7

7  Body of distinctive shape, rounded; with rows of bony tubercles along sides ................. .......................................................................................... **Cyclopteridae** (p. 493)
   Body without bony tubercles along sides; no scales ................................................. 8

8  Skin loose, prickly to touch, 1 long dorsal fin ...................................... **Liparididae** (p. 494)
   Skin smooth. Head flattened, almost triangular in shape. Single dorsal fin, short, close to tail fin .......................................................................... **Gobiesocidae** (p. 486)

9  Pelvic fins reduced to 1 long spine and a small ray, situated posterior to the base of the pectoral fins. Conspicuous row of spines anterior to dorsal fin. Without scales, but may have bony plates along sides of body .................................... **Gasterosteidae** (p. 487)
   Not as above .................................................................................................. 10

10 Pelvic fins very small, reduced to small spines just below pectoral fins. Body long and slender, with row of black spots along base of long, slender dorsal fin  .  **Pholididae** (p. 499)
   Not as above .................................................................................................. 11

11 Lips thick and pronounced. Body deep and laterally compressed. A single dorsal fin  ..... .......................................................................................... **Labridae** (p. 495)
   Not as above .................................................................................................. 12

12 Two dorsal fins, the anterior one spiny but not black in colour. Spines just in front of operculum. Without scales ........................................................ **Cottidae** (p. 492)
   Not as above .................................................................................................. 13

13  Two dorsal fins, anterior dorsal fin spiny, black in colour. Large spine on operculum. Mouth sloping sharply downwards. Scales present .......................... **Trachinidae** (p. 500)

Not as above .................................................................................................... 14

14  Two or 3 dorsal fins; anterior dorsal may be reduced to 1 long ray with row of much smaller rays in groove behind. Fins not spiny, soft. Head usually with 1 or more barbels ......................................................................... **Gadidae** (p. 484)

Single dorsal fin, long ..................................................................................... 15

15  Dorsal fin continuous with tail and anal fin ......................................... **Zoarcidae** (p. 499)

Tail fin separate from dorsal and anal fins. Dorsal fin divided into anterior and posterior portions by a depression ....................................... **Blenniidae** (p. 501)

**Family Anguillidae — eels**

Body long and narrow, slimy. Dorsal fin continuous with anal fin, anterior margin some way behind pectoral fins. No pelvic fins. Eyes small. Lower jaw longer than upper. Scales tiny and embedded in the skin.

***Anguilla anguilla*** (Linnaeus)    Common eel (Fig. 317a, b)

*Body elongate, skin slimy. Dorsal fin long, anterior margin well behind pectoral fins; continuous with tail and anal fin. Margin of pectoral fin rounded. Without pelvic fins. Lower jaw longer than upper (cf. Conger conger, below). Eyes round, small. Scales tiny, embedded in the skin. Colour varied, dorsal surface brown, black to olive-green; ventral surface yellowish. Some specimens silvery. Up to about 1 m in length.*

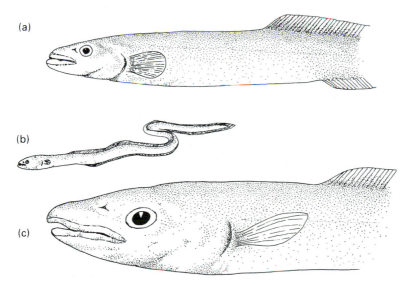

Figure 317  (a), (b) *Anguilla anguilla*. (c) *Conger conger*.

*A. anguilla* is widely distributed in the rivers and estuaries of north-west Europe and is frequently common on the shore. Part of the life-cycle is spent in fresh water and part in the sea. Males and females are believed to spend up to 12 and 20 years, respectively, in fresh water and at a length of 400–500 mm they take on a silvery colour and begin their migration to the Sargasso Sea where they spawn. Although it was believed that the leaf-like larvae (leptocephali) drifted in the plankton of the Atlantic for two to three years before reaching European coasts, it is now understood that the larvae swim actively on a journey which takes about one year. At the end of the migration the leptocephali metamorphose into transparent young eels or 'glass eels'. These darken in colour and in British waters ascend freshwater streams in large numbers in late winter and spring. At this stage they are known as elvers, and are about 50 mm in length. Both elvers and mature eels are fished commercially.

Family Congridae – conger eels
Body long and narrow, slimy. Dorsal fin continuous with anal fin, anterior margin just behind pectoral fins. No pelvic fins. Upper jaw longer than lower. Eyes oval-shaped. No scales.

*Conger conger* (Linnaeus)    Conger eel (Fig. 317c)

*Body elongate, skin slimy. Dorsal fin long, anterior margin just behind pectoral fins; continuous with tail and anal fin. Margin of pectoral fin pointed. Without pelvic fins. Upper jaw longer than lower (cf. Anguilla anguilla, above). Eyes oval. Without scales. Dorsal surface grey-brown, ventral surface paler. Up to about 2 m in length.*

*C. conger* is widely distributed in north-west Europe where it is found in rock crevices on the lower shore and in deep water. It is common on south and west coasts of Britain but much less common on the east coast. It feeds on a wide variety of fishes and crustaceans. Maturity is reached at anything between five and 15 years and the mature eels migrate to the Atlantic as far south as the Azores, where they spawn. The larvae (leptocephali) have a long larval life, drifting from the spawning grounds, to be found on the shores of north-west Europe one or two years later. They do not enter fresh water.

Family Gadidae – cod fishes
Fins soft, without spines. Usually two or three dorsal fins; one or two anal fins. Pelvic fins anterior to pectoral fins. Head usually with one or more barbels. Scales very small. A large family which includes commercially important species such as the cod, haddock and whiting. Also includes common shore fishes, the rocklings, in which the anterior dorsal fin is reduced to one long ray with a row of much smaller rays in a groove behind.

Two dorsal fins, the anterior one a single prominent ray behind which is a row of short rays in a groove. Head with 3 barbels, 1 on lower jaw and 1 close to each anterior nostril ...................................................................... *Gaidropsarus mediterraneus* (p. 486)
Two dorsal fins, the anterior a single prominent ray behind which is a row of short rays in a groove. Head with 5 barbels, 1 on lower jaw, 2 on upper jaw and 1 close to each anterior nostril ............................................................................... *Ciliata mustela* (p. 485)

**Ciliata mustela** (Linnaeus)  Five-bearded rockling (Fig. 318b)

*Rather slender body. Two dorsal fins, anterior one a single, prominent ray behind which is a row of very short rays lying in a groove. Posterior dorsal and anal fins long. Head small, with 5 barbels. Mouth extends just beyond posterior margin of orbit. Body covered with tiny scales. Dorsal surface brown, black or reddish; ventral surface paler. Length up to 250 mm.*

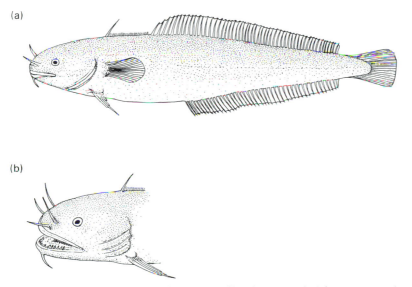

(a)

(b)

Figure 318  (a) *Gaidropsarus mediterraneus*. (b) *Ciliata mustela* (after Day, 1884).

*C. mustela* is widely distributed in north-west Europe and occurs all around Britain, where it is often common intertidally in rock pools and among seaweed. It is also found over sandy and muddy shores and extends to depths of 20 m. It feeds on a variety of invertebrates and small fishes which are believed to be located by the head barbels. Breeding has been recorded in spring and summer, and in South Wales maturity is reached in the second year. Spawning occurs offshore, but specimens with ripe ovaries have been recorded intertidally, suggesting that some spawning takes place on the lower shore. Both eggs and larvae are pelagic and young are found inshore in summer.

***Gaidropsarus mediterraneus*** (Linnaeus)    Shore rockling (Fig. 318a)

*Rather slender body. Two dorsal fins, anterior one a single prominent ray behind which is a row of very short rays lying in a groove. Posterior dorsal and anal fins long; 15–17 rays in pectoral fin. Head with 3 barbels. Mouth extends to, or just beyond, posterior margin of orbit. Body covered with tiny scales. Dorsal surface uniform in colour, brown, reddish-brown or almost black; ventral surface paler. Length usually up to 250 mm.*

G. *mediterraneus* is common on the north-west and north coasts of France and the south and west coasts of Britain, extending north to Orkney and southern Norway. It has been recorded sublittorally to 27 m but is mainly intertidal in pools and among seaweed on rocky shores. It feeds on a wide range of invertebrates and small fish. The head barbels are believed to be used to locate food and, contrary to earlier opinion, the groove and hair-like rays of the anterior dorsal fin apparently play little, if any, role in food detection. In South Wales maturity is reached in the second or third year. Breeding occurs offshore in summer and both the eggs and larvae are pelagic. Young are found inshore from late September.

***Gaidropsarus vulgaris*** (Cloquet) is widely distributed in north-west Europe. It is similar to G. *mediterraneus* in having three barbels on the head, but is mainly salmon-pink in colour. It is sublittoral, generally over rocky grounds.

Family Gobiesocidae – clingfishes
Head flattened, almost triangular in shape. Single dorsal fin, short, close to tail fin. Pelvic fins modified, in part, to form a conspicuous sucker on the ventral surface. No scales.

***Lepadogaster lepadogaster*** (Bonnaterre)    Shore clingfish, Cornish sucker (Fig. 319)

*Body flattened. Dorsal, tail and anal fins continuous. Sucker on ventral surface. Head flattened, lips thick. Flap of tissue in front of eyes. Body without scales. Pink to red in colour; conspicuous blue spot on dorsal surface just behind each eye. Length up to 65 mm.*

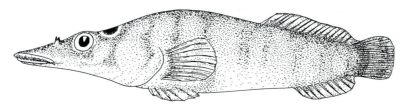

Figure 319 *Lepadogaster lepadogaster.*

*L. lepadogaster* is found on the north-west coast of France and south and west coasts of Britain. Although most common on sheltered rocky shores, where it is found clinging to the undersurface of stones and among seaweed at low water, it has also been recorded as high as the middle shore. It is carnivorous on the wide range of invertebrates found on sheltered rocky shores. On the west coast of Ireland, longevity of up to four years has been recorded, the fish reaching maturity in their second year at a body length of about 39 mm. Breeding occurs during summer. Yellow-coloured eggs are laid in sheets under stones on the shore, often as many as 250 eggs in each sheet. They are guarded by one of the parents and as development proceeds the colour changes through orange to olive-green with hatching occurring after about two weeks. Pelagic larvae have been recorded in inshore waters during summer.

Family Gasterosteidae – sticklebacks
Conspicuous row of spines anterior to dorsal fin. Dorsal and anal fins almost exactly opposite. Pelvic fins reduced to one long spine and a small ray, behind the base of the pectoral fins. Without scales but may have bony plates along sides of body.

Body somewhat elongate; 3 dorsal spines, the anterior 2 of which are long and the third small, just in front of the dorsal fin ......................... *Gasterosteus aculeatus* (p. 487)
Body long and slender; 15 (14–16) short spines in front of dorsal fin ............................
.................................................................................. *Spinachia spinachia* (p. 488)

*Gasterosteus aculeatus* Linnaeus    Three-spined stickleback (Fig. 320)

*Body somewhat elongate; 3 dorsal spines, the anterior 2 of which are long and the third small and just in front of dorsal fin. Dorsal and anal fins opposite, dorsal longer than anal. Pelvic fins reduced to long spine and small ray. Body without scales but may have bony plates along sides. Dorsal surface usually bluish-green, ventral surface silver to white. Breeding males develop red coloration on undersurface and the eyes become bright blue. Length usually up to 60 mm, occasionally longer.*

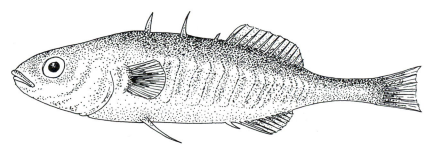

Figure 320  *Gasterosteus aculeatus.*

*G. aculeatus* is common in fresh waters but is also widely distributed throughout north-west Europe and Britain in the intertidal region, particularly in estuaries, feeding on a wide range of invertebrates. When not breeding it is often found in shoals. Breeding occurs in spring and early summer. The male develops a characteristic breeding color-ation of red belly and throat and constructs a nest out of bits of plant material held together by sticky secretions produced by the kidneys. The nest is built in a depression in the sediment. More than one female may be attracted to it and as many as 1000 eggs laid. After laying, the females are chased away by the male and play no part in caring for the eggs. The nest is guarded by the male who fans the eggs using the pectoral fins; the male also removes dead eggs. The eggs hatch after one to three weeks, the young often remaining close to the nest for a few days. Maturity is generally reached after one year and longevity is up to three years.

**Spinachia spinachia** (Linnaeus)    Fifteen-spined stickleback, Sea stickleback (Fig. 321)

*Body very slender. Fifteen (14 or 16) short spines in front of dorsal fin. Dorsal and anal fins opposite. Pelvic fins reduced to small spine and small ray. Body without scales but about 41 bony plates along sides. Dorsal and lateral surfaces green-brown with darker blotches, ventral surface paler. Dorsal and anal fins dark anteriorly. Length up to 150 mm, rarely 200 mm.*

Figure 321 *Spinachia spinachia.*

*S. spinachia* is widely distributed in north-west Europe. It is found in rock pools and shallow water among seaweeds, extending to depths of about 20 m and feeds on a wide range of crustaceans. Breeding occurs in spring and early summer. The male, which unlike that of the three-spined stickleback (above) does not take on a characteristic breed-ing coloration, constructs a nest among seaweed. The nest is clear of the ground, up to 80 mm in diameter and made of strands of weed held together by sticky secretions pro-duced by the kidneys. The nest may contain up to 200 eggs, which are laid by more than one female. After spawning, the females are chased away by the male and take no part in the care of the eggs, which are protected and aerated by the male. The eggs hatch after about two to three weeks. Longevity is usually one year but some fish are believed to survive for a second year.

Family Syngnathidae – pipefishes
Body slender, divided into bony rings along length. Head extended into tube-like snout; mouth small. Dorsal fin conspicuous; other fins reduced or absent.

1 Pectoral, anal and tail fins present ............................................................. 2
  Pectoral, anal and tail fins absent ............................................................. 3
2 Snout long, greater than ½ total length of head; 17—21 rings between base of pectoral fin and anus; 38—43 rings between anus and tail fin ................... *Syngnathus acus* (p. 490)
  Snout about ½ total length of head; 13—17 rings between base of pectoral fin and anus; 37—42 rings between anus and tail ......................... *Syngnathus rostellatus* (p. 491)
3 Snout short, upturned; dorsal fin with 24—28 rays. Dorsal surface almost black in colour .................................................................. *Nerophis lumbriciformis* (p. 489)
  Snout straight; dorsal fin with 33—34 rays. Dorsal surface green-brown in colour .........
  ...................................................................................... *Nerophis ophidion* (p. 489)

*Nerophis lumbriciformis* (Jenyns)   Worm pipefish (Fig. 322)

*Body very slender and elongate, round in section; without pectoral, anal and tail fins. Dorsal fin with 24—28 rays. Snout short, upturned. Body divided into bony rings, 17—19 rings between head and anus, 46—54 behind anus. Dorsal surface dark green, black; ventral surface with paler markings. Length up to 170 mm.*

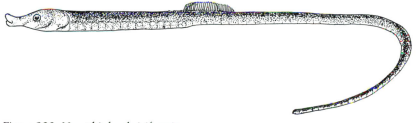

Figure 322  *Nerophis lumbriciformis.*

*N. lumbriciformis* is widely distributed in north-west Europe and is most frequently found on south and west coasts of Britain, particularly at low water on sheltered rocky shores where there is a good cover of seaweed. It is found at depths of up to about 30 m, feeding mainly on crustaceans and small fishes. Breeding has been recorded from April to October. The female transfers up to about 150 eggs to a shallow groove on the undersurface of the abdomen of the male, where fertilization and development take place. Young hatch at a body length of about 10 mm and spend some weeks in the plankton before becoming benthic at a length of about 30 mm.

*Nerophis ophidion* (Linnaeus)   Straight-nosed pipefish (Fig. 323)

*Body very slender and elongate, round in section; without pectoral, anal and tail fins. Dorsal fin with 33—34 rays. Snout about ½ total length of head, not upturned. Body divided into bony rings, 28—33 between head and anus, 68—82 rings behind anus. Dorsal surface green-brown, ventral surface paler. Females larger than males and with blue coloration. Length up to 300 mm.*

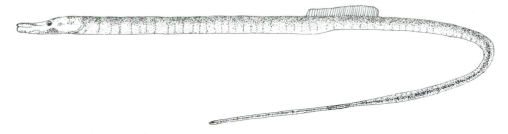

Figure 323  *Nerophis ophidion.*

*N. ophidion* has a widespread but local distribution in north-west Europe. It is predominantly sublittoral to depths of about 15 m but is occasionally found on the lower shore, especially among algae and *Zostera* (p. 85), feeding on crustaceans. It extends into estuaries. Breeding has been recorded from May to August. The female develops a skin fold on the undersurface during the breeding season, making her appear considerably larger than the male. There is a lengthy courtship period during which the female displays her blue colour and the skin fold. Up to 150 eggs are transferred to a groove on the undersurface of the male and it is here that fertilization and development take place, the embryos receiving oxygen and nutrients from the male. Development takes about a month and on hatching the young, about 12 mm in body length, are pelagic for up to four months before taking up a benthic habit. Eggs carried by the male at any one time are from a single female, but more than one brood may be carried during the breeding season.

**Syngnathus acus** Linnaeus    Greater pipefish (Fig. 324)

*Body very slender and elongate; pectoral, dorsal, anal and tail fins present. Dorsal fin with 36–45 rays. Snout long, greater than ½ length of head. Body divided into bony rings; 17–21 rings between base of pectoral fin and anus, 38–43 rings between anus and tail fin. Dorsal surface brown with darker markings. Ventral surface lighter. Length up to 460 mm.*

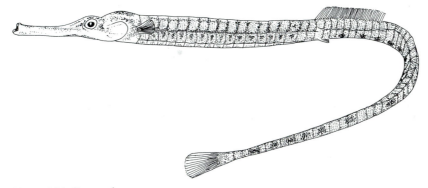

Figure 324  *Syngnathus acus.*

*S. acus* is widely distributed in north-west Europe and in Britain, where it is found in shallow water among seaweed and occasionally in rock pools on the lower shore and to depths of about 90 m. It is also found on sand and mud, extending into the mouths of estuaries. It lives among seaweed, feeding on small crustaceans and larval fishes. Breeding occurs in spring and summer. Males are mature at about 300 mm body length and about 400 eggs are deposited by the female in a brood chamber on the ventral surface of the tail of the male. The brood chamber is formed of lateral folds of skin which hold the eggs in place. Here, fertilization takes place and after about five weeks young emerge, with a body length of about 25 mm. There is no planktonic stage.

**Syngnathus rostellatus** Nilsson   Nilsson's pipefish, Lesser pipefish (Fig. 325)

*Body slender, elongate. Pectoral, dorsal, anal and tail fins present. Dorsal fin with 36–45 rays. Snout about ½ length of head. Body divided into bony rings; 13–17 rings between base of pectoral fin and anus, 37–42 rings between anus and tail. Dorsal surface brown with darker markings. Ventral surface lighter. Length up to 170 mm.*

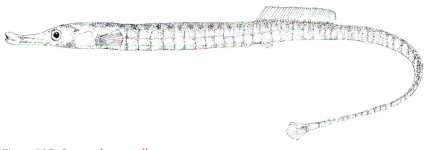

Figure 325  *Syngnathus rostellatus.*

*S. rostellatus* is widely distributed in north-west Europe over sandy and muddy deposits from low water to depths of 18 m. It extends into estuaries, where it is sometimes common, feeding on small crustaceans. Breeding occurs in summer; males are mature at about 100 mm body length and eggs are deposited by the female in a brood chamber on the ventral surface of the tail of the male. Young emerge from the brood chamber at a body length of about 14 mm and have a pelagic phase of a few weeks before assuming a benthic habit.

A third species of *Syngnathus*, **Syngnathus typhle** Linnaeus, has a wide but patchy distribution in north-west Europe, and in some areas, particularly where *Zostera* (p. 85) is present, it may be common. It can be identified by its deep, laterally flattened snout, which is more than half the length of the head.

Family Cottidae — bullheads, sea-scorpions

Stout-bodied, head large. Two dorsal fins, the anterior one spiny. Pelvic fins with one spine and two to five rays. Spines just in front of operculum. Without scales, often with spines, prickles or bony plates along lateral line.

Opercular membranes join to form single membrane under body. Largest spine on the operculum is shorter than diameter of eye ..................... *Myoxocephalus scorpius* (p. 492)
Opercular membranes join sides of body and do not form a single membrane on ventral surface. Largest spine on the operculum is greater than diameter of eye .............................
............................................................................................... *Taurulus bubalis* (p. 493)

*Myoxocephalus scorpius* (Linnaeus) (*Cottus scorpius*), Bull-rout, Short-spined sea-scorpion, Father lasher (Fig. 326a)

*Stout-bodied, 2 dorsal fins. Pectoral fins large; pelvic fins with long rays. Head large. Two conspicuous spines in front of operculum, upper one the longer but shorter than diameter of eye. Opercular membranes extend under body and join to form a single membrane. Without scales; small spines above and below lateral line. Colour varied, dorsal surface greenish-brown with darker patches; ventral surface red to orange; white spots on ventral and lateral surfaces. Length up to about 300 mm.*

(a)

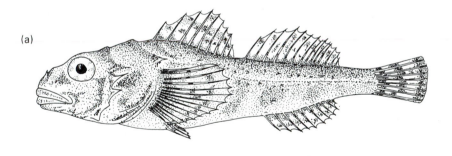

(b)

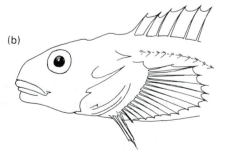

Figure 326 (a) *Myoxocephalus scorpius*. (b) *Taurulus bubalis*.

*M. scorpius* is a northern species, widely distributed in north-west Europe and extending south to the north-west coast of France. It is found all around Britain on rocky, sandy and muddy sediments to depths of about 60 m, only occurring intertidally in the north. It feeds on a variety of invertebrates, predominantly crustaceans, as well as small fishes. Breeding occurs from December to March. Orange-coloured eggs are laid in crevices and among seaweed on the shore, or in the shallow sublittoral. Egg masses are up to 80 mm across and are guarded by the male, the eggs taking from five to 12 weeks to hatch. Larvae are found in the plankton in spring and become benthic at a length of 15–20 mm. Maturity is reached in the second year and longevity is believed to be about six years.

**Taurulus bubalis** (Euphrasen)  (*Cottus bubalis*), Sea-scorpion, Long-spined sea-scorpion (Fig. 326b)

*Stout-bodied, 2 dorsal fins. Pectoral fins large; pelvic fins with 3 long rays. Head broad. Conspicuous spines in front of operculum, upper one long, longer than diameter of eye. Opercular membranes join sides of body. Without scales; small spines along lateral line. Colour varied, greenish-brown to shades of red. Length up to 170 mm.*

*T. bubalis* is widely distributed in north-west Europe, where it is found among seaweed on rocky shores. It occurs all around Britain and may be common on some shores, extending to depths of 30 m. It feeds on a variety of invertebrates and small fishes. Breeding occurs from about January to May. Orange-coloured eggs are laid in clusters on the shore under seaweed and in crevices, and hatch after about six or seven weeks. Larvae have been recorded in the plankton during spring and early summer, and they become benthic at a body length of about 12 mm. Sexual maturity is reached after two years at a body length of 100–180 mm.

Family Cyclopteridae – lumpsucker
Body rather thick, rounded; laterally compressed. Rows of bony tubercles along sides of body. Pelvic fins modified to form conspicuous sucker.

**Cyclopterus lumpus** Linnaeus   Lumpsucker (Fig. 327)

*Body thick, rounded, very distinctive in appearance. Two dorsal fins, but in adult fish first dorsal is reduced to tubercles. Three, sometimes 4, rows of bony plates and tubercles along sides of body. Large sucker on ventral surface. Dorsal surface brown to greyish, paler below. Males develop red-orange colours on ventral surface during breeding season. Length up to 600 mm.*

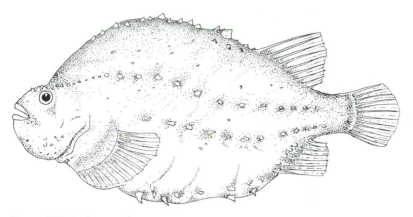

Figure 327  *Cyclopterus lumpus.*

*C. lumpus* is widely distributed in north-west Europe, where it has been recorded from the lower shore to depths of 200 m, feeding on ctenophores, crustaceans, polychaetes and other fishes. The lumpsucker migrates into shallow water to spawn, and in northern parts of Britain it spawns on the lower shore. Clusters of many thousands of bluish-red to yellow-coloured eggs, up to 300 mm across, are attached to rocks and seaweed in spring and early summer where they are guarded by the male. The eggs are fanned with the pectoral fins and this presumably aerates the egg mass. The eggs take several weeks to hatch and the larvae have a length of about 7 mm. Soon after hatching they attach to pieces of seaweed and are carried into the plankton before eventually settling on the bottom.

Family Liparididae – sea-snails
Head broad, body narrowing posteriorly. Single dorsal fin, long; anal fin long. Pelvic fins fused to form ventral sucker. No scales. Skin loose and prickly.

***Liparis montagui*** (Donovan)    Montagu's sea-snail (Fig. 328)

*Rather thick and rounded anteriorly. Long dorsal fin with 28–30 rays, anterior rays shorter than posterior. Anal fin long, 22–25 rays barely extending to the base of the tail fin. Dorsal and anal fins separate from tail fin. Pelvic fins fused to form ventral sucker. Without scales. Skin loose and prickly. Colour varied, brown, yellowish, reddish-brown, green. Length up to about 60 mm, occasionally longer.*

*L. montagui* is widely distributed on rocky shores in north-west Europe, extending into the sublittoral to depths of about 30 m. In Britain, where it is commonest in the north, it is found on the middle and lower shore on the undersurface of stones and the holdfasts

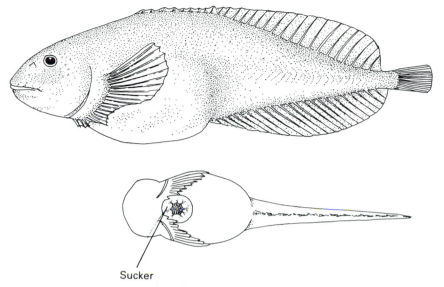

Figure 328  *Liparis montagui*, lateral and ventral views.

of laminarians, attached, often firmly, by the ventral sucker. It feeds on a variety of crustaceans. Breeding occurs in spring. Red-, pink- or orange-coloured eggs are laid in small
clusters on red seaweeds and in crevices on the lower shore. They take about six weeks
to hatch and the pelagic larvae have been recorded in the plankton during spring and
summer. Young are found on the shore when they are about 12 mm in length. Maturity
is reached in the second year and longevity is three years.

The sea-snail *Liparis liparis* (Linnaeus) is found from shallow water to depths of about
300 m around northern Britain and into Scandinavia, but is rare in the south and south-
west. It is distinguished from *Liparis montagui* by the presence of more fin rays (27–36)
in the anal fin and by the fact that the anal fin overlaps about half of the tail fin.

Family Labridae — wrasses
Body deep, laterally compressed. Lips thick and pronounced; mouth protrusible. Strong
teeth. Single dorsal fin. Dorsal and anal fins with anterior spines. Scales large. The
number of rows of scales along the lateral line is important in identification. Brightly
coloured fishes.

1  Five spines (occasionally 4 or 6) and 6–8 rays in anal fin  ...  *Centrolabrus exoletus* (p. 496)
   Three spines in anal fin ................................................................................................ 2
2  Forty-one to 47 scales along lateral line; dorsal fin with 19–20 spines anteriorly and 9–
   11 rays posteriorly ................................................................... *Labrus bergylta* (p. 498)
   Thirty-one to 39 scales along lateral line ................................................................. 3

3 Dark spot on centre of base of tail fin; 31–37 scales along lateral line ............................
.................................................................................. *Symphodus melops* (p. 496
Dark spot in dorsal position on base of tail fin; 35–39 scales along lateral line ..............
.................................................................. *Ctenolabrus rupestris* (p. 497)

**Centrolabrus exoletus** (Linnaeus)    Rock cook (Fig. 330b)

*Body deep, laterally compressed. Large dorsal fin with 18–20 spines anteriorly and 5–7 rays posteriorly. Anal fin with 5 (4–6) spines and 6–8 rays. Scales large, 32–35 along lateral line. Mouth small. Colour greenish, brown; ventral surface paler; tail fin with 1 or 2 brownish bands. Length up to 150 mm.*

*C. exoletus* is widely distributed in shallow waters around north-west Europe. It is found over rocks and *Zostera* beds (p. 85) to depths of about 25 m, feeding on small invertebrates. Breeding has been recorded in spring and summer. Longevity of up to nine years has been recorded on the west coast of Scotland.

**Symphodus melops** (Linnaeus)    (*Crenilabrus melops*), Corkwing wrasse (Fig. 329b)

*Body deep, laterally compressed. Large dorsal fin with 14–17 spines anteriorly and 8–10 rays posteriorly. Anal fin with 3 spines and 8–11 rays. Scales large, 31–37 along lateral line. Lips thick and pronounced. Region in front of operculum finely toothed. Body colour and markings vary, depending on background and age of specimen; juveniles generally olive-green or blue-green; mature males darker, often reddish-brown to deep purple-brown with bluish lines on lower part of head; mature females pale brown with dark brown lines on head. Crescent-shaped dark patch just posterior to eye; dark spot on centre of base of tail fin. Length usually up to 150 mm but can be 250 mm.*

*S. melops* is widely distributed in north-west Europe. Found among seaweed, it may be common in rock pools on the lower shore, particularly on the west coast of Britain, feeding on a wide range of molluscs and crustaceans. It extends into the sublittoral to depths of about 50 m. In spring the male builds a nest of seaweed in a crevice on the lower shore or shallow sublittoral, in preparation for spawning in summer. After a complex courtship display, eggs are laid and fertilized in the nest, which is then guarded and fanned by the male. In a single season some males build a second nest and rear another batch of eggs. The eggs hatch after about two weeks and larvae are found in the plankton off Plymouth from June to August. On English Channel coasts both sexes reach maturity in the second or third year, but the nest-building behaviour of the males apparently does not begin until they are in their fourth year. Longevity is believed to be up to nine years.

(a)

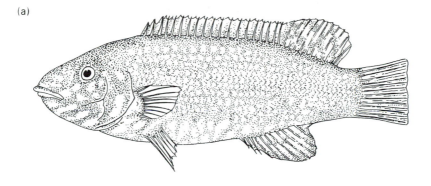

(b)

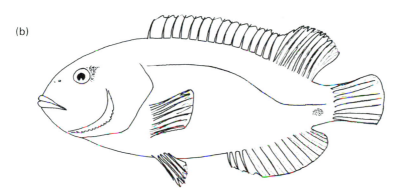

Figure 329  (a) *Labrus bergylta*. (b) *Symphodus melops* (after Du Heaume, in Wheeler, 1969).

***Ctenolabrus rupestris*** (Linnaeus)    Goldsinny (Fig. 330a)

*Body deep, laterally compressed. Large dorsal fin with 16–19 spines anteriorly and 7–10 rays posteriorly. Anal fin with 3 (occasionally 4) spines and 6–9 rays. Scales large, 35–39 along lateral line. Colour brown, green, reddish. Dark spot in dorsal position on base of tail fin and black patch on anterior of dorsal fin. Length up to 180 mm.*

C. rupestris is very widely distributed in north-west Europe and may be locally abundant in some areas around Britain, but is not common in the eastern English Channel and southern North Sea. It is found among seaweed off rocky shores to depths of about 50 m. Breeding occurs in spring and summer, and fish of up to 16 years of age have been recorded off the west coast of Scotland.

(a)

(b)

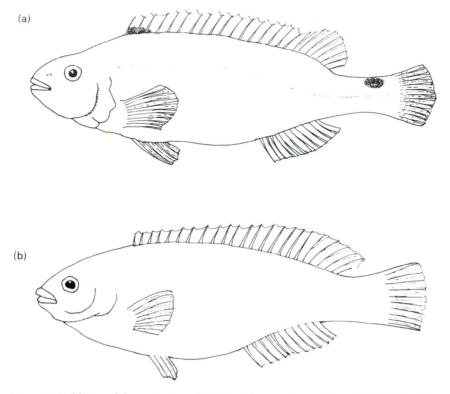

Figure 330  (a) *Ctenolabrus rupestris*. (b) *Centrolabrus exoletus*. ((a) and (b) Modified after
           Du Heaume, in Wheeler, 1969.)

***Labrus bergylta*** Ascanius    Ballan wrasse (Fig. 329a)

*Body deep, laterally compressed. Large dorsal fin with 19–20 spines anteriorly and 9–11
rays posteriorly. Anal fin with 3 spines and 8–12 rays. Lips thick and pronounced. Scales
large, usually with pale centre and dark posterior margin; 41–47 scales along lateral line.
Region in front of operculum not toothed. Colour varied, green, brown, sometimes reddish-
brown. Length up to 600 mm.*

*L. bergylta* is widely distributed in north-west Europe and Britain and is often common
among seaweed in shallow water off rocky shores, extending to depths of about 20 m.
It is also found in rock pools on the lower shore and feeds mainly on crustaceans and
mussels. The ballan wrasse undergoes a sex change. The fishes mature first as female
then many, but not all, later change sex to male. On Irish Sea coasts, females spawn
when they are five or six years old and sex change occurs at any time thereafter. Breeding
takes place in spring and summer. The eggs are believed to be laid in depressions exca-
vated by the males. Larvae hatch at a body length of about 4 mm and have been recorded

in the plankton off Plymouth from May to July. Growth is slow and specimens more than 20 years old have been recorded.

Family Zoarcidae – eelpouts
Body elongate, with relatively long head. Dorsal fin long and continuous with tail fin and long anal fin. Pelvic fins anterior to pectoral fins. Scales very small.

***Zoarces viviparus*** (Linnaeus)    Viviparous blenny, Eelpout (Fig. 331)

*Body elongate. Dorsal fin long, continuous with tail and anal fin; just anterior to tail the fin rays are short. Pectoral fins large; pelvic fins small. Head large. Scales small. Brownish colour, paler ventrally, with number of dark patches along body. Usually up to 300 mm but may reach 500 mm in length.*

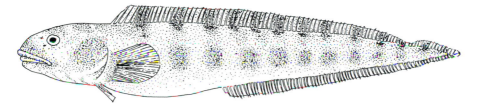

Figure 331  *Zoarces viviparus.*

*Z. viviparus* is a northern species reaching the southern limit of its distribution in the eastern English Channel. Around Britain it is found on North Sea coasts and the west coast of Scotland, where it is often common in rock pools and among seaweed on the lower shore, extending to depths of about 40 m. During summer it migrates offshore to deeper, cooler waters, and in the southern part of its range it is not common intertidally. It feeds on a wide range of invertebrates, particularly crustaceans, and small fishes. As one of the common names implies, *Zoarces* is viviparous. Mating takes place in late summer and autumn and fertilization is internal, following a brief period of copulation when sperm are deposited in the female by means of a tiny papilla on the undersurface of the male. The eggs develop inside the female; three to four months later, fully developed young are released at about 40 mm body length, generally from December to February. It has been estimated that the larger females carry as many as 300 young. Maturity is reached in the second or third year and longevity is about four years. The eelpout is fished commercially in Scandinavia, using baited pots and seine nets.

Family Pholididae – butterfish
Body long, slender. Dorsal fin long, slender, made up of large number of short spines. Pelvic fins very small, reduced to small spines. Scales tiny.

**Pholis gunnellus** (Linnaeus)     (*Centronotus gunnellus*), Butterfish, Gunnel (Fig. 332)

*Body long, slender and laterally flattened. Dorsal fin long, low, extending to tail fin; anal fin much shorter. Pelvic fins very small, reduced to small spines just below base of pectoral fins. Scales very small. Brownish in colour with darker patches; about 12 black spots, ringed with white, along base of dorsal fin. Length up to 250 mm.*

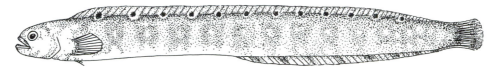

Figure 332  *Pholis gunnellus.*

*P. gunnellus* is a northern species reaching the southern limit of its distribution on the north-west coast of France. It is common around Britain in rock pools from the middle shore and into the sublittoral to depths of about 100 m. There is downshore migration during winter. It feeds on a wide variety of invertebrates. Breeding occurs during winter. The eggs, which are laid in clusters on stones, in empty shells and in crevices, are usually guarded by the female and hatch after a few weeks; larvae have been recorded in the plankton up to about May. Young are found inshore when they are about 30 mm in length and maturity is reached at a body length of 100 mm when the fish is in its second year. Longevity is up to five years.

Family Trachinidae – weevers
Body stout, mouth sloping sharply downwards. Eyes situated dorsally. Anterior dorsal fin with large spines, black in colour. Second dorsal fin and anal fin long. Pelvic fins anterior to pectoral fins. Large spine on operculum. Scales small, in conspicuous oblique rows.

**Echiichthys vipera** (Cuvier)     (*Trachinus vipera*), Lesser weever (Fig. 333)

*Body thick. Anterior dorsal fin with 5–7 spines, black in colour; posterior dorsal fin long, with 21–24 rays; anal fin long, with 24–26 rays. Margin of pectoral fin rounded. Mouth running sharply downwards. Eyes situated dorsally; no spines near eyes. Long, backwardly directed spine on operculum. Scales small. Dorsal surface yellowish-brown with dark patches; paler ventrally. Length up to 140 mm.*

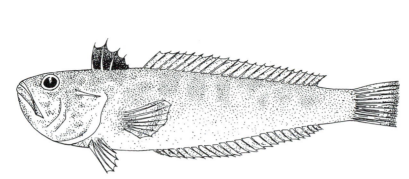

Figure 333 *Echiichthys vipera.*

*E. vipera* occurs in the southern half of the North Sea and all around Britain, where it is most abundant in the south and west. It lives on sandy sediments and lies buried with only the head and dorsal surface visible. Occasionally found on the lower shore when the tide is out, it is most common in shallow water to depths of about 50 m. It feeds on a wide range of molluscs, crustaceans and small fishes. Breeding occurs in summer; the eggs are pelagic and larvae have been recorded in the plankton off Plymouth from June to September. The spines on the operculum and the anterior dorsal fin of the weever are equipped with venom cells which can inflict a severe sting causing pronounced swelling of the tissues. The pain is described as 'intense' and 'excruciating' and may last for up to 24 hours. It is said that some relief can be obtained by immediately bathing the affected area in hot water; this apparently breaks down the toxin.

The great weever, *Trachinus draco* Linnaeus, up to 300 mm body length, is found in deeper water than *E. vipera* but is occasionally found in the shallows. It is distinguished from the latter species by the presence of two spines close to the eyes and an indented or straight margin to the pectoral fin. Like *E. vipera*, it has a poisonous sting.

Family Blenniidae – blennies
Body elongate, head rather large. Dorsal fin long and, in the species described in this text, divided into two parts by a depression. Pelvic fins with two long rays; anterior to pectoral fins. Tail fin separate from dorsal and anal fins. Without scales.

1 Flap of tissue across head, just above eyes, with mid-dorsal row of small tentacles behind ......................................................................... *Coryphoblennius galerita* (p. 502)
  Head not as above ................................................................................................ 2
2 Pair of branched tentacles between eyes ...................... *Parablennius gattorugine* (p. 504)
  Without tentacles between eyes .................................................. *Lipophrys pholis* (p. 503)

*Coryphoblennius galerita* (Linnaeus)   (*Blennius montagui*), Montagu's blenny (Fig. 334c)

*Dorsal fin long, divided into anterior and posterior parts by depression. Flap of tissue across head, just above the eyes; mid-dorsal row of small tentacles behind the flap. Anal fin long. Pelvic fins each with 2 long rays; anterior to pectoral fins. No scales. Green-brown colour with dark patches; light-coloured spots on head and body. Length up to 85 mm.*

(a)

(b)

(c)

Figure 334  (a) *Lipophrys pholis.* (b) *Parablennius gattorugine.* (c) *Coryphoblennius galerita.*
((b) After Du Heaume, in Wheeler, 1969.)

*C. galerita* is found on rocky shores on the Atlantic coast of France and south-west Britain, extending to the west coast of Ireland. It is found on the middle shore, particularly in pools with the red alga, *Corallina* (p. 61). It feeds on barnacles, nipping off the thoracic appendages as the barnacle feeds, and other small crustaceans. Eggs are laid in crevices from May to August and are guarded by the males, which are territorial.

### *Lipophrys pholis* (Linnaeus)    (*Blennius pholis*), Shanny (Fig. 334a)

*Dorsal fin long, divided into 2 by slight depression; anterior part with 11–13 rays, posterior part with 18–20 rays. Anal fin long. Pelvic fins each with 2 long rays; anterior to the pectoral fins. Body without scales. Without tentacles on head. Colour varied, green, brown, olive or yellowish with black markings. Length up to 160 mm.*

*L. pholis* is widely distributed on rocky shores, less frequently on sand and mud, in north-west Europe, but has a patchy distribution in the southern North Sea. It is essentially a rock pool species, occasionally found on the upper shore; it is highly adapted to conditions in rock pools withstanding low salinity and surviving in moist situations under stones. Small specimens are generally found under stones and in pools on the upper shore. *Lipophrys* forages over the shore when the tide is in, often returning to the same rock pool which it may occupy for many weeks, when the tide ebbs. This has been described as homing behaviour. It feeds on a wide range of crustaceans and molluscs including barnacles, littorinids and the small bivalve *Lasaea adansoni* (p. 296); the young feed extensively on the thoracic appendages of barnacles, which are nipped off by the shanny as the barnacles feed. Larger individuals often have considerable amounts of algae in the gut. There is a downshore migration in winter. Around Britain, breeding occurs in spring and summer but this shifts to winter and spring in more southerly localities. The eggs are laid in clusters under stones or in narrow crevices, where they are guarded by the male, which fans them with its tail. A single female lays up to eight batches of eggs per season, which hatch after about six to eight weeks. The larger females have been estimated to produce as many as 8000 eggs. Larvae are common in the plankton during summer, young appearing on the shore at a length of about 20 mm. Maturity is reached after two or three years and although specimens of 14 years of age have been recorded, few live more than five years.

During the winter months, specimens of *L. pholis* on the west coast of Britain are sometimes parasitized by the ectoparasitic marine leech **Oceanobdella blennii** (Knight-Jones). Small leeches up to about 4 mm in length are found under the operculum, and larger specimens (up to 14 mm) behind the pectoral fin. The mature leeches leave the host in spring and lay egg cocoons on the shore. The eggs develop slowly during summer and hatch during the following autumn/winter when the young leeches seek out a host.

*Parablennius gattorugine* (Brünnich) (*Blennius gattorugine*), Tompot blenny (Fig. 334b)

*Dorsal fin long, divided into 2 parts by slight depression; anterior part with 12–14 rays, posterior part with 17–20 longer rays. Anal fin long. Pelvic fins each with 2 long rays; anterior to the pectoral fins. Body without scales. A pair of branched tentacles between the eyes. Colour yellowish, olive, brown, with number of darker vertical bands. Length usually up to about 150 mm, may reach 300 mm.*

*P. gattorugine* is a southern species, extending into the English Channel and along the west coast of Britain as far north as Scotland. It has not been recorded from North Sea coasts. Although found occasionally in rock pools on the lower shore, it is more common in the sublittoral, extending to depths of about 35 m. It feeds on a wide range of cnidarians, molluscs, crustaceans, polychaetes and algal fragments. Breeding occurs in spring and summer. The purplish-black eggs are laid in crevices and guarded by the male. Larvae have been recorded in the plankton in summer. On the west coast of Ireland maturity is reached after one year; longevity is believed to be up to about nine years.

Family Ammodytidae – sand eels
Body elongate. Tail fin forked. Lower jaw projecting beyond upper jaw. Dorsal fin long. No pelvic fins. Scales tiny.

1 Upper jaw not extensible ................................................................................ 2
  Upper jaw extensible (Fig. 335d) ................................................................. 3
2 Black spot in front of eye .................................................. *Hyperoplus lanceolatus* (p. 506)
  Without black spot in front of eye ............................... *Hyperoplus immaculatus* (p. 506)
3 Scales in >>>>>> pattern on undersurface. Scales on base of tail fin .......................
  .................................................................................... *Ammodytes tobianus* (p. 505)
  Scales on undersurface not in >>>>> pattern. Scales not present on base of tail fin ..
  .................................................................................... *Ammodytes marinus* (p. 504)

*Ammodytes marinus* Raitt   Raitt's sand eel (Fig. 335c)

*Body elongate, eel-like. Dorsal fin long, with 55–67 rays; anal fin much shorter, 29–33 rays. Tail fin forked. Pectoral fins extend beyond anterior margin of dorsal fin. Pelvic fins absent. Lower jaw protruding beyond upper jaw; upper jaw extensible. Scales small, not in >>>> pattern on undersurface. Scales absent from base of tail fin. Dorsal surface blue-green, silvery below. Length up to 240 mm.*

*A. marinus* is essentially a sublittoral species, swimming in shoals in water of 30–150 m depth. It is often abundant and juveniles are occasionally found in much shallower water, where they burrow into sand and fine gravel. It is generally regarded as a zooplank-

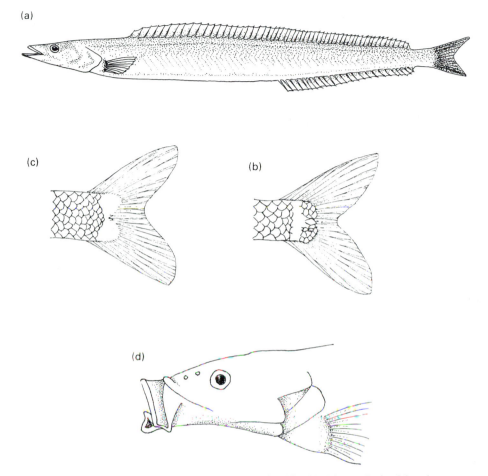

Figure 335  (a) *Ammodytes tobianus*, with detail of tail fin (b). (c) Detail of tail fin of *Ammodytes marinus*. (d) Extensible jaw of *Ammodytes* sp. ((b), (c), (d) After Stebbing, in Wheeler, 1978.)

ton feeder but there is some evidence to suggest that in the North Sea it is a serious predator of herring larvae and eggs. Breeding occurs in winter and longevity is about four years.

***Ammodytes tobianus*** Linnaeus    (*Ammodytes lancea*), Lesser sand eel (Fig. 335a, b)

*Body elongate, eel-like. Dorsal fin long, with 50–56 rays; anal fin much shorter, with 24–32 rays. Tail fin forked. Pectoral fins extend beyond anterior margin of dorsal fin. Pelvic fins absent. Lower jaw protruding beyond upper jaw; upper jaw extensible. Scales small, in >>> pattern on undersurface. Scales on base of tail fin. Dorsal surface yellow-green or bluish-green; silvery below. Length up to 200 mm.*

*A. tobianus* is widely distributed and common in north-west Europe and Britain on sandy beaches from the middle shore to depths of about 30 m. It swims in shoals just above the bottom, feeding on a wide variety of organisms, including small fishes. It is often found buried in the sand at low water when the tide is out. There are apparently two spawning groups, one spawning in spring and the other in the autumn. Eggs are attached to sand grains and the larvae are pelagic. Longevity is believed to be four years.

### *Hyperoplus immaculatus* (Corbin)    Corbin's sand eel (Fig. 336b)

*Body elongate, eel-like. Dorsal fin long, with 59–62 rays; anal fin much shorter (31–43 rays). Tail fin forked. Pectoral fins extend beyond anterior margin of dorsal fin. Pelvic fins absent. Lower jaw protruding beyond upper jaw; upper jaw not extensible. Scales small. Dorsal surface blue-green, silver below; without black spot in front of eye (cf.* Hyperoplus lanceolatus, *below). Body length up to 300 mm.*

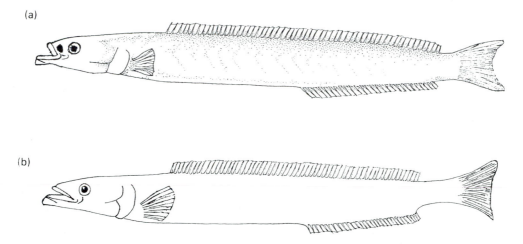

(a)

(b)

Figure 336  (a) *Hyperoplus lanceolatus.* (b) *Hyperoplus immaculatus.*

*H. immaculatus* is generally found sublittorally in waters of 50–300 m depth, but young occur close inshore on sand and gravel. It takes a wide range of food, including zooplankton and small fish. Breeding occurs in winter.

### *Hyperoplus lanceolatus* (Le Sauvage)    Greater sand eel (Fig. 336a)

*Body elongate, eel-like. Dorsal fin long, with 52–61 rays; anal fin much shorter (27–33 rays). Tail fin forked. Pectoral fins do not extend beyond anterior margin of dorsal fin. Pelvic fins absent. Lower jaw protruding beyond upper jaw; upper jaw not extensible. Scales small. Dorsal surface greenish, silver below. Conspicuous black spot in front of eye. Body length up to 320 mm.*

*H. lanceolatus* is widely distributed and common in north-west Europe, occurring sublit-torally to depths of 150 m. Young specimens are often found buried on the lower shore on sandy beaches. The greater sand eel feeds on crustaceans and other fauna associated with sandy deposits. Breeding occurs in spring and summer when the fish are about two years old. Longevity is up to five years.

Sand eels are important components in the diet of some species of birds and a number of commercial fish. They are fished commercially and have been subjected to very heavy fishing pressure.

### Family Gobiidae – gobies

Eyes large, protruding, dorso-lateral in position. Two dorsal fins; anterior dorsal short with spines, pelvic fins joined ventrally to form a crescent-shaped disc enabling the fish to attach loosely to the substratum. Body with scales. A large family of fishes found on the open coast, in estuaries and fresh water. The majority are benthic and generally of a small size, mostly less than 150 mm body length. Although identification can be diffi-cult, the species are often restricted to a particular habitat and this can be helpful in identification.

1  Large black spot at base of tail fin. Anterior dorsal fin with 7 (rarely 6 or 8) spines ........
.............................................................................. *Gobiusculus flavescens* (p. 509)
   Without black spot at base of tail fin. Anterior dorsal fin with 6 (occasionally 5–7)
   spines ......................................................................................................... 2
2  Upper rays of pectoral fins free distally, extending beyond fin membrane (Fig. 337a) ....... 3
   Upper rays of pectoral fins not free distally (Fig. 339) ...................................... 4
3  Upper rays of pectoral fins short; 32–42 scales in a line from base of pectoral fin to tail
   fin ............................................................................... *Gobius niger* (p. 507)
   Upper rays of pectoral fins long, branched; 50–57 scales in a line from base of pectoral
   fin to tail fin ......................................................... *Gobius paganellus* (p. 509)
4  (a) Fifty-five to 75 scales in a line from base of pectoral fin to tail fin ............................
.............................................................................. *Pomatoschistus minutus* (p. 511)
   *or*
   (b) Thirty-nine to 52 scales in a line from base of pectoral fin to tail fin ..............................
.............................................................................. *Pomatoschistus microps* (p. 510)
   *or*
   (c) Thirty-six to 43 scales in a line from base of pectoral fin to tail fin; 1 or 2 rows of black
   spots at base of dorsal fin ................................................... *Pomatoschistus pictus* (p. 511)

*Gobius niger* Linnaeus    Black goby (Fig. 337a)

*Anterior dorsal fin with 6 (sometimes 5–7) long spines, the central ones longest; posterior dorsal fin with 1 spine and 11–13 rays. Eyes large. Upper rays of pectoral fins short, free distally. Pelvic fins united to form crescent-shaped disc; 32–42 scales in a line from base of*

*pectoral fin to tail fin. Body various shades of brown with darker patches on sides. Black
spot on upper, anterior margin of dorsal fins. Length up to 170 mm.*

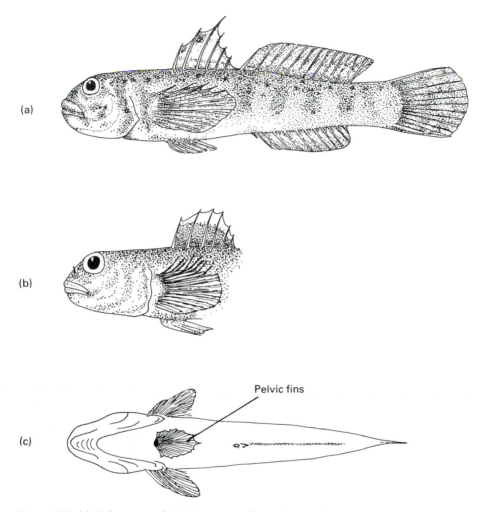

Figure 337  (a) *Gobius niger*. (b) *Gobius paganellus*, with ventral view to show pelvic fins
joined to form crescent-shaped disc (c).

*G. niger* is widely distributed throughout north-west Europe over sand and mud to
depths of about 70 m and extends into estuaries. It feeds on a wide range of invertebrates
and young fishes. Breeding occurs in spring and summer. Club-shaped eggs are laid on
stones, shells and seaweed and are guarded by the male. Larvae hatch after about two
weeks and are found in the plankton during summer, becoming benthic at a length of
about 12 mm. Maturity is reached after two or three years and longevity is believed to
be about four or five years.

*Gobius paganellus* Linnaeus   Rock goby (Fig. 337b, c)

*Anterior dorsal fin with 6 spines; posterior dorsal with 1 spine and 13–14 rays. Upper rays of pectoral fins long, branched; free distally, almost reaching the base of the anterior dorsal fin. Small branched tentacle in front of eye. Pelvic fins united to form crescent-shaped disc; 50–57 scales in a line from base of pectoral fin to tail fin. Blackish-brown in colour; pale yellowish horizontal band along upper margin of anterior dorsal fin; band is bright orange in breeding males. Length up to 130 mm.*

G. paganellus is found on the west coast of France, in the English Channel and along the west coast of Britain, and has recently been recorded from the southern North Sea. It is found under stones on rocky shores and in pools on the middle shore and below, being most common among seaweed on sheltered shores. It is found to depths of about 15 m and undergoes a downshore migration during winter. It feeds on a wide range of small crustaceans and molluscs. Breeding occurs in spring and summer. The eggs are laid in sheets under stones, but have also been found on a variety of substrata such as shells and the tubes of polychaete worms. Development takes about three weeks: the larvae have been recorded in the plankton during summer, becoming benthic at about 10 mm in length. Maturity is reached in the second or third year. Although individuals of eight and 10 years old have been recorded, most live for only two or three years.

*Gobiusculus flavescens* (Fabricius)   (*Gobius ruthensparri*), Two-spot goby (Fig. 338)

*Anterior dorsal fin with 7 spines (rarely 6 or 8); posterior dorsal with 1 spine and 9–11 rays. Upper margin of pectoral fins not free distally. Pelvic fins united to form crescent-shaped disc; 35–40 scales in a line from base of pectoral fin to tail fin. Reddish-brown colour, with pale blotches along dorsal surface; large black spot, with yellow edge, at base of tail fin; small dark spot just behind pectoral fin in male; dorsal and tail fins banded. Up to 60 mm in length.*

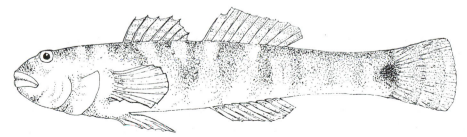

Figure 338  *Gobiusculus flavescens.*

*G. flavescens* is widely distributed in north-west Europe and is found all around Britain in rock pools and sublittorally to depths of 15 m living among seaweed, often in shoals. It is sometimes locally abundant in the south and west and is also found in sheltered bays and estuaries among *Zostera* beds (p. 85). Planktonic crustaceans form the bulk of its food. Breeding occurs in spring and summer; the eggs are laid among the holdfasts of *Sacchoriza polyschides* (p. 45) and in empty barnacle shells, and are guarded by the male.

**Pomatoschistus microps** (Krøyer)    Common goby (Fig. 339)

*Anterior dorsal fin with 6 spines, posterior dorsal with 1 spine and 8–10 rays. Pelvic fins united to form crescent-shaped disc. Eyes large. Upper rays of pectoral fins not free distally; 39–52 scales in a line from base of pectoral fin to tail fin. Greyish in colour, often with darker blotches across back and faint marks along sides of body; dark area at base of pectoral fins and tail fin; dark spot on anterior dorsal fin of males. Length up to about 65 mm.*

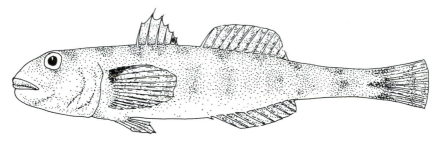

Figure 339  *Pomatoschistus microps.*

*P. microps* is widely distributed in north-west Europe and Britain on sandy and muddy sediments and has been described as 'the most important small fish of British estuaries'. It tolerates salinities from about 8 to 80‰ and is often abundant in pools and creeks on mud-flats. It feeds predominantly on small crustaceans such as *Corophium volutator* (p. 377), which are found in such habitats. When winter temperatures fall below 5 °C the goby migrates into deeper water, returning in spring. Breeding occurs in spring and summer, when a single female can produce as many as 10 batches of eggs. The pear-shaped eggs are laid under stones and shells, but apparently there is a preference for laying under empty bivalve shells. The fish removes sediment from under the shell to create a nest and the eggs are attached by the female to the inner surface of the shell which forms the roof of the nest. The shell is then anchored by being partly covered with sediment. Nest construction is carried out solely by males, which have been seen to turn shells so that the concave side faces downwards. The male guards and fans the eggs until they hatch, generally after about 14 days. Both sexes spawn a number of times during the breeding season. Larvae have been recorded in the plankton during summer and young fishes are found on the shore at a body length of 11–12 mm. Sexual maturity is reached in the first year and longevity is believed to be about two years.

**Pomatoschistus minutus** (Pallas)    (*Gobius minutus*), Sand goby (Fig. 340)

*Anterior dorsal fin with 6–7 spines, posterior dorsal with 1 spine and 10–12 rays. Pelvic fins united to form crescent-shaped disc. Eyes large. Upper rays of pectoral fins not free distally; 55–75 scales in a line from base of pectoral fin to tail fin. Pale brown with darker markings on sides; dark spot on posterior margin of anterior dorsal fin. Length up to about 110 mm but generally about 60 mm.*

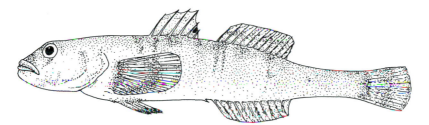

Figure 340 *Pomatoschistus minutus* (after Day, 1884).

*P. minutus* is widely distributed in north-west Europe and Britain on sand and mud from the middle shore into the sublittoral, sometimes occurring in shoals. It is also found in the mouths of estuaries. Juveniles are apparently more tolerant of reduced salinity than the adults and extend well into estuaries. The diet is predominantly small crustaceans such as *Corophium* (p. 377). There is an offshore migration into deeper water during winter. Breeding occurs in spring and summer. Pear-shaped eggs are laid in empty bivalve shells where they are guarded by the male until they hatch, usually after about two weeks. Larvae are found in the plankton throughout the summer; young become benthic at about 12 mm in length. The sand goby reaches maturity in its first year and lives for two years.

**Pomatoschistus pictus** (Malm)    (*Gobius pictus*), Painted goby (Fig. 341)

*Anterior dorsal fin with 6 spines, posterior dorsal with 1 spine and 8–9 rays. Pelvic fins united to form crescent-shaped disc. Eyes large. Upper rays of pectoral fins not free distally; 36–43 scales in a line from base of pectoral fin to tail fin. Brownish in colour with paler areas on dorsal surface and sides; 4 dark spots, sometimes double, on each side; 1 or 2 rows of black spots at base of dorsal fins. Length generally up to 60 mm.*

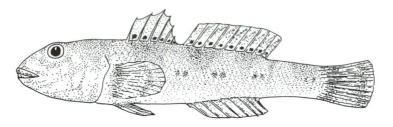

Figure 341 *Pomatoschistus pictus* (after Du Heaume, in Wheeler, 1969).

*P. pictus* is widely distributed in north-west Europe on gravel and coarse sand from the lower shore to depths of about 50 m. It is occasionally found in rock pools on the middle and lower shore. In Britain it is most common on south and west coasts. It feeds mainly on crustaceans. Breeding occurs during spring and summer. As in *P. microps* (above), pear-shaped eggs are laid in empty bivalve shells and guarded by the male, hatching after about 12 days. Larvae have been recorded in the plankton during summer. Sexual maturity is reached in the second year and longevity is believed to be two, occasionally three, years.

Two species of goby occurring in north-west Europe, **Aphia minuta** (Risso) and **Crystallogobius linearis** (von Düben), are pelagic. They can be recognized by their transparent, laterally compressed bodies.

REFERENCES
Urochordata
Alder, J. & Hancock, A. (1905–12). *The British Tunicata. An unfinished monograph.* London: Ray Society.

Berrill, N. J. (1950). *The Tunicata with an account of the British species.* London: Ray Society.

Kott, P. (1952). Observations on compound ascidians of the Plymouth area, with descriptions of two new species. *Journal of the Marine Biological Association of the United Kingdom,* 31, 65–83.

Lafargue, F. & Wahl, M. (1987). The didemnid ascidian fauna of France. *Annales de l'Institute Océanographique, Monaco,* 63, 1–46.

Millar, R. H. (1970). *British ascidians. Tunicata: Ascidiacea.* Keys and notes for the identification of the species. Synopses of the British fauna (New Series), no. 1. London: Academic Press.

Picton, B. E. (1985). *Ascidians of the British Isles. A colour guide.* Ross-on-Wye: Marine Conservation Society.


Vertebrata
Day, F. (1880–4). *The fishes of Great Britain and Ireland.* 2 vols. London: Williams & Norgate.

Hureau J. C. & Monod, Th. (eds.) (1979). *Check-list of the fishes of the north-eastern Atlantic and of the Mediterranean*, 2nd imp. with supplement, 2 vols. Paris: UNESCO.

Muus, B. J. & Dahlstrom, P. (1974). *Collins guide to the sea fishes of Britain and north-western Europe.* London: Collins.

Nelson, J. S. (1984). *Fishes of the world*, 2nd edn. New York: Wiley.

Russell, F. S. (1976). *The eggs and planktonic stages of British marine fishes.* London: Academic Press.

Wheeler, A. (1969). *The fishes of the British Isles and north-west Europe.* London: Macmillan.

Wheeler, A. (1978). *Key to the fishes of northern Europe.* London: Warne.

Whitehead, P. J. P., Bauchot, M.-L., Hureau, J.-C., Nielson, J. & Tortonese, E. (eds.) (1984–6). *Fishes of the north-eastern Atlantic and the Mediterranean*, vol. I (1984), vol. II (1986), vol. III (1986). Paris: UNESCO.

# Glossary

Words in these definitions which are italicized are listed in the glossary.

ACONTIA  thread-like structures discharged through mouth and pores in the column of sea-anemones. Armed with nematocysts.

ACRORHAGUS  wart-like structure in sea-anemones, often conspicuously coloured, bearing batteries of stinging cells.

ADDUCTOR MUSCLE  muscle which closes shell of bivalve mollusc. Adductor muscle scars are seen on inside of shell valves.

AMBULACRUM  part of surface of echinoderm with tube-feet.

ANAEROBIC  without oxygen.

ANCESTRULA  the first-formed *zooid* in a bryozoan colony.

ANNUAL  organism living for one year.

ANNULATION  ring or ring-like division.

ANTENNA  sensory appendage found on the head of some annelids and arthropods.

APOTHECIA  saucer-shaped reproductive bodies seen on the surface of lichen *thallus*.

ARISTOTLE'S LANTERN  complex feeding structure of sea-urchins; equipped with calcified teeth.

ASEXUAL REPRODUCTION  reproduction not involving a sexual process.

ATRIUM  space surrounding *pharynx* in the sea-squirts. Communicates with exterior via exhalant *siphon*.

AVICULARIUM  specialized bryozoan *zooid*, equipped with movable jaws.

BASAL LACERATION  asexual reproduction of sea-anemones by fragmentation of basal disc.

BEAK  first-formed part of bivalve shell. Usually pointed.

BENTHIC  living in or on the sea bed.

BERRIED  bearing eggs, with reference to crustaceans carrying eggs externally, e.g. crabs.

BILATERAL SYMMETRY  symmetry of organism which can be cut into two halves, each a mirror image of the other.

BIRAMOUS  having two branches or *rami*, e.g. some arthropod appendages.

BLASTOSTYLE  hydroid *polyp* specialized for reproduction.

BRACKISH  water containing less salt than normal seawater.

BROOD PLATE  see *oostegite*.

BROODING  the care of eggs, at least during the early stages of development, in a brood chamber which may be internal or external to the body.

BYSSUS  fibres produced by the foot of bivalve molluscs. Used for attachment.

CALCAREOUS  made of calcium carbonate or chalk.

CARAPACE  *exoskeleton* covering the anterior *dorsal* surface of many crustaceans.

CARDINAL TEETH  teeth on the hinge line of bivalve molluscs.

CARPOSPOROPHYTE  a stage in the life-cycle of red seaweeds. Produces carpospores which grow into the *sporophyte* generation.

CARTILAGE  firm, elastic tissue. Not calcified. Found in the skeleton of skates, rays, etc.

CEPHALOTHORAX  fusion of head and one or more thoracic segments in crustaceans.

CERATA  finger-like outgrowths on the *dorsal* surface of sea-slugs.

CHAETA (SETA)  chitinous bristle of annelids and arthropods. Usually referred to as chaetae in annelids and setae in arthropods.

CHELAE  pincers on the *pereopods* of decapod crustaceans.

CHELIFORES  first pair of appendages on the head of sea-spiders.

CHELIPEDS  first pair of *pereopods* of decapod crustaceans.

CHLOROPLAST  body in plant cell containing chlorophyll. The site of *photosynthesis*.

CHONDROPHORE  depression in which the internal ligament of bivalve shell is attached.

CHROMATOPHORE  cell containing pigment granules.

CILIUM  very small hair-like structure. Motile, creating water currents which may be used for feeding, locomotion, etc.

CIRRUS  tentacle-like appendage.

CLASS  unit of classification. A subdivision of a *phylum*.

CNIDOCYTE  cell containing capsule-like structure (cnida) with long thread which is discharged on stimulation. Characteristic of *phylum* Cnidaria.

COLUMELLA  central pillar of a gastropod shell.

COMMENSAL  organism living in close association with another of a different *species*.

CONCEPTACLE  flask-shaped invagination opening by tiny pore onto the surface of seaweed *frond*. Contains *gametes*.

CRENULATE  scalloped or indented margin.

CRUSTOSE  crust-like *thallus* of lichens; growth form of red seaweeds.

CUTICLE  external, non-cellular layer on the outside of plants and animals.

DACTYLOZOOID  *polyp* specialized for the defence of hydrozoan colony.

DEPOSIT FEEDING  feeding on bottom deposits and *detritus*.

DETRITUS  decomposing plant and animal material.

DIATOMS  microscopic, unicellular algae with cell walls of silica. Main constituent of *phytoplankton*.

DICHOTOMOUS BRANCHING  repeatedly dividing into two branches.

DIGITATE  divided into finger-like segments.

DIOECIOUS  separate sexes.

DIPLOID  chromosomes in pairs in nucleus.

DIRECT DEVELOPMENT  development in which a *larval* stage is lacking.

DISTAL  away from point of attachment, e.g. the distal part of a limb.

DORSAL  the upper side of a plant or animal.

ECDYSIS  the periodic shedding (moulting) of the *cuticle* by arthropods.

ECTOPARASITE  organism living *parasitically* on the outer surface of another organism.

EMERSION  describes the period when an intertidal organism is out of water.

ENDEMIC  native to a country.

ENDOPARASITE  organism living *parasitically* within another organism.

ENDOSKELETON  an internal skeleton, e.g. the skeleton of fishes.

ENZYME  a substance which promotes a chemical reaction without itself being used up
    in the reaction.

EPIDERMIS  outer layer of cells of plants and animals.

EPIFAUNA  animals living on the surface of other plants and animals, but not living
    *parasitically*.

EPIPHYTE  plant living on the surface of another plant or animal, but not living
    *parasitically*.

EPIPODIAL TENTACLE  tentacle on the lateral lobes of the foot of prosobranch
    molluscs.

EPITOKE  specialized polychaete adapted for reproduction.

EQUINOCTIAL TIDE  tide of large amplitude occurring at time of spring and autumn
    equinoxes.

EQUIVALVE  valves of shell of bivalve mollusc similar in size and shape.

ERRANT  animal moving freely on or in substratum, with particular reference to
    polychaetes.

EULITTORAL ZONE  middle region of shore, lying between the upper limit of the
    laminarian seaweeds and the upper limit of the acorn barnacles.

EXOSKELETON  skeleton covering the outside of an animal, e.g. the shell of a crab.

FAMILY  unit of classification. A subdivision of an *order*.

FILAMENTOUS  thread-like chain of cells with reference to seaweed *thallus*.

FOLIOSE  leaf-like *thallus* of lichens.

FOOT  muscular organ of molluscs.

FROND  that part of a seaweed excluding the *holdfast*.

FRUTICOSE  upright or pendant lichen *thallus*, attached at base.

GAMETE  reproductive cell, male or female.

GAMETOPHYTE  stage in the life-cycle of seaweeds; produces *gametes*.

GASTROZOOID (HYDRANTH)  hydroid *polyp* specialized for feeding.

GENUS  unit of classification. A subdivision of a *family*.

GIRDLE  margin of muscular tissue surrounding valves of chitons.

GNATHOPOD  anterior thoracic appendage of amphipods, usually used for feeding and
    grasping.

GONOTHECA  protective cup surrounding the *gonozooid* of a hydroid.

GONOZOOID  hydroid *polyp* specialized for reproduction.

HAPLOID  single set of unpaired chromosomes in nucleus.

HERBIVORE  animal eating plant material.

HERMAPHRODITE  organism producing both male and female *gametes*.

HETEROMORPHIC  morphologically dissimilar *gametophyte* and *sporophyte* generations of seaweeds.

HOLDFAST  attachment organ of seaweeds.

HYBRID  offspring resulting from cross between two different *species*.

HYDRANTH  see *gastrozooid*.

HYDROTHECA  protective cup around the *hydranth* of a hydroid.

INEQUIVALVE  valves of bivalve mollusc dissimilar in size and/or shape.

INTERSTITIAL FAUNA  fauna living in space between sand grains.

INTROVERT  narrow, anterior region of sipunculans.

ISOMORPHIC  morphologically similar *gametophyte* and *sporophyte* generations of seaweeds.

LANCEOLATE  spear-like in shape.

LARVA  immature, pre-adult form in which some animals hatch from the egg. Usually different in form and way of life from adult. Often an important dispersal phase for animals.

LATERAL LINE  system of sense organs arranged along sides of fish. Sensitive to pressure changes in water.

LATERAL TEETH  teeth on hinge line of shell of bivalve mollusc.

LITTORAL FRINGE  region of shore lying between upper limit of barnacles and upper limit of periwinkles and black lichens, i.e. between *eulittoral zone* and *terrestrial fringe*.

LOPHOPHORE  group of ciliated tentacles encircling the mouth in phoronids and bryozoans.

LUNULE  depressed area, often heart-shaped, lying in front of the *beaks* in some bivalve molluscs. Usually of different sculpture and/or colour from rest of shell.

MANDIBLE  anterior crustacean mouthpart.

MANTLE (PALLIUM)  fold of tissue covering whole or part of the body of a mollusc. Secretes the shell. Also used to describe the flaps of tissue inside the *opercular aperture* of barnacles.

MANTLE CAVITY  cavity enclosed by *mantle* in which gills and other structures are housed.

MAXILLA  crustacean mouthpart. Two pairs, maxillae 1 and maxillae 2.

MAXILLIPED  posterior crustacean mouthpart(s).

MEDUSA  free-swimming (jellyfish) stage in the life-cycle of hydrozoans and scyphozoans.

MEMBRANOUS  thin sheet of cells, with reference to seaweed *thallus*.

MEIOSIS  cell division in which the number of chromosomes is halved.

METAMERISM  repetition of *segments* of similar structure.

METAMORPHOSIS  the process by which a *larva* changes into the adult form.

METAPODIAL TENTACLE  tentacle on the posterior lobes of the foot of prosobranch molluscs.

MIDRIB  thickened central axis of a seaweed *frond*.

MITOSIS  cell division in which the number of chromosomes is not reduced.

MONOSIPHONOUS  single row of siphon-like cells, with reference to seaweed *frond*.

NACRE  the inner layer of the shell of molluscs. The nacreous layer.

NATENT  swimming, with reference to decapods.

NEAP TIDE  tide with the smallest difference between the heights of high and low water.

NEMATOPHORE  hydroid *polyp* armed with nematocysts, specialized for the defence of a hydroid colony.

NEMATOTHECA  protective cup surrounding *nematophore* of a hydroid.

NEUROPODIUM  ventral part of *parapodium* of polychaetes.

NOTOCHORD  dorsal, flexible rod between nerve cord and gut; found at some stage in the life-cycle of vertebrates.

NOTOPODIUM  dorsal part of *parapodium* of polychaetes.

OMNIVORE  animal eating both plant and animal material.

OOSTEGITE (BROOD PLATE)  plate arising from the base of thoracic appendages of females of some crustacean groups. Oostegites form brood chamber in which eggs are held.

OPERCULAR APERTURE  opening at the apex of acorn barnacle through which feeding appendages emerge. The aperture is closed by paired plates, the terga and scuta.

OPERCULUM  structure which closes the opening of, for example, polychaete tube, molluscan shell, bryozoan *zooid*, etc. Also the gill cover of a fish.

ORDER  unit of classification. A subdivision of a *class*.

OSCULUM  opening of sponge through which water leaves the body.

OVIGEROUS LEG  appendage of sea-spiders, used to carry eggs.

OVOVIVIPAROUS  embryos develop within female but are nourished from yolk of egg. Young emerge from parent.

PALEA  broad, flattened *chaeta* of polychaetes.

PALLIAL LINE  line on inside of shell of bivalve mollusc, marking position of attachment of *mantle* to shell.

PALLIAL SINUS  indentation of *pallial line* marking area occupied by siphons of bivalve mollusc.

PALLIUM  see *mantle*.

PALMATE  divided like a hand into finger-like lobes.

PALPS  paired projections arising from the head of polychaetes and arthropods.

PARAGNATHS  chitinous teeth on the *proboscis* of polychaetes.

PARAPODIA  paired lateral outgrowths from body *segments* of polychaetes.

PARTHENOGENESIS  development of an egg without fertilization.

PARASITE  organism living in or on another organism (the host) from which it obtains nourishment at the expense of the host.

PAXILLAE  modified spines of asteroid. Have brush-like margin.

PEDICELLARIA  pincer-like structure of echinoderms, borne on long stalk or sessile.

PELAGIC  swimming or floating in the sea.

PELAGIC DEVELOPMENT  development in which there is a *pelagic* stage.

PENTAMEROUS  symmetry in which the body can be divided into five parts arranged round centre, e.g. a starfish.

PERENNIAL  plant that continues to grow year after year.

PEREOPOD  appendage of thoracic *segment* of crustaceans.

PERIOSTRACUM  outer layer of the molluscan shell.

PERISARC  thin, skeletal structure surrounding hydroid.

PERITHECIUM  flask-shaped reproductive body embedded in *thallus* of lichen.

PHARYNGEAL CLEFTS  openings leading from the cavity of the *pharynx* to the exterior. Sometimes known as gill slits.

PHARYNX  anterior region of alimentary canal.

PHOTOSYNTHESIS  synthesis of organic compounds by green plants from water and carbon dioxide, using energy from sunlight.

PHYLUM  major division of animal kingdom.

PHYTOPLANKTON  plants (usually microscopic) found in *plankton*.

PINNATE  branched on either side of central axis, feather-like.

PLANKTON  plants and animals which drift in the sea. Most are microscopic.

PLEOPOD  appendage of abdominal *segment* of crustaceans.

PLUMOSE  feather-like, with reference to arthropod *setae*.

PODIUM  see *tube-foot*.

POLYMORPHISM  occurrence of two or more different forms of the same species, e.g. in cnidarians and bryozoans.

POLYP  *sedentary* stage in the life-cycle of cnidarians. Usually has single opening, the mouth, surrounded by tentacles.

POLYSIPHONOUS  with several rows of siphon-like cells, with reference to seaweed frond.

PROBOSCIS  specialized, tubular structure, usually associated with the anterior part of the gut of some animals, e.g. the polychaetes.

PROTANDROUS HERMAPHRODITE  *hermaphrodite* which matures first as a male.

PROXIMAL  towards point of attachment, e.g.the proximal part of limb.

RADIAL SYMMETRY  symmetry of an organism in which the parts are symmetrically arranged around a central axis, e.g. a jellyfish.

RADIOLES  tentacle-like projections borne on head of some polychaetes and forming crown or fan.

RADULA  narrow, ribbon-like structure bearing teeth and used for rasping food. Occurs in some groups of molluscs.

RAMUS  branch of appendage of arthropod.

RECEPTACLE  specialized reproductive structure on seaweed *frond* bearing
  *conceptacles*.

REPTANT  crawling, with reference to decapods.

RHIZOME  underground stem of angiosperms.

ROSTRUM  forward projection of the *carapace* or head between the eyes in crustaceans,
  e.g. prawns.

SALINITY  measure of the salt content of water. Expressed as parts per thousand.
  'Normal' seawater is 35‰.

SEDENTARY  animal not moving freely on or in substratum. Often living in burrow or
  tube.

SCYPHISTOMA  *polyp* stage in the life-cycle of a jellyfish.

SEGMENT  unit of body, usually bearing appendages; clearly seen in annelids and
  arthropods. Term used to describe divisions of arthropod limb.

SEMI-DIURNAL TIDE  tidal system giving two high waters and two low waters in
  approximately 24 hours.

SEPTUM  partition.

SESSILE  organism living attached to the substratum.

SETA  see *chaeta*.

SEXUAL REPRODUCTION  reproduction involving fusion of male and female *gametes*.

SIPHON  tube carrying water current into or out of body, e.g. in bivalve molluscs,
  sea-squirts.

SPATULATE  having broad end.

SPECIES  smallest unit of classification normally used. Organisms in a species are able
  to breed among themselves but not with other individuals from another species.

SPERMATHECA  organ in female (or *hermaphroditic* animal) in which sperm is stored.

SPERMATOPHORE  packet of sperm.

SPINE  stout *seta*.

SPOROPHYTE  stage in the life-cycle of seaweeds, producing spores.

SPRING TIDE  tide with the greatest difference between heights of high and low water.

STATOCYST  organ of balance.

STIPE  stalk-like part of seaweed.

STOLON  root-like structure which anchors colony to substratum, e.g. in hydroids.

SUBCHELATE  claw of *pereopod* closes against part of preceding segment.

SUBLITTORAL  lowest zone on the shore, extending below the level of the lowest
  *spring tides*. Its upper limit is the upper limit of laminarian seaweeds.

SUBMERSION  describes the period when an intertidal organism is covered by water.

SUSPENSION FEEDING  feeding on organic particles, e.g. *plankton* or *detritus*
  suspended in water.

SYMBIOSIS  close association between two individuals of different *species*, to the
  benefit of both.

TAXODONT  arrangement of teeth on hinge line of bivalve mollusc in which there are
  large numbers of small teeth and alternating sockets.

TELSON  terminal, usually flattened, region of abdomen, seen in some crustaceans. Telson and *uropods* often form a tail fan.

TERRESTRIAL FRINGE  region of shore above the *littoral fringe*.

TEST  hard, *calcareous* covering of a sea-urchin. The jelly-like coating of a sea-squirt.

THALLUS  the entire body of a seaweed or lichen.

TORSION  twisting of the visceral mass through 180° during development of gastropods.

TRIRAMOUS  having three branches or *rami*, e.g. some arthropod appendages.

TUBE-FOOT (PODIUM)  blind ending projection from the water vascular system of echinoderms.

UMBILICUS  cavity in the *columella* of a gastropod shell.

UMBO  raised area of bivalve shell, just behind the *beak*.

UNIRAMOUS  having one branch or *ramus*, e.g. some arthropod appendages.

UROPOD  appendage of posterior abdominal segment(s) of crustaceans.

VEGETATIVE REPRODUCTION  asexual reproduction in both plants and animals, by the separation of part of the body which develops into a complete organism.

VENTRAL  the underside of a plant or animal.

VERTEBRAE  small bones or cartilages in chain-like arrangement to form a backbone. Characteristic of vertebrates.

VIBRACULUM  specialized *zooid* of bryozoans, equipped with a *seta* which sweeps over surface and keeps it clear of *detritus*.

VISCERAL MASS  region of gastropod mollusc in which most internal organs are found.

VIVIPARITY  embryos develop within the body of the female. Nutrition provided directly by mother. Young emerge.

WATER VASCULAR SYSTEM  fluid-filled canals functioning as a hydraulic system. Unique to echinoderms.

ZONATION  distribution of shore organisms in zones or bands on the shore.

ZOOID  individual member of a colony of animals, e.g. ascidians and bryozoans.

ZOOPLANKTON  animals of the *plankton*.

ZYGOTE  fertilized egg.

# Bibliography

The following references are a selection of those used in preparing the text. The list is not comprehensive.

GENERAL REFERENCES

Barnes, R. S. K., Calow, P. & Olive, P. J. W. (1993). *The invertebrates: a new synthesis*, 2nd edn. Oxford: Blackwell Scientific.

Barrett, J. H. & Yonge, C. M. (1958). *Collins pocket guide to the sea shore*. London: Collins.

Boyden, C. R., Crothers, J. H., Little, C. & Mettam, C. (1977). The intertidal invertebrate fauna of the Severn estuary. *Field Studies*, 4, 477–554.

Bruce, J. R., Coleman, J. S. & Jones, N. S. (eds.) (1963). *Marine fauna of the Isle of Man*. Liverpool: Liverpool University Press.

Campbell, A. C. (1976). *The Country Life guide to the seashore and shallow seas of Britain and Europe*. Feltham: Country Life Books.

Crothers, J. H. (ed.) (1966). *Dale Fort marine fauna*, 2nd edn. London: Field Studies Council.

Green, J. (1968). *The biology of estuarine animals*. London: Sidgwick & Jackson.

Hayward, P. J. (1988). *Animals on seaweed*. Richmond: Richmond Publishing.

Hayward, P. J. (1994). *Animals on sandy shores*. Slough: Richmond Publishing.

Hayward, P. J. & Ryland, J. S. (eds.) (1990). *The marine fauna of the British Isles and north-west Europe*, vols. 1 and 2. Oxford: Clarendon Press.

Hayward, P. J. & Ryland, J. S. (eds.) (1995). *Handbook of the marine fauna of north-west Europe*. Oxford: Oxford University Press.

Howson, C. M. (ed.) (1987). *Directory of the British marine fauna and flora. A coded checklist of the marine fauna and flora of the British Isles and its surrounding seas*. Ross-on-Wye: Marine Conservation Society.

Laverack, M. S. & Blackler, M. (eds.) (1974). *Fauna and flora of St. Andrews Bay*. Edinburgh: Scottish Academic Press.

Laverack, M. S. & Dando, J. (1987). *Lecture notes on invertebrate zoology*, 3rd edn. Oxford: Blackwell Scientific.

Marine Biological Association of the UK (1957). *Plymouth marine fauna*, 3rd edn. Plymouth: Marine Biological Association of the United Kingdom.

Newell, G. E. (1954). The marine fauna of Whitstable. *Annals and Magazine of Natural History, Series 12*, 7, 321–50.

Quigley, M. & Crump, R. (1986). *Animals and plants of rocky shores*. Oxford: Basil Blackwell.

Ruppert, E. E. & Barnes, R. D. (1994). *Invertebrate zoology*, 6th edn. Philadelphia: Saunders.

Russell, F. S. & Yonge, C. M. (1975). *The seas*, 4th edn. London: Frederick Warne.

Sims, R. W., Freedman, P. & Hawksworth, D. L. (1988). *Key works to the fauna and flora of the British Isles and north-western Europe*, 5th edn. Systematics Association Special Volume No. 33. Oxford: Clarendon Press.

The Marine Fauna of the Cullercoats District. Papers published at intervals in *Reports of the Dove Marine Laboratory, Cullercoats.*


ALGAE

Brawley, S. H. (1992). Fertilization in natural populations of the dioecious brown alga *Fucus ceranoides* and the importance of the polyspermy block. *Marine Biology*, 113, 145–57.

Chapman, A. R. O. (1979). *Biology of seaweeds: levels of organisation.* London: Edward Arnold.

Chapman, A. R. O. (1986). Population and community ecology of seaweeds. *Advances in Marine Biology*, 23, 1–161.

Dring, M. J. (1982). *The biology of marine plants.* London: Edward Arnold.

Edyvean, R. G. J. & Ford, H. (1987). Growth rates of *Lithophyllum incrustans* (Corallinales, Rhodophyta) from south west Wales. *British Phycological Journal*, 22, 139–46.

Geiselman, J. A. & McConnell, O. J. (1981). Polyphenols in brown algae *Fucus vesiculosus* and *Ascophyllum nodosum*: chemical defenses against the marine herbivorous snail, *Littorina littorea*. *Journal of Chemical Ecology*, 7, 1115–33.

Gibb, D. C. (1957). The free-living forms of *Ascophyllum nodosum* (L.) Le Jol. *Journal of Ecology*, 45, 49–83.

Grahame, J. & Hanna, F. S. (1989). Factors affecting the distribution of the epiphytic fauna of *Corallina officinalis* (L.) on an exposed rocky shore. *Ophelia*, 30, 113–29.

Kain, J. M. (1979). A view of the genus *Laminaria*. *Oceanography and Marine Biology. An Annual Review*, 17, 101–61.

Kitching, J. A. (1987). The flora and fauna associated with *Himanthalia elongata* (L.) S. F. Gray in relation to water current and wave action in the Lough Hyne marine nature reserve. *Estuarine, Coastal and Shelf Science*, 25, 663–76.

Lobban, C. S., Harrison, P. J. & Duncan, M. J. (1985). *The physiological ecology of seaweeds.* Cambridge: Cambridge University Press.

Norton, T. A. (1969). Growth form and environment in *Saccorhiza polyschides*. *Journal of the Marine Biological Association of the United Kingdom*, 49, 1025–45.

Norton, T. A. & Burrows, E. M. (1969). Studies on marine algae of the British Isles. 7. *Saccorhiza polyschides* (Lightf.) Batt. *British Phycological Journal*, 4, 19–53.

Oswald, R. C., Telford, N., Seed, R. & Happey-Wood, C. M. (1984). The effect of encrusting bryozoans on the photosynthetic activity of *Fucus serratus* L. *Estuarine, Coastal and Shelf Science*, 19, 697–702.

Roberts, M. (1970). Studies on marine algae of the British Isles. 8. *Cystoseira tamariscifolia* (Hudson) Papenfuss. *British Phycological Journal*, 5, 201–10.

South, G. R. & Burrows, E. M. (1967). Studies on marine algae of the British Isles. 5. *Chorda filum* (L.) Stackh. *British Phycological Bulletin*, 3, 379–402.

## LICHENES

Boney, A. D. (1961). A note on the intertidal lichen *Lichina pygmaea* Ag. *Journal of the Marine Biological Association of the United Kingdom*, **41**, 123–6.

Fletcher, A. (1973). The ecology of marine (littoral) lichens on some rocky shores of Anglesey. *Lichenologist*, **5**, 368–400.

Fletcher, A. (1973). The ecology of maritime (supralittoral) lichens on some rocky shores of Anglesey. *Lichenologist*, **5**, 401–22.

Hale, M. E. (1983). *The biology of lichens*, 3rd edn. London: Edward Arnold.


## MAGNOLIOPSIDA (ANGIOSPERMAE)

Cox, P. A., Laushman, R. H. & Ruckelshaus, M. H. (1992). Surface and submarine pollination in the sea grass *Zostera marina*. *Botanical Journal of the Linnean Society*, **109**, 281–91.

Hubbard, J. C. E. (1965). *Spartina* marshes in southern England. VI. Pattern of invasion in Poole Harbour. *Journal of Ecology*, **53**, 799–813.

Muehlstein, L. K., Porter, D. & Short, F. T. (1988). *Labyrinthula* sp., a marine slime mold producing the symptoms of wasting disease in eelgrass, *Zostera marina*. *Marine Biology*, **99**, 465–72.

Portig, A. A., Mathers, R. G., Montgomery, W. I. & Govier, R. N. (1994). The distribution and utilisation of *Zostera* species in Strangford Lough, Northern Ireland. *Aquatic Botany*, **47**, 317–28.

Thompson, J. D. (1991). The biology of an invasive plant. What makes *Spartina anglica* so successful? *BioScience*, **41**, 393–401.

Turk, S. M. (1986). The three species of eelgrass *(Zostera)* on the Cornish coast. *Cornish Studies*, **14**, 15–22.

Vergeer, L. H. T. & den Hartog, C. (1991). Occurrence of wasting disease in *Zostera noltii*. *Aquatic Botany*, **40**, 155–63

Wilson, D. P. (1949). The decline of *Zostera marina* L. at Salcombe and its effects on the shore. *Journal of the Marine Biological Association of the United Kingdom*, **28**, 395–412.


## PORIFERA

Bergquist, P. R. (1978). *Sponges.* London: Hutchinson.

Fell, P. E. (1974). Porifera. In *Reproduction of marine invertebrates*, vol. 1, ed. A. C. Giese & J. S. Pearse, pp. 51–132. New York: Academic Press.

Florkin, M. & Scheer, B. T. (eds.) (1968). *Chemical zoology. Vol. 2. Porifera, Coelenterata and Platyhelminthes.* New York: Academic Press.

Peattie, M. E. & Hoare, R. (1981). The subittoral ecology of the Menai Strait II. The sponge *Halichondria panicea* (Pallas) and its associated fauna. *Estuarine, Coastal and Shelf Science*, **13**, 621–35.

Solé-Cava, A. M. & Thorpe, J. P. (1986). Genetic differentiation between morphotypes of the marine sponge *Suberites ficus* (Demospongiae: Hadromerida). *Marine Biology*, **93**, 247–53.

Stone, A. R. (1970). Growth and reproduction of *Hymeniacidon perleve* (Montagu) (Porifera) in Langstone Harbour, Hampshire. *Journal of Zoology, London*, **161**, 443–59.

CNIDARIA

Brace, R. C. & Quicke, D. L. J. (1985). Further analysis of individual spacing within aggregations of the anemone, *Actinia equina*. *Journal of the Marine Biological Association of the United Kingdom*, **65**, 35–53.

Brace, R. C. & Quicke, D. L. J. (1986). Dynamics of colonization by the beadlet anemone, *Actinia equina*. *Journal of the Marine Biological Association of the United Kingdom*, **66**, 21–47.

Brace, R. C. & Sauter, S-J. (1991). Experimental habituation of aggression in the sea anemone *Actinia equina*. *Hydrobiologia*, **216/217**, 533–7.

Bucklin, A. (1985). Biochemical genetic variation, growth and regeneration of the sea anemone, *Metridium*, of British shores. *Journal of the Marine Biological Association of the United Kingdom*, **65**, 141–57.

Carter, M. A. & Miles, J. (1989). Gametogenic cycles and reproduction in the beadlet sea anemone *Actinia equina* (Cnidaria: Anthozoa). *Biological Journal of the Linnean Society*, **36**, 129–55.

Cheng, L. (1975). Marine pleuston: animals at the sea–air interface. *Oceanography and Marine Biology. An Annual Review*, **13**, 181–212.

Christiansen, B. O. (1972). The hydroid fauna of the Oslo Fiord in Norway. *Norwegian Journal of Zoology*, **20**, 279–310.

Donaghue, A. M., Quicke, D. L. J. & Brace, R. C. (1985). Biochemical–genetic and acrorhagial characteristics of pedal disc colour phenotypes of *Actinia equina*. *Journal of the Marine Biological Association of the United Kingdom*, **65**, 21–33.

Dorsett, D. A. & Turner, R. (1986). Colour morph dominance and avoidance responses of *Anemonia sulcata*. *Marine Behaviour and Physiology*, **12**, 115–23.

Edwards, C. & Harvey, S. M. (1975). The hydroids *Clava multicornis* and *Clava squamata*. *Journal of the Marine Biological Association of the United Kingdom*, **55**, 879–86.

Gashout, S. E. & Ormond, R. F. G. (1979). Evidence for parthenogenetic reproduction in the sea anemone *Actinia equina* L. *Journal of the Marine Biological Association of the United Kingdom*, **59**, 975–87.

Gröndahl, F. (1989). Evidence of gregarious settlement of planula larvae of the scyphozoan *Aurelia aurita*: an experimental study. *Marine Ecology Progress Series*, **56**, 119–25.

Hamond, R. (1959). Notes on the Hydrozoa of the Norfolk coast. *Journal of the Linnean Society of London, Zoology*, **43**, 294–324.

Hancock, D. A., Drinnan, R. E. & Harris, W. N. (1956). Notes on the biology of *Sertularia argentea* L. *Journal of the Marine Biological Association of the United Kingdom*, **35**, 307–25.

Hartnoll, R. G. (1975). The annual cycle of *Alcyonium digitatum*. *Estuarine and Coastal Marine Science*, **3**, 71–8.

Haylor, G. S., Thorpe, J. P. & Carter, M. A. (1984). Genetic and ecological differentiation between sympatric colour morphs of the common intertidal sea anemone *Actinia equina*. *Marine Ecology Progress Series*, **16**, 281–9.

Heuwinkel, U. B. (1988). Brooding behaviour in the marine hydropolyp *Myriothela cocksi* (Vigors) by means of special tentacles. *Acta Biologica Benrodis*, **1**, 57–67.

Hiscock, K. & Howlett, R. M. (1976). The ecology of *Caryophyllia smithi* Stokes and Broderip on south-western coasts of the British Isles. In *Underwater research*, ed. A. Drew, J. N. Lythgoe & J. D. Woods, pp. 319–34. London: Academic Press.

Hughes, R. G. (1983). The life-history of *Tubularia indivisa* (Hydrozoa: Tubulariidae) with observations on the status of *T. ceratogyne*. *Journal of the Marine Biological Association of the United Kingdom*, **63**, 467–79.

Hughes, R. G. (1986). Differences in the growth, form and life-history of *Plumularia setacea* (Ellis & Solander) (Hydrozoa: Plumulariidae) in two contrasting habitats. *Proceedings of the Royal Society of London, Series B*, **288**, 113–25.

Manton, S. M. (1944). On the hydrorhiza and claspers of the hydroid *Myriothela cocksi* (Vigurs). *Journal of the Marine Biological Association of the United Kingdom*, **25**, 143–50.

Möller, H. (1980). Population dynamics of *Aurelia aurita* medusae in Kiel Bight, Germany (FRG). *Marine Biology*, **60**, 123–8.

Orr, J., Thorpe, J. P. & Carter, M. A. (1982). Biochemical genetic confirmation of the asexual reproduction of brooded offspring in the sea anemone *Actinia equina*. *Marine Ecology Progress Series*, **7**, 227–9.

Pyefinch, K. A. & Downing, F. S. (1949). Notes on the general biology of *Tubularia larynx* Ellis & Solander. *Journal of the Marine Biological Association of the United Kingdom*, **28**, 21–43.

Quicke, D. L. J., Donoghue, A. M., Keeling, T. F. & Brace, R. C. (1985). Littoral distributions and evidence for differential post-settlement selection of the morphs of *Actinia equina*. *Journal of the Marine Biological Association of the United Kingdom*, **65**, 1–20.

Robins, M. W. (1969). The marine flora and fauna of the Isles of Scilly: Cnidaria and Ctenophora. *Journal of Natural History*, **3**, 329–43.

Ross, D. M. (1967). Behavioural and ecological relationships between sea anemones and other invertebrates. *Oceanography and Marine Biology. An Annual Review*, **5**, 291–316.

Schmidt, G. H. & Warner, G. F. (1991). The settlement and growth of *Sertularia cupressina* (Hydrozoa: Sertulariidae) in Langstone Harbour, Hampshire, U.K. *Hydrobiologia*, **216/217**, 215–19.

Solé-Cava, A. M. & Thorpe, J. P. (1987). Further genetic evidence for the reproductive isolation of the green sea anemone *Actinia prasina* Gosse from common intertidal beadlet anemone *Actinia equina* (L.). *Marine Ecology Progress Series*, **38**, 225–9.

Solé-Cava, A. M. & Thorpe, J. P. (1992). Genetic divergence between colour morphs in populations of the common intertidal sea anemones *Actinia equina* and *A. prasina* (Anthozoa: Actiniaria) in the Isle of Man. *Marine Biology*, **112**, 243–52.

Solé-Cava, A. M., Thorpe, J. P. & Kaye, J. G. (1985). Reproductive isolation with little genetic divergence between *Urticina* (=*Tealia*) *felina* and *U. eques* (Anthozoa: Actiniaria). *Marine Biology*, **85**, 279–84.

Turk, S. M. (1982). Influx of warm-water oceanic drift animals into the Bristol and English Channels, summer 1981. *Journal of the Marine Biological Association of the United Kingdom*, **62**, 487–9.

Williams, G. B. (1965). Observations on the behaviour of the planulae larvae of *Clava squamata*. *Journal of the Marine Biological Association of the United Kingdom*, **45**, 257–73.

Williams, R. B., Cornelius, P. F. S., Hughes, R. G. & Robson, E. A. (eds.) (1991). *Coelenterate biology: recent research on Cnidaria and Ctenophora. Fifth International Conference on Coelenterate Biology, 1989. Hydrobiologia*, **216/217**.

Yund, P. O., Cunningham, C. W. & Buss, L. W. (1987). Recruitment and postrecruitment interactions in a colonial hydroid. *Ecology*, **68**, 971–82.

CTENOPHORA

Frank, K. T. (1986). Ecological significance of the ctenophore *Pleurobrachia pileus* off southwestern Nova Scotia. *Canadian Journal of Fisheries and Aquatic Science*, **43**, 211–22.

Fraser, J. H. (1970). The ecology of the ctenophore *Pleurobrachia pileus* in Scottish waters. *Journal du Conseil International pour l'exploration de la mer*, **33**, 149–68.

Harbison, G. R. (1985). On the classification and evolution of the Ctenophora. In *The origins and relationships of lower invertebrates*, ed. S. C. Morris, J. D. George, R. Gibson & H. M. Platt, pp. 78–100. Systematics Association Special Volume No. 28. Oxford: Clarendon Press.

Pianka, H. D. (1974). Ctenophora. *In Reproduction of marine invertebrates*, vol. 1, ed. A. C. Giese & J. S. Pearse, pp. 201–65. New York: Academic Press.

Swanberg, N. (1974). The feeding behaviour of *Beroe ovata*. *Marine Biology*, **24**, 69–76.


PLATYHELMINTHES

Henley, C. (1974). Platyhelminthes (Turbellaria). In *Reproduction of marine invertebrates*, vol. 1, ed. A. C. Giese & J. S. Pearse, pp. 267–343. New York: Academic Press.

Sluys, R. (1987). The taxonomy of three species of the genus *Procerodes* (Platyhelminthes: Tricladida: Maricola). *Journal of the Marine Biological Association of the United Kingdom*, **67**, 373–84.


NEMERTEA

Bartsch, I. (1973). Zur Nahrungsaufnahme von *Tetrastemma melanocephalum* (Nemertini). *Helgoländer wissenschaftliche Meeresuntersuchungen*, **25**, 326–31.

Gibson, R. (1972). *Nemerteans*. London: Hutchinson.

McDermott, J. J. & Roe, P. (1985). Food, feeding behaviour and feeding ecology of nemerteans. *American Zoologist*, **25**, 113–25.

Riser, N. W. (1974). Nemertinea. In *Reproduction of marine invertebrates*, vol. 1, ed. A. C. Giese & J. S. Pearse, pp. 359–89. New York: Academic Press.

Sundberg, P. (1984). Multivariate analysis of polymorphism in the hoplonemertean *Oerstedia dorsalis* (Abildgaard, 1806). *Journal of Experimental Marine Biology and Ecology*, **78**, 1–22.

Sundberg, P. & Andersson, S. (1995). Random amplified polymorphic DNA (RAPD) and intraspecific variation in *Oerstedia dorsalis* (Hoplonemertea, Nemertea). *Journal of the Marine Biological Association of the United Kingdom*, **75**, 483–90.

Sundberg, P. & Janson, K. (1988). Polymorphism in *Oerstedia dorsalis* (Abildgaard, 1806) revisited. Electrophoretic evidence for a species complex. *Hydrobiologia*, **156**, 93–8.

Vernet, G. & Bierne, T. (1993). The influence of light and sea water temperature on the reproductive cycle of *Lineus ruber* (Heteronemertea). *Hydrobiologia*, **266**, 267–71.


PRIAPULA

Hammond, R. A. (1970). The burrowing of *Priapulus caudatus*. *Journal of Zoology, London*, **162**, 469–80.

Lang, K. (1948). Contribution to the ecology of *Priapulus caudatus* Lam. *Arkiv för Zoologi*, **41**, 1–12.

ANNELIDA

Bass, N. R. & Brafield, A. E. (1972). The life-cycle of the polychaete *Nereis virens. Journal of the Marine Biological Association of the United Kingdom*, **52**, 701–26.

Bedford, A. P. & Moore, P. G. (1985). Macrofaunal involvement in the sublittoral decay of kelp debris: the polychaete *Platynereis dumerilii* (Audouin and Milne-Edwards) (Annelida: Polychaeta). *Estuarine, Coastal and Shelf Science*, **20**, 117–34.

Bentley, M. G., Olive, P. J. W., Garwood, P. R. & Wright, N. H. (1984). The spawning and spawning mechanism of *Nephtys caeca* (Fabricius, 1780) and *Nephtys hombergi* Sauvigny, 1818 (Annelida: Polychaeta). *Sarsia*, **69**, 63–8.

Bhaud, M. R. (1991). Larval release from the egg mass and settlement of *Eupolymnia nebulosa* (Polychaeta, Terebellidae). *Bulletin of Marine Science*, **48**, 420–31.

Bhaud, M. R., Grémare, A., Lang, F. & Retière, C. (1987). Étude comparée des caractères reproductifs du Terebellien *Eupolymnia nebulosa* (Montagu) (Annélide-Polychète) en deux points de son aire géographique. *Comptes Rendus-Académie des Sciences, Paris*, Série 3, **304**, 119–22.

Bosence, D. W. J. (1979). The factors leading to aggregation and reef formation in *Serpula vermicularis* L. In *Biology and systematics of colonial organisms*, ed. G. Larwood & B. R. Rosen, Systematics Association Special Volume No. 11, pp. 299–318. London: Academic Press.

Cadman, P. S. & Nelson-Smith, A. (1990). Genetic evidence for two species of lugworm *(Arenicola)* in south Wales. *Marine Ecology Progress Series*, **64**, 107–12.

Cram, A. & Evans, S. M. (1980). Shell entry behaviour in the commensal ragworm *Nereis fucata. Marine Behaviour and Physiology*, **7**, 57–64.

Crisp, D. J. & Ekaratne, K. (1984). Polymorphism in *Pomatoceros. Zoological Journal of the Linnean Society*, **80**, 157–75.

Dales, R. P. (1950). The reproduction and larval development of *Nereis diversicolor* O. F. Müller. *Journal of the Marine Biological Association of the United Kingdom*, **29**, 321–60.

Dales, R. P. (1955). Feeding and digestion in terebellid polychaetes. *Journal of the Marine Biological Association of the United Kingdom*, **34**, 55–79.

Dales, R. P. (1957). The feeding mechanism and structure of the gut of *Owenia fusiformis* Delle Chiaje. *Journal of the Marine Biological Association of the United Kingdom*, **36**, 81–9.

Daly, J. M. (1972). The maturation and breeding biology of *Harmothoë imbricata* (Polychaeta: Polynoidae). *Marine Biology*, **12**, 53–66.

Daly, J. M. (1973). Behavioural and secretory activity during tube construction by *Platynereis dumerilii* Aud & M. Edw. (Polychaeta: Nereidae). *Journal of the Marine Biological Association of the United Kingdom*, **53**, 521–9.

Daly, J. M. (1978). The annual cycle and the short term periodicity of breeding in a Northumberland population of *Spirorbis spirorbis* (Polychaeta: Serpulidae). *Journal of the Marine Biological Association of the United Kingdom*, **58**, 161–76.

Daly, J. M. & Golding, D. W. (1977). A description of the spermatheca of *Spirorbis spirorbis* (L.) (Polychaeta: Serpulidae) and evidence for a novel mode of sperm transmission. *Journal of the Marine Biological Association of the United Kingdom*, **57**, 219–27.

Daro, M. H. & Polk, P. (1973). The autecology of *Polydora ciliata* along the Belgian coast. *Netherlands Journal of Sea Research*, **6**, 130–40.

Dauer, D. M. & Ewing, M. R. (1991). Functional morphology and feeding behaviour of *Malacoceros indicus* (Polychaeta: Spionidae). *Bulletin of Marine Science*, **48**, 395–400.

Dean, D., Chapman, S. R. & Chapman, C. S. (1987). Reproduction and development of the sabellid polychaete *Myxicola infundibulum*. *Journal of the Marine Biological Association of the United Kingdom*, **67**, 431–9.

de Silva, P. H. D. H. (1962). Experiments on choice of substrata by *Spirorbis* larvae (Serpulidae). *Journal of Experimental Biology*, **39**, 483–90.

de Silva, P. H. D. H. (1967). Studies on the biology of Spirorbinae (Polychaeta). *Journal of Zoology, London*, **152**, 269–79.

Dorsett, D. A. (1961). The reproduction and maintenance of *Polydora ciliata* (Johnst.) at Whitstable. *Journal of the Marine Biological Association of the United Kingdom*, **41**, 383–96.

Dorsett, D. A. (1961). The behaviour of *Polydora ciliata* (Johnst.): tube building and burrowing. *Journal of the Marine Biological Association of the United Kingdom*, **41**, 577–90.

Emson, R. H. (1977). The feeding and consequent role of *Eulalia viridis* (O. F. Müller) (Polychaeta) in intertidal communities. *Journal of the Marine Biological Association of the United Kingdom*, **57**, 93–6.

Enders, H. E. (1909). A study of the life-history and habits of *Chaetopterus variopedatus*, Renier et Claparede. *Journal of Morphology*, **20**, 479–531.

Fauchald, K. & Jumars, P. A. (1979). The diet of worms: a study of polychaete feeding guilds. *Oceanography and Marine Biology. An Annual Review*, **17**, 193–284.

Fischer, A. & Pfannenstiel, H-D. (eds.) (1984). *Polychaete reproduction: progress in comparative reproductive biology*. Stuttgart: Gustav Fischer.

George, J. D. (1964). The life history of the cirratulid worm, *Cirriformia tentaculata*, on an intertidal mud flat. *Journal of the Marine Biological Association of the United Kingdom*, **44**, 47–65.

Gibbs, P. E. (1968). Observations on the population of *Scoloplos armiger* at Whitstable. *Journal of the Marine Biological Association of the United Kingdom*, **48**, 225–54.

Gibbs, P. E. (1971). A comparative study of reproductive cycles in four polychaete species belonging to the family Cirratulidae. *Journal of the Marine Biological Association of the United Kingdom*, **51**, 745–69.

Gillet, P. (1990). Variation intraspécifique des paragnathes chez *Nereis diversicolor* (Annélides, Polychètes) de L'Atlantique Nord-est. *Vie Milieu*, **40**, 297–303.

Gilpin-Brown, J. B. (1959). The reproduction and larval development of *Nereis fucata* (Savigny). *Journal of the Marine Biological Association of the United Kingdom*, **38**, 65–80.

Gilpin-Brown, J. B. (1969). Host-adoption in the commensal polychaete *Nereis fucata*. *Journal of the Marine Biological Association of the United Kingdom*, **49**, 121–7.

Grant, A. (1989). The reproductive cycle of *Platynereis dumerilii* (Audouin & Milne-Edwards) (Polychaeta: Nereidae) from the Firth of Clyde. *Sarsia*, **74**, 79–84.

Grassle, J. P. (1984). Speciation in the genus *Capitella* (Polychaeta, Capitellidae). In *Polychaete reproduction: progress in comparative reproductive biology*, ed. A. Fischer & H-D. Pfannenstiel, pp. 293–8. Stuttgart: Gustav Fischer.

Grémare, A. (1986). A comparative study of reproductive energetics in two populations of the terebellid polychaete *Eupolymnia nebulosa* Montagu with different reproductive modes. *Journal of Experimental Marine Biology and Ecology*, **96**, 287–302.

Gruet, Y. & Lassus, P. (1983). Contribution à l'étude de la biologie reproductive d'une population naturelle de l'annélide polychète *Sabellaria alveolata* (Linné). *Annales de l'Institut Océanographique, Paris*, **59**, 127–40.

Gudmundsson, H. (1985). Life history pattern of polychaete species of the family Spionidae. *Journal of the Marine Biological Association of the United Kingdom*, **65**, 93–111.

Guérin, J-P. & Kerambrun, P. (1984). Role of reproductive characters in the taxonomy of spionids and elements of speciation in the '*Malacoceros fuliginosus* complex'. In *Polychaete reproduction: progress in comparative reproductive biology*, ed. A. Fischer & H-D. Pfannenstiel, pp. 317–33. Stuttgart: Gustav Fischer.

Hamond, R. (1969). On the preferred foods of some autolytoides (Polychaeta, Syllidae). *Cahiers de Biologie Marine*, **10**, 439–45.

Harley, M. B. (1953). The feeding habits of *Nereis diversicolor* (O. F. Müller). *The British Journal of Animal Behaviour*, **1**, 88.

Hateley, J. G., Grant, A., Taylor, S. M. & Jones, N. V. (1992). Morphological and other evidence on the degree of genetic differentiation between populations of *Nereis diversicolor*. *Journal of the Marine Biological Association of the United Kingdom*, **72**, 365–81.

Howie, D. I. D. (1984). The reproductive biology of the lugworm, *Arenicola marina* L. In *Polychaete reproduction: progress in comparative reproductive biology*, ed. A. Fischer & H-D. Pfannenstiel, pp. 247–63. Stuttgart: Gustav Fischer.

Irlinger, J. P., Gentil, F. & Quintino, V. (1991). Reproductive biology of the polychaete *Pectinaria koreni* (Malmgren) in the Bay of Seine (English Channel). *Ophelia* Supplement, **5**, 343–50.

Joyner, A. (1962). Reproduction and larval life of *Nerine cirratulus* (Delle Chiaje) Family Spionidae. *Proceedings of the Zoological Society of London*, **138**, 655–66.

Kent, R. M. L. (1979). The influence of heavy infestations of *Polydora ciliata* on the flesh content of *Mytilus edulis*. *Journal of the Marine Biological Association of the United Kingdom*, **59**, 289–97.

Knight-Jones, P. & Thorp, C. H. (1984). The opercular brood chambers of Spirorbidae. *Zoological Journal of the Linnean Society*, **80**, 121–33.

Kristensen, E. (1984). Life cycle, growth and production in estuarine populations of the polychaetes, *Nereis virens* and *N. diversicolor*. *Holarctic Ecology*, **7**, 249–56.

Lewis, D. B. (1968). Some aspects of the ecology of *Fabricia sabella* (Ehr.) (Annelida, Polychaeta). *Journal of the Linnean Society of London, Zoology*, **47**, 515–26.

Lewis, D. B. (1968). Feeding and tube-building in the Fabriciinae (Annelida, Polychaeta). *Proceedings of the Linnean Society of London*, **179**, 37–49.

Mettam, C. (1980). On the feeding habits of *Aphrodita aculeata* and commensal polynoids. *Journal of the Marine Biological Association of the United Kingdom*, **60**, 833–4.

Mustaquim, J. (1986). Morphological variation in *Polydora ciliata* complex (Polychaeta: Annelida). *Zoological Journal of the Linnean Society*, **86**, 75–88.

Newell, G. E. (1949). The later larval life of *Arenicola marina* L. *Journal of the Marine Biological Association of the United Kingdom*, **28**, 635–9.

Nicol, E. A. T. (1931). The feeding mechanism, formation of the tube, and physiology of digestion in *Sabella pavonina*. *Transactions of the Royal Society of Edinburgh*, **56**, 537–98.

Nicolaidou, A. (1983). Life history and productivity of *Pectinaria koreni* Malmgren (Polychaeta). *Estuarine, Coastal and Shelf Science*, **17**, 31–43.

Nicolaidou, A. (1988). Notes on the behaviour of *Pectinaria koreni*. *Journal of the Marine Biological Association of the United Kingdom*, **68**, 55–9.

O'Connor, B. (1982). *Pomatoceros lamarcki* (Quatrefages) (Polychaeta: Serpulidae). A polychaete new to Irish waters. *Irish Naturalists' Journal*, **20**, 401.

Olive, P. J. W. (1970). Reproduction of a Northumberland population of the polychaete *Cirratulus cirratus*. *Marine Biology*, **5**, 259–73.

Olive, P. J. W. (1975). Reproductive biology of *Eulalia viridis* (Müller) (Polychaeta: Phyllodocidae) in the north eastern U.K. *Journal of the Marine Biological Association of the United Kingdom*, **55**, 313–26.

Olive, P. J. W. (1978). Reproduction and annual gametogenic cycle in *Nephtys hombergii* and *N. caeca* (Polychaeta: Nephtyidae). *Marine Biology*, **46**, 83–90.

Olive, P. J. W. & Garwood, P. R. (1981). Gametogenic cycle and population structure of *Nereis* (*Hediste*) *diversicolor* and *Nereis* (*Nereis*) *pelagica* from north-east England. *Journal of the Marine Biological Association of the United Kingdom*, **61**, 193–213.

Olive, P. J. W. & Pillai, G. (1983). Reproductive biology of the polychaete *Kefersteinia cirrata* Keferstein (Hesionidae). II. The gametogenic cycle and evidence for photoperiodic control of oogenesis. *International Journal of Invertebrate Reproduction*, **6**, 307–15.

Olive, P. J. W., Garwood, P. R., Bentley, M. G. & Wright, N. (1981). Reproductive success, relative abundance and population structure of two species of *Nephtys* in an estuarine beach. *Marine Biology*, **63**, 189–96.

Oyenekan, J. A. (1983). Production and population dynamics of *Capitella capitata*. *Archiv fuer Hydrobiologie*, **98**, 115–26.

Pearson, M. & Pearson, T. H. (1991). Variation in populations of *Capitella capitata* (Fabricius, 1780) (Polychaeta) from the west coast of Scotland. *Ophelia* Supplement, **5**, 363–70.

Pearson, T. H. (1969). *Scionella lornensis* sp. nov., a new terebellid (Polychaeta: Annelida) from the west coast of Scotland, with notes on the genus *Scionella* Moore, and a key to the genera of the Terebellidae recorded from European waters. *Journal of Natural History*, **3**, 509–16.

Pettibone, M. H. (1963). *Marine polychaete worms of the New England region*. Washington: Smithsonian Institution.

Plate, S. & Husemann, E. (1991). An alternative mode of larval development in *Scoloplos armiger* (O. F. Müller, 1776) (Polychaeta, Orbiniidae). *Helgoländer Meeresuntersuchungen*, **45**, 487–92.

Sach, G. (1975). Zur Fortpflanzung des Polychaeten *Anaitides mucosa*. *Marine Biology*, **31**, 157–60.

Schiedges, K-L. (1980). Morphological and systematic studies of an *Autolytus* population (Polychaeta, Syllidae, Autolytinae) from the Oosterschelde Estuary. *Netherlands Journal of Sea Research*, **14**, 208–19.

Schroeder, P. C. & Hermans, C. O. (1975). Annelida: Polychaeta. In *Reproduction of marine invertebrates*, vol. III, ed. A. C. Giese & J. S. Pearse, pp. 1–213. New York: Academic Press.

Theede, H., Schaudinn, J. & Saffé, F. (1973). Ecophysiological studies on four *Nereis* species of the Kiel Bay. *Oikos Supplementum*, **15**, 246–52.

Tsutsumi, H. & Kikuchi, T. (1984). Study on the life history of *Capitella capitata*

(Polychaeta: Capitellidae) in Amakusa, south Japan including a comparison with other geographical regions. *Marine Biology*, **80**, 315–21.

Upton, B. (1953). The escape from the cocoon of *Scoloplos armiger*. *The British Journal of Animal Behaviour*, **1**, 87.

Vedel, A. & Riisgård, H. A. (1993). Filter-feeding in the polychaete *Nereis diversicolor*: growth and biogenetics. *Marine Ecology Progress Series*, **100**, 145–52.

Warren, L. M. (1977). The ecology of *Capitella capitata* in British waters. *Journal of the Marine Biological Association of the United Kingdom*, **57**, 151–9.

de Wilde, P. A. W. J. & Farke, H. (1983). The lugworm *Arenicola marina*. In *Ecology of the Wadden Sea*, I, ed. W. J. Wolff, pp. 4/111–4/113. Rotterdam: A. A. Balkema.

Wilson, D. P. (1932). On the mitraria larva of *Owenia fusiformis* Delle Chiaje. *Transactions of the Royal Society, Series B*, **221**, 231–334.

Wilson, D. P. (1948). The larval development of *Ophelia bicornis* Savigny. *Journal of the Marine Biological Association of the United Kingdom*, **27**, 540–53.

Wilson, D. P. (1971). *Sabellaria* colonies in Duckpool, North Cornwall, 1961–70. *Journal of the Marine Biological Association of the United Kingdom*, **51**, 509–80.

Wilson, W. H. (1991). Sexual reproductive modes in polychaetes: classification and diversity. *Bulletin of Marine Science*, **48**, 500–16.


MOLLUSCA: GASTROPODA

Barnes, R. S. K. (1991). On the distribution of the northwest European species of the gastropod *Hydrobia*, with particular reference to *H. neglecta*. *Journal of Conchology*, **34**, 59–62.

Barnett, P. R. O., Hardy, B. L. S. & Watson, J. (1980). Substratum selection and egg-capsule deposition in *Nassarius reticulatus* (L.). *Journal of Experimental Marine Biology and Ecology*, **45**, 95–103.

Blackmore, D. T. (1969). Studies on *Patella vulgata* L. 1. Growth, reproduction and zonal distribution. *Journal of Experimental Marine Biology and Ecology*, **3**, 200–13.

Boyle, P. R. (1977). The physiology and behaviour of chitons (Mollusca: Polyplacophora). *Oceanography and Marine Biology. An Annual Review*, **15**, 461–509.

Byers, B. A. (1990). Shell colour polymorphism associated with substrate colour in the intertidal snail *Littorina saxatilis* Olivi (Prosobranchia: Littorinidae). *Biological Journal of the Linnean Society*, **40**, 3–10.

Carefoot, T. H. (1987). *Aplysia*: its biology and ecology. *Oceanography and Marine Biology. An Annual Review*, **25**, 167–284.

Chia, F. S. (1971). Oviposition, fecundity and larval development of three saccoglossan opisthobranchs from the Northumberland coast, England. *Veliger*, **13**, 319–25.

Chipperfield, P. N. J. (1951). The breeding of *Crepidula fornicata* (L.) in the River Blackwater, Essex. *Journal of the Marine Biological Association of the United Kingdom*, **30**, 49–71.

Clare, A. S. (1990). Laboratory-induced spawning of the gastropod *Gibbula cineraria* as an indicator of field spawning. *Marine Ecology Progress Series*, **63**, 303–4.

Crothers, J. H. (1985). Dog-whelks: an introduction to the biology of *Nucella lapillus* (L.). *Field Studies*, **6**, 291–360.

Desai, B. N. (1966). The biology of *Monodonta lineata* (da Costa). *Proceedings of the Malacological Society of London*, **37**, 1–17.

Dussart, G. B. J. (1977). The ecology of *Potamopyrgus jenkinsi* (Smith) in north west England with a note on *Marstoniopsis scholtzi* (Schmidt). *Journal of Molluscan Studies*, **43**, 208–16.

Fish, J. D. (1972). The breeding cycle and growth of open coast and estuarine populations of *Littorina littorea*. *Journal of the Marine Biological Association of the United Kingdom*, **52**, 1011–19.

Fish, J. D. (1979). The rhythmic spawning behaviour of *Littorina littorea* (L.). *Journal of Molluscan Studies*, **45**, 172–7.

Fish, J. D. & Fish, S. (1974). The breeding cycle and growth of *Hydrobia ulvae* in the Dovey Estuary. *Journal of the Marine Biological Association of the United Kingdom*, **54**, 685–97.

Fish, J. D. & Fish, S. (1977). The veliger larva of *Hydrobia ulvae* with observations on the veliger of *Littorina littorea* (Mollusca: Prosobranchia). *Journal of Zoology, London*, **182**, 495–503.

Fish, J. D. & Fish, S. (1981). The early life-cycle stages of *Hydrobia ventrosa* and *Hydrobia neglecta* with observations on *Potamopyrgus jenkinsi*. *Journal of Molluscan Studies*, **47**, 89–98.

Fish, J. D. & Sharp, L. (1985). The ecology of the periwinkle, *Littorina neglecta* Bean. In *The ecology of rocky coasts*, ed. P. G. Moore & R. Seed, pp. 143–56. London: Hodder & Stoughton.

Fretter, V. & Graham, A. (1954). Observations on the opisthobranch mollusc *Acteon tornatilis* (L.). *Journal of the Marine Biological Association of the United Kingdom*, **33**, 565–85.

Fretter, V. & Graham, A. (1994). *British prosobranch molluscs: their functional anatomy and ecology*, revised and updated edition. London: Ray Society.

Fretter, V. & Manly, R. (1977). Algal associations of *Tricolia pullus*, *Lacuna vincta* and *Cerithiopsis tubercularis* (Gastropoda) with special reference to the settlement of their larvae. *Journal of the Marine Biological Association of the United Kingdom*, **57**, 999–1017.

Fretter, V. & Manly, R. (1979). Observations on the biology of some sublittoral prosobranchs. *Journal of Molluscan Studies*, **45**, 209–18.

Garwood, P. R. & Kendall, M. A. (1985). The reproductive cycles of *Monodonta lineata* and *Gibbula umbilicalis* on the coast of mid-Wales. *Journal of the Marine Biological Association of the United Kingdom*, **65**, 993–1008.

Goodwin, B. J. (1978). The growth and breeding cycle of *Littorina obtusata* (Gastropoda: Prosobranchiata) from Cardigan Bay. *Journal of Molluscan Studies*, **44**, 231–42.

Goodwin, B. J. (1979). The egg mass of *Littorina obtusata* and *Lacuna pallidula* (Gastropoda: Prosobranchia). *Journal of Molluscan Studies*, **45**, 1–11.

Goodwin, B. J. & Fish, J. D. (1977). Inter- and intraspecific variation in *Littorina obtusata* and *L. mariae* (Gastropoda: Prosobranchia). *Journal of Molluscan Studies*, **43**, 241–54.

Graham, A. & Fretter, V. (1947). The life history of *Patina pellucida* (L.). *Journal of the Marine Biological Association of the United Kingdom*, **26**, 590–601.

Grahame, J. (1970). Shedding of the penis in *Littorina littorea*. *Nature, London*, **221**, 976.

Hall, S. J. & Todd, C. D. (1986). Growth and reproduction in the aeolid nudibranch *Aeolidia papillosa* (L.). *Journal of Molluscan Studies*, **52**, 193–205.

Havenhand, J. N. & Todd, C. D. (1988). Physiological ecology of *Adalaria proxima* (Alder et Hancock) and *Onchidoris muricata* (Müller) (Gastropoda: Nudibranchia). 1.

Feeding, growth and respiration. *Journal of Experimental Marine Biology and Ecology*, 118, 151–72. 2. Reproduction. *Journal of Experimental Marine Biology and Ecology*, 118, 173–89.

Hayashi, I. (1980). The reproductive biology of the ormer, *Haliotis tuberculata*. *Journal of the Marine Biological Association of the United Kingdom*, 60, 415–30.

Hayashi, I. (1980). Structure and growth of a shore population of the ormer, *Haliotis tuberculata*. *Journal of the Marine Biological Association of the United Kingdom*, 60, 431–7.

Hughes, R. N. (1980). Population dynamics, growth and reproductive rates of *Littorina nigrolineata* Gray from a moderately sheltered locality in North Wales. *Journal of Experimental Marine Biology and Ecology*, 44, 211–28.

Hughes, R. N. & Roberts, D. J. (1980). Growth and reproductive rates of *Littorina neritoides* (L.) in North Wales. *Journal of the Marine Biological Association of the United Kingdom*, 60, 591–9.

Johannesson, K. & Johannesson, B. (1990). Genetic variation within *Littorina saxatilis* (Olivi) and *Littorina neglecta* Bean: Is *L. neglecta* a good species? *Hydrobiologia*, 193, 89–97.

Jones, H. D. (1984). Shell cleaning behaviour of *Calliostoma zizyphinum*. *Journal of Molluscan Studies*, 50, 245–7.

Lebour, M. V. (1933). The eggs and larvae of *Turritella communis* Lamarck and *Aporrhais pes-pelicani* (L.). *Journal of the Marine Biological Association of the United Kingdom*, 18, 499–506.

Lebour, M. V. (1937). The eggs and larvae of the British prosobranchs with special reference to those living in the plankton. *Journal of the Marine Biological Association of the United Kingdom*, 22, 105–66.

Lewis, J. R. & Bowman, R. S. (1975). Local habitat-induced variations in the population dynamics of *Patella vulgata* L. *Journal of Experimental Marine Biology and Ecology*, 17, 165–203.

McGrath, D. (1992). Recruitment and growth of the blue-rayed limpet, *Helcion pellucidum* (L.), in south east Ireland. *Journal of Molluscan Studies*, 58, 425–31.

Morton, J. E. (1954). The crevice faunas of the upper intertidal zone at Wembury. *Journal of the Marine Biological Association of the United Kingdom*, 33, 187–224.

Morton, J. E. (1955). The functional morphology of *Otina otis*, a primitive marine pulmonate. *Journal of the Marine Biological Association of the United Kingdom*, 34, 113–50.

Muus, B. J. (1967). The fauna of Danish estuaries and lagoons. *Meddelelser fra Danmarks Fiskeri-og Havundersøgelser*, Ny Serie, 5, 1–316.

Orton, J. H. & Southward, A. J. (1961). Studies on the biology of limpets IV. The breeding of *Patella depressa* Pennant on the north Cornish coast. *Journal of the Marine Biological Association of the United Kingdom*, 41, 653–62.

Patil, A. M. (1958). The occurrence of a male of the prosobranch *Potamopyrgus jenkinsi* (Smith) var. *carinata* Marshall in the Thames at Sonning, Berkshire. *Annals and Magazine of Natural History*, 1, 232–40.

Phorson, J. E. (1990). Some observations on the development of juvenile shells of *Trivia*. *The Conchologists' Newsletter*, no. 113, 279–83.

Phorson, J. E. (1991). Some observations on juveniles and growth series of *Leucophytia bidentata* (Montagu 1808) and *Ovatella myosotis* (Draparnaud 1801). *The Conchologists' Newsletter*, no. 117, 366–70.

Potts, G. W. (1970). The ecology of *Onchidoris fusca* (Nudibranchia). *Journal of the Marine Biological Association of the United Kingdom*, **50**, 269–92.

Raffaelli, D. (1982). Recent ecological research on some European species of *Littorina*. *Journal of Molluscan Studies*, **48**, 342–54.

Reid, D. G. (1989). The comparative morphology, phylogeny and evolution of the gastropod family Littorinidae. *Philosophical Transactions of the Royal Society of London B*, **324**, 1–110.

Reimchen, T. E. (1979). Substratum heterogeneity, crypsis, and colour polymorphism in an intertidal snail (*Littorina mariae*). *Canadian Journal of Zoology*, **57**, 1070–85.

Seed, R. (1979). Distribution and shell characteristics of the painted topshell *Calliostoma zizyphinum* (L.) (Prosobranchia; Trochidae) in County Down, N. Ireland. *Journal of Molluscan Studies*, **45**, 12–18.

Siegismund, H. R. (1982). Life cycle and production of *Hydrobia ventrosa* and *H. neglecta* (Mollusca: Prosobranchia). *Marine Ecology Progress Series*, **7**, 75–82.

Smith, D. A. S. (1973). The population biology of *Lacuna pallidula* (da Costa) and *Lacuna vincta* (Montagu) in north-east England. *Journal of the Marine Biological Association of the United Kingdom*, **53**, 493–520.

Smith, S. M. (1982). A review of the genus *Littorina* in British and Atlantic waters (Gastropod: Prosobranchia). *Malacologia*, **22**, 539–9.

Smith, S. T. (1967). The ecology and life-history of *Retusa obtusa* (Montagu) (Gastropoda, Opisthobranchia). *Canadian Journal of Zoology*, **45**, 397–405.

Smith, S. T. (1967). The development of *Retusa obtusa* (Montagu) (Gastropoda, Opisthobranchia). *Canadian Journal of Zoology*, **45**, 737–64.

Southgate, T. (1982). Studies on an intertidal population of *Rissoa parva* (Gastropoda: Prosobranchia) in south-west Ireland. *Journal of Natural History*, **16**, 183–94.

Spence, S. K., Bryan, G. W., Gibbs, P. E., Masters, D., Morris, L. & Hawkins, S. J. (1990). Effects of TBT contamination on *Nucella* populations. *Functional Ecology*, **4**, 425–32.

Steneck, R. S. & Watling, L. (1982). Feeding capabilities and limitation of herbivorous molluscs: a functional group approach. *Marine Biology*, **68**, 299–319.

Tallmark, B. (1980). Population dynamics of *Nassarius reticulatus* (Gastropoda, Prosobranchia) in Gullmar Fjord, Sweden. *Marine Ecology Progress Series*, **3**, 51–62.

Taylor, D. L. (1968). Chloroplasts as symbiotic organelles in the digestive gland of *Elysia viridis* (Gastropoda: Opisthobranchia). *Journal of the Marine Biological Association of the United Kingdom*, **48**, 1–15.

Thompson, G. B. (1979). Distribution and population dynamics of the limpet *Patella aspera* (Lamarck) in Bantry Bay. *Journal of Experimental Marine Biology and Ecology*, **40**, 115–35.

Thompson, T. E. & Slinn, D. J. (1959). On the biology of the opisthobranch *Pleurobranchus membranaceus*. *Journal of the Marine Biological Association of the United Kingdom*, **38**, 507–24.

Todd, C. D. (1979). The population ecology of *Onchidoris bilamellata* (L.) (Gastropoda: Nudibranchia). *Journal of Experimental Marine Biology and Ecology*, **41**, 213–55.

Todd, C. D. (1981). The ecology of nudibranch molluscs. *Oceanography and Marine Biology. An Annual Review*, **19**, 141–234.

Underwood, A. J. (1972). Observations on the reproductive cycles of *Monodonta lineata*, *Gibbula umbilicalis* and *G. cineraria*. *Marine Biology*, **17**, 333–40.

Underwood, A. J. (1972). Spawning, larval development and settlement behaviour of

*Gibbula cineraria* (Gastropoda: Prosobranchia) with a reappraisal of torsion in gastropods. *Marine Biology*, **17**, 341–9.

Vahl, O. (1971). Growth and density of *Patina pellucida* (L.) (Gastropoda: Prosobranchia) on *Laminaria hyperborea* (Gunnerus) from western Norway. *Ophelia*, **9**, 31–50.

Waite, M. E., Waldock, M. J., Thain, J. E., Smith, D. J. & Milton, S. M. (1991). Reductions in TBT concentrations in UK estuaries following legislation in 1986 and 1987. *Marine Environmental Research*, **32**, 89–111.

Wallace, C. (1985). On the distribution of the sexes of *Potamopyrgus jenkinsi* (Smith). *Journal of Molluscan Studies*, **51**, 290–6.

Watson, D. C. & Norton, T. A. (1985). Dietary preferences of the common periwinkle, *Littorina littorea* (L.). *Journal of Experimental Marine Biology and Ecology*, **88**, 193–211.

Wigham, G. D. (1975). The biology and ecology of *Rissoa parva* (da Costa) (Gastropoda: Prosobranchia). *Journal of the Marine Biological Association of the United Kingdom*, **55**, 45–67.

Wigham, G. D. (1975). Environmental influences upon the expression of shell form in *Rissoa parva* (da Costa) (Gastropoda: Prosobranchia). *Journal of the Marine Biological Association of the United Kingdom*, **55**, 425–38.

Williams, G. R. (1990). The comparative ecology of the flat periwinkles, *Littorina obtusata* (L.) and *L. mariae* Sacchi et Rastelli. *Field Studies*, **7**, 469–82.

Williamson, P. & Kendall, M. A. (1981). Population age structure and growth of the trochid *Monodonta lineata* determined from shell rings. *Journal of the Marine Biological Association of the United Kingdom*, **61**, 1011–26.

Yonge, C. M. (1937). The biology of *Aporrhais pes-pelecani* (L.) and *A. serresiana* (Mich). *Journal of the Marine Biological Association of the United Kingdom*, **21**, 687–703.

Yonge, C. M. (1946). On the habits of *Turritella communis* Risso. *Journal of the Marine Biological Association of the United Kingdom*, **26**, 377–80.

Yonge, C. M. & Thompson, T. E. (1976). *Living marine molluscs*. London: Collins.

Yonow, N. & Ryland, J. S. (1992). Growth and life history parameters in *Acteon tornatilis* (L.) (Opisthobranchia: Cephalaspidea). In *Marine eutrophication and population dynamics, with a special section on the Adriatic Sea. 25th European Marine Biological Symposium*, ed. G. Columbo, I. Ferrari, V. U. Ceccherelli & R. Rossi, pp. 271–6. Fredensburg: Olsen & Olsen.

## MOLLUSCA: BIVALVIA

Allen, J. A. (1969). Observations on size composition and breeding of Northumberland populations of *Zirphaea crispata* (Pholadidae: Bivalvia). *Marine Biology*, **3**, 269–75.

Ansell, A. D. (1961). Reproduction, growth and mortality of *Venus striatula* (da Costa) in Kames Bay, Millport. *Journal of the Marine Biological Association of the United Kingdom*, **41**, 191–215.

Ansell, A. D. (1961). The functional morphology of the British species of Veneracea (Eulamellibranchia). *Journal of the Marine Biological Association of the United Kingdom*, **41**, 489–515.

Ansell, A. D. (1972). Distribution, growth and seasonal changes in biochemical composition for the bivalve *Donax vittatus* (da Costa) from Kames Bay, Millport. *Journal of Experimental Marine Biology and Ecology*, 10, 137–50.

Ansell, A. D. & Trevallion, A. (1967). Studies on *Tellina tenuis* da Costa 1. Seasonal growth and biochemical cycle. *Journal of Experimental Marine Biology and Ecology*, 1, 220–35.

Bayne, B. L. (1964). Primary and secondary settlement in *Mytilus edulis* L. (Mollusca). *Journal of Animal Ecology*, 33, 513–23.

Bayne, B. L. (ed.) (1976). *Marine mussels: their ecology and physiology*. International Biological Programme 10. Cambridge: Cambridge University Press.

Board, P. A. (1970). Some observations on the tunnelling of shipworms. *Journal of Zoology, London*, 161, 193–201.

Boyden, C. R. (1971). A comparative study of the reproductive cycles of the cockles *Cerastoderma edule* and *C. glaucum*. *Journal of the Marine Biological Association of the United Kingdom*, 51, 605–22.

Boyden, C. R. (1971). A note on the nomenclature of two European cockles. *Zoological Journal of the Linnean Society*, 50, 307–10.

Boyden, C. R. (1972). The behaviour, survival and respiration of the cockles *Cerastoderma edule* and *C. glaucum* in air. *Journal of the Marine Biological Association of the United Kingdom*, 52, 661–80.

Boyden, C. R. & Russell, P. J. C. (1972). The distribution and habitat range of the brackish water cockle (*Cardium (Cerastoderma) glaucum*) in the British Isles. *Journal of Animal Ecology*, 41, 719–34.

Brand, A. R., Paul, J. D. & Hoogesteger, J. N. (1980). Spat settlement of the scallops *Chlamys opercularis* (L.) and *Pecten maximus* (L.) on artificial collectors. *Journal of the Marine Biological Association of the United Kingdom*, 60, 379–90.

Cerrato, R. M., Wallace, H. V. E. & Lightfoot, K. G. (1991). Tidal and seasonal patterns in the chondrophore of the soft-shell clam *Mya arenaria*. *Biological Bulletin*, 181, 307–11.

Conan, G. & Shafee, M. S. (1978). Growth and biannual recruitment of the Black Scallop *Chlamys varia* (L.) in Lanveoc area, Bay of Brest. *Journal of Experimental Marine Biology and Ecology*, 35, 59–71.

Crisp, D. J. & Standen, A. (1988). *Lasaea rubra* (Montagu) (Bivalvia: Erycinacea), an apomictic crevice-living bivalve with clones separated by tidal level preference. *Journal of Experimental Marine Biology and Ecology*, 117, 27–45.

Dixon, D. R. & Flavell, N. (1986). A comparative study of the chromosomes of *Mytilus edulis* and *Mytilus galloprovincialis*. *Journal of the Marine Biological Association of the United Kingdom*, 66, 219–28.

Duval, D. M. (1962). Observations on the annual cycle of *Barnea candida* (Class Lamellibranchiata, Family Pholadidae). *Proceedings of the Malacological Society of London*, 35, 101–2.

Gage, J. (1966). Observations on the bivalves *Montacuta substriata* and *M. ferruginosa*, 'commensals' with spatangoids. *Journal of the Marine Biological Association of the United Kingdom*, 46, 49–70.

Gage, J. (1966). The life-histories of the bivalves *Montacuta substriata* and *M. ferruginosa*, 'commensals' with spatangoids. *Journal of the Marine Biological Association of the United Kingdom*, 46, 499–511.

Gardner, J. P. A. (1992). *Mytilus galloprovincialis* (Lmk) (Bivalvia, Mollusca): the taxonomic status of the Mediterranean mussel. *Ophelia*, 35, 219–43.

Gosling, E. M. & McGrath, D. (1990). Genetic variability in exposed-shore mussels, *Mytilus* spp., along an environmental gradient. *Marine Biology*, **104**, 413–18.

Green, J. (1957). The growth of *Scrobicularia plana* (da Costa) in the Gwendraeth estuary. *Journal of the Marine Biological Association of the United Kingdom*, **36**, 41–7.

Hodgson, A. N. (1982). Studies on wound healing, and an estimation of the rate of regeneration, of the siphon of *Scrobicularia plana* (da Costa). *Journal of Experimental Marine Biology and Ecology*, **62**, 117–28.

Hodgson, A. N. & Bernard, R. T. F. (1986). Observations on the ultrastructure of the spermatozoon of two mytilids from the south-west coast of England. *Journal of the Marine Biological Association of the United Kingdom*, **66**, 385–90.

Holme, N. A. (1951). The identification of British species of the genus *Ensis* Schumacher (Lammellibranchiata). *Journal of the Marine Biological Association of the United Kingdom*, **29**, 639–47.

Holme, N. A. (1954). The ecology of British species of *Ensis*. *Journal of the Marine Biological Association of the United Kingdom*, **33**, 145–72.

Holme, N. A. (1959). The British species of *Lutraria* (Lammellibranchia), with a description of *L. angustior* Philippi. *Journal of the Marine Biological Association of the United Kingdom*, **38**, 557–68.

Holme, N. A. (1961). Notes on the mode of life of the Tellinidae (Lamellibranchia). *Journal of the Marine Biological Association of the United Kingdom*, **41**, 699–703.

Holme, N. A. (1961). Shell form in *Venerupis rhomboides*. *Journal of the Marine Biological Association of the United Kingdom*, **41**, 705–22.

Hughes, R. N. (1969). A study of feeding in *Scrobicularia plana*. *Journal of the Marine Biological Association of the United Kingdom*, **49**, 805–23.

Hughes, R. N. (1971). Reproduction of *Scrobicularia plana* da Costa (Pelecypoda: Semelidae) in North Wales. *Veliger*, **14**, 77–81.

Hunter, W. R. (1948). The structure and behaviour of *Hiatella gallicana* (Lamarck) and *H. artica* (L.), with special reference to the boring habit. *Proceedings of the Royal Society of Edinburgh, Series B*, **63**, 271–89.

Ivell, R. (1979). The biology and ecology of a brackish lagoon bivalve, *Cerastoderma glaucum* Bruguière in an English lagoon, the Widewater, Sussex. *Journal of Molluscan Studies*, **45**, 383–400.

Johannessen, O. H. (1973). Population structure and individual growth of *Venerupis pullastra* (Montagu) (Lammellibranchia). *Sarsia*, **52**, 97–116.

Jones, A. M. (1979). Structure and growth of a high-level population of *Cerastoderma edule* (Lamellibranchiata). *Journal of the Marine Biological Association of the United Kingdom*, **59**, 277–87.

Kingston, P. F. (1974). Studies on the reproductive cycles of *Cardium edule* and *C. glaucum*. *Marine Biology*, **28**, 317–23.

Kristensen, E. S. (1979). Observations on growth and life cycle of the shipworm *Teredo navalis* L. (Bivalvia, Mollusca) in the Isefjord, Denmark. *Ophelia*, **18**, 235–42.

Lane, D. J. W., Beaumont, A. R. & Hunter, J. R. (1985). Byssus drifting and the drifting threads of the young post-larval mussel *Mytilus edulis*. *Marine Biology*, **84**, 301–8.

Lebour, M. V. (1938). Notes on the breeding of some lamellibranchs from Plymouth and their larvae. *Journal of the Marine Biological Association of the United Kingdom*, **23**, 119–44.

Lin, J. & Hines, A. H. (1994). Effect of suspended food availability on the feeding mode and burial depth of the Baltic clam, *Macoma bathica*. *Oikos*, 69, 28–36.

Macdonald, S. A. & Thomas, M. L. H. (1980). Age determination of the soft-shell clam *Mya arenaria* using shell internal growth lines. *Marine Biology*, 58, 105–9.

Macleod, J. A. A., Thorpe, J. P. & Duggan, N. A. (1985). A biochemical genetic study of population structure in queen scallop (*Chlamys opercularis*) stocks in the Northern Irish Sea. *Marine Biology*, 87, 77–82.

Mason, J. (1958). The breeding of the scallop, *Pecten maximus* (L.), in Manx waters. *Journal of the Marine Biological Association of the United Kingdom*, 37, 653–71.

Mason, J. (1972). The cultivation of the European mussel, *Mytilus edulis* Linnaeus. *Oceanography and Marine Biology. An Annual Review*, 10, 437–60.

McGrath, D. & O'Foighil, D. (1986). Population dynamics and reproduction of hermaphroditic *Lasaea rubra* (Montagu) (Bivalvia, Galeommatacea). *Ophelia*, 25, 209–19.

McGrath, D., King, P. A. & Gosling, E. M. (1988). Evidence for the direct settlement of *Mytilus edulis* larvae on adult mussel beds. *Marine Ecology Progress Series*, 47, 103–6.

Meadows, P. S. & Shand, P. (1989). Experimental analysis of byssus thread production by *Mytilus edulis* and *Modiolus modiolus* in sediments. *Marine Biology*, 101, 219–26.

Morton, B. (1978). Feeding and digestion in shipworms. *Oceanography and Marine Biology. An Annual Review*, 16, 107–44.

Morton, J. E., Boney, A. D. & Corner, E. D. S. (1957). The adaptations of *Lasaea rubra* (Montagu), a small intertidal lamellibranch. *Journal of the Marine Biological Association of the United Kingdom*, 36, 383–405.

Newell, C. R. & Hidu, H. (1982). The effects of sediment type on growth rate and shell allometry in the soft shelled clam *Mya arenaria* L. *Journal of Experimental Marine Biology and Ecology*, 65, 285–95.

Ockelmann, K. W. (1964). *Turtonia minuta* (Fabricius), a neotenous veneracean bivalve. *Ophelia*, 1, 121–46.

Petraitis, P. S. (1987). Immobilization of the predatory gastropod, *Nucella lapillus*, by its prey, *Mytilus edulis*. *Biological Bulletin*, 172, 307–14.

Purchon, R. D. (1955). The structure and function of the British Pholadidae (rock-boring Lamellibranchia). *Proceedings of the Zoological Society of London*, 124, 859–911.

Quayle, D. B. (1951). The rate of growth of *Venerupis pullastra* (Montagu) at Millport, Scotland. *Proceedings of the Royal Society of Edinburgh, Series B*, 64, 384–406.

Reddiah, K. (1962). The sexuality and spawning of Manx pectinids. *Journal of the Marine Biological Association of the United Kingdom*, 42, 683–703.

Seed, R. (1980). Reproduction and growth in *Anomia ephippium* (L.) (Bivalvia: Anomiidae) in Strangford Lough, Northern Ireland. *Journal of Conchology*, 30, 239–45.

Seed, R. (1992). Systematics evolution and distribution of mussels belonging to the genus *Mytilus*: an overview. *American Malacological Bulletin*, 9, 123–37.

Seed, R. & Brown, R. A. (1977). A comparison of the reproductive cycles of *Modiolus modiolus* (L.), *Cerastoderma* (=*Cardium*) *edule* (L.), and *Mytilus edulis* L. in Strangford Lough, Northern Ireland. *Oecologia, Berlin*, 30, 173–88.)

Seed, R. & O'Connor, R. J. (1980). Shell shape and seasonal changes in population structure in *Lasaea rubra* (Bivalvia: Erycinidae). *Journal of Molluscan Studies*, 46, 66–73.

Taylor, A. C. & Venn, T. J. (1978). Growth of the queen scallop, *Chlamys opercularis*, from the Clyde Sea area. *Journal of the Marine Biological Association of the United Kingdom*, 58, 687–700.

Thompson, I., Jones, D. S. & Dreibelbis, D. (1980). Annual internal growth banding and life history of the ocean quahog *Arctica islandica* (Mollusca: Bivalvia). *Marine Biology*, 57, 25–34.

Trevallion, A. (1971). Studies on *Tellina tenuis* da Costa. III. Aspects of general biology and energy flow. *Journal of Experimental Marine Biology and Ecology*, 7, 95–122.

Willis, G. L. & Skibinski, D. O. F. (1992). Variation in strength of attachment to the substrate explains differential mortality in hybrid mussel (*Mytilus galloprovincialis* and *M. edulis*) populations. *Marine Biology*, 112 403–8.

Wilson, J. G. (1976). Dispersion of *Tellina tenuis* from Kames Bay, Millport, Scotland. *Marine Biology*, 37, 371–6.

Yonge, C. M. (1959). On the structure, biology and systematic position of *Pharus legumen* (L.). *Journal of the Marine Biological Association of the United Kingdom*, 38, 277–90.

Yonge, C. M. (1960). *Oysters*. London: Collins.

Zwarts, L. (1986). Burying depth of the benthic bivalve *Scrobicularia plana* (da Costa) in relation to siphon-cropping. *Journal of Experimental Marine Biology and Ecology*, 101, 25–39.


ARTHROPODA

Allen, J. A. (1963). Observations on the biology of *Pandalus montagui* (Crustacea: Decapoda). *Journal of the Marine Biological Association of the United Kingdom*, 43, 665–82.

Antonopoulou, E. & Emson, R. H. (1992). Aspects of the population dynamics of *Palaemonetes varians* (Leach). In *Marine eutrophication and population dynamics, with a special section on the Adriatic Sea. 25th European Marine Biological Symposium*, ed. G. Columbo, I. Ferrari, V. U. Ceccherelli & R. Rossi, pp. 157–64. Fredensburg: Olsen & Olsen.

Barnes, H. & Stone, R. L. (1973). The general biology of *Verruca stroemia* (O. F. Müller). *Journal of Experimental Marine Biology and Ecology*, 12, 279–97.

Burrows, M. T., Hawkins, S. J. & Southward, A. J. (1992). A comparison of reproduction in co-occurring chthamalid barnacles, *Chthamalus stellatus* (Poli) and *Chthamalus montagui* Southward. *Journal of Experimental Marine Biology and Ecology*, 160, 229–49.

Cheng, L. & Frank, J. H. (1993). Marine insects and their reproduction. *Oceanography and Marine Biology. An Annual Review*, 31, 479–506.

Choy, S. C. (1986). Natural diet and feeding habits of the crabs *Liocarcinus puber* and *L. holsatus* (Decapoda, Brachyura, Portunidae). *Marine Ecology Progress Series*, 31, 87–99.

Crisp, D. J., Southward, A. J. & Southward, E. C. (1981). On the distribution of the intertidal barnacles *Chthamalus stellatus*, *Chthamalus montagui* and *Euraphia depressa*. *Journal of the Marine Biological Association of the United Kingdom*, 61, 359–80.

Crothers, J. H. (1967). The biology of the shore crab, *Carcinus maenas* (L.). 1. The background — anatomy, growth and life-history. *Field Studies*, 2, 407–34.

Crothers, J. H. (1968). The biology of the shore crab, *Carcinus maenas* (L.). 2. The life of the adult crab. *Field Studies*, 2, 579–614.

Davies, A. J. (1981). A scanning electron microscope study of the praniza lava of *Gnathia maxillaris* Montagu (Crustacea, Isopoda, Gnathiidae), with special reference to the mouthparts. *Journal of Natural History*, 15, 545–54.

Davies, L. & Richardson, J. (1970). Distribution in Britain and habitat requirements of *Petrobius maritimus* (Leach) and *P. brevistylis* Carpenter (Thysanura). *Entomologist*, 103, 97–114.

Edwards, E. (1978). *The edible crab and its fishery in British waters*. Farnham: Fishing News Books.

Elwood, R. W. & Stewart, A. (1987). Reproduction in the littoral hermit crab *Pagurus bernhardus*. *Irish Naturalists' Journal*, 22, 252–5.

Fincham, A. A. (1971). Ecology and population studies of some intertidal and sublittoral sand-dwelling amphipods. *Journal of the Marine Biological Association of the United Kingdom*, 51, 471–88.

Fish, J. D. & Fish, S. (1972). The swimming rhythm of *Eurydice pulchra* Leach and a possible explanation of intertidal migration. *Journal of Experimental Marine Biology and Ecology*, 8, 195–200.

Fish, J. D. & Mills, A. (1979). The reproductive biology of *Corophium volutator* and *C. arenarium* (Crustacea: Amphipoda). *Journal of the Marine Biological Association of the United Kingdom*, 59, 355–68.

Fish, J. D. & Preece, G. S. (1970). The annual reproductive patterns of *Bathyporeia pilosa* and *Bathyporeia pelagica* (Crustacea: Amphipoda). *Journal of the Marine Biological Association of the United Kingdom*, 50, 475–88.

Fish, S. (1970). The biology of *Eurydice pulchra* (Crustacea: Isopoda). *Journal of the Marine Biological Association of the United Kingdom*, 50, 753–68.

Franke, H-D. (1993). Mating system of the commensal marine isopod *Jaera hopeana* (Crustacea). I The male-manca (I) amplexus. *Marine Biology*, 115, 65–73.

Furman, E. R. & Yule, A. B. (1991). *Balanus improvisus* Darwin in British estuaries: gene-flow and recolonisation. In *Estuaries and coasts: spatial and temporal intercomparisons*, ed. M. Elliott & J-P. Ducrotoy, pp. 273–6. Fredensborg: Olsen & Olsen.

Haahtela, I. & Naylor, E. (1965). *Jaera hopeana*, an intertidal isopod new to the British fauna. *Journal of the Marine Biological Association of the United Kingdom*, 45, 367–71.

Hamond, R. (1967). The Amphipoda of Norfolk. *Cahiers de Biologie Marine*, 8, 113–52.

Hartnoll, R. G. (1963). The biology of Manx spider crabs. *Proceedings of the Zoological Society of London*, 141, 423–96.

Hartnoll, R. G. (1968). Reproduction in the burrowing crab, *Corystes cassivelaunus* (Pennant, 1777) (Decapoda, Brachyura). *Crustaceana*, 15, 165–70.

Hartnoll, R. G. (1972). The biology of the burrowing crab, *Corystes cassivelaunus*. *Bijdragen tot de Dierkunde*, 42, 139–55.

Harvey, C. E. (1968). Distribution and seasonal population changes of *Campecopea hirsuta* (Isopoda: Flabellifera). *Journal of the Marine Biological Association of the United Kingdom*, 48, 761–7.

Harvey, C. E. (1969). Breeding and distribution of *Sphaeroma* (Crustacea: Isopoda) in Britain. *Journal of Animal Ecology*, 38, 399–406.

Healey, B. & O'Neill, M. (1984). The life cycle and population dynamics of *Idotea pelagica* and *I. granulosa* (Isopoda: Valvifera) in south-east Ireland. *Journal of the Marine Biological Association of the United Kingdom*, **64**, 21–33.

Heath, D. J. (1974). Seasonal changes in frequency of the 'yellow' morph of the isopod, *Sphaeroma rugicauda. Heredity*, **32**, 299–307.

Heath, D. J. (1975). Geographical variation in populations of the polymorphic isopod, *Sphaeroma rugicauda. Heredity*, **35**, 99–107.

Hewett, C. J. (1974). Growth and moulting in the common lobster (*Homarus vulgaris* Milne-Edwards). *Journal of the Marine Biological Association of the United Kingdom*, **54**, 379–91.

Høeg, J. T. (1984). Size and settling behaviour in male and female cypris larvae of the parasitic barnacle *Sacculina carcini* Thompson (Crustacea: Cirripedia: Rhizocephala). *Journal of Experimental Marine Biology and Ecology*, **76**, 145–56.

Høeg, J. T. (1995). The biology and life cycle of the rhizocephala (Cirripedia). *Journal of the Marine Biological Association of the United Kingdom*, **75**, 517–50.

Hogarth, P. J. (1978). Variation in the carapace pattern of juvenile *Carcinus maenas. Marine Biology*, **44**, 337–43.

Holdich, D. M. (1968). Reproduction, growth and bionomics of *Dynamene bidentata* (Crustacea: Isopoda). *Journal of Zoology, London*, **156**, 137–53.

Holdich, D. M. (1970). The distribution and habitat preferences of the Afro-European species of *Dynamene* (Crustacea: Isopoda). *Journal of Natural History*, **4**, 419–38.

Holdich, D. M. (1971). Changes in physiology, structure and histochemistry occurring during the life-history of the sexually dimorphic isopod *Dynamene bidentata* (Crustacea: Peracarida). *Marine Biology*, **8**, 35–47.

Howard, A. E. & Bennett, D. B. (1979). The substrate preference and burrowing behaviour of juvenile lobsters (*Hommarus gammarus* (L.)). *Journal of Natural History*, **13**, 433–8.

Jarvis, J. H. & King, P. E. (1972). Reproduction and development in the pycnogonid *Pycnogonum littorale. Marine Biology*, **13**, 146–54.

Jones, L. T. (1963). The geographical and vertical distribution of British *Limnoria* (Crustacea: Isopoda). *Journal of the Marine Biological Association of the United Kingdom*, **43**, 589–603.

Jones, M. B. & Naylor, E. (1971). Breeding and bionomics of the British members of the *Jaera albifrons* group of species (Isopoda: Asellota). *Journal of Zoology, London*, **165**, 183–99.

Kamermans, P. & Huitema, H. J. (1994). Shrimp (*Crangon crangon* L.) browsing upon siphon tips inhibits feeding and growth in the bivalve *Macoma balthica* (L.). *Journal of Experimental Marine Biology and Ecology*, **175**, 59–75.

King, P. E. & Jarvis, J. H. (1970). Egg development in a littoral pycnogonid *Nymphon gracile. Marine Biology*, **7**, 294–304.

King, P. E., Pugh, P. J. A., Fordy, M. R., Love, N. & Wheeler, S. A. (1990). A comparison of some environmental adaptations of the littoral collembolans *Anuridella marina* (Willem) and *Anurida maritima* (Guérin). *Journal of Natural History*, **24**, 673–88.

Lauckner, G. (1983). Diseases of Mollusca: Bivalvia. In *Diseases of marine animals*, Vol. II, ed. O. Kinne, pp. 477–961. Hamburg: Biologische Anstalt Helgoland.

Laval, P. (1980). Hyperiid amphipods as crustacean parasitoids associated with gelatinous zooplankton. *Oceanography and Marine Biology. An Annual Review*, **18**, 11–56.

Mauchline, J. (1971). The biology of *Neomysis integer* (Crustacea, Mysidacea). *Journal of the Marine Biological Association of the United Kingdom*, **51**, 347–54.

Mauchline, J. (1971). The biology of *Praunus flexuosus* and *P. neglectus* (Crustacea, Mysidacea). *Journal of the Marine Biological Association of the United Kingdom*, 51, 641–52.

McGaw, I. J., Kaiser, M. J., Naylor, E. & Hughes, R. N. (1992). Intraspecific morphological variation related to the moult-cycle in colour forms of the shore crab *Carcinus maenas*. *Journal of Zoology, London*, 228, 351–9.

Mettam, C. (1989). The life cycle of *Bathyporeia pilosa* Lindström (Amphipoda) in a stressful, low salinity environment. *Topics in Marine Biology*, 53, 543–50.

Metz, P. (1967). On the relations between *Hyperia galba* Montagu (Amphipoda, Hyperiidae) and its host *Aurelia aurita* in the Isefjord area (Sjaelland, Denmark). *Videnskabelige Meddelelser fra Dansk naturhistorisk Forening i Kjøbenhavn*, 130, 85–108.

Moore, P. G. (1983). The apparent role of temperature in breeding initiation and winter population structure in *Hyalle nilssoni* Rathke (Amphipoda): field observtions 1972–83. *Journal of Experimental Marine Biology and Ecology*, 71, 237–48.

Moore, P. G. & Francis, C. H. (1986). Notes on breeding periodicity and sex ratio of *Orchestia gammarellus* (Pallas) (Crustacea: Amphipoda) at Millport, Scotland. *Journal of Experimental Marine Biology and Ecology*, 95, 203–9.

Nair, K. K. C. & Anger, K. (1980). Seasonal variation in population structure and biochemical composition of *Jassa falcata* (Crustacea, Amphipoda) off the island of Helgoland (North Sea). *Estuarine and Coastal Marine Science*, 11, 505–13.

Norman, C. P. & Jones, M. B. (1992). Influence of depth, season and moult stage on the diet of the velvet swimming crab *Necora puber* (Brachyura: Portunidae). *Estuarine and Coastal Shelf Science*, 34, 71–83.

Norman, C. P. & Jones, M. B. (1993). Reproductive ecology of the velvet swimming crab, *Necora puber* (Brachyura: Portunidae) at Plymouth. *Journal of the Marine Biological Association of the United Kingdom*, 73, 379–89.

O'Riordan, R. M., Myers, A. A. & Cross, T. F. (1992). Brooding in the intertidal barnacles *Chthamalus stellatus* (Poli) and *Chthamalus montagui* Southward in south-western Ireland. *Journal of Experimental Marine Biology and Ecology*, 164, 135–45.

Pike, R. B. & Williamson, D. I. (1959). Observations on the distribution and breeding of British hermit crabs and the stone crab (Crustacea: Diogenidae, Paguridae and Lithodidae). *Proceedings of the Zoological Society of London*, 132, 551–67.

Rangeley, R. W. & Thomas, M. L. H. (1987). Predatory behaviour of juvenile shore crab *Carcinus maenas* (L.). *Journal of Experimental Marine Biology and Ecology*, 108, 191–7.

Salemaa, H. (1986). Breeding biology and microhabitat utilization of the intertidal isopod *Idotea granulosa* Rathke in the Irish Sea. *Estuarine, Coastal and Shelf Science*, 22, 335–55.

Sheader, M. (1974). North Sea hyperiid amphipods. *Proceedings of the Challenger Society, London*, 4, 247.

Sheader, M. (1977). The breeding biology of *Idotea pelagica* (Isopoda: Valvifera) with notes on the occurrence and biology of its parasite *Clypeoniscus hanseni* (Isopoda: Epicaridea). *Journal of the Marine Biological Association of the United Kingdom*, 57, 659–74.

Simpson, A. C., Howell, B. R. & Warren, P. J. (1970). Synopsis of biological data on the shrimp *Pandalus montagui* Leach, 1814. *F.A.O. Fisheries Reports*, no. 57, 4, 1225–49.

Smaldon, G. (1972). Population structure and breeding biology of *Pisidia longicornis* and *Porcellana platycheles*. *Marine Biology*, 17, 171–9.

Sneli, J-A. (1983). Larvae of *Lepas anatifera* L., 1758, in the North Sea (Cirripedia). *Crustaceana*, 45, 306–8.

Tiews, K. (1970). Synopsis of biological data on the common shrimp *Crangon crangon* (Linnaeus, 1758). *F.A.O. Fisheries Reports*, no. 57, 4, 1167–224.

Ugolini, A., Scapini, F. & Pardi, L. (1986). Interaction between solar orientation and landscape visibility in *Talitrus saltator* (Crustacea: Amphipoda). *Marine Biology*, 90, 449–60.

Upton, N. P. D. (1987). Asynchronous male and female life cycles in the sexually dimorphic, harem-forming isopod *Paragnathia formica* (Crustacea: Isopoda). *Journal of Zoology, London*, 212, 677–90.

Upton, N. P. D. (1987). Gregarious larval settlement within a restricted intertidal zone and sex differences in subsequent mortality in the polygynous saltmarsh isopod *Paragnathia formica* (Crustacea: Isopoda). *Journal of the Marine Biological Association of the United Kingdom*, 67, 663–78.

Vader, W. J. M. (1965). Intertidal distribution of haustoriid amphipods in the Netherlands. *Botanica Gothoburgensia*, 3, 233–46.

Walker, G. (1987). Further studies concerning the sex ratio of the larvae of the parasitic barnacle, *Sacculina carcini* Thompson. *Journal of Experimental Marine Biology and Ecology*, 106, 151–63.

Warner, G. F. (1977). *The biology of crabs*. London: Paul Elek (Scientific Books).

Wildish, D. J. (1987). Estuarine species of *Orchestia* (Crustacea: Amphipoda: Talitroidea) from Britain. *Journal of the Marine Biological Association of the United Kingdom*, 67, 571–83.

Williams, J. A. (1978). The annual pattern of reproduction of *Talitrus saltator* (Crustacea: Amphipoda: Talitridae). *Journal of Zoology, London*, 184, 231–44.

Willows, R. I. (1987). Population dynamics and life history of two contrasting populations of *Ligia oceanica* (Crustacea: Oniscidea) in the rocky supralittoral. *Journal of Animal Ecology*, 56, 315–30.

Willows, R. I. (1987). Intrapopulation variation in the reproductive characteristics of two populations of *Ligia oceanica* (Crustacea: Oniscidea). *Journal of Animal Ecology*, 56, 331–40.

Witteveen, J., Verhoef, H. A. & Huipen, T. E. A. M. (1988). Life history strategy and egg diapause in the intertidal collembolan *Anurida maritima*. *Ecological Entomology*, 13, 443–51.

Wyer, D. & King, P. E. (1974). Feeding in British littoral pycnogonids. *Estuarine and Coastal Marine Science*, 2, 177–84.


SIPUNCULA

Kristensen, J. H. (1969). Irrigation in the sipunculid *Phascolion strombi* (Mont.). *Ophelia*, 7, 101–12.

Kristensen, J. H. (1970). Fauna associated with the sipunculid *Phascolion strombi* (Montagu), especially the parasitic gastropod *Menestho diaphana* (Jeffreys). *Ophelia*, 7, 257–76.

BRYOZOA

Cancino, J. M. & Hughes, R. N. (1987). The effects of water flow on growth and reproduction of *Celleporella hyalina* (L.) (Bryozoa: Cheilostomata). *Journal of Experimental Marine Biology and Ecology*, 112,109–30.

Cancino, J. M. & Hughes, R. N. (1988). The zooidal polymorphism and astogeny of *Celleporella hyalina* (Bryozoa: Cheilostomata). *Journal of Zoology, London*, 215, 167–81.

Cancino, J. M., Hughes, R. N. & Ramirez, C. (1991). Environmental cues and the phasing of larval release in the bryozoan *Celleporella hyalina* (L.). *Proceedings of the Royal Society of London, Series B*, 246, 39–45.

Eggleston, D. (1972). Patterns of reproduction in the marine Ectoprocta of the Isle of Man. *Journal of Natural History*, 6, 31–8.

Hayward, P. J. (1980). Invertebrate epiphytes of coastal marine algae. In *The shore environment. Vol. 2. Ecosystems*. Systematics Association Special Volume No. 17b, ed. J. H. Price, D. E. G. Irvine & W. F. Farnham, pp. 761–87. London: Academic Press.

Hayward, P. J. & Ryland, J. S. (1975). Growth, reproduction and larval dispersal in *Alcyonidium hirsutum* (Fleming) and some other Bryozoa. *Pubblicazioni della Stazione Zoologica di Napoli*, 39, 226–41.

Hughes, D. J. (1987). Gametogenesis and embryonic brooding in the cheilostome bryozoan *Celleporella hyalina*. *Journal of Zoology, London*, 212, 691–711.

Hughes, D. J. & Hughes, R. N. (1986). Life history variation in *Celleporella hyalina* (Bryozoa). *Proceedings of the Royal Society of London, Series B*, 228, 127–32.

Owrid, G. M. A. & Ryland, J. S. (1991). Sexual reproduction in *Alcyonidium hirsutum* (Bryozoa: Ctenostomata). *Bulletin Société des Sciences Naturelles de L'Ouest de la France*, 1, 317–326.

Ryland, J. S. (1976). Physiology and ecology of marine bryozoans. *Advances in Marine Biology*, 14, 285–443.

Stebbing, A. R. D. (1971). The epizoic fauna of *Flustra foliacea* (Bryozoa). *Journal of the Marine Biological Association of the United Kingdom*, 51, 283–300.

Stebbing, A. R. D. (1971). Growth of *Flustra foliacea* (Bryozoa). *Marine Biology*, 9, 267–73.

Thorpe, J. P. & Ryland, J. S. (1979). Cryptic speciation detected by biochemical genetics in three ecologically important intertidal bryozoans. *Estuarine and Coastal Marine Science*, 8, 395–8.

Thorpe, J. P., Beardmore, J. A. & Ryland, J. S. (1978). Genetic evidence for cryptic speciation in the marine bryozoan *Alcyonidium gelatinosum*. *Marine Biology*, 49, 27–32.

Woollacott, R. M. & Zimmer, R. L. (eds.) (1977). *Biology of bryozoans*. New York: Academic Press.


ECHINODERMATA

Barker, M. F. & Nichols, D. (1983). Reproduction, recruitment and juvenile ecology of the starfish, *Asterias rubens* and *Marthasterias glacialis*. *Journal of the Marine Biological Association of the United Kingdom*, 63, 745–65.

Bourgoin, A. & Guillou, M. (1990). Variations in the reproductive cycle of *Acrocnida brachiata* (Echinodermata: Ophiuroidea) according to environment in the Bay of

Douarnenez (Brittany). *Journal of the Marine Biological Association of the United Kingdom*, 70, 57–66.

Buchanan, J. B. (1966). The biology of *Echinocardium cordatum* (Echinodermata: Spatangoidea) from different habitats. *Journal of the Marine Biological Association of the United Kingdom*, 46, 97–114.

Byrne, M. (1990). Annual reproductive cycles of the commercial sea urchin *Paracentrotus lividus* from an exposed intertidal and a sheltered subtidal habitat on the west coast of Ireland. *Marine Biology*, 104, 275–89.

Christensen, A. M. (1970). Feeding biology of the sea-star *Astropecten irregularis* Pennant. *Ophelia*, 8, 1–134.

Comely, C. A. & Ansell, A. D. (1989). The reproductive cycle of *Echinus esculentus* L. on the Scottish west coast. *Estuarine, Coastal and Shelf Science*, 29, 385–407.

Costelloe, J. (1985). The annual reproductive cycle of the holothurian *Aslia lefevrei* (Dendrochirota: Echinodermata). *Marine Biology*, 88, 155–65.

Costelloe, J. (1988). Reproductive cycle, development and recruitment of two geographically separated populations of the dendrochirote holothurian *Aslia lefevrei*. *Marine Biology*, 99, 535–45.

Costelloe, J. & Keegan, B. F. (1984). Feeding and related morphological structures in the dendrochirote *Aslia lefevrei* (Holothuroidea: Echinodermata). *Marine Biology*, 84, 135–42.

Crapp, G. B. & Willis, M. E. (1975). Age determination in the sea urchin *Paracentrotus lividus* (Lamarck), with notes on the reproductive cycle. *Journal of Experimental Marine Biology and Ecology*, 20, 157–78.

Emson, R. H. & Crump, R. G. (1979). Description of a new species of *Asterina* (Asteroidea), with an account of its ecology. *Journal of the Marine Biological Association of the United Kingdom*, 59, 77–94.

Emson, R. H. & Crump, R. G. (1984). Comparative studies on the ecology of *Asterina gibbosa* and *A. phylactica* at Lough Ine. *Journal of the Marine Biological Association of the United Kingdom*, 64, 35–53.

Emson, R. H. & Whitfield, P. J. (1989). Aspects of the life history of a tide pool population of *Amphipholis squamata* (Ophiuroidea) from south Devon. *Journal of the Marine Biological Association of the United Kingdom*, 69, 27–41.

Emson, R. H. & Wilkie, I. C. (1980). Fission and autotomy in echinoderms. *Oceanography and Marine Biology. An Annual Review*, 18, 155–250.

Emson, R. H., Jones, M. B. & Whitfield, P. J. (1989). Habitat and latitude differences in reproductive pattern and life-history in the cosmopolitan brittle-star *Amphipholis squamata* (Echinodermata). In *Reproduction, genetics and distributions of marine organisms. 23rd European Marine Biology Symposium*, ed. J. S. Ryland & P. A. Tyler, pp. 75–82. Fredensborg: Olsen & Olsen.

Feder, H. M. (1981). Aspects of the feeding biology of the brittle star *Ophiura texturata*. *Ophelia*, 20, 215–35.

Fontaine, A. R. (1965). The feeding mechanisms of the ophiuroid *Ophiocomina nigra*. *Journal of the Marine Biological Association of the United Kingdom*, 45, 373–85.

Frid, C. L. J. (1992). Foraging behaviour of the spiny starfish *Marthasterias glacialis* in Lough Ine, Co. Cork. *Marine Behaviour and Physiology*, 19, 227–39.

Gage, J. D. (1990). Skeletal growth bands in brittle stars: microstructure and significance as age markers. *Journal of the Marine Biological Association of the United Kingdom*, 70, 209–24.

Gage, J. D. (1991). Skeletal growth zones as age-markers in the sea urchin *Psammechinus miliaris. Marine Biology*, 110, 217–28.

Gage, J. D. (1992). Growth bands in the sea urchin *Echinus esculentus*: results from tetracycline-mark/recapture. *Journal of the Marine Biological Association of the United Kingdom*, 72, 257–60.

Hancock, D. A. (1974). Some aspects of the biology of the sunstar *Crossaster papposus* (L.). *Ophelia*, 13, 1–30.

Holme, N. A. (1984). Fluctuations of *Ophiothrix fragilis* in the western English Channel. *Journal of the Marine Biological Association of the United Kingdom*, 64, 351–78.

Jangoux, M. (ed.) (1980). *Echinoderms: present and past.* Rotterdam: A. A. Balkema.

Jangoux, M. & Lawrence, J. M. (1982). *Echinoderm nutrition.* Rotterdam: A. A. Balkema.

Jensen, M. (1969). Age determination of echinoids. *Sarsia*, 37, 41–4.

Jensen, M. (1969). Breeding and growth of *Psammechinus miliaris* (Gmelin). *Ophelia*, 7, 65–78.

Jones, M. B. & Smaldon, G. (1989). Aspects of the biology of a population of the cosmopolitan brittlestar *Amphipholis squamata* (Echinodermata) from the Firth of Forth, Scotland. *Journal of Natural History*, 23, 613–25.

Lahaye, M-C. & Jangoux, M. (1985). Post-spawning behaviour and early development of the comatulid crinoid, *Antedon bifida.* In *Echinodermata. Proceedings of the Fifth International Echinoderm Conference, Galway*, ed. B. F. Keegan & B. D. S. O'Connor, pp. 181–4. Rotterdam: Balkema.

Lahaye, M-C. & Jangoux, M. (1985). Functional morphology of the podia and ambulacral grooves of the comatulid crinoid *Antedon bifida* (Echinodermata). *Marine Biology*, 86, 307–18.

La Touche, R. W. (1978). The feeding behaviour of the featherstar *Antedon bifida* (Echinodermata: Crinoidea). *Journal of the Marine Biological Association of the United Kingdom*, 58, 877–90.

Lawrence, J. (1987). *A functional biology of echinoderms.* London: Croom Helm.

Minchin, D. (1987). Sea-water temperature and spawning behaviour in the seastar *Marthasterias glacialis. Marine Biology*, 95, 139–43.

Nauen, C. E. (1978). The growth of the sea star, *Asterias rubens*, and its role as benthic predator in Kiel Bay. *Kieler Meeresforschungen*, 4, 68–81.

Nichols, D. (1979). A nationwide survey of the British sea-urchin, *Echinus esculentus.* In *Progress in underwater science*, Vol. 4, ed. J. C. Gamble & J. D. George, pp. 161–87. Plymouth: Pentech Press.

Nichols, D. (1994). Reproductive seasonality in the comatulid crinoid *Antedon bifida* (Pennant) from the English Channel. *Philosophical Transactions of the Royal Society of London B*, 343, 113–34.

Nichols, D. & Barker, M. F. (1984). Growth of juvenile *Asterias rubens* L. (Echinodermata: Asteroidea) on an intertidal reef in southwestern Britain. *Journal of Experimental Marine Biology and Ecology*, 78, 157–65.

Nichols, D. & Barker, M. F. (1984). Reproductive and nutritional periodicities in the starfish, *Marthasterias glacialis*, from Plymouth Sound. *Journal of the Marine Biological Association of the United Kingdom*, 64, 461–70.

Nichols, D. & Barker, M. F. (1984). A comparative study of reproductive and nutritional periodicities in two populations of *Asterias rubens* (Echinodermata: Asteroidea) from the English Channel. *Journal of the Marine Biological Association of the United Kingdom*, **64**, 471–84.

Nichols, D., Sime, A. A. T. & Bishop, G. M. (1985). Growth in populations of the sea-urchin *Echinus esculentus* L. (Echinodermata: Echinoidea) from the English Channel and Firth of Clyde. *Journal of Experimental Marine Biology and Ecology*, **86**, 219–28.

Rowe, F. W. E. (1970). A note on the British species of cucumarians, involving the erection of two new nominal genera. *Journal of the Marine Biological Association of the United Kingdom*, **50**, 683–7.

Rowe, F. W. E. (1971). The marine flora and fauna of the Isles of Scilly Echinodermata. *Journal of Natural History*, **5**, 233–8.

Sime, A. A. T. & Cranmer, G. J. (1985). Age and growth of North Sea echinoids. *Journal of the Marine Biological Association of the United Kingdom*, **65**, 583–8.

Sloan, N. A. (1980). Aspects of the feeding biology of asteroids. *Oceanography and Marine Biology. An Annual Review*, **18**, 57–124.

Sloan, N. A. & Aldridge, T. H. (1981). Observations on an aggregation of the starfish *Asterias rubens* L. in Morecambe Bay, Lancashire, England. *Journal of Natural History*, **15**, 407–18.

Tuwo, A. & Conand, C. (1992). Reproductive biology of the holothurian *Holothuria forskali* (Echinodermata). *Journal of the Marine Biological Association of the United Kingdom*, **72**, 745–58.

Warner, G. F. (1971). On the ecology of a dense bed of the brittle-star *Ophiothrix fragilis*. *Journal of the Marine Biological Association of the United Kingdom*, **51**, 267–82.

Warner, G. F. (1979). Aggregation in echinoderms. In *Biology and systematics of colonial organisms*, ed. G. Larwood & B. R. Rosen, Systematics Association Special Volume No. 11, pp. 375–96. London: Academic Press.

Warner, G. F. & Woodley, J. D. (1975). Suspension-feeding in the brittle-star *Ophiothrix fragilis*. *Journal of the Marine Biological Association of the United Kingdom*, **55**, 199–210.

Webb, C. M. & Tyler, P. A. (1985). Post larval development of the common north-west European brittle stars *Ophiura ophiura, O. albida* and *Acrocnida brachiata* (Echinodermata: Ophiuroidea). *Marine Biology*, **89**, 281–92.

CHORDATA: ASCIDIACEA

Berrill, N. J. (1975). Chordata: Tunicata. In *Reproduction of marine invertebrates*, vol. II, ed. A. C. Giese & J. S. Pearse, pp. 241–82. New York: Academic Press.

Dybern, B. I. (1965). The life cycle of *Ciona intestinalis* (L.) f. *typica* in relation to the environmental temperature. *Oikos*, **16**, 109–31.

Dybern, B. I. (1967). The distribution and salinity tolerance of *Ciona intestinalis* (L.) f. *typica* with special reference to the waters around southern Scandinavia. *Ophelia*, **4**, 207–26.

Dybern, B. I. (1969). Distribution and ecology of the tunicate *Ascidiella scabra* (Müller) in the Skagerak–Kattegat area. *Ophelia*, **6**, 183–201.

Dybern, B. I. (1969). Distribution and ecology of ascidians in Kviturdvikpollen and Våagsböpollen on the west coast of Norway. *Sarsia*, **37**, 21–40.

Millar, R. H. (1952). The annual growth and reproductive cycle in four ascidians. *Journal of the Marine Biological Association of the United Kingdom*, 31, 41–61.

Millar, R. H. (1954). The annual growth and reproductive cycle of the ascidian *Dendrodoa grossularia* (Van Beneden). *Journal of the Marine Biological Association of the United Kingdom*, 33, 33–48.

Millar, R. H. (1958). The breeding season of some littoral ascidians in Scottish waters. *Journal of the Marine Biological Association of the United Kingdom*, 37, 649–52.

Millar, R. H. (1971). The biology of ascidians. *Advances in Marine Biology*, 9, 1–100.

Rowe, F. W. E. (1972). The marine flora and fauna of the Isles of Scilly Enteropneusta, Ascidiacea, Thaliacea, Larvacea and Cephalochordata. *Journal of Natural History*, 6, 207–13.

Stoecker, D. (1980). Chemical defenses of ascidians against predators. *Ecology*, 61, 1327–34.

Svane, I. & Lundälv, T. (1981). Reproductive patterns and populations dynamics of *Ascidia mentula* O. F. Müller on the Swedish west coast. *Journal of Experimental Marine Biology and Ecology*, 50, 163–82.


CHORDATA: VERTEBRATA

Azzarello, M. Y. (1991). Some questions concerning the Syngnathidae brood pouch. *Bulletin of Marine Science*, 49, 741–7.

Barron, R. J. C. (1976). The occurrence of the rock goby, *Gobius paganellus* L. 1758 and the two-spot goby, *Chaparrudo flavescens* (Fabricius, 1779) in the Blackwater estuary, Essex (S.E. England). *Journal of Fish Biology*, 8, 93–5.

Deady, S. & Fives, J. M. (1995). The diet of corkwing wrasse, *Crenilabrus melops*, in Galway Bay, Ireland, and in Dinard, France. *Journal of the Marine Biological Association of the United Kingdom*, 75, 635–649.

Deady, S. & Fives, J. M. (1995). Diet of ballan wrasse, *Labrus bergylta*, and some comparisons with the diet of corkwing wrasse, *Crenilabrus melops*. *Journal of the Marine Biological Association of the United Kingdom*, 75, 651–65.

Dipper, F. (1981). The strange sex lives of British wrasse. *New Scientist*, 90, 444–5.

Dipper, F. A. & Pullin, R. S. V. (1979). Gonochorism and sex-inversion in British Labridae (Pisces). *Journal of Zoology, London*, 187, 97–112.

Dipper, F. A., Bridges, C. R. & Menz, A. (1977). Age, growth and feeding in the ballan wrasse *Labrus bergylta* Ascanius 1767. *Journal of Fish Biology*, 11, 105–20.

Douglas, D. J. (1989). Broad nosed pipe fish *Syngnathus typhle* (L.). *Irish Naturalists' Journal*, 23, 115.

Dunne, J. (1977). Littoral and benthic investigations on the west coast of Ireland – VII. (Section A: Faunistic and ecological studies.) The biology of the shanny, *Blennius pholis* L. (Pisces) at Carna, Connemara. *Proceedings of the Royal Irish Academy, Section B*, 77, 207–26.

Dunne, J. (1978). Littoral and benthic investigations on the west coast of Ireland – IX. (Section A: Faunistic and ecological studies.) The biology of the rock-goby, *Gobius paganellus* L., at Carna. *Proceedings of the Royal Irish Academy, Section B*, 78, 179–91.

Dunne, J. (1981). A contribution to the biology of Montagu's sea snail, *Liparis montagui* Donovan (Pisces). *Irish Naturalists' Journal*, 20, 217–22.

Dunne, J. & Byrne, P. (1979). Notes on the biology of the tompot blenny, *Blennius gattorugine* Brunnich. *Irish Naturalists' Journal*, 19, 414–8.

Gauld, J. A. & Hutcheon, J. R. (1990). Spawning and fecundity in the lesser sandeel, *Ammodytes marinus* Raitt, in the north-western North Sea. *Journal of Fish Biology*, 36, 611–13.

Gibson, R. N. (1969). The biology and behaviour of littoral fish. *Oceanography and Marine Biology. An Annual Review*, 7, 367–410.

Gibson, R. N. (1969). Powers of adhesion in *Liparis montagui* (Donovan) and other shore fish. *Journal of Experimental Marine Biology and Ecology*, 3, 179–90.

Gibson, R. N. (1982). Recent studies on the biology and intertidal fishes. *Oceanography and Marine Biology. An Annual Review*, 20, 363–414.

Gibson, R. N. (1986). Intertidal teleosts: life in a fluctuating environment. In *The behaviour of teleost fishes*, ed. T. J. Pitcher, pp. 388–408. London: Croom Helm.

Gibson, R. N. & Tong, L. J. (1969). Observations on the biology of the marine leech *Oceanobdella blennii*. *Journal of the Marine Biological Association of the United Kingdom*, 49, 433–8.

Healey, M. C. (1971). The distribution and abundance of sand gobies, *Gobius minutus*, in the Ythan estuary. *Journal of Zoology, London*, 163, 177–229.

Healey, M. C. (1972). On the population ecology of the common goby in the Ythan estuary. *Journal of Natural History*, 6, 133–45.

Henderson, P. A. (1989). On the structure of the inshore fish community of England and Wales. *Journal of the Marine Biological Association of the United Kingdom*, 69, 145–63.

Henderson, P. A. & Holmes, R. H. A. (1990). Population stability over a ten-year period in the short-lived fish *Liparis liparis* (L.). *Journal of Fish Biology*, 37, 605–15.

Jacquet, T. N. & Raffaelli, D. (1989). The ecological importance of the sand goby *Pomatoscistus minutus* Pallas). *Journal of Experimental Marine Biology and Ecology*, 128, 147–56.

Kaiser, M. J. & Croy, M. I. (1991). Population structure of the fifteen-spined stickleback, *Spinachia spinachia* (L.). *Journal of Fish Biology*, 39, 129–131.

King, P. A. (1989). Littoral and benthic investigations on the west coast of Ireland – XXII. The biology of a population of shore clingfish *Lepadogaster lepadogaster* (Bonaterre, 1788) at Inishbofin, Co. Galway. *Proceedings of the Royal Irish Academy, Section B*, 89, 47–58.

Lecomte-Finiger, R. (1992). Growth history and age at recruitment of European glass eels (*Anguilla anguilla*) as revealed by otolith microstructure. *Marine Biology*, 114, 205–10.

Magnhagen, C. (1992). Alternative reproductive behaviour in the common goby, *Pomatoschistus microps*: an ontogenetic gradient? *Animal Behaviour*, 44, 182–4.

Miller, P. J. (1961). Age, growth and reproduction of the rock goby, *Gobius paganellus* L., in the Isle of Man. *Journal of the Marine Biological Association of the United Kingdom*, 41, 737–69.

Miller, P. J. (1974). A new species of *Gobius* (Teleostei: Gobiidae) from the western English Channel, with a key to related species in the British and Irish fauna. *Journal of Zoology, London*, 174, 467–80.

Miller, P. J. (1975). Age-structure and life-span in the common goby. *Pomatoschistus microps*. *Journal of Zoology, London*, 177, 425–48.

Milton, P. (1983). Biology of littoral blenniid fishes on the coast of south-west England. *Journal of the Marine Biological Association of the United Kingdom*, 63, 223–37.

Nolan, C. P. (1985). Observations on the breeding season of the worm pipefish, *Nerophis lumbriciformis* (Jenyns). *Irish Naturalists' Journal*, 21, 422.

O'Farrel, M. M. & Fives, J. M. (1990). The feeding relationships of the shanny, *Lipophrys pholis* (L.). and Montagu's blenny, *Coryphoblennius galerita* (L.) (Teleostei: Blenniidae). *Irish Fisheries Investigations, Series B (Marine)*, 36. Dublin: Stationery Office.

Potts, G. W. (1974). The colouration and its behavioural significance in the corkwing wrasse, *Crenilabrus melops. Journal of the Marine Biological Association of the United Kingdom*, 54, 925–38.

Potts, G. W. (1984). Parental behaviour in temperate marine teleosts with special reference to the development of nest structures. In *Fish reproduction: strategies and tactics*, ed. G. W. Potts & R. J. Wootton, pp. 230–44. London: Academic Press.

Potts, G. W. (1985). The nest structure of the corkwing wrasse, *Crenilabrus melops* (Labridae: Teleostei). *Journal of the Marine Biological Association of the United Kingdom*, 65, 531–46.

Potts, G. W. & Swaby, S. E. (1993). *Marine and estuarine fishes of Wales*. Countryside Council for Wales Contract No. FC-73-01-53.

Qasim, S. Z. (1957). The biology of *Centronotus gunnellus* (L.) (Teleostei). *Journal of Animal Ecology*, 26, 389–401.

Rosenqvist, G. (1993). Sex role reversal in a pipefish. *Marine Behaviour and Physiology*, 23, 219–30.

Russell, F. E. (1965). Marine toxins and venomous and poisonous marine animals. *Advances in Marine Biology*, 3, 256–384.

Sawyer, R. T. (1970). The juvenile anatomy and post-hatching development of the marine leech *Oceanobdella blennii* (Knight-Jones, 1940). *Journal of Natural History*, 4, 175–88.

Treasurer, J. W. (1994). The distribution, age and growth of wrasse (Labridae) in inshore waters of west Scotland. *Journal of Fish Biology*, 44, 905–18.

Vestergaard, K. (1976). Nest building behaviour in the common goby *Pomatoschistus microps* (Krøyer) (Pisces, Gobiidae). *Videnskabelige Meddelelser fra Dansk naturhistorisk Forening i Kjøbenhavn*, 139, 91–108.

Webb, C. J. (1980). Systematics of the *Pomatoschistus minutus* complex (Teleostei: Gobioidei). *Proceedings of the Royal Society of London, Series B*, 291, 201–39.

Wootton, R. J. (1976). *The biology of the sticklebacks*. London: Academic Press.

Wyttenbach, A. & Senn, D. G. (1993). Intertidal habitat: does the shore level affect the nutritional condition of the shanny (*Lipophrys pholis*, Teleostei, Blenniidae)? *Experientia*, 49, 725–8.

# Index